Instant Notes
Vegetable Science Objective Type

NIPA® GENX ELECTRONIC RESOURCES & SOLUTIONS P. LTD.
New Delhi-110 034

About the Authors

Nagendra Rai began professional career as Junior Scientist in May 1993 and went up the ladder through Senior Scientist (Horticulture) at ICAR, Research Complex for NEH Region, Meghalaya and Principal Scientist (Vegetable Science) at Indian Institute of Vegetable Research, Varanasi. Taught several graduate/ post-graduate level courses in horticulture (Vegetable Crops) and guided research of many students which led to award of M.Sc. and Ph.D. degrees in various disciplines of Vegetable Science. Developed more than 25 varieties/ lines/ hybrids of vegetable crops like tomato, brinjal, cowpea, French bean, Indian bean and turmeric. Also generated production technologies in Vegetable Crops for various part of the country like Udham Singh Nagar of Uttrakhand, Raipur of Chhattisgarh and Umiam of Meghalaya. Fellows of National Academy of Agricultural Sciences, International Society for Noni Sciences, Society for Horticultural Research and Development, Uttar Pradesh Academy of Agricultural Sciences and Indian Society of Vegetable Science. Recipient of Kirti Singh Lifetime Achievement Award, Harbhajan Singh Memorial Award, Research Paper Reviewer Award, Year of the Scientist Award, Distinguished Scientist Award, Young Scientist Associate Award and Crop Research Award. Publications 275 comprising research papers, books, book chapters, popular articles, technical bulletins/ manuals/ souvenirs and extension folders related to vegetable crops. Editor of Current Horticulture, Progressive Agriculture- An International Journal & Legume Research and Joint -Editor of Vegetable Science. Leadership in both fund mobilization and human resource development.

B.S. Asati possesses a meritorious academic record. He has an illustrious university career in agriculture, with an array of medals, prizes and scholarships. He has completed M.Sc. (Ag) Horticulture from Indira Gandhi Agricultural University, Raipur obtaining first position in the department of Horticulture. He served for four years as Research Associate in ICAR Research Complex for NEH Region, Umiam (Meghalaya). Presently, he is Associate Professor, Rani Avanti Lodhi College of Agriculture and Research Station, Chhuikhadan, KCG under Indira Gandhi Krishi Vishwavidyalaya, Chhattisgarh. He has number of research papers, popular articles, book chapters and technical bulletin to his credit. He has also a life membership of various journals and magazines.

 **Krishna Kumar Rai** has completed his M.Sc. in Biotechnology from, Amity University, Noida. He has served 5 years as Senior Research Fellow (SRF) in DBT/ICAR funded projects at Indian Institute of Vegetable Research, Varanasi, where he has worked on deciphering biochemical and molecular mechanisms linked to biotic and abiotic stress tolerance in tomato and Indian bean plants. He has published quality research papers and book chapters in the field of plant physiology, biochemistry and molecular biology. He has completed his Ph.D. degree from Institute of Sciences, Department of Botany, Banaras Hindu University, Varanasi.

Instant Notes
Vegetable Science Objective Types
(ARS, NET, SET, SRF, JRF and PhD Entrance Examinations)

Nagendra Rai
Principal Scientist (Vegetable Science)
ICAR- Indian Institute of Vegetable Research
Jakhini (Shahanshahpur)
Varanasi, Uttar Pradesh

Bhagwat Saran Asati
Associate Professor
Rani Avanti Lodhi College of Agriculture and Research Station
Indira Gandhi Krishi Vishwavidyalaya
Chhuikhadan, KCG, (Chhattisgarh)

Krishna Kumar Rai
Research Associate
Institute of Sciences, Department of Botany
Banaras Hindu University
Varanasi, Uttar Pradesh

NIPA® GENX ELECTRONIC RESOURCES & SOLUTIONS P. LTD.
New Delhi-110 034

**NIPA® GENX ELECTRONIC
RESOURCES & SOLUTIONS P. LTD.**

101,103, Vikas Surya Plaza, CU Block
L.S.C. Market, Pitam Pura, New Delhi-110 034
Ph : +91 11 27341616, 27341717, 27341718
E-mail:newindiapublishingagency@gmail.com
Web: www.nipabooks.com

For customer assistance, please contact
Phone: + 91-11-27 34 17 17
Fax: + 91-11-27 34 16 16
E-Mail: feedbacks@nipabooks.com

© 2024, Publisher

ISBN: 978-93-87973-83-1

All rights reserved. No part of this publication may be reproduced, stored in a retrieval system or transmitted in any form or by any means, including electronic, mechanical, photocopying recording or otherwise without the prior written permission of the publisher or the copyright holder.

This book contains information obtained from authentic and highly reliable sources. Reasonable efforts have been made to publish reliable data and information, but the author/s, editor/s and publisher cannot assume responsibility for the validity, accuracy or completeness of all materials or information published herein or the consequences of their use. The work is published with the understanding that the publisher and author/s are not attempting to render any professional services. The author/s, editor/s and publisher have attempted to trace and acknowledge the copyright holders of 7all material reproduced in this publication and apologize to copyright holders if permission and/or acknowledgements to publish in this form have not been taken. If any copyrighted material has not been acknowledged, please write to us and let us know so that we may rectify the error, in subsequent reprints.

Trademark Notice: NIPA®, the NIPA® logos and their presentations (the way they are written/ presented) in this book are the trademarks of the publisher and hence may not be used without written permission, if copied or used without authorization, the infringer will be prosecuted as per law.

NIPA® also publishes books in a variety of electronic formats. Some content that appears in print may not be available in electronic books, and vice versa.

Composed and Designed by NIPA®.

Preface

Vegetable is considered to be one of the most important commodities in agriculture and an impressive growth has been recorded in vegetable production since independence, which was merely 5.8 tonnes per hectare. However, there is need of to increase the vegetable productivity to meet the increasing demand of the growing populations. Challenges of shrinking land and water resources with a threat of climate change need an effective support in term of human resources *vis-a-vis* financial support for the development of technologies. Development of vegetable hybrids embedded with gene for higher yield and resistance to biotic/ abiotic stresses; production and protection technologies for mitigating the problematic soil, use of less water and nutrient and safe management of pests and diseases, reduction in post -harvest losses and value addition are need of the hour. Updated information's on Vegetable Science is interest for teachers, researchers and scholars.

The book entitled Vegetable Science *Objective Type* will fulfil a long felt need of Indian students for graduate and post- graduate of Olericulture. This book has been designed to cover the courses offered by various universities/institutes for seeking degree in Vegetable Science and also the syllabi of various competitive examinations for the requirements of those students whose are to compete for ARS, NET, SRF, JRF, SET, and PhD entrance examinations. The book is based on multiple choice questions, difference in between, fill in the blanks and true and false which covers vegetable crops based production, breeding and biotechnology, statistics, biometrics and application of computer. In this book authors have also given latest information in agriculture as special chapter *General Agriculture* for updating knowledge in agriculture for readers. Memory based questions which have been asked last 22 years (from 1997-2019) in competitive examinations of JRF/ SRF/ NET in Vegetable Science have also been given in this book.

The authors are thankful to family members for their cooperation during the writing of this book.

Valuable suggestions are welcome from readers especially the teaching fraternity and students for making this book a much better one.

Nagendra Rai
Bhagwat Saran Asati
Krishna Kumar Rai

Contents

Preface .... *vii*

1. **Vegetable Production** 1
   1.1. Multiple Choice Questions 1
      1.1.1. General 1
      1.1.2. Vegetable Crops Based 32
         1.1.2.1. Bulb Vegetables 32
         1.1.2.2. Cole Crops 44
         1.1.2.3. Cucurbitaceous Vegetables 57
         1.1.2.4. Fruit Vegetables 81
         1.1.2.5. Leafy Vegetables 107
         1.1.2.6. Legume Vegetables 116
         1.1.2.7. Perennial Vegetables 130
         1.1.2.8. Root Vegetables 137
         1.1.2.9. Tuber Vegetables 149
         1.1.2.10. Spices Vegetables 161
         1.1.2.11. Underutilized Vegetable Crops 178
   1.2. Differentiate in Between 186
   1.3. Fill in the Blanks 188
   1.4. True and False 210

2. **Breeding, Biotechnology and Seed Technology** 227
   2.1. Multiple Choice Questions 227
   2.2. Differentiate in Between 280
   2.3. Fill in the Blanks 285
   2.4. True and False 298

3. **Statistics, Biometrics and Application of Computers** 308
   3.1. Multiples Choice Questions 308
   3.2. Differenciate in Between 328

	3.3. Fill in the Blanks	335
	3.4. True and False	339
4.	**General Agriculture Question Bank**	**345**
	4.1. Multiple Choice Questions	345
	4.2. Fill in the Blanks	424
	4.3. Current Affairs	432
5.	**Memory Based Question (JRF, SRF, NET) 1997 to 2019**	**450**
	Appendices	**514**
	1. Abbreviations used in Vegetable Science	514
	2. First in Vegetable Science	517
	3. Latest Released Varieties/ Hybrids	520
	Glossary	**531**

1
Vegetable Production

1.1. Multiple Choice Questions

1.1.1. General

(1) Vegetables combat under nourishment and are known to be the cheapest source of ——————.
    (a) Artificial protective food   (b) Natural protective food
    (c) a and b both   (d) Only vitamins
    (e) None of the above

(2) As per nutritionists, per capita requirements of vegetable should be ——————.
    (a) 100g, in which 75 g leafy vegetable and 25 g root vegetable
    (b) 150 g, in which 75g leafy vegetables and 75 g other vegetables
    (c) 500g, in which 250g leafy vegetables, 100g root vegetables and 150 g other vegetables
    (d) 250 g, in which 125 g root vegetable and 125 g leafy vegetables
    (e) 300g, in which 115g leafy vegetables, 70 g root vegetables and 115 g others.

(3) The deficiency of —————— causes night blindness.
    (a) Vitamin A   (b) Vitamin C
    (c) Vitamin D   (d) Vitamin E
    (e) None of the above

(4) The deficiency of vitamin 'B_1' causes ——————
    (a) Night blindness   (b) Rickets
    (c) Scurvey   (d) Beriberi
    (e) Non e of the above

(5) Scurvey disease is due to deficiency of——————.
    (a) Vitamin D   (b) Vitamin E
    (c) Vitamin C   (d) Vitamin K
    (e) Vitamin B

(6) —————— is also known as antisterility vitamin.
    (a) Vitamin C   (b) Vitamin E
    (c) Vitamin K   (d) Vitamin A
    (e) Vitamin B

(7) The deficiency of ———— in human diet causes disorders like rough skin, abnormality in the tongue and cellular respiration.
   (a) Magnesium   (b) Sodium
   (c) Cupper   (d) Niacin
   (e) None of the above

(8) ———— is essential for prevention of rickets, osteomalacia and dental diseases.
   (a) Vitamin A   (b) Vitamin C
   (c) Vitamin D   (d) Vitamin K
   (e) Vitamin E

(9) The synthesis of prothrombin and normal blood cloting regulate by the ————.
   (a) Vitamin A   (b) Vitamin C
   (c) Vitamin D   (d) Vitamin K
   (e) None of the above

(10) The deficiency of ———— in body causes ricket and osteomalacia.
   (a) Sodium   (b) Calcium
   (c) Phosphorous   (d) Mangenes
   (e) All of the above

(11) ————is essential in human diet for cell multiplication of bones and soft tissue.
   (a) Iron   (b) Sodium
   (c) Magnesium   (d) Sulphur
   (e) Phosphorous

(12) The 'goitre' disease in human is due to deficiency of ————.
   (a) Iron   (b) Sodium
   (c) Iodine   (d) Calcium
   (e) None of the above

(13) For good health, the requirement of vitamin 'A' per day is ————.
   (a) 200 IU   (b) 500 IU
   (c) 1000 IU   (d) 2000 IU
   (e) None of the above

(14) The requirement of vitamin 'B' per capita/day for good health is————.
   (a) 0.05 mg   (b) 0.01 mg
   (c) 0.02 mg   (d) Above 0.17 mg
   (e) None of the above

(15) For good health per capita/day requirement of vitamin 'C' is ————.
   (a) Above 1 mg   (b) Above 5 mg
   (c) Above 20 mg   (d) Above 15 mg
   (e) In between 10-15 mg

(16) ———————requirement per capita/day for good health is above 3.0 mg.
   (a) Calcium   (b) Iron
   (c) Sodium   (d) Phosphorous
   (e) None of the above

(17) ——————— requirement per capita/day is above 20 mg.
   (a) Iron   (b) Calcium
   (c) Magnessium   (d) Sodium
   (e) None of the above

(18) Mangnesium is implicated to have role in———————.
   (a) Cardiovascular disease   (b) Skin disease
   (c) Stomach disease   (d) Throat disease
   (e) None of the above

(19) ———————deficiency in human body leads to growth failure and poor development of body growth.
   (a) Iron   (b) Zinc
   (c) Sodium   (d) Managese
   (e) All of the above

(20) ———————deficiency in body leads to impaired glucose tolerance.
   (a) Sodium   (b) Phosphorous
   (c) Chromium   (d) Potassium
   (e) Nitrogen

(21) ——————— deficiency in body leads to abnormality in skeletal bone mineralization.
   (a) Magnessium   (b) Iron
   (c) Chromium   (d) Sodium
   (e) Manganese

(22) Excess molybednum intake in human body may increase the risk of———————.
   (a) Beri-beri   (b) Gout
   (c) Hemophelia   (d) Rickets
   (e) None of the above

(23) Selenium deficiency in human body is also implicated as a risk factor in———————.
   (a) Brain hammerage   (b) Heart disease
   (c) Cancer   (d) Joundice
   (e) None of above

(24) Megaloblastic anaemia in living organism is due to deficiency of———————.
   (a) Vitamin E   (b) Vitamin B_6
   (c) Vitamin C   (d) Vitamin B_{12}
   (e) None of the Above

(25) Inadequate intake of vitamin like ———————— results in soreness of the tongue (glossaries), cracking at the angles of mouth (angular stomatitis), redness of the eye and burning sensation in eyes, scaliness of the skin in the region between the nose and the angles of the lips (seborroic dermatitis).
   (a) Thiamin (b) Naicin
   (c) Riboflavin (d) Uracil

(26) Psychomotor development in children may be impaired in ————————.
   (a) Vitamin A (b) Vitamin C
   (c) Vitamin D (d) Vitamin B
   (e) Vitamin E

(27) ———————— is incorporated in rhodopsin (Eye pigment).
   (a) Vitamin K (b) Vitamin E
   (c) Vitamin A (d) Vitamin D
   (e) None of the Above

(28) ———————— is the richest source of carbohydrate (38.1 g/ 100 g edible part).
   (a) Tomato (b) Sweet potato
   (c) Cucumber (d) Tapioca
   (e) None of the Above

(29) ———————— is the richest source of fiber (6.8 g / 100 g edible part).
   (a) Brinjal (b) Pea
   (c) Chilli (d) Tomato
   (e) Potato

(30) Pea is the richest source of ————————.
   (a) Carbohydrate (b) Protein
   (c) Vitamin (d) Fat
   (e) Minerals

(31) Parsley is the richest source of ————————.
   (a) Vitamin C (b) Vitamin B
   (c) Vitamin D (d) Vitamin E
   (e) None of the Above

(32) ———————— is the richest source of vitamin A (14190 µg/100 g edible parts).
   (a) Carrot (b) Tomato
   (c) Potato (d) Sweet potato
   (e) None of the above

(33) Research on vegetable crops was started by ICAR with the sanctioning of nucleus Plant Introduction at the IARI, New Delhi in ——————————.
    (a) 1947-48   (b) 1957-58
    (c) 1967-68   (d) 1984

(34) The vegetable breeding station at Katrain in Kulu Valley, Himachal Pradesh for production seeds of temperate vegetables was established in ——————————.
    (a) 1947   (b) 1957-58
    (c) 1949   (d) 1984

(35) The Division of Horticulture at the IARI, New Delhi was established in——————.
    (a) 1947-48   (b) 1958
    (c) 1956   (d) 1984

(36) Recommendation by quin- quennial review team (QRT) of ICAR to upgrade the AICVIP at the level of project Directorate of Vegetable Research (PDVR) was in ——————.
    (a) 1947   (b) 1957
    (c) 1967   (d) 1984

(37) The head quarter for AICRP(VC) was established in——————————.
    (a) 1968   (b) 1988
    (c) 68   (d) 1987

(38) At present the head quarter of AICRP(VC) is situated in——————————.
    (a) Banglore   (b) New Delhi
    (c) Pune   (d) Varanasi

(39) Presently, in AICVIP, the corrdinated research programme is going on —————— main ————— subcentre ————— and voluntary centres.
    (a) 10,8, 6   (b) 8, 18, 31
    (c) 25, 20, 10   (d) None of them

(40) The All india corrdinated research programme was started during—————————————— by merging the on going adhoc projects running in Agricultural universities
    (a) 1969   (b) 1958
    (c) 1968   (d)1984

(41) The all India Coordinated Project on spices was initiated in ——————————.
    (a) 1975   (b) 1971
    (c) 68   (d) 1984

(42) At present there are ——————————————————— State Agricultural Universities, where researches on vegetable crops are going on.
    (a) 31   (b) 38
    (c) 25   (d) None of them

(43) Project Directorate of Vegetable Research was upgraded to Indian Institute of Vegetable Research in ――――――.
  (a) 1992 (b) 1999
  (c) 2001 (d) 1984

(44) ―――――― was the founder Director of Indian Institute of Vegetable Research
  (a) Dr. SP Ghosh (b) Dr. G Kalloo
  (c) Dr. K. L Chadha (d) Dr. Kirti Singh

(45) National Seed Project for production of Breeder Seeds of Vegetable Crops was initiated in ――――――.
  (a) 1991 (b) 1999
  (c) 1994 (d) None of them

(46) Central University of Agricultural at Imphal came in to existence in ――――――.
  (a) 1993 (b) 1999
  (c) 1885 (d) None of them

(47) India, the first report of hybrid vigour in tomato appeared as early as ――――――.
  (a) 1908 (b) 1934
  (c) 1956 (d) 1965
  (e) None of the above

(48) At the National level, first report of hybrid vigour appeared in 1933 in ――――――――――― at Indian Agricultural Research Institute (IARI), New Delhi.
  (a) Tomato (b) Cucumber
  (c) Chilli (d) Pumpkin
  (e) All of the Above

(49) Indo-American Hybrid Seed Company, Bangalore took the lead in the release of first tomato hybrid "Karnataka" and first capsicum hybrid "Bharat" in ――――――.
  (a) 1978 (b) 1973
  (c) 1989 (d) 1992
  (e) 1952

(50) Realizing potential of vegetable hybrid technology in India, Indian Council of Agricultural Research (ICAR) initiated a network project entitled "Promotion of Hybrid Research in Vegetable Crops" during ――――――――――.
  (a) 1978-79 (b) 1985-1986
  (c) 1995-96 (d) 2000-2001 (e) None of the above

(51) ―――――――――― of F_1 hybrid seed is one of the major handicaps of successful hybrid vegetable technology.
  (a) Non availiability of seeds (b) High cost
  (c) Lack of techanical gudiance (d) a and b both
  (e) All of the above

(52) ———— and ———— hybrids of onion developed by using cytoplasmic male sterility.
   (a) Pusa Red and Patana Red   (b) Pusa Nasik
   (c) Arka Pitamber and Arka Kirtiman   (d) N-53
   (e) Udaipur-201

(53) The criteria as to which element should be considered as essential element was given by Arnon and Stout in ————.
   (a) 1942   (b) 1945
   (c) 1939   (d) 1922
   (e) 1932

(54) Nitrogen, phosphorus and sulphur are considered as ————.
   (a) Balance elements   (b) Growth promoting elements
   (c) Protoplasmic elements   (d) Growth inhibiting elements
   (e) All of the above

(55) Calcium, magnesium and potassium counteract the toxic effect of other minerals by causing ionic balance. Hence, these elements are called as ————.
   (a) Balance elements   (b) Protoplasmic elements
   (c) Growth promoting elements   (d) Growth inhibiting elements
   (e) All of the above

(56) Carbon, hydrogen and oxygen are considered as ————.
   (a) Growth promoting elements   (b) Growth inhibiting elements
   (c) Protoplasmic elements   (d) Framework elements'
   (e) Balance elements

(57) The most important of all known Ca^{2+} modulated is 'Calmodulin', a highly conserved, heat stable, small molecular weight protein found in all ————
   (a) Prokaryotic cells   (b) Eukaryotic cells
   (c) Multicells   (d) None of the above

(58) 'Calmodulin' has been isolated from lower and higher plants, the ————. *Chalmydomonas* and *Mougetia*,
   (a) Fungi   (b) Green algae
   (c) Bacteria   (d) Virus
   (e) None of the above

(59) Manganese deficiency occurs in plants, when in the leaf manganese concentration is ————.
   (a) Less than 50 ppm on dry weight basis
   (b) More than 50 ppm on dry weight basis
   (c) Less than 20 ppm on dry weight basis
   (d) Less than 30 ppm on dry weight basis
   (e) Less than 10 ppm on dry weight basis.

(60) Zinc is responsible for the biosynthesis of plant auxin ——————.
   (a) ABA (b) IAA
   (c) NAA (d) 2,4-D
   (e) None of the above

(61) —————————— reduces auxin content in plants through its involvement in the synthesis of tryptophan a precursor of auxin.
   (a) Mg (b) Fe
   (c) Zn (d) Na
   (e) K

(62) Warington (1923) first demonstrated the essential role of boron in ——————
   (a) Cowpea (b) Sem
   (c) Broad bean (d) Lima bean
   (e) All of the above

(63) Boron deficiency occurs in most the vegetable crops, if the level is ————.
   (a) More 30 ppm on dry weight basis
   (b) Less than 30 ppm on dry weight basis
   (c) Less than 20 ppm on dry weight basis
   (d) Less than 10 ppm on dry weight basis
   (e) Less than 5 ppm on dry weight basis

(64) 'Brown heart', 'Heart rot' or 'Hollow centre' occurs in the roots of radish, turnip and beet, respectively is due to deficiency of ——————————————.
   (a) Mo (b) Boron
   (c) Iron (d) Nitrogen
   (e) All of the above

(65) —————————— acts as a component of phenolases, laccase and ascorbic acid oxidase.
   (a) Iron (b) Sodium
   (c) Nitrogen (d) Copper
   (e) None of the above

(66) Whiptail of cauliflower is due to deficiency of ——————————.
   (a) Boron (b) Iron
   (c) Molybdenum (d) Copper
   (e) Sodium

(67) Molybdenum deficiencies occur in vegetable crops when grown on————————, or when molybdenum is unavailable, or on soils where molybdenum is fixed by secondary soil mineral, or on very well drained alkaline soils.
   (a) Very acidic soils (b) Alkali soil
   (c) Saline soil (d) B and C both
   (e) None of the above

(68) The concept of plant growth substances was started during the mid-1800s by the famous German plant physiologist ———————— suggested that plant form is attained through the action of specific "Organ-farming" substances such as "leaf-forming" substances, "root-forming" substances and "flower-forming" substances. The triggering substances referred to as phytohormones, which initiate biochemical reaction and changes in chemical composition within the plant.
   (a) Charls Darwin (b) W. C. Went
   (c) Gillbert (d) Julis Von Sachs
   (e) All of the above

(69) ———————— wrote a book entitled "The Power of Movement in Plants".
   (a) Charles Darwin (b) W. C. Went
   (c) Julis Von Sachs (d) W. C. Went
   (e) None of the above

(70) The term auxin (Greek *auxein,* "to increase") was first used by Frits Went, who, as a graduate student in Holland in ————————, discovered that some unidentified compound probably caused curvature of oat coleoptiles toward light.
   (a) 1932 (b) 1926
   (c) 1934 (d) 1954
   (e) None of the above

(71) 2-methyl-4-chlorophenoxyacetic acid (MCPA) is ———————— compound.
   (a) Synthetic GA (b) Synthetic Auxin
   (c) Synthetic cytokinin (d) Synthetic retardants
   (e) Synthetic inhibitor

(72) Tryptophan is a ———————— precursor.
   (a) GA (b) Auxin
   (c) Cytokinin (d) A and C both
   (e) None of the above

(73) Slow movement, polar transports and requires metabolic energy are three important features of ——————————.
  (a) GAs
  (b) Cytokinins
  (c) Auxins
  (d) Growth retardants
  (e) Growth inhibitors

(74) TIBA (2,3,5-triiodobenzoic acid and NPAC (a-naphthythalamic acid) compounds are called ——————————.
  (a) Anti-auxins
  (b) Anti-Gibberellins
  (c) Anti-retardants
  (d) Anti-inhibitors
  (e) None of the above

(75) 2,4-D, 2,4,5-T, MCPA and derivatives of picolinic acid such as pictoram (solid under the trade name Tordon) are considered as ——————————.
  (a) Auxin rootin
  (b) Auxin growth promoting
  (c) Auxin herbicides
  (d) Auxin flowering
  (e) None of the above

(76) The derivative of benzoic acid such as —————————— also has auxin activity and is more effective than the others against deep-rooted perennial weeds.
  (a) Salicylic acid
  (b) Nicotonic acid
  (c) Pictoram
  (d) Dicamba
  (e) Picolinic acid

(77) Modern hypothesis regarding auxins herbicides suggested that these compounds alter —————————— so greatly that enzymes needed for coordinated growth are not produced properly.
  (a) DNA translocation and RNA transcription
  (b) DNA transcription and RNA translocation
  (c) DNA transcription and RNA transcription
  (d) DNA translocation and RNA translocation
  (e) None of the above

(78) Gibberellins were first discovered in —————————— in the 1930s from studies with diseased rice plants that grew excessively tall. These plants often could not support themselves and extentually died from combined weakness and parasite damage. As early as 1890s, the Japanese called this the "bakbane" (foolish seedling) disease which is caused by fungus (*Gibberella fujikuroi* (the asexual or imperfect stage is *Fusarium moniliformae*).
  (a) Italy
  (b) Japan
  (c) Turkey
  (d) USA
  (e) England

(79) As of 1990, ———————— gibberellins had been discovered in various fungi and plants. Of these, 73 occur in higher plants, 25 in the *Gibberella* fungus and 14 in both. Seed of the cucurbit *Sechium edule* contain at least 20 gibberellins, and seeds of the French bean (*Phaseolus vulgaris*) contain at least 16.

    (a) 56   (b) 78
    (c) 84   (d) 100
    (e) None of the above

(80) Geranylgeranyl pyrophosphate, a ———————— carbon compound, serves as the donor for all gibberellin carbon atoms.

    (a) 30   (b) 20
    (c) 10   (d) 15
    (e) 40

(81) Kaurene is the precursor of ————————————.

    (a) Gibberellins   (b) Auxins
    (c) Cytokinins   (d) Ethylene
    (e) None of the above

(82) Certain commercial ———————— that inhibit stem elongation and cause overall stunting does so in part because they inhibit gibberellin synthesis. These products include phosphon D, Amo-1618, CCC or cycocel, ancymidol and paclobutrazol. The first two, block conversion of geranyl geranyl pyrophosphate to copalylpyrophosphate. Phosphon D also inhibits subsequent formation of kaurene, whereas ancymidol and paclobutrazol block oxidation reactions between kaurene and kaurenoic acid. However, phosphon D, Amo 1618 and CCC inhibit sterol synthesis.

    (a) Growth retardants   (b) Growth inhibitors
    (c) Gibberellins   (d) Auxins
    (e) Ethylene

(83) Three actions of gibberellins are.

    (a) Stimulation of cell fructose   (b) Increase growth stem
    (c) Sucrose into glucose   (d) All of the above
    (e) Increase cell-wall plasticity

(84) Gibberellins (especially GA_4 and GA_7) cause ———————————— development in some vegetable crops.

    (a) Parthenocarpic fruit   (b) Stimulate growth
    (c) Stimulate rooting   (d) Ripening of fruits
    (e) None of the above

(85) Gottlief Haberlandt discovered cytokinin in Austria in ————————————.

    (a) 1920   (b) 1932
    (c) 1884   (d) 1913
    (e) 1987

(86) Cytokinins are as substituted ———————— compounds that promote cell division.
   (a) Adenine
   (b) Thiamine
   (c) Uracil
   (d) Guanine
   (e) None of the above

(87) The most commonly known synthetic cytokinins are ————————.
   (a) Alar and MH
   (b) CCC and Paclobutrazol
   (c) NAA and NOA
   (d) IAA and IBA
   (e) BA and kinetin

(88) Promalin is a ————————, constituting two molecules of gibberellins (GA_4 and GA_7) and one molecule of BA.
   (a) Growth inhibitor
   (b) Growth promoter
   (c) Fruit ripener
   (d) Rootner
   (e) None of the above

(89) The Russian physiologist Dimitry N. Neljubow first established that ethylene affects plant growth. In ————————, he identified ethylene in illuminating gas and showed that causes a triple response on pea seedlings, inhibited stem elongation, increased stem thickening and a horizontal growth habit.
   (a) 1888
   (b) 1901
   (c) 1922
   (d) 1864
   (e) None of the above

(90) Ethephon (2-chloroethyl phosphoric acid) is best known synthetic ———————— compound.
   (a) Auxin
   (b) Gibbrellin
   (c) Ethylene
   (d) Cytokinin
   (e) None of the above

(91) Plant growth retardants counteract the ———————— biosynthesis.
   (a) Auxin
   (b) ABA
   (c) GA
   (d) MH
   (e) None of the above

(92) Paclobutrazol (PP_{333}, cultar, bonzi), triapenthenol (RSW 0411) and uniconazole (S-3307; XE-1019) are plant ————————.
   (a) Auxins
   (b) Gibberellins
   (c) Growth inhibitors
   (d) Growth retardants
   (e) Fruit ripners

(93) Flowering hormone is known as ——————————. This term was given by Soviet plant physiologist M. Kh. Chailakhyan.
    (a) Anti auxin   (b) Anti gibbrellin
    (c) Florigen   (d) A and b both
    (e) None of the above

(94) Triacontanol is a plant ——————————. Triacontanol has been known as a constituent of bee's wax and the cuticle of many leaves. It is a 30-carbon, saturated primary alcohol first isolated from shoots of alfalfa. It is insoluble in water (less than 2×10^{-16} M or 9×10^{-14} g L^{-1}).
    (a) Herbicides   (b) Fungicides
    (c) Growth regulators   (d) Nematicides
    (e) Bio - agents

(95) ——————————, a growth retardant at 125 ppm has beneficial effect on the osmotic adjustment, plant biomass and yield of tomato plants grown under water stress conditions.
    (a) Alar   (b) Cycocel
    (c) Mepiquatchloride   (d) Paclobutrazol
    (e) Promalin

(96) —————————— are compounds in yam plants (*Dioscorea batata*) that seem to cause dormancy in bulbils (vegetative reproductive structures) that arise from swelling of the aerial lateral buds.
    (a) Batasins   (b) Mepiquatchloride
    (c) Paclobutrazol   (d) Uniconzol
    (e) None of the above

(97) Brassins or brassinosteroids is steroid —————————— first isolated from pollen grains of rape plants but now known to be present in several other species as well.
    (a) Weedicide   (b) Herbicides
    (c) Growth retardants   (d) Growth promoter
    (e) None of the above

(98) Salicylic acid (2-hydroxybenzoic acid) is a —————————— important for some known physiological responses.
    (a) Insecticides   (b) Virocide
    (c) Plant harmone   (d) Nematicide
    (e) None of the above

(99) Turgorins are also plant ——————————.
    (a) Growth regulators   (b) Bio- agent
    (c) Fungicide   (d) Nematicide
    (e) None of the above

14   Vegetable Science: Objective Type

(100) In ————————, abscisic acid was first identified and chemically characterized in California by Frederick T. Addicott and his co-workers.
- (a) 1963
- (b) 1988
- (c) 1945
- (d) 1962
- (e) 1965

(101) ABA is a ———————— carbon sesquiterpenoid synthesized partly in chloroplasts and other plastids by the mevalonic acid pathway. Thus early reaction in ABA synthesis are identical to those of isoprenoids such as gibberellins, sterols and carotenoids.
- (a) 10
- (b) 15
- (c) 20
- (d) 30
- (e) 40

(102) Abscisic acid (ABA) was originally described as ———————— inducing and abscision-accelerating substance.
- (a) Sprouting
- (b) Rooting
- (c) Dormancy
- (d) A and C both
- (e) None of the Above

(103) Abscisic acid may act on plant system as
- (i) ABA does not exert a general inhibition of DNA transcription but may inhibit the synthesis of a limited number of specific mRNA species, e.g., the a-amylase mRNA of the cells of the aleurone layer in barley seeds.
- (ii) ABA exert its main effect by blocking protein synthesis at a post transcriptional level such as mRNA processing or translation rather than by preventing DNA transcription, e.g., ABA has been shown to inhibit the synthesis of protease and isocitrate lyase in the embryos of cotton seeds even though appropriate mRNAs are present.
- (iii) The other major effect is inhibition of RNA and protein synthesis. Thus inhibits the synthesis of specific enzymes for e.g. inhibition of a number of enzymes like -amylase, protease a-glucanase and ribonuclease has been reported to be suppressed by ABA. In most of the studies, ABA has been shown to inhibit the accumulation of labelled RNA precursors into all fractions of RNA as also in specific RNA fractions. This could be due to an increased activity of nucleases, decreased availability of template or inhibition of polymerase activity.

- (a) Only (i)
- (b) Both (i) and (ii)
- (c) All (i), (ii) and (iii)
- (d) Only (iii)
- (e) None of the above

(104) The History of Vegetable production under protected cultivation started from Roman emperor Tibenus Caeser during ———————— AD who had advised to eat one cucumber every day by royal doctors
- (a) First century
- (b) Second century
- (c) Third century
- (d) None of them

(105) The first use of polyhouse as a green house cover was in ─────── when Professor Emery Myers Emmett at University of Kentuckey used the less expensive material in place of more expensive glass.

    (a) 1948   (b) 1890

    (c) 1968   (d) 1977

(106) ─────── is very old green house cultivation

    (a) Germany   (b) China

    (c) Holland   (d) Turkey

(107) ─────── are the largest users of protected cultivation in Asia as well as rest of the world

    (a) India   (b) Israel

    (c) China and Japan   (d) None of them

(108) India has entered into the arena of protected cultivation users more recently. The green house is increasing at about ─────── per cent per year.

    (a) 30   (b) 10

    (c) 20   (d) 15

(109) Organic production refers to organically grown crops which are not exposed to any chemical right from the stage of seed treatment to the final post harvest handelling and processing. It is based on the recycling of

    (a) Only natural organic matter   (b) Only crop rotation

    (c) Both a and b   (d) None of them

(110) Nature forming was developed in Japan in ─────── by Mokichi Okada who later formed the "Mokichi Okada Association (MOA).

    (a) 1930   (b) 1945

    (c) 1965   (d) 1985

(111) Poneers in organic farming include names like ───────.

    (a) Sir Albert Howard and Lady Eva Balfour

    (b) Dr. R. K. Pathak   (c) Vandana Shiva

    (d) All of them

(112) ─────── in short can be summerised up as: putting ones energies into supporting the good, rather than fighting the bad.

    (a) Biodynamic   (b) Biopesticide

    (c) Biological control   (d) None of them

(113) The concept of biodynamic is based on polarity of planets

    (a) Only Sun   (b) Only Moon

    (c) All the planets   (d) None of them

(114) Cow Pat Pit (CPP) is a biodynamic field preparation also called as ───────.

    (a) Soil ammender   (b) Soil Shampoo

    (c) Soil reactor   (d) None of them

(115) Vermiculture biotechnology is an aspect involving the use of ———————— as versatile natural bioreactor for effective recycling of non-toxic organic wastes to the soil.
    (a) Earthworm   (b) Bioagents
    (c) Biopesticides   (d) All of the above

(116) The ability to fix nitrogen by certain spirilla (*Spirillum lipoferum*) was first recorded by Beijermck in ————————.
    (a) 1930   (b) 1925
    (c) 1945   (d) 1965

(117) It is estimated that 1800 worms which is an ideal population for one square meter can feed on ———————— tonnes of humus per year.
    (a) 19   (b) 50
    (c) 60   (d) 80

(118) Organic farming practices includes the use of ————————
1. Balanced rotation systems.
2. Organic materials, farmyard manure, composted wastes and green mnaures.
3. Specific cultivation techniques- mulching stale and seed beds.
4. Management of natural habitats.

    (a) Only 1 and 2   (b) Only 3 and 4
    (c) Only 2 and 3   (d) All 1, 2, 3, 4,

(119) With the passage of the Organic Food Production Act 1990, the USDA began the process of developing federal standards for organic foods in ————————
    (a) 1999   (b) 1968
    (c) 1990   (d) None of them

(120) IFOAM stands ————————
    (a) International Federation of Association of Management
    (b) International Federation of Organic Agriculture Movements
    (c) International Farmer Associsation and Management
    (d) None of them

(121) IOFGA stands ————————
    (a) Indian Organic Farmers and Growers Association
    (b) International Association Organic Farmers and Growers
    (c) Irish Organic Farmers and Growers Association
    (d) Organic Farmers and Growers Association

(122) OFPA stands ————————.
    (a) Organic Foods Production Act   (b) Organic Fertilizer Production Act.
    (c) Organic Fish Production Act   (d) Organic Fruit Production Act
    (e) None of the above

(123) NOSB stands ─────────────.
    (a) National Observation   (b) National Organic Standard Board
        Standard Board
    (c) A and b both   (d) None of above

(124) India is the world's largest producer of certified ─────────────.
    (a) Organic ginger   (b) Organic tea
    (c) Organic apple   (d) None of them

(125) The headquarter of Institute for Market Ecology (IME) is in ─────────.
    (a) Brazil   (b) Switzerland
    (c) Tiawan   (d) None of them

(126) ───────────── is a joint creation of the FAO and the WHO to implement the joints FAO / WHO standards programme.
    (a) Irish Organic Farmers and Growers Association
    (b) Codex Alimentaius Commission
    (c) Both a & b   (d) None of above

(127) NPOP stands ─────────────.
    (a) National Program for Production   (b) National Produce Office
    (c) National Program for Organic Production   (d) None of the above

(128) Nicotine sulphate, sabadilla, rotenone, neem and pyrethrum/ pyrethrins are .
    (a) Chemical pesticide   (b) Botanical pesticide
    (c) Nematicide   (d) None of them

(129) Sulphur and lime sulphur are ─────────────.
    (a) Mineral based pesticides   (b) Natural pesticide
    (c) Organic pesticide   (d) None of them

(130) ───────────── is considered as organic zone in India.
    (a) Western UP   (b) NEH Region
    (c) Punjab   (d) Uttarakhand

(131) Symcha Blass, an Israli engineer, first visualized vegetable production through micro- irrigation in ─────────.
    (a) 1968   (b) 1959
    (c) 1972   (d) None of them

(132) The experiment on vegetables indicates that micro-irrigation can save water up to ───────── per cent with increase yield up to ───────── per cent, respectively.
    (a) 79 and 88   (b) 30 and
    (c) 50 and 60   (d) None of them

(133) The net gain in return under drip irrigation system was found considerably higher in ——————.

(a) Brinjal (b) Tomato
(c) Bottle gourd (d) None of them

(134) The potentiality of drip irrigation in India is about ——————.

(a) 15 % (b) 27%
(c) 80% (d) None of them

(135) It has been observed that soil solarization increases —————— percent yields over control in corky root disease in tomato.

(a) 1 to 2 (b) 10.0 to 35.0
(c) 5 to 10 (d) 10 to 15
(e) 15 to 20

(136) In biological contol of diseses, the antagonists like *T. virens* produces ————.

(a) Gliotoxin (b) Trichdermin
(c) Only a (d) Both a & b
(e) None of the above

(137) In biocontrol of —————— disease of vegetables, Trichoderma has been extensively exploited because of its inhabitant nature.

(a) Soil born (b) Air born
(c) Seed born (d) Both seed and air born
(e) None of the above

(138) Multiplication of antagonists depend upon to pH, temperature and substrate. The best pH is 5.5 to 6.0 but it can grow from pH range of 5.0 -9.0. The most favourable temperature is between.

(a) 5 to 10°c (b) 15 to 20°c
(c) 35 to 40°c (d) 25 to 30°c
(e) None of abobe

(139) Many countries like —————— have under taken commercial use of bio- pesticides in the management of soil born disease.

(a) Only Israel (b) Only Bulgaria
(c) Only USA (d) Only Australia
(e) All of the above

(140) Antagonists are —————— to particular soil and location specific.

(a) Tolerant (b) Resistant
(c) Succeptible (d) None of above

(141) Insect –pests are one of the major factors in minimizing productivity, causing more than ———————————— percent of yield loss.
    (a) More than 10   (b) More than 20
    (c) More than 30   (d) More than 40
    (e) None of above

(142) The concept of IPM was formulated by Stern and Co- worker in —————.
    (a) 1959   (b) 1969
    (c) 1979   (d) 1989.
    (e) 1949

(143) Macrcos (1988) suggested that IPM is ecological complexities —————.
    (a) Only population   (b) Only community
    (c) Only ecosystems   (d) All a,b and c
    (e) None of above

(144) In African marigold, tight bud stage functions a good source to trap the adults of _____.
    (a) Aphid   (b) Gandhi bug
    (c) Fruit borer   (d) Jassid
    (e) None of above

(145) Used as a trap crop _____ along with cabbage has been successfully utilized for the control of diamond back moth, aphid and leaf webber.
    (a) Castor   (b) Maize
    (c) Jawar   (d) Mustard
    (e) None of above

(146) Among egg parasites, *Trichogramma* spp has been utilized to some extent for control of _____
    (a) Leaf curl virus   (b) Leaf minor
    (c) Tomato fruit borer   (d) RKN
    (e) All of the above

(147) ———————————— is the most extensively used as biocontrol agent in vegetables
    (a) Azatobactor   (b) Aspergillous
    (c) Bt   (d) None of above

(148) On the basis of eggs, economic threshold for tomato, fruit borer has been determined ———————————— for spray initiation with 0.07 per cent endosulfan .
    (a) 8 egg /15 plants   (b) 18 egg /15 plants
    (c) 28 egg /25 plants   (d) 38 egg /35 plants
    (e) 48 egg /45 plants

(149) Reduction on yield of vegetables due to weeds ranges from _____ in potato.
- (a) 25-35%
- (b) 70-80%
- (c) 67%
- (d) 6-82%
- (e) 42-71%

(150) On an average weeds extract two times more N and Ca and ─────── more potassium than the crops.
- (a) 35%
- (b) 45%
- (c) 55%
- (d) 25%
- (e) 15%

(151) Pendimethalin and butachlor are ─────── type of weedicide.
- (a) Post emergence
- (b) Pre-emergence
- (c) Early post emergence
- (d) Emergence
- (e) None of the above

(152) Fluchloralin & Trifluralin are applied in vegetable production as _____
- (a) Pre emergence
- (b) Post emergence
- (c) Pre- plant incorporation
- (d) Early post emergence
- (e) None of the above

(153) Recommended dose of pendimethalin is 0.65 to 1.0 kg a.i./ha, while butachlor applied at ─────── a.i./ ha in vegetables.
- (a) 2.0 kg
- (b) 4.0 kg
- (c) 8.0 kg
- (d) 10.0 kg
- (e) 1 2.0 kg

(154) The critical stage for weed control in onion and garlic are ───────.
- (a) 25 to 30
- (b) 30 to 75
- (c) 30 to 45
- (d) 15 to 20
- (e) None of the above

(155) The critical stage for weed control in potato and radish are ───────.
- (a) 25 to 30
- (b) 30 to 75
- (c) 30 to 45
- (d) 15 to 20
- (e) None of the above

(156) The critical stage for weed control in tomato and chilli are ───────.
- (a) 25 to 30
- (b) 30 to 75
- (c) 30 to 45
- (d) 15 to 20
- (e) None of the above

(157) The critical stage for weed control in carrot and beans are —————————.
   (a) 25 to 30   (b) 30 to 75
   (c) 30to45   (d) 15 to 20
   (e) None of the above

(158) Pectin is used as a thickening agent in the preparation of —————————.
   (a) Squash   (b) syrup
   (c) Ketchup   (d) Jelly
   (e) Both

(159) Post-harvest losses are noticed _____ in India
   (a) 30-40 per cent   (b) 15-20 per cent
   (c) 40-50 per cent   (d) 20-25 per cent
   (e) None of the Above

(160) Most commonly used antioxidants in vegetable products are _____
   (a) Butylated hydroxy anisole (BHA)   (b) Butylated hydroxy toluene (BHT)
   (c) Ascorbic Acid   (d) Tocopherols and prophyl gallate (PG)
   (e) All of above

(161) Food additives are used for are main reasons. (i) To maintain product consistency. (ii) To improve or maintain nutritional value. (iii)To maintain palatability and wholesomeness. (iv) To provide leavening or control acidity/alkalinity. (v) To enhance flavour or impart desired colour.
   (a) Only (i)   (b) Only (v)
   (c) Both (ii) and (iv)   (d) All (i), (ii), (iii), (iv) and (v)
   (e) None of the above

(162) ————————— present in the vegetables such as potato, amaranths, spinach, celery, sugar beets and colocasia etc. Causing the corrosion in the human mouth and gastrointestinal tract, gastric hemorrhages, renal failure, renal colic, bloody urine etc.
   (a) Citric acid   (b) Oxalic acid
   (c) Pectin   (d) Nicotinic acid
   (e) None of the above

(163) *Clostridium botulinum* is common infection in canned vegetable products and could be over comed by heating can to ————————— °C for 5 minutes or 115 °C for 15 minutes.
   (a) 165   (b) 104
   (c) 98   (d) 121
   (e) None of the Above

(164) —————————— is a technique of identifying and monitoring those processing points where contamination can best be controlled.

(a) PFAA (b) HACCP
(c) FPO (d) IQF
(e) All of them

(165) Main advantages of primary processing are (i) Reduction of bulk. (ii) Longer shelf life. (iii) Less transportation cost. (iv) Conventional handling and (v) Employment generation

(a) Only (i) (b) Only (ii)
(c) Only (iii) (d) All (i), (ii), (iii), (iv) and (v)
(e) None of the above

(166) —————————— is the cooling of fresh produce in order to remove field heat and is device to enhance the shelf life of vegetables.

(a) Pasteurization (b) Precooling
(c) Sterilization (d) B and C both
(e) None of the above

(167) Cold water treatment by immersion or showering of water is ——————.

(a) Hydrocooling (b) MAS
(c) Blanching (d) Lacquering
(e) None of the above

(168) In Modified Atmosphere Storage (MAS) the build up of respiratory gases such as —————— inside packed vegetables for extending the shelf life of storage fresh produce is called as MAS.

(a) CO_2 (b) H_2O
(c) H (d) N
(e) None of the above

(169) Removal of unwanted organisms by heat, radiation, chemicals or by filtration is known as ——————.

(a) Sterilization (b) Pasteurization
(c) Blanching (d) Bleaching
(e) None of the above

(170) In pasteurization, reduction of the number of microorganism in milk by maintaining it in a holder at a temperature of from —————————— for 30 minutes.

(a) 62.8 °C to 65.5 °C (b) 72.8 °C to 75.5 °C
(c) 62.8 °C to 85.5 °C (d) 42.8 °C to 65.5 °C
(e) 52.8 °C to 65.5 °C

(171) Factors, which influence the rate of heat transfer during vegetables freezing are (i) Thermal conductivity of the vegetables. (ii) Area of vegetables available for heat transfer. (iii) Distance that the heat must travel through the vegetables. (iv) Temperature difference between vegetables and freezing medium. (v) Insulating effect of the boundary film of air surrounding the vegetables.

    (a) Only (i)   (b) Only (ii)
    (c) Both (i) and (ii)   (d) All (i), (ii), (iii), (iv) and (v)
    (e) None of the above

(172) Undesirable changes to some vegetables occur when the temperature is reduced below a specific optimum level and such changes are collectively known as ——————.

    (a) Chilling injury   (b) Freezing injury
    (c) Heat injury   (d) Sun burning
    (e) None of the above

(173) —————————————— refers to freezing of water present in intercellular spaces resulting in the formation of ice crystals. These ice crystals then penetrate the cells and result in their damage.

    (a) Chilling injury   (b) Pasteurization
    (c) Freezing injury   (d) A and C both
    (e) None of the above

(174) —————————— commonly used plastic packaging materials for fresh vegetables are: (i) Plastic field boxes (ii) Plastic boxes (iii) Corrugated plastic cartons (iv) Rigid plastic trays (v) Polyethylene bags

    (a) Only (v)   (b) Only (i)
    (c) All (i), (ii), (iii), (iv) and (v)   (d) Both (i) and (ii)
    (e) None of the above

(175) Commonly used plastic packaging materials for processed vegetable and are: (i) Polyethylene bags (ii) High density polyethylene containers (iii) PVC bottles/jars/tube/can etc. (iv) PET bottles/jars/containers

    (a) Only (i)   (b) Both (i) and (ii)
    (c) Only (iii)   (d) All (i), (ii), (iii) and (iv)
    (e) None of the above

(176) A process involves exposure to ionizing radiation, such as gamma rays from radioactive source and electron beams an X-ray generated by machines. When such rays pass through cells, they knock electrons off atoms and molecules, breaking chemical bonds and creating ions and free radicals is known as ———.

    (a) Irradiation   (b) Sterilization
    (c) Pasteurization   (d) Dehydration
    (e) Sun drying

(177) Inactivation of genes involved in ethylene synthesis was first achieved in ——————— Two small multigene families encoding ACC synthase (ACS) and ACC oxidase (ACO) control the biosynthesis of ethylene.

(a) Brinjal   (b) Tomato
(c) Pea   (d) Watermelon
(e) None of the above

(178) The first genetically modified plant food to enter the market place was the "Flavr Savr" ———————. This is a low PG fresh market ———————, first sold by Calgene in 1994 in the USA.

(a) Tomato   (b) Chilli
(c) Okra   (d) Spinach
(e) None of the above

(179) A low ethylene variety produced by sense-suppression of ACS was marketed in the USA in ———————. Reduction in ethylene is likely to be beneficial for climateric vegetables like tomato.

(a) 1965   (b) 1978
(c) 1995   (d) 1998
(e) None of the above

(180) Different pre-drying treatments, which are common for vegetables drying, are— ——————— (i) Selection and sorting for size and soundness (ii) Washing (iii) Peeling by hand, lye solution or abrasions (iv) Cutting into slices, cubes etc and (v) Blanching.

(a) Only (i)   (b) Only (ii)
(c) All (i), (ii), (iii), (iv) and (v)   (d) Only (v)
(e) None of the above

(181) The objectives of blanching treatments are ——————— (i) To inactivate enzyme. (ii) To expel respiratory air (gases). (iii) To remove raw flavour. (iv) Fixation of colour. (v) Softening of the tissue to reduce bulk storage. (vi) Reduces microbial load.

(a) Only (i)   (b) Only (ii)
(c) Only (v)   (d) All (i), (ii), (iii), (iv) (v) and (vi)
(e) None of the above

(182) In exhausting, removal of air by passing through exhausting box where steam circulation around it. Exhausting is done at 190 °F for about ———————.

(a) 15 minutes   (b) 10-20 minutes
(c) 5-7 minutes   (d) 1-2 minutes
(e) None of the above

(183) Main disadvantages of solar drying are ─────── (i) Inability to control. (ii) Large area requirement. (iii) Degradation of product by bio-chemical and microbiological reaction. (iv) Uneven drying. (v) Long drying time.
    (a) Only (i)   (b) Only (ii)
    (c) Only (iii)   (d) All (i), (ii), (iii), (iv) and (v)
    (e) None of the above

(184) Advantages of microwave drying are ─────── (i) A penetrating quality that leads to uniform drying. (ii) Selective absorption by liquid water, which leads to uniform moisture profile within the particle. (iii) Case of control due to rapid response of such heating.
    (a) Only (ii) and (iii)   (b) Only (i)
    (c) All (i), (ii) and (iii)   (d) Only (iii)
    (e) None of the above

(185) Freeze drying is also called as ─────────────, is characterized by heating (i.e. conduction, radiation, radio frequency, microwaves etc.) a moist frozen product in a vacuum chamber maintained at an absolute pressure below the vapour pressure of ice within the product. When the material is heated, the ice sublimes.
    (a) Lyophilization   (b) Iceing
    (c) Precooling   (d) Pasteurization
    (e) None of the above

(186) Main advantages of freeze drying are ───────.
    (a) High flavour, aroma and colour retention   (b) High retention of nutritional value
    (c) Both A and B   (d) Only A
    (e) None of the above

(187) Osmotic dehydration happens in the immersion of cut vegetables in concentrated solution of ───────.
    (a) Salts or sugars   (b) Citric acid
    (c) Acetic acid   (d) Sodium benzoate
    (e) Potassium meta bi sulphide

(188) Importance of fermented vegetables is ─────── (i) Preservation of vegetables. (ii) Develop characteristic flavour, aroma and texture. (iii) Destroy naturally occurring toxins and undesirable components in raw material (iv) Improve digestibility. (v) Enrich products with desired microbial metabolites. (vi) Create new products for new markets. (vii) Enhance dietary value.
    (a) Only (iv)   (b) Only (i)
    (c) Only (v)   (d) All (i), (ii), (iii), (iv), (v), (iv), (vi) and (vii)
    (e) None of the above

(189) Important fermented vegetable products are ———— (i) Sauerkraut (ii) Carrot pickles (iii) Cucumber pickles (iv) Sweet turnip pickles and (v) Cauliflower pickles.

    (a) Only (v)   (b) Only (iii)

    (c) All (i), (ii), (iii), (iv) and (v)   (d) Only (i)

    (e) None of the above

(190) Quality of vegetables is characterized by ————————————.

    (a) Shape   (b) Size

    (c) shape, size, flavour, firmness, colour, composition, freeness from diseases etc   (d) Only A

    (e) Only C

(191) Non-destructive techniques used for quality evaluation of vegetables are ————————————.

    (a) DLE   (b) NMR

    (c) NIS   (d) Ultrasonic

    (e) All of the above

(192) General Food Texturometers and Universal Testing machines are two important ———————————— measurement equipments for vegetables.

    (a) Shape   (b) Size

    (c) Texture   (d) Colour

    (e) Thickness

(193) ———————————— is an edible product preserved and flavoured in a solution of common salt and vinegar spices and oils are also added.

    (a) Ketchup   (b) Sauce

    (c) Pickle   (d) Puree

    (e) None of the above

(194) ———————————— may be defined as the food, which can be eaten as such, without dehydration, and is self-stable without refrigeration or thermal processing.

    (a) IMF   (b) FPO

    (c) PFAA   (d) IQF

    (e) None of the above

(195) The parameters like water activity, high temperature, low temperature, acidification (pH), redox potential and preservatives are called ————————————.

    (a) Acidification   (b) Pasteurization

    (c) Sterilization   (d) Hurdles

    (e) None of the above

(196) ———————————— can be defined as a pressure driven membrane process that can be used in the fractionation, purification and concentration of the substances having a molecular weight between 10^3 to 10^6 daltons.
   (a) Ultrafiltration (b) Pasteurization
   (c) Sterilization (d) Sulphitation
   (e) None of the above

(197) Blanched vegetables are sulphited by placing them in ———————————— solution of potassium metabisulphite for sufficient time.
   (a) 1% (b) 2%
   (c) 0.5% (d) 5%
   (e) None of the above

(198) ISO-9001, ISO-9002, ISO-9003 are used for ———————— assurance purpose in contractual situations.
   (a) Internal quality of food (b) External quality
   (c) Both A and B (d) None of the above

(199) ISO-9004 gives guidance to all organization for ———————— management.
   (a) Internal (b) External quality
   (c) A and B both (d) None of the above

(200) 'Greening' of potato is a physiological disorder and caused by————————.
   (a) Nitrogen deffiency (b) Boron defficency
   (c) Potash defficency (d) Exposure of the tuber to the direct sun light
   (e) None of the above

(201) Elimination of fruits under sized by manually or mechanically is known as ————————.
   (a) Presizing (b) Grading
   (c) Screening (d) None o f the above

(202) Vegetables are sorted by quality into two or more grades as per the standards available in that country is known as ————————.
   (a) Storing (b) Waxing
   (c) Grading (d) All of above

(203) In case of potato, post harvest application of a growth regulator CIPC (Isopropyl N-3 chlorophenyl carbamate) is very effective in reducing sprouting during storage at a comparatively higher temperature around ————————.
   (a) 15 to 20 °C (b) 5 to 10 °C
   (c) 25 to 30 °C (d) 35 to 40 °C
   (e) None of the above

(204) —————— is defined as the ratio of the vapour pressure (VP) of a solution to that of pure water at a specified temperature.
   (a) Water potential   (b) Water activity
   (c) Water retention   (d) All of the above

(205) In —————— strictly controlled with certain specific gaseous concentrations of N_2, CO_2 and O_2 inside the storage package material is termed as controlled atmosphere (CA) storage.
   (a) Atmospheric storage   (b) Cold stroage
   (c) Zero Cool energy Chamber   (d) Controlled atmosphere storage

(206) —————— is a distinct type of product which contains the whole tomato paste inclusive of skin and seeds and thus differs from tomato puree and paste.
   (a) Tomato puree   (b) Tomato crush
   (c) Tomato paste   (d) Tomato ketchup

(207) —————— should contain minimum percentage of acidity of one per cent as acetic acid and minimum total soluble solids of 25 per cent (w/w).
   (a) Tomato crush   (b) Tomato paste
   (c) Tomato juice   (d) Tomato ketchup

(208) Tomato sauce should contain minimum percentage of acidity of 1.2 per cent as acetic acid and minimum total soluble solids of —————— per cent (w/w).
   (a) 20   (b) 25
   (c) 15   (d) 35
   (e) None of the above

(209) Tomato juice should contain minimum percentage of total soluble solids as 5 per cent whereas tomato soup should contain minimum of —————— per cent total soluble solids.
   (a) 10   (b) 7
   (c) 15   (d) 22
   (e) 20

(210) —————— can be made by the 'cold process' or by the 'hot process'
   (a) Ketchup   (b) Jelly
   (c) Syrup   (d) Jam
   (e) None of the above

(211) —————— are used in canning of vegetables.
   (a) Sugar   (b) Oil
   (c) Spice   (d) Brine
   (e) None of the Above

(212) The Brix or Balling hydrometer gives directly the ——————— percentage of by weight, in the syrup.
   (a) Salt (b) Oil
   (c) Spices (d) Sugar
   (e) None of the above

(213) Strength of brine is measured by ———————.
   (a) Salometer (b) Hydrometer
   (c) Hygrometer (d) Thermometer
   (e) None of the above

(214) The salometer is calibrated from 0 to 100 degrees at ———————.
   (a) 15°C (b) 30°C
   (c) 20°C (d) 10°C
   (e) None of the above

(215) In swell ——————— can, the ends are tightly bulged. The bulge is due to the formation of carbon dioxide or other gases inside the can as a result of decomposition of the contents caused by microorganism.
   (a) Swell (b) Springer
   (c) Flipper (d) Flat sour
   (e) None of above

(216) A mild swell at one or both ends of a can called a ———————.
   (a) Springer (b) Swell
   (c) Flat sour (d) A and B both
   (e) None of the Above

(217) A can with a mild positive pressure is called a ———————.
   (a) Swell (b) Flat sour
   (c) Flipper (d) Springer
   (e) None of the above

(218) The product inside the can has a sour odour, and its acidity will be much higher than that of the normal product. This type of spoilage is mostly caused in non-acid vegetables by microorganism is known as ———————.
   (a) Flipper (b) Flat sour
   (c) Swell (d) Springer
   (e) Swell

(219) Tomato pulp can be prepared either 'hot' process or 'cold' process.
   (a) Tomato pulp (b) Tomato ketchup
   (c) Tomato sauce (d) Tomato puree
   (e) All of the above

(220) Advantages of glacial acetic acid, used in tomato products are (i) It is cheaper than vinegar. (ii) It does not impart any colour of its own to the product. (iii) Can be added at the end of boiling without causing any loss of flavour or acidity.

    (a) Only (i)   (b) Only (ii)

    (c) All (i), (ii) and (iii)   (d) Only (iii)

    (e) None of the Above

(221) Food colours used are mostly the permitted coaltar dyes, which are designated as synthetic colours by —————.

    (a) UNO   (b) WHO

    (c) UNICEF   (d) IFO

    (e) None of the above

(222) Cherry tomato is preferred for export to —————.

    (a) Asia   (b) Europe

    (c) China   (d) Japan

    (e) None of above

(223) The export quality tomato producing area is —————.

    (a) Nasik and Pune in Maharashtra

    (b) Banglore in Karnataka

    (c) Western Uttar Pradesh

    (d) Central India

    (e) a and b both

(224) In processed products of tomato especially _____ have great demand for export.

    (a) Tomato juice   (b) Tomato puree and paste

    (c) Tomato Ketchup   (d) Tomato concentrates

    (e) Tomato cocktail

(225) The Netherland is the leading exporter of —————, contribute about 21% of world export.

    (a) Tomato   (b) Chilli

    (c) Onion   (d) Potato

    (e) None of the above

(226) ————— is the biggest onion market in India.

    (a) Gujarat   (b) Maharashtra

    (c) Punjab   (d) South India

    (e) None of the above

(227) —————— export is of big onion varieties, viz. N-53, N-2-4-1, Bellary Red, Patana red, Agrifound Dark red, Agrifound Light red, Pusa red, Pusa Madhvi and Arka Niketan.

    (a) 15%   (b) 25%
    (c) 5%   (d) 75%
    (e) 90%

(228) G-282 variety of —————— is suitable for export purpose.

    (a) Onion   (b) Okra
    (c) Garlic   (d) Tomato
    (e) Chilli

(229) For fresh fruit export, okra fruits should be green, tender, —————— long.

    (a) 3-6 cm   (b) 6-9 cm
    (c) 16-19 cm   (d) 26-29 cm
    (e) 10-20 cm

(230) Pusa Sawani, Parbhani Kranti, Versha Uphar and Pusa A-4 are varieties of —————— suitable for export.

    (a) Chilli   (b) Okra
    (c) Pea   (d) Frenchbean
    (e) None of above

(231) —————— major producer and exporter of fenugreek seed spice

    (a) China   (b) Japan
    (c) India   (d) USA
    (e) None of the above

Answer Sheet

1	b	2	e	3	a	4	d	5	c	6	b	7	d	8	c	9	d	10	b
11	e	12	c	13	d	14	d	15	c	16	b	17	b	18	a	19	b	20	c
21	e	22	b	23	c	24	d	25	c	26	d	27	c	28	d	29	c	30	b
31	a	32	d	33	a	34	c	35	c	36	d	37	d	38	d	39	b	40	a
41	b	42	a	43	b	44	b	45	c	46	a	47	a	48	c	49	b	50	c
51	b	52	c	53	c	54	c	55	a	56	d	57	b	58	b	59	a	60	b
61	c	62	c	63	b	64	b	65	d	66	c	67	a	68	d	69	a	70	b
71	b	72	b	73	a	74	a	75	c	76	d	77	b	78	b	79	c	80	b
81	a	82	a	83	d	84	a	85	d	86	a	87	e	88	b	89	b	90	c
91	c	92	d	93	c	94	c	95	c	96	a	97	d	98	c	99	a	100	a
101	b	102	c	103	c	104	a	105	a	106	c	107	b	108	a	109	c	110	a
111	d	112	a	113	d	114	b	115	a	116	b	117	d	118	d	119	c	120	b
121	c	122	a	123	b	124	b	125	b	126	b	127	c	128	b	129	a	130	b
131	b	132	a	133	b	134	b	135	b	136	a	137	a	138	d	139	e	140	b
141	d	142	a	143	d	144	c	145	d	146	c	147	c	148	a	149	d	150	d
151	b	152	a	153	a	154	b	155	b	156	c	157	d	158	e	159	b	160	e
161	d	162	b	163	d	164	b	165	d	166	b	167	a	168	a	169	a	170	a
171	d	172	a	173	c	174	c	175	d	176	a	177	b	178	a	179	c	180	c
181	d	182	c	183	d	184	c	185	a	186	c	187	a	188	d	189	c	190	e
191	e	192	c	193	c	194	a	195	d	196	a	197	-	198	b	199	a	200	d
201	a	202	c	203	a	204	b	205	d	206	b	207	d	208	c	209	b	210	c
211	d	212	d	213	a	214	a	215	a	216	a	217	c	218	b	219	a	220	c
221	b	222	b	223	e	224	b	225	c	226	b	227	e	228	c	229	b	230	b
231	c																		

1.1.2. Vegetable Crops Based

1.1.2.1. Bulb Vegetables

(1) Garlic (*Alium sativum* L.) belongs to the family Alliaceae, is the second most important bulb crop in India, the first ———.

(a) Leek
(b) Onion
(c) Shallot
(d) Multiplier onion
(e) None of the Above

(2) Garlic bulb is a multiple or compound bulb consisted of small bulblets or segments or popularly known as ——————— which are covered with a protective thin membranous sheath.

(a) Bulb
(b) Clove
(c) Tuber
(d) Rhizome
(e) None of the above

(3) Among most of cultivars available, ——————— cultivar of garlic having largest bulb size.

(a) Jamnagar Local
(b) Agrifound White
(c) Yamuna Safed-2
(d) Agrifound Parvat
(e) None of the above

(4) Garlic contains a colourless and odourless amino acid allicin of it, is ─────.
   (a) Diallyldisulphide   (b) Anthocyanin
   (c) Cucurbitacin   (d) Lycopene
   (e) None of the above

(5) Garlic is ───────── season crop
   (a) Hot   (b) Cool
   (c) Both hot and cool   (d) Rainy
   (e) None of the above

(6) Sometimes, garlic bulbs sprout in the field near harvesting time, which is a disorder, which affects yield and quality of bulbs. This disorder is well attributable due to [1] fluctuation in temperature during later stages of crop growth may enhance sometimes sprouting of cloves in some garlic strains, [2] the sprouting in cloves may also be due to some physiological changes occurs in later stages of bulbs formation, [3] excessive soil moisture in the later stage of crop and supply of nitrogen and [4] further, the degeneration of planting bulbs is also attributable.
   (a) Only 1   (b) 1 and 2 both
   (c) All 1, 2, 3 and 4   (d) Only 4
   (e) None of the above

(7) Agrifound White (G-41), Yamuna Safed-K (G-1), Yamuna Safed-2 (G-50), G-282 and Agrifound Parvati (G-313) are newly released varieties of ─────.
   (a) Onion   (b) Garlic
   (c) Multiplier onion   (d) Potato
   (e) Sweet potato

(8) Agrifound Parvati (G-313) variety of garlic is tolerant to ───── disease
   (a) Soft rot   (b) Purple blotch
   (c) Stemphyllum blight   (d) Smut
   (e) None of the above

(9) The average yield of garlic ranges ─────.
   (a) 150-200 q/ha   (b) 250-300 q/ha
   (c) 450-500 q/ha   (d) 650-700 q/ha
   (e) 50-100 q/ha.

(10) In India, major garlic growing states are ─────.
   (a) Uttar Pradesh and Bihar   (b) Madhya Pradesh and Chittisgarh
   (c) Andhra Pradesh and Orissa   (d) Gujarat, Orissa, and Madhya Pradesh
   (e) None of the above

(11) The inhalation of ———— oil or juice has generally been recommended by doctors in cases of pulmonary tuberculosis, rheumatism, sterility, impotency, cough and red eyes.
- (a) Garlic
- (b) Onion
- (c) Tomato
- (d) Sarsoo
- (e) None of the above

(12) ———— is the largest producer of garlic.
- (a) Australia
- (b) Spain
- (c) Nigeria
- (d) USA
- (e) None of the above

(13) Origin place of garlic is said to be ———— especially Mediterranean region.
- (a) South America
- (b) Central Asia and Southern Europe
- (c) India
- (d) China
- (e) None of the above

(14) Two distinct types of garlic namely Fawari and Rajalle Gaddi with slightly bigger bulbs are grown in the Bellary District of ————.
- (a) Maharashtra
- (b) Gujarat
- (c) Bihar
- (d) MP
- (e) South India

(15) ———— promoted secondary growth and suppressed bulbing in garlic. Treatment of long day (16 hour light) markedly accelerated bulbing and senescence.
- (a) Short day
- (b) Long day
- (c) Day netural
- (d) Both short day and long day
- (e) None of the above

(16) About 350-500 kg of cloves are required if planting is done at spacing of ————.
- (a) 20X 20
- (b) 15 X 10
- (c) 30 X 30
- (d) 45X60
- (e) None of the above

(17) Recovery of cloves in the garlic bulbs ranges from ————.
- (a) 45%
- (b) 35%
- (c) 90%
- (d) 65%
- (e) None of the above

(18) Carl Linnaeus gave the name *Allium sativum* to garlic in ————.
- (a) 1854
- (b) 1754
- (c) 1954
- (d) 1944
- (e) None of the above

(19) Onion belongs to family ―――――.
    (a) Araceae   (b) Alliaceae
    (c) Lilaceae   (d) Cucurbitaceae
    (e) None of the above

(20) Multiplier onion is botanically known as ―――――
    (a) *Benincasa hispada*   (b) *Allium cepa* var. sativum
    (c) *Allium cepa*   (d) *Allium cepa* var. *aggregatum*
    (e) *Allium cepa* var. *porum*

(21) The basic chromosome number of cultivated *Allium* is x = ―――――.
    (a) 8   (b) 16
    (c) 12   (d) 10
    (e) None of the above

(22) The losses due to purple blotch in onion is ranged ――――― per cent.
    (a) 10%   (b) 5%
    (c) 38-62%   (d) 15%
    (e) None of the above

(23) The genotype ――――― has been found resistant to purple blotch.
    (a) IIHR-56-1   (b) Naphad 2-4-1
    (c) Arka kalyan   (d) All of these
    (e) None of the above

(24) The varieties N-53 and Pusa Ratnar are resistant to ―――――.
    (a) Thrips.   (b) Stemphylium blight
    (c) Head borer   (d) Root borer
    (e) All of these

(25) ――――― variety is suited for both *kharif* and. *rabi* seasons.
    (a) Pusa Red   (b) Arka Niketan
    (c) Agrifound Dark Red   (d) All of these
    (e) None of the above

(26) Arka Kalyan variety of onion is local collection of ―――――.
    (a) IIHR-193   (b) IIHR56-1
    (c) Pusa Red   (d) Patana Red
    (e) IIHR-145

(27) ――――― variety of onion which can tolerate salinity.
    (a) Hisar-2   (b) Arka Pragati
    (c) Arka Kalyan   (d) Kalyanpur Red Round
    (e) All of the above

(28) ———————— variety of onion suited for salad purposes.
   (a) N-53   (b) Pusa Madhvi
   (c) VL-3   (d) Udaipur-101
   (e) None of the above

(29) Onion is richest source of ————————.
   (a) Calcium   (b) Magnesium
   (c) Vanadium   (d) Sodium
   (e) Cupper

(30) Pungency in onion is due to ————————.
   (a) Glycoalkaloids   (b) Cucurbitacin
   (c) Capsacin   (d) Allylpropyl disulfide
   (e) None of the above

(31) Onion contains an enzyme called ————————.
   (a) Lactase   (b) Hydrolase
   (c) Maltase   (d) Allinase
   (e) All of them

(32) The optimum temperature required for onion bulb development is ————————.
   (a) 30°C   (b) 15.5°C to 21°C
   (c) below 10°C   (d) Above 35°C
   (e) None of the above

(33) The cultivars grown in plains of India are ————————.
   (a) Day netural   (b) Long day
   (c) Short day   (d) a and b
   (e) None of the above

(34) For checking post emergence "damping off", drenching of the nursery with ———————— has been found most effective ————————.
   (a) 0. 1% malthion   (b) 0.5 % endosulfan
   (c) 0.1 per cent brassical/copper oxychloride (d) All of them
   (e) None of them

(35) In hilly area of India, onion seeds are sown in nursery bed in the end of February to end of May while in north India, the best time of nursery for *kharif* onion is ————————.
   (a) Feburary- March   (b) October- November
   (c) August-September (d) All of them
   (e) May-June

(36) About 10-12 kg seed in *rabi* and 12-15 kg seed in *kharif* is required to raise seedling for planting in ————.
    (a) One acre   (b) Two acre
    (c) One hectare   (d) Two hectare
    (e) None of them

(37) Best planting density for onion cultivation is ————.
    (a) 30 X 20cm   (b) 15 X 8-10 cm
    (c) 45 X 30 cm   (d) 45 X 15 cm
    (e) All of them

(38) Onion seedlings of 0.8 to 0.9 cm in diameter and about ———— in height are ready for transplanting.
    (a) 5-10cm   (b) 60-45cm
    (c) 10-15cm   (d) 20-35 cm.
    (e) None of them

(39) The origin place of onion is ————.
    (a) Central Asia   (b) South America
    (c) India   (d) China
    (e) Europe

(40) The edible portion of onion is a modified stem known as
    (a) Suckers   (b) Rhizome
    (c) Root   (d) Bulb
    (e) Leaf

(41) Onion is mainly exported from India to other countries are ————.
    (a) Only Malaysia   (b) Only Russia
    (c) Only Kuwait   (d) Dubai, Malaysia, Russia and Kuwait
    (e) Only D.

(42) Dehydrator No-8, White Imperial Spinach, Rivrina Late Brown, Country Queen, Verma Jaint and Punjab-48 varieties of onion are suitable for ————.
    (a) Salad   (b) Pickle
    (c) Juice   (d) Dehydration
    (e) All of the above

(43) ———— is the second largest producer of onion in world.
    (a) China   (b) America
    (c) India   (d) Malaysia
    (e) Kuwait

(44) ———— is a variety of onion.
    (a) Pusa Sawani (b) Punjab Naroya
    (c) Pusa Ruby (d) Punjab Padmani
    (e) None of the above

(45) Improved Mako variety of onion is tolerant for————.
    (a) Nematodes (b) Salanity
    (c) Blight (d) Thrips
    (e) Neck rot

(46) Black mould is a serious storage disease of ————.
    (a) Onion (b) Tomato
    (c) Chilli (d) Okara
    (e) Pumpkin

(47) In India, ———— is the leading state for onion production accounting for more than 20 per cent in area and 25 per cent of production.
    (a) Gujarat (b) Punjab .
    (c) Maharashtra (d) Andhra Pradesh
    (e) Madhya Pradesh

(48) ———— species of onion is considered as pink/red nodding onion.
    (a) *A. cepa* (b) *A. sativum*
    (c) *A. porum* (d) *A. cernuum*
    (e) None of the them

(49) ———— species of onion has highly fragrant yellow pendulous inflorescences.
    (a) *A. flavum* (b) *A. cernuum*
    (c) *A. sativum* (d) *A. aggregatum*
    (e) *A. cepa*

(50) *A. karataviense* species of onion preferred in ————.
    (a) Salad (b) Pickle
    (c) Rock garden (d) Dehydrastion
    (e) None of them

(51) ———— species of onion has purple, scented, pendulous inflorescence.
    (a) *A. pulchellum* (b) *A. cepa*
    (c) *A. Sativum* (d) *A. flavum*
    (e) All of the above

(52) Rakkyo is botanically known as ————.
    (a) *A. aggregatum* (b) *Allium chinese*
    (c) *A. sativum* (d) *A. flavum*
    (e) None of them

(53) Conjumatlan is cultivar of onion in which bulb formation may occurs under ──────── condition.
   (a) Long day   (b) Day netural
   (c) a and b both   (d) Only a
   (e) Short day

(54) The bulb formation in onion is governed by many factors like ────────
   (a) Photoperiod   (b) Temperature
   (c) Light quality and intensity   (d) a, b and c
   (e) None of them

(55) Early Grano and Bermudo Yellow are yellow skinned cultivar of ────────.
   (a) Onion   (b) Turnip
   (c) Carrot   (d) Radish
   (e) None of them

(56) About ──────── temperature is optimum for onion seed germination.
   (a) 15°C   (b) Above 30°C
   (c) 20-25°C   (d) Below 10°C
   (e) None of them

(57) In onion, harvesting is ────────.
   (a) Planting season   (b) Cultivars
   (c) Market price   (d) Condition of crops
   (e) All of them

(58) *Aspergillus* spp. and *Penicillium* rot are the most important ──────── diseases of onion.
   (a) Nursery stage   (b) Crop growing stage
   (c) Storage   (d) Maturity stage
   (e) None of them

(59) Yellow dwarf disease of onion is ──────── a disease.
   (a) Fungal   (b) Viral
   (c) Bacterial   (d) Mycoplasma
   (e) None of them

(60) Thrips is most destructive insect of ──────── in India.
   (a) Radish   (b) Muskmelon
   (c) Tomato   (d) Onion
   (e) Carrot

(61) The factor which influences for production of onion as seed crop are ————.
   (a) Cultivars
   (b) Bulb weight
   (c) soil and climate
   (d) Spacing, fertilizer application and date of planting
   (e) All of the above

(62) The anti fungal factor in onion is phenolic compound known as ————.
   (a) Catechol
   (b) Caffeic acid
   (c) Caumeric acid
   (d) a and b
   (e) All of the above

(63) ———————— can also be used against sun strokes (loo).
   (a) Tomato
   (b) Water melon
   (c) Onion
   (d) Cucumber
   (e) None of them

(64) The colour of the outer skin of onion bulb is due to ————.
   (a) Lycopene
   (b) Xanthophyll
   (c) Quercetin
   (d) a and b
   (e) None of them

(65) It is possible to grow onion in ————————.
   (a) Only rabi
   (b) Only in kharif
   (c) Both rabi and kharif
   (d) Only c
   (e) None of the them

(66) Bolting in onion is harmful for bulb production. Bolting may be due to the interaction of more than one factor ————.
   (a) Temperature
   (b) Cultivar
   (c) Time of planting and age of seedling
   (d) Avaliability of nutrients
   (e) None of the them

(67) Poor nitrogen application in onion may results ————.
   (a) Bolting
   (b) Poor growth
   (c) Highly pungent
   (d) All of them
   (e) None of them

(68) Onion grows better in soil pH 5.8 and 6.5 but cultivars like ———— is comparatively salt tolerant.
   (a) Pusa Red
   (b) Patna Red
   (c) Arka niketan
   (d) Hisar-2
   (e) All of the them

(69) High nitrogen application may ———————— the storage life of onion bulbs.
   (a) Enhance storage life
   (b) Reduce storage life
   (c) Moderate storage life
   (d) Only a
   (e) None of them

(70) The critical stages for irrigation in onion is ————————.
   (a) 15 days after transplanting
   (b) 30 days after transplanting
   (c) At bulb formation and enlargement
   (d) At maturity stage
   (e) None of them

(71) Curing of onion bulbs before storage i.e. after harvesting is essential for ————.
   (a) For drying
   (b) For dehydration
   (c) For better storage
   (d) a and b both .
   (e) None of the them

(72) Onion is a cross-pollinated crop. Therefore about ———— meters isolation distance is kept for foundation seed and ———— meters for certified seed between two cultivars for seed production.
   (a) 250 m and 5000
   (b) 10 m and 20 m
   (c) 1000 m and 400 m
   (d) 1600 m and 1000 m
   (e) None of the above

(73) Seed production of onion is done by seed to seed and bulb to seed method. Though, there are fair chances of getting higher seed yields in seed to seed method but quality seed is obtained by ————————.
   (a) Seed to seed
   (b) Both seed to seed and bulb to bulb
   (c) Only bulb toseed
   (d) only seed to seed
   (e) None of them

(74) Flower initiation and higher yields may be expected with the treatment of onion bulbs by ————————.
   (a) Alar
   (b) Cytokinin
   (c) Paclobutrazol
   (d) GA3
   (e) None of above

(75) Onion seeds (true seed) stored in sealed containers may remain viable for ————.
   (a) 6 months
   (b) 9 months
   (c) One years
   (d) 3-4 years
   (e) None of the above

(76) In onion, the first two F_1 hybrids ———— and ———— have been developed using gene cytoplasmic male sterility.
   (a) Hisar -2 and N-53
   (b) Pantan Red
   (c) Arka Niketan and Arka Kalyan
   (d) Arka Pitamber and Arka Kirtaman
   (e) None of them

(77) In onion, dehiscence of flower usually occurs between ─────────.
   (a) 6-10 AM   (b) 11 AM- 3 PM
   (c) 9.00 AM to 5.00 PM   (d) Night
   (e) Early in the morning

(78) An inflorescence may have flowers opening for a period of two weeks or more and a plant may be in bloom for more than ─────────.
   (a) 30 days   (b) 75 days
   (c) 45 days   (d) 60 days
   (e) None of the them

(79) Red Creole, Granex, Bombay Red, Red Kano, Patna Red, Texas Grano, Yellow Bermuda and Early Cape Flat varieties of onion are suitable for ─────────.
   (a) Salad   (b) Processing
   (c) Export   (d) Spices
   (e) None of them

(80) The onion varieties 'Italian Red' and 'Local Brazilian' are observed resistant to purple blotch *(Alternaria porii)*, Granex and Excel -35 are reported resistant to.
   (a) Damping off   (b) Pink rot
   (c) Purple blotch   (d) Phytphthora
   (e) None of them

(81) Special characteristics of onion cultivar for dehydration are ─────────.
   (a) Onion bulb should be white, round without any green patches, and with small but tight neck and small root zone.
   (b) Bulbs should have TSS 15 to 20 brix and low ratio of reducing to non-reducing sugar
   (c) Bulb should have good keeping quality and resistant to pest and disease.
   (d) All of the above
   (e) None of the above

(82) Onion cultivars suitable for *kharif* season are ─────────.
   (a) Arka Niketan and Arka Kalyan   (b) Patana Red and White Red
   (c) N-53 and Agrifound Dark Red   (d) All of them
   (e) None of the them

(83) Brown Spanish, Cream Gold, White Spanish and Early Lockyar Brown cultivars are suitable for ─────────.
   (a) Hilly   (b) Plain
   (c) Both plain and hilly   (d) All of them
   (e) None of them

(84) *Allium cepa* var. *ascalonicum* (2n=2x=16) is a perennial onion, which rarely produces seed and which is perpetuated each year by replanting some of the bulbs which form in clusters on the surface of the soil is known as ─────.

    (a) Shallot   (b) Onion

    (c) Garlic   (d) Multiplier onion

    (e) Potato onion

(85) *A. cepa* var. aggregattum (2n=2x=16) is also also known ───────── which grows as closely packed clusters of bulbs underground rather than on the surface like shallot.

    (a) Common onion   (b) Shallot

    (c) Potato onion   (d) Multiplier onion

    (e) None of the above

(86) *A. cepa* var. *viviparum/proliferum* (2n=2x=16) is a viviparous plant that grows as a underground has leaves similar to onion and produces clusters of bulblets at the top of the stem in place of inflorescence is known as.

    (a) Tree onion or Egyptian tree onion   (b) Potato onion

    (c) Multiplier onion   (d) Garlic

    (e) None of them

(87) Storage losses in onion can be reduced: (a) By checking sprouting in storage which can be achieved by field application of ───── 75-90 days after transplanting. Similarly, 4-6 kr gamma irradiation also check sprouting. (b) By maintaining sanitation in around store –house during disinfectants like formalin @ 0.65 per cent. (c) Treating bulbs soon after harvesting with 0.1 per cent bavistin solution.

    (a) GA3 50ppm   (b) Alar 3000ppm

    (c) MH 2500 ppm   (d) BA 50 ppm

    (e) None of them

Answer Sheet

1	b	2	b	3	a	4	a	5	b	6	c	7	b	8	b	9	a	10	d
11	a	12	b	13	b	14	e	15	a	16	b	17	c	18	b	19	b	20	d
21	a	22	c	23	a	24	a	25	b	26	e	27	a	28	d	29	c	30	d
31	d	32	b	33	c	34	c	35	e	36	c	37	b	38	d	39	a	40	d
41	e	42	d	43	c	44	b	45	e	46	a	47	c	48	d	49	a	50	c
51	a	52	d	53	e	54	d	55	a	56	c	57	e	58	c	59	b	60	d
61	d	62	a	63	c	64	c	65	c	66	e	67	a	68	d	69	b	70	c
71	c	72	c	73	c	74	d	75	d	76	d	77	c	78	a	79	c	80	b
81	d	82	c	83	a	84	c	85	a	86	a	87	c						

1.1.2.2. Cole Crops

(1) Origin place of sprouting broccoli is ―――――――.
   (a) Italy (b) India
   (c) Europe (d) China
   (e) None of them

(2) Chinese broccoli is botanically known as ―――――――
   (a) *Brassica oleracea* (b) *B. campestris*
   (c) *B. alboglara* (d) *Raphanus sativus*
   (e) None of them

(3) The 2n-chromosome number in sprouting broccoli is ―――――――.
   (a) 12 (b) 14
   (c) 32 (d) 18
   (e) 10

(4) USA is the leading producer of sprouting broccoli. In India, it is widely grown in ――――――――, which occupies an area of about 50 hectares with an annual production of 600-750 tonnes
   (a) Uttar Pradesh (b) Madhya Pradesh
   (c) Himachal Pradesh (d) Bihar
   (e) Gujarat

(5) Broccoli can be stored successfully at 0 °C and 95-100 per cent RH for ―――.
   (a) 10 days (b) 15 days
   (c) One month (d) Two months
   (e) 2-4 weeks

(6) Besides vitamin and minerals, broccoli is also rich source of ―――――――, a compound associated with reducing the risk of cancer―――――――.
   (a) Lycopene (b) Vanadium
   (c) Sulphoraphane (d) All of them
   (e) None of them

(7) Cultivars like Waltham -29, Green Mountain and Coastal Atlantic are ――――――― ――――――― and can be grown as early or late crop. Green Sprouting Medium take 100 days to maturity while late strain like Green Sprouting Late is biennial.
   (a) Heat tolerant (b) Frost tolerant
   (c) Less sensitive to buttoning (d) Cold tolerant
   (e) None of the above

(8) Like other crops there are some hybrids Viz.; Southern Comet, Premium Crop, Clipper, Laser (extra early and early) Dorsair, Excalibut, Cruiser Emerald Corona (mid season), Late Corona Stiff, Kayak and Green Surf are marketed by seed companies in ――――――――――.

(a) Japan, the United States and Europe (b) China
(c) Malaysia (d) India
(e) Thailand

(9) The average yield of Brussels sprout is ―――――.
(a) 250 q/ha (b) 150 q/ha
(c) 60-80 q/ha (d) 125 q/ha
(e) None of them

(10) The seed production of Brussels sprout is done in the Himalayas at ―――― height.
(a) 250 m (b) 1000 m
(c) 750 m (d) 500 m
(e) 1200 m and above

(11) Brussels sprouts can be stored at 0 to $1°C$ temperature and 90-95 per cent RH for
(a) 15 days (b) 30 days
(c) Two months (d) Four months
(e) 3-5 weeks

(12) 'Amager Market' and 'Danish Prize' are varieties of ―――――.
(a) Cauliflower (b) Cabbage
(c) Knol-khol (d) Brussels sprouts
(e) None of the above

(13) Catskill, Early Dwarf, Dwarf Gem, Long Island Improved are dwarf types varieties of ―――――.
(a) Sprouting broccoli (b) Radish
(c) Turnip (d) Tomato
(e) Brussels sprouts

(14) Amager Market and Danish Prize are ―――――.
(a) Intermediate type (b) Short type
(c) Tall types (d) All of them
(e) None of them

(15) Flavour in cabbage leaves is due to the glucoside ―――――.
(a) Capsicinoids (b) Sinigrin
(c) Cucurbitacin (d) Xanthophyll
(e) None of them

(16) Edible part of cabbage is botanically known as ―――――.
(a) Curd (b) Head.
(c) Infloresence (d) All of them
(e) None of them

(17) Optimum temperature for seed germination of cabbage is ―――――.
    (a) 12-16 °C (b) 20-26 °C.
    (c) 30-35 °C (d) 5-10 °C (e) None of them

(18) Early varieties of cabbage take ――――― for harvesting.
    (a) 100-110 days (b) 60-70 days
    (c) 45-60 days (d) 120-130 days
    (e) None of them

(19) 'Pride of India' variety of cabbage has ―――――――――.
    (a) Cauliflower (b) Cabbage
    (c) Knol-khol (d) Radish
    (e) None of them

(20) White cabbage is botanically known as ―――――――.
    (a) *Brassica oleracea* var. *gongylodes* (b) *Brassica rapa*
    (c) *B. oleracea* sub sub group *rubra* (d) *B. oleracea* sub group *alba*
    (e) None of the them

(21) 'Jersy Wakefield' variety of cabbage has head ―――――――――.
    (a) Round type (b) Flate type
    (c) Oval type (d) Conical type
    (e) None of them

(22) Chinese cabbage and kale are found resistant to ―――――――――.
    (a) Powdery mildew (b) Anthracnose
    (c) Downy mildew (d) Leaf spot
    (e) None of them

(23) Premature seedling or bolting in cabbage is ―――――――――.
    (a) Early seed sowing (b) Warm winter
    (c) Extreme change in temperature (d) All of the above ―――――.
    (e) None of the above

(24) Affected plants show light brown canker on stem and they enlarge until the stem is girdled, such situation is due to infestation of disease ―――――――.
    (a) Powdery mildew (b) Anthracnose
    (c) Black leg (d) Root rot
    (e) Downey mildew

(25) To maintain genetic purity, the isolation distances for cabbage seed production of foundation and certified seeds are ――――― and ――――― meter, respectively.
    (a) 200 m and 400 m (b) 800 m and 1000 m
    (c) 1000 and 1600 m (d) 25 m and 50 m
    (e) None of them

(26) All the cole crops are originated from a common parent ―――――.
    (a) *Brassica oleracea* var. *sylvestris* (b) *Brassica oleracea* var.*rapa*
    (c) *Brassica compestris*   (d) *Raphanus sativus*
    (e) None of them

(27) Cabbage juice can be used for remedy of ―――――.
    (a) Heart disease   (b) Dysentry
    (c) Stomach ache   (d) Poisonous snakes
    (e) None of them

(28) 'Pusa Mukta' variety of cabbage is cross of EC-24855 x EC-10109 and tolerant to ―――――.
    (a) Buttoning   (b) Reiceyness
    (c) Whiptail   (d) Salinity
    (e) None of the them

(29) 'Pusa Drum Head' variety of cabbage is resistant to ―――――.
    (a) Downey mildew   (b) Leaf spot
    (c) Powdery mildew   (d) Damping off
    (e) Black leg

(30) 'Chieftain' is variety of ――――― belongs in to savoy group
    (a) Cauliflower   (b) Knol khol
    (c) Cabbage   (d) Broccoli
    (e) Brussels sprouts

(31) ――――― is botanically known as *B. oleracea* sub group *rubra*.
    (a) Red cabbage   (b) Savoy cabbage
    (c) Cauliflower   (d) Knol khol
    (e) None of the above

(32) The word 'Cole' is abbreviated from ―――――.
    (a) Cole   (b) Cole wart
    (c) Cabbage   (d) Caulis
    (e) None of the above

(33) Common biofertilizers used for enhancing cabbage yield are ―――――.
    (a) PSB   (b) Blue algae
    (c) Rhizobium   (d) Only *Azospirillum*
    (e) *Azospirillum* and *Azatobactor*

(34) For planting one hectare area, ――――― seed is required.
    (a) 1kg   (b) 125-250g
    (c) 200-500 g   (d) 10-12kg
    (e) None of the above

(35) Planting of cabbage is done in the month of August-December in ―――――.
   (a) South India
   (b) Plains of India
   (c) NEH Region
   (d) Central India
   (e) None of the above

(36) The basic chromosome number in cabbage is ―――――――――.
   (a) 12
   (b) 16
   (c) 21
   (d) 9
   (e) 10

(37) ――――――― is botanically known as *B. oleracea* sub group *sabauda*.
   (a) Red cabbage
   (b) Savoy cabbage
   (c) Cabbage
   (d) Turnip
   (e) Cauliflower

(38) Cabbage originated from Western Europe and Mediterranean region. It was introduced to India from ――――――.
   (a) England
   (b) Turkey
   (c) South America
   (d) Asia
   (e) Western Europe

(39) ―――――― is botanically known as *B. napus* subgroup *napobrassica* (n= 19).
   (a) Cabbage
   (b) Turnip
   (c) Rutabaga
   (d) Radish
   (e) None of the above

(40) 'Sri Ganesh Gol' is the hybrid of ――――――.
   (a) Cauliflower
   (b) Cabbage
   (c) Radish
   (d) Turnip
   (e) None of the above

(41) ――――――― is botanically known as *B. campestris* group *pekinensis*.
   (a) Chinese cabbage
   (b) White cabbage
   (c) Cauliflower
   (d) Brussels sprouts
   (e) Broccoli

(42) Japan is first country which developed cabbage hybrid and Nagaoka was first hybrid released from Japan in ――――――.
   (a) 1961
   (b) 1951
   (c) 1971
   (d) 1981
   (e) 1944

(43) Hybrid vigour in cabbage was observed as early as ——————.
    (a) 1945   (b) 1935
    (c) 1920   (d) 1944
    (e) None of the above

(44) Problems in exploiting self incompatible line in production of cabbage hybrid ——————.

(a) Continuous inbreeding in many brassica crops may lead to complete loss of inbred lines. Sibmating for maintenance of these lines may prove useful.
(b) Pseudo incompatibility may lead to self in an otherwise hybrid seed.
(c) Reduction in strength of incompatibility by environment factors i.e. evaluation of temperature may weakens may even break it down.
(d) Hybrid seed would be expensive if SI lines are difficult to maintain, produce low in volume and weak seed.
(e) Restriction of pollination within parental lines. It is found that bees do not visit the plants randomly. This results in more of sib pollination and less of out crossing.

    (a) Only a   (b) Only e
    (c) Both b and c   (d) All a, b, c, d and e
    (e) Both d and b

(45) Methods used for identifying of self-compatible plants are ——————.
    (a) Only molecular method   (b) Only seed set method
    (c) Only pollen tube growth mehod (d) All a, b and c
    (e) None of the above

(46) —————— class is a useful gene resistance against cabbage butterfly and diamond back moth of cabbage.
    (a) Cry 1A   (b) Cry 1B
    (c) Cry 1C   (d) Cry 1D
    (e) None of the above

(47) Methods used for seed production of cabbage are ——————.
    (a) Only seed to seed method   (b) Only head to seed method
    (c) Only stump method   (d) All a, b and c
    (e) None of the above

(48) Grow cabbage for seed where chilling temperature —————— is available at head maturity stage for over wintering which helps in initiation of flower primordia.
    (a) -10 to -12 °C   (b) 10 to 16 °C
    (c) - 3 to -8 °C   (d) 1 to 2 °C
    (e) 3 to 8 °C

(49) Red Acre is cultivar of ——————————.
    (a) Red cabbage   (b) Savoy cabbage
    (c) Cauliflower   (d) Radish
    (e) None of the above

(50) Copenhagen Resistant, Wisconsin Copenhagen, Marion Market are varieties of cabbage resistant to ——————————.
    (a) Aphid   (b) DBM
    (c) Cabbage yellow   (d) Damping off
    (e) None of the above

(51) Cabbage can be stored at 0 °C and 90-95 per cent RH for about ——————————.
    (a) 15days   (b) One month
    (c) 2 to 8 months   (d) 45 days
    (e) None of the above

(52) —————————— is botanically known as *B. oleracea* group *acephala*.
    (a) Cabbage   (b) Cauliflower
    (c) Knol-khol   (d) Kale
    (e) None of the above

(53) —————————— is botanically known as *Brassica oleracea* var. *botrytis* ——————.
    (a) Cabbage   (b) Kale
    (c) Turnip   (d) Cauliflower
    (e) None of the above

(54) Edible part of cauliflower is known as ——————————.
    (a) Head   (b) Leaf
    (c) Curd   (d) Stem
    (e) None of the above

(55) The optimum temperature for curd initiation is——————————.
    (a) 17-20 °C   (b) 10-15 °C
    (c) 7-10 °C   (d) 25-30°C
    (e) None of the above

(56) —————————— is a self-blanched variety of cauliflower——————————.
    (a) Golden Acre   (b) 'Hisar-1'
    (c) Pant Gobhi-4   (d) Pusa Ageti
    (e) None of the above

(57) For planting of one-hectare cauliflower, it requires —————————— seeds for early varieties.
    (a) 100-125g   (b) 200-400g
    (c) 500-1000 g   (d) 5kg
    (e) None of the above

(58) Blanching operation is done in cauliflower for obtaining better quality curd and its period ranges ——————— in cool weather.
   (a) 10-15 days (b) 3-5 days
   (c) 20-25 days (d) 1-3 days
   (e) None of the above

(59) The water soaked areas are developed in the stem and brown colour develops in the centre. This situation in cauliflower is due to ——————— deficiency.
   (a) Iron (b) Nitrogen
   (c) Boron (d) Calcium
   (e) None of the above

(60) Cauliflower having leaf-blade strap-like and severely savoyed and sometimes only the midribs develop. Such type of situation in cauliflower occurs due to deficiency of———————.
   (a) Boron (b) Phosphorous
   (c) Molybdenum (d) Potassium
   (e) None of the above

(61) Cauliflower curds give full velvety appearance due to elongation of pedicels and peduncles in flowers. This situation in cauliflower is due to ———————.
   (a) High nitrogen (b) High moisture regime
   (c) Fluctuation in temperature (d) Low moisture regime
   (e) None of the above

(62) A few thin leaves come out from the cauliflower curds is known as ———————.
   (a) Blanching (b) Browning
   (c) Whiptail (d) Pinking
   (e) Leafiness

(63) Cauliflower curds show yellowish, white, pink colour. This situation occurs due to physiological disorder ———————.
   (a) Leafiness (b) Pinking
   (c) Browning (d) Whiptail
   (e) None of the above

(64) ——————— in cauliflower is due to insufficient nitrogen application, use of wrong varieties and high temperature.
   (a) Browning (b) Buttoning
   (c) Pinking (d) Leafiness
   (e) None of the above

(65) ——————— in cauliflower is due to molybdenum deficiency.
   (a) Pinking (b) Leafiness
   (c) Browning (d) Buttoning
   (e) Whiptail

(66) In studies in India on cauliflower, it has been found cauliflower varieties like Snowball, MGS-2-3, 1-6-1-4, 1-6-1-2, 12 etc. are highly resistant to ——————— under field conditions.
   (a) Powdery mildew
   (b) Downey mildew
   (c) Anthracnose
   (d) Cabbage yellow
   (e) None of the above

(67) Cauliflower is susceptible to downy mildew, while ——————— are found to be resistant.
   (a) Cabbage
   (b) Red cabbage
   (c) Savoy cabbage
   (d) Chinese cabbage and Kale
   (e) None of the above

(68) ——————— disease of cauliflower is due to *Xanthomonas campestris*
   (a) Club rot
   (b) Root rot
   (c) Black rot
   (d) Downey mildew
   (e) None of the above

(69) The scientific name of ——————— is *Diacrisia obliqua* Walker.
   (a) DBM
   (b) Aphid
   (c) Beetle
   (d) Bihar hairy caterpillar
   (e) None of the above.

(70) Most of the temperate type of cauliflower's are grown in the temperature of ———————.
   (a) 25-30 °C
   (b) 35-40 °C
   (c) 15-20 °C
   (d) 5-10 °C
   (e) 1-2 °C

(71) The temperature higher or lower than the optimum required for curd formation of the cultivar may causes physiological disorders ———————.
   (a) Burning effect
   (b) Good growth
   (c) Physiological disorder
   (d) Good seed setting
   (e) None of the above

(72) The optimum soil pH for cauliflower cultivation is ———————.
   a) 6-7
   (b) 4.5-5
   (c) 8-9
   (d) 10-12
   (e) None of the above

(73) Most of the early Indian varieties of cauliflower are called by the name of month in which the curds mature and become ready for market in the month of September - October. This cultivation is known as ———————.
   (a) Maghi
   (b) Poss
   (c) Aghani
   (d) Kunwari
   (e) None of the above

(74) 'Pusa Deepali' is variety of cauliflower which requires 20-25 °C temperature for curd development and almost free from —————.
    (a) Browning   (b) Pinking
    (c) Self blanching   (d) Riceyness
    (e) None of the above

(75) Spraying of 1-1.5 kg/ha of sodium or ammonium molybdate per hectare can control ————— physiological disorder of cauliflower
    (a) Browning   (b) Pinking
    (c) Whiptail   (d) Riceyness
    (e) None of the above

(76) The present ————— developed as results of inter crossing between European and Cornish type —————.
    (a) Cabbage   (b) Indian cauliflower
    (c) Kale   (d) Knol-khol
    (e) None of the above

(77) Florescence Microscopy technique used for distinguishing compatible and incompatible population in —————.
    (a) Cabbage   (b) Kale
    (c) Indian Cauliflower   (d) Turnip
    (e) None of the above

(78) Pusa Shubra is the variety of —————.
    (a) Cauliflower   (b) Cabbage
    (c) Kale   (d) Tomato
    (e) None of the above

(79) Sporophytic self-incompatibility is used for commercial hybrid production in —————.
    (a) Tomato   (b) Cabbage
    (c) Kale   (d) Radish
    (e) Cauliflower

(80) The chromosome number (2n) in cauliflower is —————.
    (a) 18   (b) 22
    (c) 12   (d) 14
    (e) None of the above

(81) Origin place of cauliflower is —————.
    (a) Asia   (b) China
    (c) Japan   (d) Italy
    (e) Eastern Mediterranean region

(82) The variety viz; Improved Japanese, Pusa Hybrid-2, Pusa Sharad and Pant Gobhi-4, the temperature requirement for curd initiation and development is —————.
   (a) 16-20 °C (b) 6-10 °C
   (c) 26-30 °C (d) 30-40 °C
   (e) None of the above

(83) On the basis of maturity of curd, cauliflower can be classified in ————— groups —————.
   (a) Two (b) Three
   (c) Four (d) Five
   (e) None of the above

(84) Cauliflower is a cross-pollinated crop, therefore maintain ————— meters isolation distance between two cultivars.
   (a) 400 (b) 200
   (c) 800 (d) 1600
   (e) None of the above

(85) From one-hectare seed crop of cauliflower, about seed is obtained —————.
   (a) 500g (b) 2kg
   (c) 500-600 kg (d) 15-20 kg
   (e) None of the above

(86) Blindness in cauliflower means —————.
   (a) Without leaves (b) Without stem
   (c) Without terminal bud (d) Without root
   (e) None ot he above

(87) Alpha, Danish Giant and Mechlin varieties of ————— belong in summer and autumn temperate.
   (a) Cabbage (b) Knol- khol
   (c) Radish (d) Turnip
   (e) Cauliflower

(88) Snow Queen, Snow King, Snow Crown, White Contessa, Meigetsu are hybrids of _____ cauliflower.
   (a) Radish (b) Turnip
   (c) Cauliflower (d) Cabbage
   (e) Tomato

(89) Tropical cultivars of cauliflower show growth even at —————.
   (a) 35 °C (b) 25 °C
   (c) 15 °C (d) 30 °C
   (e) None of the above

(90) Kale Patta is variety of ——————.
 (a) Spinach   (b) Amranthus
 (c) Cauliflower   (d) Cabbage
 (e) Methi

(91) Dr. Jenson, a botanist from 'KEW' introduced —————— in 1822 and grows it in Saharanpur.
 (a) Cauliflower   (b) Tomato
 (c) Radish   (d) Chilli
 (e) None of the above

(92) Dr. Harbhajan Singh (1955-57), who contributed greatly to the standardization of seed production technology of cole crops and in 1958, the —————— seed was produced for the first time.
 (a) Late cauliflower   (b) Late onion
 (c) Late radish   (d) Late turnip
 (e) None of the above

(93) 'U's triangle reported by U.N. in ——————.
 (a) 1945   (b) 1948
 (c) 1935   (d) 1905
 (e) 1925

(94) Cauliflower can be stored successfully at 0 °C and 90-95 per cent relative humidity for 2-4 weeks.
 (a) 6 months   (b) 1 weeks
 (c) Two months   (d) Two – four months
 (e) None of the above

(95) Varieties like Svale' Fanolz, Bandone-Z and Fano II-2 are bred to meet specific requirement like frost and cold tolerance freedom from bolting, better storability and quality of ——————.
 (a) Curry leaf   (b) Chow-chow
 (c) Cauliflower   (d) Tomato
 (e) Radish

(96) Some of the —————— hybrids viz. Himani Gardian (Ando-American), Candid Charm, White Flash, Cashmere (Sakata), Serrano (Sandoz), Early Himlata, Early Himangiri (Century), Nath Ujwala (Nath), Namdhari-84 (Namdhari), Pusa Hybrid-2 (IARI) are also commercially grown in India.
 (a) Radish   (b) Kale
 (c) Cauliflower   (d) Cabbage
 (e) Turnip

(97) 'Takii-2' variety of ———————— is suitable for cold storage; 'Green Comet', 'Green Hornets' and 'Premium Crop' are recommended for fresh market and freezing and 'Neptune F1' hybrid for canning and fresh use.
    (a) Chilli   (b) Capsicum
    (c) Radish   (d) Bitter gourd
    (e) Cauliflower

(98) Knol khol is also known as ————————————.
    (a) Khol rabi   (b) Savoy cabbage
    (c) Cauliflower   (d) Red cabbage
    (e) None of the above

(99) About _____ seeds will be sufficient to raise seedlings for planting of one-hectare area of Knol khol.
    (a) Five kg   (b) Six kg
    (c) One kg   (d) Two kg
    (e) None of the above

(100) Botanical name of ———————————— is *Brassica oleracea* var. *gongylodes* ————————.
    (a) Cabbage   (b) Turnip
    (c) Radish   (d) Knol - khol
    (e) None of the above

(101) Knol khol can be stored successfully at 0°C and 95-100 per cent RH for 25-30 days.
    (a) Three months   (b) One months
    (c) Two months   (d) 15 days
    (e) None of the above

(102) Edible part of Knol- khol is known as ————————.
    (a) Curd   (b) Head
    (c) Knob   (d) Leaves
    (e) None of the above

Answer Sheet

1	a	2	c	3	d	4	c	5	e	6	c	7	c	8	a	9	c	10	e
11	e	12	d	13	e	14	c	15	b	16	b	17	a	18	b	19	b	20	d
21	d	22	c	23	d	24	c	25	c	26	a	27	d	28	d	29	e	30	c
31	a	32	d	33	e	34	c	35	b	36	d	37	b	38	e	39	c	40	b
41	a	42	b	43	b	44	d	45	d	46	a	47	d	48	e	49	a	50	c
51	c	52	d	53	d	54	c	55	a	56	b	57	c	58	b	59	c	60	c
61	c	62	e	63	b	64	b	65	e	66	b	67	d	68	c	69	d	70	c
71	c	72	a	73	d	74	d	75	c	76	b	77	c	78	a	79	e	80	a
81	e	82	a	83	d	84	d	85	c	86	c	87	e	88	c	89	a	90	c
91	a	92	a	93	c	94	d	95	c	96	c	97	e	98	a	99	c	100	d
101	b	102	c																

1.1.2.3. Cucurbitaceous Vegetables

(1) Pusa Vishesh is a variety of ——————————.
    (a) Brinjal   (b) Bitter gourd
    (c) Tomato   (d) Okra
    (e) None of the above

(2) For raising one-hectare crop of bitter gourd, —————————— of seeds per hectare is sufficient.
    (a) 10-12kg   (b) 500-600g
    (c) 15-20kg   (d) 4.5 to 6 kg
    (e) None of the above

(3) Optimum temperature requirement for bitter gourd cultivation is ——————————.
    (a) 24-27 °C   (b) 14-17 °C
    (c) 10-15 °C   (d) 30-40 °C
    (e) None of the above

(4) —————————— is botanically known as *Momordica charantia* Linn.
    (a) Pointed gourd   (b) Ridge gourd
    (c) Bitter gourd   (d) Bottle gourd
    (e) None of the above

(5) Short days helps in increasing female flower production in ——————————.
    (a) Spinach   (b) Chow-chow
    (c) Bitter gourd   (d) Bhat karela
    (e) None of the above

(6) —————————— is also known as balsam pear.
    (a) Bottle gourd   (b) Ridge gourd
    (c) Ash gourd   (d) Snake gourd
    (e) Bitter gourd

(7) Bitter gourd fruit contains ———— of vitamin A per 100 g edible portion.
   (a) 210 IU (b) 110 IU
   (c) 440 IU (d) 1100 IU
   (e) 510 IU

(8) Cucurbitacin-bitter glucosides in preventing spoilage of cooked vegetable of ————.
   (a) Tomato (b) Onion
   (c) Leek (d) Bitter gourd
   (e) None of the above

(9) In West Bengal and South India, bitter gourd is sown in ————.
   (a) June - July (b) Feburary - March
   (c) October - November (d) August - September
   (e) None of the above

(10) The origin place of bitter gourd is ————.
   (a) Asia (b) Indo- Burma
   (c) China (d) Europe
   (e) None of the above

(11) The crop will take about 55 to 110 days from seed sowing to reach first harvest.
   (a) Pumpkin (b) Bitter gourd
   (c) Radish (d) Sweet potato
   (e) None of the above

(12) In bitter gourd, flowering starts ———— days after planting.
   (a) 25-30 (b) 65-75
   (c) 45-55 (d) 10-15
   (e) None of the above

(13) The chromosome number (2n) in bitter gourd is ————.
   (a) 24 (b) 22
   (c) 14 (d) 32
   (e) 42

(14) About 100-150 quintals of fruits are obtained in one hectare.
   (a) Tomato (b) Brinjal
   (c) Radish (d) Bitter gourd
   (e) None of the above

(15) The vegetable made from ———— is beneficial to diabetic patient.
   (a) Tomato (b) Radish
   (c) Spinach (d) Bitter gourd
   (e) None of the above

(16) Optimum soil pH requirement for bitter gourd cultivation is ———————.
    (a) 4.5-5   (b) 5.5-6.8
    (c) 7.5-8.5   (d) 9.5-10.5
    (e) None of the above

(17) ——————— is least affected by anthracnose disease.
    (a) Bottle gourd   (b) Chilli
    (c) Bitter gourd   (d) Radish
    (e) None of the above

(18) Red pumpkin beetles comparatively less harm ———————.
    (a) Pumpkin   (b) Cucumber
    (c) Musk melon   (d) Bitter gourd
    (e) None of the above

(19) The alkaloid ——————— imparts the bitter taste of the fruit.
    (a) Cucurbitacin   (b) Capsacin
    (c) Momordicasoides   (d) Luffein
    (e) All of them

(20) In bitter gourd, anthesis starts by ——————— and completed by 9 AM.
    (a) 10 AM   (b) 4 PM
    (c) 11 AM   (d) 4 AM
    (e) None of the above

(21) Green Smooth, White Rough and White Smooth varieties of bitter gourd are found resistant to ———————.
    (a) Red pumpkin beetle   (b) Powdery mildew
    (c) Fruit fly   (d) Aphid
    (e) None of the above

(22) Morordicin is found in ———————.
    (a) Muskmelon   (b) Bitter gourd
    (c) Watermelon   (d) Ridged gourd

(23) The origin place of bottle gourd is ———————.
    (a) South America   (b) Europe
    (c) Asia   (d) Africa and India
    (e) None of the above

(24) ——————— variety of bottle gourd is a cross of PSPL x Selection- 2.
    (a) Pusa Naveen   (b) Pusa Manjari
    (c) Pusa Komal   (d) Arka Bahar
    (e) Pusa Meghdoot

(25) Basic chromosome number in bottle gourd is —————.
    (a) 12   (b) 10
    (c) 11   (d) 24
    (e) 7

(26) Bottle gourd is a ————— crop.
    (a) Self-pollinated   (b) Cross-pollinated
    (c) Often cross-pollinated   (d) None of the above

(27) Arka Bahar is a variety of —————.
    (a) Tinda   (b) Pumpkin
    (c) Bottle gourd   (d) Bitter gourd
    (e) None of the above

(28) ————— is botanically known as *Lagenaria siceraria*.
    (a) Ridged gourd   (b) Sponge gourd
    (c) Bottle gourd   (d) Pumpkin
    (e) None of the above

(29) About ————— of seed is enough for raising one hectare of bottle gourd.
    (a) 10-12 kg   (b) 6-8 kg
    (c) 25-35 kg   (d) 45-50 kg
    (e) None of the above

(30) Bottle gourd belongs to family —————.
    (a) Onion   (b) Tomato
    (c) Radish   (d) Bottle gourd
    (e) None of the above

(31) ————— is also known as white flowered gourd.
    (a) Kundru   (b) Parwal
    (c) Bottle gourd   (d) Snake gourd
    (e) None of the above

(32) In bottle gourd, anthesis has been observed in between ————— under Punjab conditions.
    (a) 5-8 AM   (b) 11 AM
    (c) 5-8 PM   (d) 2-3 PM
    (e) None of the above

(33) Bees are the pollinators of —————.
    (a) Tomato   (b) Spinach
    (c) Brinjal   (d) Cow-chow
    (e) Bottle gourd

(34) *Lagenaria sphaerica* (Sand) Naud is other relative of ——————.
  (a) Luffa (b) Pumpkin
  (c) Ash gourd (d) Bottle gourd
  (e) None of the above

(35) —————————— is a family of frost sensitive predominately tendril bearing vines which are found in subtropical and tropical regions around the globes
  (a) Solanaceae (b) Leguminosae
  (c) Cucurbitaceae (d) Malvaceae

(36) Ecologiaclly the family cucurbitaceae is ——————.
  (a) Dichotomous (b) Monochotomous
  (c) Polychotomous (d) None of them

(37) The term cucurbits was coined by ——————.
  (a) Liberty Hyde Bailey (b) Banthem and Hooker
  (c) Muller (d) None of them

(38) Family cucurbitaceae consists ———— genera and ———— species.
  (a) 20 and 50 (b) 118 and 825
  (c) 300 and 500 (d) None of the them

(39) —————————— applied to the family and various of its members are gourd, melon, cucumber, squash and pumpkin
  (a) Leguminosae (b) Cucurbitaceae
  (c) Solanaceae (d) None of them

(40) ————— is also known as Khira.
  (a) Watermelon (b) Muskmelon
  (c) Cucumber (d) Bitter gourd
  (e) None of the above

(41) ————— is botanically known as *Cucumis sativus* Linn.
  (a) Muskmelon (b) Watermelon
  (c) Cucumber (d) Ash gourd
  (e) None of the above

(42) Average daily temperature of ————— is most favourable for cucumber growth.
  (a) 18 to 24 °C (b) 10 to 14 °C
  (c) 28 to 30 °C (d) 5 to 10 °C
  (e) 8 to 14 °C

(43) Low humidity accelerates the appearance of staminate flowers of cv. Nerosinge in —————.
  (a) Cucumber (b) Pumpkin
  (c) Watermelon (d) Muskmelon
  (e) None of the above

(44) ———— is the chief pollinating agent in cucumber.
    (a) Housefly   (b) Bumble bee
    (c) Honey bee   (d) None of the above

(45) Cucumber is ———— in nature.
    (a) Monoecious   (b) Dioecious
    (c) Gynomonoecious   (d) Andromonoecious
    (e) None of the above

(46) For seed production of cucumber, maintain ————meter isolation distance between two cultivars.
    (a) 1600   (b) 800
    (c) 1000   (d) 400
    (e) 20

(47) The origin place of cucumber is ————.
    (a) Brazil   (b) Peru
    (c) Asia Minor   (d) India
    (e) China

(48) Basic chromosome number in cucumber is ————.
    (a) 11   (b) 10
    (c) 12   (d) 7
    (e) 11

(49) Straight Eight is a variety of ————.
    (a) Pumpkin   (b) Cucumber
    (c) Tinda   (d) Bottle gourd
    (e) None of the above

(50) ———— kilogram of cucumber seeds required for raising one hectare crop.
    (a) Three to four   (b) Four to five
    (c) Eight to nine   (d) Ten to Twelve
    (e) None of the above

(51) The average yield of cucumber fruits per hectare is _____ quintals.
    (a) 300-400 quintals   (b) 500-600 quintals
    (c) 80-100 quintals   (d) 200-300 quintals

(52) High humidity hastens formation of pistillate flowers in ———— cultivation.
    (a) Tomato   (b) Spinach
    (c) Brinjal   (d) Cucumber
    (e) None of the above

(53) Lower fertility, high temperature and longer light period induces ———— in cucumber.
- (a) Maleness
- (b) Femaleness
- (c) Vegetative growth
- (d) Fruit growth
- (e) None of the above

(54) While eating cucumber, the bitterness in cucumber can be reduced to minimum when the upper portion of fruits cut upto 1 cm horizontally, and a pinch off of ———————————— is sprinkled on the cut surface and is rubbed with the same cut piece gently for about 20 to 30 second.
- (a) HCl
- (b) NaCl
- (c) NaOH
- (d) KMS
- (e) None of the above

(55) *Cucumis hardwicki*; *C. callosus*; *C. melo*; *C. propnaterum* and *C. dispaceus* are relatives of ————.
- (a) Cucumber
- (b) Muskmelon
- (c) Watermelon
- (d) Ridge gourd

(56) Long melon, Snap melon, Mango melon, are the members of genus ————.
- (a) *Cucumis*
- (b) *Cucurbita*
- (c) *Luffa*
- (d) *Citrullus*
- (e) *Tricosanthus*

(57) *Ecbalium elaterium* is also known as ————.
- (a) Pickling cucumber
- (b) Squiriting cucumber
- (c) Snapmelon
- (d) Roundmelon
- (e) None of them

(58) Parthenocarpy in cucumber and squash is promoted by ————.
- (a) Only low temperature
- (b) Only short day length
- (c) Only old plant age
- (d) All of the above

(59) In muskmelon, the TSS content varies from ———— per cent.
- (a) 8-10
- (b) 5-6
- (c) 8-17
- (d) 3-4
- (e) None of them

(60) Hara Madhu, Pusa Madhuras and Arka Rajhans are varieties of ————.
- (a) Watermelon
- (b) Muskmelon
- (c) Snapmelon
- (d) Longmelon
- (e) None of the above

(61) —————— is also known as kharbooz.
   (a) Watermelon (b) Muskmelon
   (c) Bittergourd (d) Bottle gourd
   (e) Snapmelon

(62) —————— variety of muskmelon is cross of Hara Madhu x Edisto.
   (a) Sugar Baby (b) Arka Manik
   (c) Punjab Sunehari (d) Hara Madhu
   (e) None of the above

(63) Swarna is a hybrid of ——————.
   (a) Pumpkin (b) Watermelon
   (c) Muskmelon (d) Bottle gourd
   (e) None of the above

(64) Sharbat-e Anar and Bathesa are local cultivars of muskmelon, which are cultivated widely in ——————.
   (a) MP (b) AP
   (c) UP (d) Bihar
   (e) Orissa

(65) Muskmelon seed does not germinate at temperature lower than ——————.
   (a) 18°C (b) 24°C
   (c) 30°C (d) 35°C
   (e) 25°C

(66) In South India, muskmelon is sown in the month of ——————.
   (a) June-July (b) Septumber-October
   (c) October-November (d) February -March
   (e) None of them

(67) —————— kilogram seed per hectare is required.
   (a) 10-15 (b) 4-6
   (c) 15-20 (d) 8-10

(68) Removal of all secondary growth upto —————— nodes in Hara Madhu variety of muskmelon has been reported to enhance fruit yield.
   (a) 7th nodes (b) 8th nodes
   (c) 9th nodes (d) 10th nodes
   (e) 11th nodes

(69) The muskmelon crop is ready for harvesting in about —————— days after seed sowing, depending upon the variety and season.
   (a) 50-60 (b) 70-90
   (c) 100-110 (d) 50-60
   (e) None of the above

(70) Formation of female flowers early as well as more in number can also be obtained with the application of growth regulators like ———— at 25 ppm applied at 2-leaf and 4-leaf stages of plant growth.
    (a) CCC   (b) Alar
    (c) TIBA   (d) Ethephon
    (e) None of the above

(71) ———— is botanically known as *Cucumis melo* L.
    (a) Cucumber   (b) Watermelon
    (c) Muskmelon   (d) Longmelon
    (e) Snapmelon

(72) The maturity in ———— can be determined from the change in outer colour to yellow, green or brown and the fruit also slip from the vine.
    (a) Bottle gourd   (b) Watermelon
    (c) Muskmelon   (d) Longmelon
    (e) None of the above

(73) Basic chromosome in muskmelon is ————.
    (a) 24   (b) 20
    (c) 12   (d) 10
    (e) 7

(74) ———— variety of muskmelon is cross of Kutana x Cantaloupe.
    (a) Pusa Sunehari   (b) Hara Madhu
    (c) Pusa Sharbati   (d) Sugar Baby
    (e) None of the above

(75) ———— is the primary centre of origin of muskmelon.
    (a) Japan   (b) China
    (c) Asia   (d) Europe
    (e) None of them

(76) Full slip stage of muskmelon contains ———— amount of sugar.
    (a) Minimum   (b) Maximum
    (c) Optimum   (d) Average
    (e) None of them

(77) ———— is also known as parwal.
    (a) Snake gourd   (b) Little gourd
    (c) Pointed gourd   (d) Ridge gourd
    (e) None of them

(78) ————— is original home of parwal.
    (a) Bihar   (b) Orissa
    (c) UttarPradesh   (d) Bengal
    (e) None of them

(79) ————— is considered as 'king of gourds'.
    (a) Ridged gourd   (b) Snake gourd
    (c) Little gourd   (d) Pointed gourd
    (e) Sponge gourd

(80) For planting of one-hectare area of parwal about ————— cuttings are required.
    (a) 200-500   (b) 500-1000
    (c) 5000-10000   (d) 2000-2500
    (e) None of them

(81) Under good management, once planted crop of pointed gourd give economic return for at least ————— years.
    (a) 2   (b) 3
    (c) 4   (d) None of the them

(82) Pointed gourd is commercially propagated by —————.
    (a) Seed   (b) Root
    (c) Rhizome   (d) Cutting
    (e) Tuber

(83) ————— is botanically known as *Trichosanthes dioica*.
    (a) Snake gourd   (b) Pointed gourd
    (c) Ridged gourd   (d) Ash gourd
    (e) None of them

(84) ————— is easily digestable, diuretic and laxative.
    (a) Tomato   (b) Brinjal
    (c) Okra   (d) Pointed gourd
    (e) None of them

(85) ————— is a dioecious, perennial, climbing or trailing in habit.
    (a) Ash gourd   (b) Pointed gourd
    (c) Watermelon   (d) Spinach
    (e) None of the them

(86) Pointed gourd is most favourite vegetable in —————.
    (a) Maharashtra   (b) Assam
    (c) Gujarat   (d) Uttar Pradesh and Bihar
    (e) None of the them

(87) The vines of pointed gourd come in dormancy in ─────────.
    (a) Spring season   (b) Rainy season
    (c) Winter season   (d) Summer season
    (e) None of them

(88) Parwal is a good crop for ───────── cultivation.
    (a) Murshy land   (b) Rainfed
    (c) Riverbed   (d) Hilly area
    (e) None of them

(89) In West Bengal, optimum-transplanting time of pointed gourd is ─────────.
    (a) July - August   (b) October- November
    (c) February - March   (d) June - July
    (e) None of the above

(90) The origin place of pointed gourd is ─────────.
    (a) Ethopia   (b) Asia
    (c) China   (d) India
    (e) South America

(91) The fruits of pointed gourd are ready for harvesting ───────── days after planting.
    (a) 45-60   (b) 70-75
    (c) 80-90   (d) 35-45
    (e) None of them

(92) Average fruit yield of parwal varies from 60-80 q/ha during first year and ───────── q/ha during second year, respectively.
    (a) 30-40   (b) 50-60
    (c) 70-80   (d) 100-140
    (e) None of them

(93) Basic chromosome number in parwal is ─────────.
    (a) 14   (b) 48
    (c) 22   (d) 32
    (e) None of the them

(94) Fruits after harvesting can be stored under ordinary conditions for about ───────── days.
    (a) 7-10   (b) 10-15
    (c) 3-4   (d) 15-20
    (e) None of them

(95) Parwal belongs to family ―――――――.
    (a) Solanaceae (b) Cucurbitaceae
    (c) Cruciferae (d) Leguminaceae
    (e) None of them

(96) ――――――― enhanced post harvest life of parwal.
    (a) GA (b) NAA
    (c) Salicylic Acid (d) Alar
    (e) None of them

(97) The other relatives of pointed gourd is ―――――――.
    (a) *Momordica cochinensis* (b) *T. japonica*
    (c) *Citrullus vulgaris* (d) *Momordica charantia*
    (e) None of the avove

(98) Bitterness in parwal is due to ―――――――.
    (a) Lutein (b) Trichosanthin
    (c) Momordiacin (d) None of the above

(99) Pumpkin is a monoecious ――――――― climber.
    (a) Annual (b) Perennial
    (c) Biennial (d) Woody
    (e) None of the above

(100) The pumpkin flowers make an excellent preparation so called as ―――――――.
    (a) Curry (b) Pakora
    (c) Juice (d) Cooked vegetable
    (e) None of the above

(101) The pumpkin fruits contains ――――――― mg vitamin A per 100 g edible portion.
    (a) 25 (b) 30
    (c) 84 (d) 60
    (e) None of the above

(102) Optimum temperature for pumpkin cultivation is ―――――――.
    (a) 10-14 °C (b) 18-24 °C
    (c) 8-12 °C (d) 30-34 °C
    (e) None of the above

(103) Arka Chandan is a variety of ―――――――
    (a) Bottle gourd (b) Bitter gourd
    (c) Snake gourd (d) Luffa
    (e) Pumpkin

(104) The origin place of pumpkin is ———————————————.
    (a) Asia   (b) China
    (c) Tropical America   (d) Africa
    (e) None of the above

(105) Basic chromosome number in pumpkin is ———————————
    (a) 14   (b) 32
    (c) 20   (d) 48
    (e) None of the above

(106) Botanical name of ———————————— is *Cucurbita moschata* Poir.
    (a) Tomato   (b) Pumpkin
    (c) Cucumber   (d) Ridge Gourd
    (e) Snake gourd

(107) In plains, pumpkin is grown ——————————— in a year.
    (a) Once   (b) Twice
    (c) Three time   (d) None of the above

(108) ———————————— seeds are required for pumpkin cultivation.
    (a) 8-10 kg/ha   (b) 6-8 kg/ha
    (c) 3-4 kg/ha   (d) 10-12 kg/ha
    (e) None of the above

(109) The pumpkin fruits matured ————————— after seed depending upon variety, season and other conditions.
    (a) 50-55 days   (b) 60-65 days
    (c) 75-80 days   (d) 85-90 days
    (e) None of the above

(110) Pusa Viswas, Pusa Vikas and Pusa Hybrid-1 are hybrid varieties of ———————
    (a) Pumpkin.   (b) Bitter gourd
    (c) Snake gourd   (d) Ridge Gourd
    (e) None of the above

(111) Arka Suruamukhi is fruit fly resistant variety of ———————————
    (a) Ridge gourd   (b) Bitter gourd
    (c) Snake gourd   (d) Pumpkin
    (e) None of the above

(112) Ethephon is best plant growth regulators in ————————— used for enhancing fruiting.
    (a) Tomato   (b) Okra
    (c) Pumpkin   (d) Brinjal
    (e) None of the above

(113) —————— is also known as kashiphal.
    (a) Tomato   (b) Okra
    (c) Pumpkin   (d) Brinjal
    (e) None of the above

(114) —————— are also known as marrow or musky gourd.
    (a) Bitter gourd   (b) Spong gourd
    (c) Squashes   (d) Kundru
    (e) Snake gourd

(115) The other relatives of pumpkin is ——————
    (a) *Cucurbita pepo* L   (b) *Momordica charantia*
    (c) *Citrullus vulgaris*   (d) *Benincasa hispida*
    (e) None of the above

(116) Lutein alkaloid is found in ——————.
    (a) Parwal   (b) *Cucurbita maxima*
    (c) *Cucurbita moschata*   (d) None of the above

(117) —————— is useful in cough and for improving blood circulation.
    (a) Tomato   (b) Bitter gourd
    (c) Pumpkin   (d) Ash gourd
    (e) Round gourd

(118) —————— is also known as round gourd or squash melon or Indian squash.
    (a) Round melon   (b) Bottle gourd
    (c) Pumpkin   (d) Ash gourd
    (e) None of the above

(119) Arka Tinda is variety of ——————.
    (a) Watermelon   (b) Pumpkin
    (c) Indian squash   (d) Chow-chow
    (e) Muskmelon

(120) The optimum temperature for germination of round gourd seed is ——————.
    (a) 17 °C   (b) 27 °C
    (c) 20 °C   (d) 37 °C
    (e) 10 °C

(121) The origin place of round gourd is ——————.
    (a) South India   (b) Western India
    (c) North India   (d) NEH Region
    (e) None of the above

(122) The planting of tinda is done February-March and June-July in north Indian plains and ——————— on hills, respectively
  (a) October – November   (b) March – April
  (c) May – June   (d) November – December
  (e) None of the above

(123) Three to seven kilogram seeds are required to sow one hectare of _____.
  (a) Onion   (b) Pea
  (c) French bean   (d) Round gourd
  (e) None of the above

(124) Under ordinary conditions, round gourd fruits can be kept for _____ only.
  (a) 20   (b) 15
  (c) 10   (d) 3-4
  (e) None of the above

(125) The fruit of round gourd has cooling effect and contain———————.
  (a) Vitamin C   (b) Vitamin B
  (c) Vitamin A   (d) Vitamin E
  (e) None of the above

(126) *Citrullus vulgaris* Schard var. *fistulosus* is a botanical name of ———————.
  (a) Round gourd   (b) Watermelon
  (c) Muskmelon   (d) Bitter gourd
  (e) None of the above

(127) The average yield of tinda is ——————— quintals per hectare.
  (a) 30-40   (b) 40-50
  (c) 60-80   (d) 80-120
  (e) None of the above

(128) ——————— the most effective plant growth regulator which enhanced femaleness in tinda.
  (a) GA   (b) NAA
  (c) MH   (d) ABA
  (e) Ethrel

(129) The origin place of snake gourd is ———————.
  (a) Asia   (b) China
  (c) India   (d) Ethopia
  (e) None of the above

(130) ——————— is also known as chichinda, pallakaya and attlakaya.
  (a) Bitter gourd   (b) Snake gourd
  (c) Ash gourd   (d) Pumpkin
  (e) None of the above

(131) Basic chromosome number in snake gourd is —————.
    (a) 14   (b) 20
    (c) 24   (d) 34
    (e) 42

(132) Snake gourd is a popular vegetable of —————.
    (a) North India   (b) South India
    (c) Central India   (d) Western India
    (e) None of the above

(133) *Trichosanthes anguina* is botanical name of —————.
    (a) Snakegourd   (b) Bittergourd
    (c) Bottlegourd   (d) Parwal
    (e) Pumpkin

(134) Fruit length of snake gourd goes upto ————— cm in length.
    (a) 45   (b) 60
    (c) 90   (d) 75
    (e) 150

(135) Snake gourd may not be successfully grown above an altitude of ————— meters.
    (a) 300   (b) 1000
    (c) 1500   (d) 500
    (e) None of the above

(136) Under ordinary conditions, harvested fruits of snakegourd can easily be kept for about ————— days.
    (a) 10   (b) 7
    (c) 15   (d) 3
    (e) 9

(137) ————— weeks after planting, snake gourd fruits becomes harvesting for vegetable purposes.
    (a) Seven to eight   (b) Three to four
    (c) One to Two   (d) Four to five
    (e) None of the above

(138) ————— seeds per hectare are needed for snake gourd cultivation.
    (a) Five to six kilogram   (b) One to two kilogram
    (c) Three to four kilogram   (d) Ten to twelve kilogram
    (e) None of the above

(139) The average yield of snake gourd is about ———————— quintals per hectare under good crop management.
  (a) 70-80 (b) 100-110
  (c) 70-90 (d) 150-200
  (e) None of the above

(140) Snakegourd belongs to family ————————.
  (a) Solanaceae (b) Cruciferae
  (c) Cucubitaceae (d) Leguminoaceae
  (e) None of the above

(141) Snake gourd is ———————— in nature.
  (a) Monoecious (b) Dioecious
  (c) Gynoecious (d) Androecious
  (e) None of the above

(142) ———————— is known as Chinese snake gourd
  (a) *Trichosanthus ovigera* (b) *Trichosanthus kirilowii*
  (c) *Trichosanthus angunia* (d) *Cucumis humifructus*
  (e) None of the above

(143) Pusa Nasdar is a variety of ————————.
  (a) Ridge gourd (b) Bitter gourd
  (c) Bottle gourd (d) Sponge gourd
  (e) None of the above

(144) A hermaphrodite variety of ridge gourd is known as satputia widely cultivated in ———————— state.
  (a) Gujarat (b) Bihar
  (c) Madhya Pradesh (d) Delhi
  (e) None of the above

(145) Jhingle, Yard Long and Pocha's Jhinga Torai are ridge gourd cultivars with ———————— in nature.
  (a) Monoecious (b) Dioecious
  (c) Androecious (d) Gynoecious
  (e) None of the above

(146) The origin place of ridge gourd is ————————
  (a) Ethopia (b) India
  (c) China (d) Brazil
  (e) None of the above

(147) —————————— is also known as kali torai.
    (a) Bottle gourd   (b) Pumpkin
    (c) Muskmelon   (d) Watermelon
    (e) Ridge gourd or ribbed gourd

(148) The ridge gourd flowers opening ——————————.
    (a) Early in the morning   (b) At noon
    (c) At evening   (d) Mid night
    (e) None of the above

(149) *Luffa acutangula* Roxb is botanical name of ——————.
    (a) Sponge gourd   (b) Ridge gourd
    (c) Snake gourd   (d) Round gourd
    (e) None of the above

(150) The average yield of sponge gourd is —————————— quintal per hectare.
    (a) 50-60   (b) 60-70
    (c) 80-90   (d) 100-120
    (e) 150-200

(151) About 4-5 kg seed is required for planting of one-hectare area of ——————.
    (a) Sponge gourd   (b) Bitter gourd
    (c) Ridged gourd   (d) Pumpkin
    (e) None of the above

(152) Kali torai belongs to family ——————.
    (a) Solanaceae   (b) Malavaceae
    (c) Leguminoaceae   (d) Cucurbitaceae
    (e) None of the above

(153) —————————— is also known as four-angled gourd or angled loofah.
    (a) Ridge gourd   (b) Sponge gourd
    (c) Bitter gourd   (d) Pointed gourd
    (e) None of the above

(154) Luffaculin alkaloid is found in ——————————.
    (a) *Luffa operculata*   (b) *Trichosanthus*
    (c) *Cucurbita maxima*   (d) None of the above

(155) —————————— is also known as dish rag gourd and nenua.
    (a) Smooth gourd   (b) Ridged gourd
    (c) Snake gourd   (d) Bitter gourd
    (e) None of the above

(156) Basic chromosome number in sponge gourd is ——————————.
    (a) 10   (b) 12
    (c) 13   (d) 16
    (e) 20

(157) Sponge gourd is strictly ———————— in nature.
    (a) Hermaphrodite   (b) Gynoecious
    (c) Monoecious   (d) Andromonoecious
    (e) None of the above

(158) Pusa Chikni is a variety of ————————.
    (a) Sponge gourd   (b) Ridged gourd
    (c) Snake gourd   (d) Bitter gourd
    (e) Pumpkin

(159) Sponge gourd is a ———————— crop.
    (a) Cross-pollinated   (b) Self pollinated
    (c) Often cross-pollinated (d) Wind pollinated
    (e) None of the above

(160) Flower colour of sponge gourd is ————————.
    (a) Deep white   (b) Bluish
    (c) Deep yellow   (d) Pinkish
    (e) None of the above

(161) *Luffa cylindrica* is botanical name of ————————.
    (a) Ridged gourd   (b) Sponge gourd
    (c) Snake gourd   (d) Bitter gourd
    (e) None of the above

(162) In West Bengal, sponge gourd is sown in the month of ————————.
    (a) September-October (b) March-April
    (c) November-January   (d) June-July
    (e) None of the above

(163) Long day promotes ———————— in sponge gourd.
    (a) Maleness   (b) Femaleness
    (c) Hermaphordite   (d) No effect on sex
    (e) None of the above

(164) The crop will ready for harvest in ———————— days after seed sowing.
    (a) 30-40   (b) 50-60
    (c) 60-90   (d) 45-60
    (e) None of the above

(165) The average yield of sponge gourd varies from ———————— q/ha.
- (a) 50-60
- (b) 40-50
- (c) 60-80
- (d) 100-120
- (e) 300-400

(166) The fruits of sponge gourd attained marketable maturity in ———————— days after anthesis.
- (a) 2-3
- (b) 3-4
- (c) 5-7
- (d) 10-12
- (e) 15-17

(167) The origin place of sponge gourd is ————————.
- (a) Asia
- (b) Africa
- (c) China
- (d) South America
- (e) India

(168) Sponge gourd is propagated by ————————.
- (a) Cutting
- (b) Tuber
- (c) Seed
- (d) Root
- (e) None of the above

(169) Sponge gourd contains ———————— mg Vitamin C per 100 g edible portion.
- (a) 10
- (b) 5
- (c) 0.2
- (d) 18
- (e) None of the above

(170) In summer, sponge gourd is sown during ———————— in Northern India.
- (a) June - July
- (b) Feburary - March
- (c) November - December
- (d) September - October
- (e) None of the above

(171) Recommended NPK ratio for sponge gourd is ———————— in Punjab condition.
- (a) 40:40:60
- (b) 100:60:60
- (c) 100:80:80
- (d) 120:100:150
- (e) None of the above

(172) Ethrel 200 ppm and NAA 200 ppm sprayed on very young seedlings (2-4-leaf stage) on sponge gourd stimulated the production of ———————— flowers.
- (a) Male
- (b) Female
- (c) Hermaphordite
- (d) Dioecious
- (e) None of the above

(173) Pusa Bedana is a seedless variety of —————————————.
  (a) Muskmelon (b) Watermelon
  (c) Sponge gourd (d) Round melon
  (e) None of the above

(174) ————————— variety of watermelon is cross of Tetra-2 x Pusa Rasal.
  (a) Arka Joyti (b) Sugar Baby
  (c) Arka Manik (d) Pusa Bedana
  (e) None of the above

(175) ————————— is a powdery mildew resistant variety of watermelon.
  (a) Arka Manik (b)) Sugar Baby
  (c) Pusa Bedana (d) Shipper
  (e) None of the above

(176) *Citrullus lanatus* Mansf is botanical name of ————————— .
  (a) Muskmelon (b) Watermelon
  (c) Pumpkin (d) Round Melon
  (e) None of the above

(177) Centre of origin of watermelon is —————————————.
  (a) Tropical Asia (b) India
  (c) South America (d) Tropical Africa
  (e) None of the above

(178) Watermelon is rich source of —————————————————.
  (a) Potassium (b) Sodium
  (c) Iron (d) Calcium
  (e) None of the above

(179) A temperature of 24-27 °C is the best temperature plant growth of —————————
  (a) Tomato (b) Watermelon
  (c) Radish (d) Pea
  (e) Onion

(180) Arid regions of ————————— are best for production of quality fruits of watermelon.
  (a) Bihar (b) UttarPradesh
  (c) Rajasthan (d) Gujarat
  (e) MP

(181) Mateera cultivar of watermelon is widely grown in —————————
  (a) Rajasthan (b) UttarPradesh
  (c) Coimbatore (d) Gujarat
  (e) None of the above

(182) Germination of watermelon seed is ———————————————
- (a) Hypogeal
- (b) Epigeal
- (c) Semi-hypogeal
- (d) Both hypogeal and epigeal
- (e) None of the above

(183) In watermelon, ——————————— performs pollination.
- (a) Wind
- (b) Wasp
- (c) Butterfly
- (d) Honeybee
- (e) None of the above

(184) Watermelon is —————————— annual.
- (a) Monoecious
- (b) Andomonoecious
- (c) Dioecious
- (d) Trioecious
- (e) None of the above

(185) Watermelons are sown in the month of December-February in plains and ——————————— in hills, respectively.
- (a) June - July
- (b) November- December
- (c) January - Febuary
- (d) March - April
- (e) None of the above

(186) Basic chromosome number in watermelon is ———————————————.
- (a) 14
- (b) 11
- (c) 12
- (d) 24
- (e) 32

(187) Watermelon, crop is ready for harvesting in about ——————————— days after sowing depending upon cultivar and season.
- (a) 45-60
- (b) 70-75
- (c) 75-100
- (d) 100-120
- (e) None of the above

(188) In watermelon, apical shoots are removed and side shoots are allowed to grow when ——————————— side shoots are allowed to grow, it gives significantly higher fruit yield than unprunned plants.
- (a) 1-2
- (b) 2-3
- (c) 3-4
- (d) 6-7
- (e) 8-10

(189) A well-maintained crop can yield ——————————— quintals of watermelon of high quality from one-hectare area.
- (a) 100-120
- (b) 150-200
- (c) 400-600
- (d) 200-300
- (e) None of the above

(190) Micronutrients like boron and molybdenum at ———————— ppm proved effective in sex expression in watermelon.
    (a) 20   (b) 1
    (c) 3   (d) 10
    (e) 15

(191) Varieties 'Clatham Gray', 'Summit', 'Shipper', 'White Hope' and 'Verons' of ———————— are resistant to *Fusarium oxysporum* f. sp. *Niveum*.
    (a) Snapmelon   (b) Muskmelon
    (c) Watermelon   (d) Roundmelon
    (e) None of the above

(192) Line PI-189225 and PI-271778 of ———————— are resistant to *Mycosphaerella citrullina* and 'Fairfax' to *Alternaria cucumerina*.
    (a) Snake gourd   (b) Round melon
    (c) Watermelon   (d) Ridged gourd
    (e) None of the above

(193) Cucubitocitrin is found in seeds of ————————.
    (a) Bitter gourd   (b) Snapmelon
    (c) Watermelon   (d) Pumpkin

(194) ———————— is botanically known as *Sicana odorifera*.
    (a) Bottle gourd   (b) Ridge gourd
    (c) Pumpkin   (d) Casabanana

(195) Petha kadu is known as ————————.
    (a) Pumpkin   (b) Ash gourd
    (c) Smooth gourd   (d) Round gourd
    (e) Little gourd

(196) Wax gourd is mainly grown in ————————.
    (a) Tamilnadu   (b) Kerala
    (c) North India   (d) North east Hill Region
    (e) None of the above

(197) Ash gourd is considered good for people suffering from ————————.
    (a) Dysentry   (b) Stomach ache
    (c) Nervousness and debility   (d) High blood pressure
    (e) None of the above

(198) Short days, low night temperature and humid climate are good for production of ———————— flowers in ash gourd.
    (a) Male   (b) Female
    (c) Hermaphordite   (d) Both male and female
    (e) None of the above

(199) Optimum temperature requirement for cultivation is ─────────.
    (a) 24-30 °C   (b) 14-20 °C
    (c) 34-40 °C   (d) 4-10 °C
    (e) None of the above

(200) The recommended seed rate per hectare for ash gourd is _____.
    (a) 2-3   (b) 1-2
    (c) 5-7   (d) 10-12
    (e) None of the above

(201) Agra Petha, famous all over India, is prepared from ─────────.
    (a) Bottle gourd   (b) Pumpkin
    (c) Ash gourd   (d) Muskmelon
    (e) Watermelon

(202) The crop of ash gourd will be about ───────── days after seed sowing to reach marketable maturity.
    (a) 45-60   (b) 35-45
    (c) 55-65   (d) 75-120
    (e) 120-150

(203) Basic chromosome number in wax gourd is ─────────.
    (a) 10   (b) 12
    (c) 14   (d) 16
    (e) 20

(204) Under good crop management, an average yield of ash gourd is ───── q/ha per hectare is obtained.
    (a) 50-60   (b) 100-200
    (c) 300-400   (d) 600-1000
    (e) None of the above

(205) Wax gourd is ───────── and annual climber.
    (a) Monoecious   (b) Dioecious
    (c) Trioecious   (d) Gynodioecious
    (e) None of the above

(206) The seed of wax gourd germinate within ───────── days when sufficient soil moisture is available.
    (a) 2-3   (b) 3-4
    (c) 8-10   (d) 1-2
    (e) None of the above

(207) *Benincasa hispida* Cogn is botanical name of ─────────
    (a) Snapmelon   (b) Round melon
    (c) Pumpkin   (d) Watermelon   (e) Wax gourd

(208) Hairy melon, winter melon, ash pumpkin, white pumpkin Chinese preserving melon are synonymous of wax gourd.
  (a) Watermelon
  (b) Pumpkin
  (c) Snapmelon
  (d) Long melon
  (e) Wax gourd

Answer Sheet

1	b	2	d	3	a	4	c	5	c	6	e	7	a	8	d	9	c	10	b
11	b	12	c	13	b	14	d	15	d	16	b	17	c	18	d	19	c	20	d
21	c	22	b	23	d	24	e	25	c	26	b	27	c	28	c	29	b	30	d
31	c	32	C	33	e	34	d	35	c	36	a	37	a	38	b	39	b	40	c
41	c	42	a	43	a	44	c	45	a	46	b	47	d	48	d	49	d	50	b
51	c	52	d	53	a	54	b	55	a	56	a	57	b	58	d	59	c	60	b
61	b	62	c	63	c	64	b	65	a	66	c	67	b	68	a	69	b	70	c
71	c	72	c	73	c	74	c	75	b	76	b	77	c	78	d	79	d	80	d
81	c	82	d	83	b	84	d	85	b	86	d	87	c	88	c	89	b	90	d
91	c	92	d	93	c	94	c	95	b	96	c	97	b	98	b	99	a	100	b
101	c	102	b	103	e	104	c	105	c	106	b	107	b	108	b	109	c	110	a
111	d	112	c	113	c	114	c	115	a	116	b	117	e	118	a	119	c	120	b
121	c	122	c	123	c	124	d	125	c	126	a	127	d	128	c	129	c	130	b
131	c	132	b	133	a	134	e	135	c	136	d	137	a	138	a	139	d	140	c
141	a	142	b	143	a	144	b	145	a	146	b	147	e	148	c	149	b	150	e
151	a	152	d	153	a	154	a	155	a	156	c	157	c	158	a	159	a	160	c
161	b	162	c	163	b	164	c	165	d	166	c	167	e	168	c	169	d	170	b
171	c	172	b	173	b	174	d	175	a	176	b	177	d	178	c	179	b	180	c
181	a	182	b	183	d	184	a	185	d	186	b	187	c	188	c	189	c	190	c
191	c	192	c	193	c	194	d	195	b	196	c	197	c	198	b	199	a	200	c
201	c	202	d	203	b	204	c	205	a	206	c	207	e	208	e				

1.1.2.4. Fruit Vegetables

(1) ———————— is also known as 'Eggplant' or 'Baigan'.
  (a) Tomato
  (b) Brinjal
  (c) Chilli
  (d) Okra
  (e) None of the above

(2) Bitter taste in brinjal is due to ————————.
  (a) Cucurbitacin
  (b) Luffin
  (c) Momordicin
  (d) Glycoalkaloids
  (e) None of the above

(3) ———————— variety of brinjal is suited for ratoon cropping.
  (a) Pant Samara
  (b) Hisar Shyamal
  (c) Hissar Jamuni
  (d) Pusa Purple Long
  (e) None of the above

(4) Brinjal seeds germinate in ——————————days.
    (a) 1-2   (b) 5-6
    (c) 7-8   (d) 12-18
    (e) None of the above

(5) One gram seeds of brinjal having about —————— seeds.
    (a) 50-60   (b) 100-125
    (c) 150-200   (d) 250-300
    (e) None of the above

(6) Brinjal borne —————— types of flower on the basis of length of style
    (a) 4   (b) 5
    (c) 3   (d) 6
    (e) 1

(7) —————— is scientifically known as *Epliachna vigintioctopunctata*.
    (a) Hondda beetle   (b) Butterfly
    (c) Moth   (d) Jassid
    (e) None of the above

(8) 'Florida Market' variety of brinjal is resistant to ——————.
    (a) Damping off   (b) Bacterial wilt
    (c) Fusarium wilt   (d) Anthracnose
    (e) None of the above

(9) Little leaf of brinjal is due to ——————.
    (a) Fungus   (b) Bacteria
    (c) Mycoplasma   (d) Root knot nematode
    (e) None of the above

(10) For seed production, the isolation distance for foundation and certified are —————— meters, respectively.
    (a) 50 and 25   (b) 100 and 200
    (c) 200 and 100   (d) 200 and 400
    (e) None of the above

(11) African brinjal is botanically known as——————.
    (a) *Solanum gilo*   (b) *Solanum macrocarpon*
    (c) *Solanum torvum*   (d) *Solanummelongena*
    (e) *Solanum incanum*

(12) Brinjal is an ——————crop.
    (a) Self- pollinated   (b) Cross -pollinated
    (c) Often cross-pollinated   (d) Protandry
    (e) None of the above

(13) In brinjal, fruit setting is usually in the flowers having ———————— type style.
    (a) Long and medium   (b) Short and medium
    (c) Pseudo-short and medium   (d) Long and short
    (e) None of the above

(14) ———————————————— colour brinjal is good for diabetic patient.
    (a) Purple   (b) Green
    (c) White   (d) Black
    (e) None of the above

(15) ———————— is botanically known as *Solanum quitoense*.
    (a) Tree tomato   (b) Tree bean
    (c) Naranjillo   (d) Brinjal
    (e) None of the above

(16) Basic chromosome number in brinjal is ————————————.
    (a) 12   (b) 20
    (c) 10   (d) 16
    (e) 14

(17) ———————— is botanically known as *Solanum muricatum*.
    (a) Pepino or malon pear   (b) Egg plant
    (c) Musk melon   (d) Water melon
    (e) None of the above

(18) Discolouration in ———————— after cutting the fruits is due to orthodihydroxy phenolic compounds.
    (a) Cucumber   (b) Brinjal
    (c) Tomato   (d) Ridged gourd
    (e) None of the above

(19) ———————— disease of brinjal is the most serious disease, causing crop damage in field ranging from 40 to 80 per cent.
    (a) Damping off   (b) Little leaf
    (c) Bacterial wilt   (d) Fusarium wilt
    (e) None of the above

(20) Little leaf disease of brinjal is transmitted by an insect vector ————————
    (a) Aphid   (b) Leaf hopper
    (c) Butterfly   (d) Housefly
    (e) None of the above

(21) *Solanum sisymbrifolium* species of brinjal is resistant to ———— of brinjal.
   (a) Bacterial wilt
   (b) Fusarum wilt
   (c) Little leaf
   (d) Nematode
   (e) None of the above

(22) 'Vijay' and 'Black Beauty' cultivar of ———— are tolerant to root knot nematode.
   (a) Tomato
   (b) Brinjal
   (c) Chilli
   (d) Okra
   (e) None of the above

(23) The wild species *Solanum khasianum* is resistant to ————
   (a) Jassid
   (b) Shoot and fruit borer
   (c) Beetle
   (d) White fly
   (e) None of the above

(24) ———— variety of brinjal is resistant to jassids.
   (a) Pusa purple Long
   (b) Manjri Gota'
   (c) Pusa Purple round
   (d) Arka sheel
   (e) None of the above

(25) ———— is variety of brinjal which moderately resistant to bacterial wilt.
   (a) Arka Sheel
   (b) Pusa Purple Round
   (c) Pusa Purple Long
   (d) Pusa Purple Cluster
   (e) None of the above

(26) ———— is the variety of brinjal resistant to both, blight and bacterial wilt under field conditions.
   (a) Pant Rituraj
   (b) Hisar Shyamal
   (c) Hisar Pragarti
   (d) Pant Samrat
   (e) None of the above

(27) Optimum growing temperature for brinjal is ————.
   (a) 22 to 30 °C
   (b) 12 to 20 °C
   (c) 32 to 40 °C
   (d) 18 to 20 °C
   (e) None of the above

(28) ———— is botanically known as *Solanum nigrum*.
   (a) Gardenhuckle berry
   (b) Goose berry
   (c) Indian goose berry
   (d) Raspbeery
   (e) None of the above

(29) In brinjal about ———————— g of seeds are sufficient for raising seedlings for planting of one hectare of land.
    (a) 125-250   (b) 200-300
    (c) 200-500   (d) 1000
    (e) None of the above

(30) In 60 x 60 cm spacing, per hectare plant accommodates————————.
    (a) 27778   (b) 37778
    (c) 47778   (d) 57778
    (e) 67778

(31) Brinjal is a long duration crop occupying the land for nearly ———————— months.
    (a) 3-4   (b 5-6
    (c) 6-8   (d) 10-12
    (e) None of the above

(32) ———————— is botanically known as *Erianthus arundinaceus*.
    (a) Doob grass   (b) Sarkanda grass
    (c) Motha grass   (d) Napier grass
    (e) None of the above

(33) Brinjal can be store for ———————— days in fairly good condition at 7.2 °C to 10 °C with 85-95 per cent RH.
    (a) 1-3   (b) 3-4
    (c) 4-5   (d) 7-10
    (e) None of the above

(34) The wild species of brinjal namely *S. khasianum* and *S. aviculare* are important sources of ————————
    (a) Nicotine   (b) Solasodine
    (c) Naicin   (d) Valanine
    (e) None of the above

(35) In brinjal chance of cross-pollination is more in ————————.
    (a) Long style   (b) Medium
    (c) Pseudo-short   (d) Short
    (e) None of the above

(36) 'Pusa Purple Long' and 'Pusa Purple Round' are moderately resistant to ————
    (a) Jassid   (b) Shoot and fruit borer
    (c) Fusarium wilt   (d) Bacterial wilt
    (e) None of the above

(37) ———— variety of brinjal is cross of Pusa Purple Long and Hyderpur Round.
    (a) Pusa Bindu   (b) Pant Rituraj
    (c) Pusa Anmol   (d) Pusa Pragati
    (e) None of the above

(38) ———— is variety of brinjal resistant to phomopsis blight.
    (a) Pusa Purple Long   (b) Pusa Bhairav
    (c) Pant Rituraj   (d) Pant Samarat
    (e) None of the above

(39) C-1, RAH-51, Pragati, Pusa Bindu and CO-1 varieties of brinjal are resistant to ————.
    (a) Salt   (b) Bacterial wilt
    (c) Fusarium wilt   (d) Damping off
    (e) None of the above

(40) A daily mean temperature of ———— is most favourable for better growth and yield.
    (a) 22-30 °C   (b) 12-20 °C
    (c) 8-10 °C   (d) 15-30 °C
    (e) None the above

(41) Inverted 'V' shaped interveinal chlorosis is found in brinjal due to———— deficiency.
    (a) Ca   (b) Mn
    (c) Mg   (d) Cu
    (e) None of the above

(42) Wilting, stunting and yellowing of the foliage followed by collapse of the entire plant are symptom of ————.
    (a) Bacterial wilt   (b) Fusarium wilt
    (c) Bacterial blight   (d) Little leaf
    (e) None of the above

(43) Thomas and Krishnaswamy were first person who reported little leaf disease of brinjal in India in ————.
    (a) 1939   (b) 1945
    (c) 1930   (d) 1925
    (e) None of the above

(44) Nagai and Kida first reported hybrid vigour in brinjal in ————.
    (a) 1930   (b) 1945
    (c) 1955   (d) 1965
    (e) None of the above

(45) *S. xantocarpum, S. indicum, S. gilo, S. khasianum, S. nigrum* and *S. sisymbrifolium* of wild species of brinjal are highly resistant to ——————.
    (a) Bacterial wilt   (b) Fusarium wilt
    (c) Phomopsis blight   (d) Damping off
    (e) None of the above

(46) On an average each crossed fruit of brinjal yields —————— g seed.
    (a) 20   (b) 15
    (c) 3   (d) 10
    (e) None of the above

(47) —————— is botanically known as *C. annuum*.
    (a) Bird eye chilli   (b) Pimento pepper
    (c) Black pepper   (d) Sweet pepper
    (e) None of the above

(48) The capsaicin is present in —————— walls and a placenta.
    (a) Core or septa   (b) Epicarp
    (c) Mesocarp   (d) Seeds
    (e) None of the above

(49) The chilli is originated from ——————.
    (a) India   (b) Brazil
    (c) Tropical America   (d) China
    (e) None of the above

(50) The most important chilli growing states in India is ——————.
    (a) Gujarat   (b) Andhra Pradesh
    (c) Madhya Pradesh   (d) Uttar Pradesh
    (e) Bihar

(51) —————— is botanically known as *C. frutescens*.
    (a) Black pepper   (b) Sweet pepper
    (c) Hot pepper   (d) Cardamom
    (e) None of the above

(52) Chilli can be grown from sea level to an altitude of —————— m.
    (a) 200   (b) 400
    (c) 800   (d) 1000
    (e) 2000

(53) For chilli cultivation, the maximum temperature ranging from —————— and minimum temperature 10 °C.
    (a) 20-30 °C   (b) 10-20 °C
    (c) 30-40 °C   (d) 5-10 °C
    (e) None of the above

(54) Chilli thrives in areas having moderate rainfall with the range of ———
 (a) 30-40 cm
 (b) 60-120 cm
 (c) 60-70 cm
 (d) 60-80 cm
 (e) None of the above

(55) ——— kilogram seeds of chilli required for planting in one-hectare.
 (a) Two
 (b) One
 (c) Three
 (d) One and half
 (e) None of the above

(56) One gram chilli contains ——— seeds.
 (a) 50-60
 (b) 60-80
 (c) 120-170
 (d) 200-300
 (e) None of the above

(57) The 3 x3m area nursery bed accommodates about ——— chilli seedlings.
 (a) 3000
 (b) 4000
 (c) 5000
 (d) 6000
 (e) None of the above

(58) Chilli is transplanted mainly in ——— if grown under rainfed conditions.
 (a) August- September
 (b) October- November
 (c) June - July
 (d) May- June
 (e) None of the above

(59) Basic chromosome number in chilli is ———.
 (a) 14
 (b) 16
 (c) 12
 (d) 20
 (e) 24

(60) In sweet peppers, anthesis starts at ——— and continues upto 11.15 A.M.
 (a) 7.15 P.M
 (b) 7.15 A.M
 (c) 3 A.M
 (d) 5 A.M
 (e) None of the above

(61) The sweet pepper stigma becomes receptive from the day of anthesis and remains receptive upto ——— days after anthesis.
 (a) Four
 (b) Five
 (c) Two
 (d) One
 (e) None of the above

(62) Punjab Lal is the variety of ––––––––––––––––––––––––––––––––.
    (a) Tomato   (b) Chilli
    (c) Sweet pepper   (d) Sweet potato
    (e) None of the above

(63) Pant C-1 variety of chilli is resistant to –––––––––––––––––––––––––.
    (a) Damping off   (b) Fruit rot
    (c) Mosaic and leaf curl virus   (d) Fusarium wilt
    (e) None of the above

(64) –––––––––––––––––––– is a variety of chilli, resistant to multi diseases.
    (a) Pusa Jawala   (b) NP-46
    (c) KA-2   (d) Pusa Sadabahar
    (e) None of the above

(65) –––––––––––––––––––– are botanically known as *Capsicum anuum* var. *cerasiforme*.
    (a) Cherry peppers   (b) Hot peppers
    (c) Sweet pepper   (d) Black pepper
    (e) None of the above

(66) –––––––––––––––––––– is the largest exporter of chilli.
    (a) India   (b) China
    (c) Kuwat   (d) Bangladesh
    (e) None of the above

(67) Andhra Pradesh, Karnataka and Maharashtra account for –––––––––––––– per cent net area of the country chilli and its production.
    (a) 20   (b) 30
    (c) 40   (d) 50
    (e) 75

(68) *C. annuum x C. pubescens* are completely ––––––––––––––––––––––––––.
    (a) Sef- fertile   (b) Cross- fertile
    (c) Self- incompatable   (d) Cross - compatable
    (e) None of the above

(69) In chilli, low light temperature (8-10 °C and 15 °C) increases the fruit set and –––––––––––––––––––––––– development of fruits.
    (a) Parthenocary   (b) Parthenogenesis
    (c) Stenosperomocary   (d) Fruit cracking
    (e) None of the above

(70) Short day conditions (9-10 hr light) stimulates plant growth and increased the productivity by ———————— per cent in chilli.
   (a) 5-10
   (b) 10-15
   (c) 2-4
   (d) 21-24
   (e) None of the above

(71) MDU-1 is a variety of chilli developed through ————————————.
   (a) Pure line selection
   (b) Mutation breeding
   (c) Mass selection
   (d) Pedigree method
   (e) Back cross method

(72) ———————— variety of chilli is cross of NP 46A x Puri Red.
   (a) Pant C-1
   (b) Jawahar -218
   (c) Pusa Jawala
   (d) Pusa Sadabahar
   (e) None of the above

(73) ———————— in chilli are due to *Colletotrichum capsici*.
   (a) Damping off
   (b) Leaf curl virus
   (c) Root knot nematode
   (d) Bacterial wilt
   (e) Fruit rot

(74) *Meloidogyne arenaria* is the most harmful species of nematode for ———— crop.
   (a) Okra
   (b) Chilli
   (c) Pumpkin
   (d) Cowpea
   (e) None of the above

(75) The future breeding programmes for improvement of chilli are:
   1. Evolution of cultivars to suit both internal and export markets.
   2. Wider adaptibility and quality viz; better pod colour and shape, high capsaicin, carotene and ascorbic acid contents.
   3. Cultivars, which are early, determinate and compact with shorter flowering phase, have to be developed.
   4. Resistant breeding needs emphasis for breeding lines resistant to viruses TMV, leaf curl, potato mosaic etc, thrips, fruit rot and nematodes and expliotation of hybrid vigour.
   (a) 1 and 2
   (b) Only 3 and 4
   (c) Only 4 and 5
   (d) 1,2,3,4 and 5
   (e) None of the above

(76) Callifornia Wonder, Chinese Giant and World Beater are varieties of ————
   (a) Chilli
   (b) Capsicum
   (c) Okra
   (d) Tomato
   (e) None of the above

(77) Bharat is first hybrid of capsicum released by ——————
   (a) MAHYCO (b) IAHS
   (c) Bio-Seeds (d) Sugrow
   (e) Ganesha Seeds

(78) The isolation distance between two cultivars of capsicum should be kept ——— meters for foundation seed and ——— meters for certified seed production.
   (a) 50 and 25 (b) 800 and 400
   (c) 200 and 100 (d) 1000 and 1600
   (e) None of the above

(79) In Kullu Velly, the average seed yield of the California Wonder cultivar is ——————— kg/ha.
   (a) 400 (b) 500
   (c) 100 (d) 225
   (e) None of the above

(80) Sweet or bell pepper, capsanthin accounts for about ——— of the total carotenoid content.
   (a) 10 (b) 20
   (c) 36 (d) 15
   (e) None of the above

(81) The treatment with 2,4-D at ——————— ppm fortnightly interval during the flowering season of winter and summer grown sweet pepper increased flowering in the winter crop.
   (a) 5 (b) 10
   (c) 2 (d) 20
   (e) None of the above

(82) Sweet peppers can be kept in good condition for at least ——————— days at 0 °C and RH of 95-98 per cent.
   (a) 10 (b) 15
   (c) 20 (d) 30
   (e) 40

(83) Tobacco Pepper and Rocoto are relative of ——————
   (a) Sweet pepper (b) Black pepper
   (c) Chilli (d) Cucumin
   (e) None of the above

(84) ———————————— belongs to family malavaceae.
   (a) Tomato (b) Brinjal
   (c) Okra (d) Radish
   (e) None of the above

(85) Eating fresh raw okra fruits on an empty stomach every morning nourishes the body and increases the ——————————————————— content.
- (a) Fat
- (b) Semen
- (c) Blood sugar
- (d) Colestrol
- (e) None of the above

(86) The powedered roots of okra is given with sugar for curing of———————.
- (a) Joindice
- (b) Leucorrhoea backache
- (c) Head-ache
- (d) Dysentry
- (e) None of the above

(87) Okra fruits are excellent source of——————————————————.
- (a) Calcium
- (b) Magnesium
- (c) Manganese
- (d) Iodine
- (e) None of the above

(88) Bland mucilage obtained from——————fruits is also used as a clarifying agent in the perparation of gur.
- (a) Tomato
- (b) Cotton
- (c) Okra
- (d) Jute
- (e) None of the above

(89) In okra, 2n chromosome number range from———————————.
- (a) 32-64
- (b) 20-44
- (c) 72-144
- (d) 150-160
- (e) None of the above

(90) ——————————— is also known as lady's finger.
- (a) Chilli
- (b) Okra
- (c) Guar
- (d) Pea
- (e) None of the above

(91) Okra seeds fail to germinate below ———————————temperatutre.
- (a) 10 °C
- (b) 5 °C
- (c) 20 °C
- (d) 2 °C
- (e) Noneof the above

(92) The optimum temperature for seed germination of okra is ———————.
- (a) 29 °C
- (b) 20 °C
- (c) 12 °C
- (d) 39 °C
- (e) 19 °C

(93) ———————variety of okra is cross of Pusa Makhmali x IC-1542.
- (a) Gujarat Bhindi-1
- (b) Pusa Makhmali
- (c) Pusa Sawani
- (d) VRO-10
- (e) Noneof the above

(94) In okra, if temperature goes above——————— may cause flower drop.
    (a) 22 °C　　　　　　　　　(b) 32 °C
    (c) 40 °C　　　　　　　　　(d) 42 °C
    (e) None of the above

(95) The rainy season, okra crop is sown in the month of ———————through out India.
    (a) May-June　　　　　　　(b) June-July
    (c) August-September　　　　(d) September-October
    (e) None of the above

(96) A flower bud of okra takes about ——————— days from initiation to full bloom.
    (a) 35-40　　　　　　　　　(b) 22-26
    (c) 50-60　　　　　　　　　(d) 60-70
    (e) None of the above

(97) Okra is ———————
    (a) Self - pollinated　　　　　(b) Cross- pollinated
    (c) Often cross-pollinated　　(d) Hermaphordite
    (e) None of the above

(98) Okra pods become ready for first harvesting after ——————— days of sowing.
    (a) 25　　　　　　　　　　　(b) 35
    (c) 45　　　　　　　　　　　(d) 65
    (e) None of the above

(99) Okra green pods can stored for ——————— days in 400 gauge polythene bags under room temperature (32 ± 2 °C) and 70-75 per cent RH as against 2-3 days without package.
    (a) 4　　　　　　　　　　　(b) 5
    (c) 7　　　　　　　　　　　(d) 9
    (e) None of the above

(100) Weeds in okra may result——————— per cent reduction in the yield.
    (a) 5　　　　　　　　　　　(b) 10
    (c) 15　　　　　　　　　　　(d) 20
    (e) None of the above

(101) Cultivated okra is ———————in nature.
    (a) Polyploidy　　　　　　　(b) Triloidy
    (c) Tetraploidy　　　　　　　(d) Autoploidy
    (e) None of above

(102) Sterility in okra is due to. ─────────
    (a) Chromosomal   (b) Genic defferentiation.
    (c) Mutation   (d) a and b both
    (e) Only C

(103) Yield loss in okra due to yellow vein mosaic virus is between ─────── per cent.
    (a) 2-30   (b) 30-40
    (c) 50-60   (d) 50-90
    (e) None of the above

(104) The origin place of okra is ───────────────.
    (a) South America   (b) Tropical Africa
    (c) Asia   (d) India
    (e) None of the above

(105) ─────────────────variety of bhindi is recommended from Collage of Agriculture, Solan for hilly areas under the name Harbhajan.
    (a) Hissar Unnat   (b) Pusa Makhamali
    (c) Perkins Long Green   (d) Pusa Sawani
    (e) None of the above

(106) ─────────────────of okra is scientifically known as *Earias vittela* Fab.
    (a) Jassid   (b) Shoot and fruit borer
    (c) Aphid   (d) Beetle
    (e) None of the above

(107) Okra is an often-cross pollinated crops, therfore, maintain isolation distance of ─────── meters for foundation seed and ─────── for certified seed, respectively.
    (a) 100 and 200   (b) 400 and 200
    (c) 800 and 400   (d) 200 and 100
    (e) None of the above

(108) The related *species A. manihot var. pungens, A.crinitus, A. panduraeformis* and *A. vitifolius* and varieties Sel-4 and Arka Anamika are resistant to YVMV disease of ──────────────.
    (a) Okra   (b) Amranths
    (c) Tomato   (d) Cowpea
    (e) None of the above

(109) Variety Long Green Smooth of okra has showed high resistance to ───────.
    (a) Yellow vein mosaic   (b) Shoot and fruit borer
    (c) White fly   (d) Jassid
    (e) Nematode

(110) Lines 'AE-22' 'AE-79' and Red Wonder lines of okra showed resistance to———.
    (a) Fruit borer   (b) Whitefly
    (c) Jassid   (d) Leaf spot
    (e) None of the above

(111) Lines IC-18960, IC-15055, Round selection, Walgaon, EC-41292 and IC -1542 showed resistance to———————.
    (a) Leaf spot   (b) White fly
    (c) Yellow vein mosaic   (d) Jassid
    (e) None of the above

(112) ——————————is considerd as 'Poor man's Orange' in India, while 'Love of Apple' in England.
    (a) Brinjal   (b) Tomato
    (c) Chow- chow   (d) Potato
    (e) Muskmelon

(113) Cultivated tomato originated from ———————.
    (a) *Lycopersicon esculentum* var. *cerasaforme*
    (b) *Lycopersicon esculentum* var. *pyriformi*
    (c) *Lycopersicon esculentum* var. *chesmani*
    (d) *Lycopersicon esculentum* var. *glabaratum*
    (e) None of the above

(114) Tomato originated from———————.
    (a) North America   (b) Central America
    (c) South America   (d) India
    (e) Brazil

(115) Lycopene in tomato is the highest at ————————.
    (a) 70-75 ^{0}F   (b) 60-70 ^{0}F
    (c) 100-125 ^{0}F   (d) 90-95 ^{0}F
    (e) None of the above

(116) In tomato, locular jelly may not fill the locular cavity which may lead ———.
    (a) Cracking   (b) Puffiness
    (c) Cat facing   (d) Sun scalding
    (e) None of the above

(117) ————————————— is a rainfed tomato variety.
    (a) Pusa Shetal   (b) Arka Meghali
    (c) NDT-120   (d) Punjab Chhuhara
    (e) None of the above

(118) The red colour in tomato is due to pigment————.
- (a) Anthocyanin
- (b) Quercetin
- (c) Lycopene
- (d) Xanthophyll
- (e) None of the above

(119) Karnataka is a hybrid of————.
- (a) Tomato
- (b) Brinjal
- (c) Okra
- (d) Muskmelon
- (e) None of the above

(120) Dr. C.M. Rick and Dr. G. Kalloo are well known in improvement of————.
- (a) Brinjal
- (b) Chilli
- (c) Tomato
- (d) Okra
- (e) Watermelon

(121) Metribuzin 0.5 kg/ha and alachlor 2.3 kg/ha are most effective weedicides in ————.
- (a) Okra
- (b) Brinjal
- (c) Water melon
- (d) Indian bean
- (e) Tomato

(122) For distant market, tomato fruits should be harvested at mature ———— stage.
- (a) Half ripe stage
- (b) Mature stage
- (c) Green stage
- (d) Pink stage
- (e) Red ripe stage

(123) The tomato fruits are graded as specified by ISI ———— in category.
- (a) 3
- (b) 4
- (c) 5
- (d) 6
- (e) 2

(124) *Heliothis armigera* is the scientific name of————.
- (a) Jassid
- (b) White fly
- (c) Fruit borer
- (d) Gross hopper
- (e) None of the above

(125) White fly (*Bemisia tabaci*) is vector of————.
- (a) Tomato leaf curl virus
- (b) Little leaf
- (c) Cucumber mosaic
- (d) Buttoning
- (e) None of the above

(126) Hisar Lalit and SL-120 varieties of tomato are resistant to————.
- (a) Fusarium wilt
- (b) Bacterial wilt
- (c) Root knot nematode
- (d) Leaf curl virus
- (e) None of the above

(127) Aroma of green leaf of tomato is due to ————————————.
    (a) Alkaloid   (b) Xanthophyll
    (c) Isovaleraldehyde and hexyl alcohol   (d) Chlorophyll
    (e) All of the above

(128) Tomato fruit aroma is due to ————————————————.
    (a) Sulfonium   (b) Curcurmin
    (c) Allicin   (d) Alcohol
    (e) None of the above

(129) One-gram tomato seed contains ———————————— seeds.
    (a) 100   (b) 200
    (c) 300   (d) 1000
    (e) None of them

(130) A net area of about 225 sqm is required to raise the seedling for ———— hectare.
    (a) One   (b) Two
    (c) Three   (d) Four
    (e) None of the above

(131) CCC at the rate of ———————————— ppm sprayed in the nursery two to three days before transplanting reduces the incidence of leaf curl virus.
    (a) 100   (b) 250
    (c) 500   (d) 50
    (e) 25

(132) The plant growth is continued and there is less initiation of flowers and fruits on the stem. The lateral buds always exist to continue vegetative growth, such type of growth habit in tomato is known as————————————————.
    (a) Determinate   (b) Indeterminate
    (c) Semideterminate   (d) Erect
    (e) None of the above

(133) The inforesence of tomato is ————————————.
    (a) Dicotomous   (b) Trimerous
    (c) Tetramerous   (d) Pentamerous
    (e) None of the above

(134) The fruits of tomato is ————————————.
    (a) Berry   (b) Pome
    (c) Pepo   (d) Drup
    (e) None of the above

(135) —————————— is botanically known as *Phsalis ixocarpa*.
    (a) Tree tomato   (b) Tomatillo
    (c) Pomato   (d) Tomato
    (e) None of the above

(136) In tomato, fertilization take place after ————————— of pollination.
    (a) 1-3 hrs   (b) 4-5 hrs
    (c) 10-30 hrs   (d) 10-15 hrs
    (e) None of the above

(137) In tomato, maximum acidity is found when fruits are harvested at ———— stage.
    (a) Green   (b) Pink
    (c) Ripe   (d) Red ripe
    (e) None of the above

(138) In tomato, seed germination is inhibited due to presence of ———————————.
    (a) Caffic acid and ferulic acid   (b) Mallic
    (c) Juglone   (d) Asperatic
    (e) None of the above

(139) Tomato fruit contains ———————— mg/100g fruit weight lycopene.
    (a) 20-30   (b) 20-50
    (c) 20-40   (d) 10-20
    (e) None of the above

(140) ——————————— is steroidal glycoalkaloid.
    (a) Tomatine   (b) Lycopene
    (c) Caffein   (d) Ferrulic
    (e) None of the above

(141) In tomato, asexual propagation is employed in order to overcome ———— barrior.
    (a) Incompatability   (b) Sterility
    (c) Fast multiplication   (d) TLCV testing
    (e) Both b and d

(142) Tomato produces ———————— flowers in cluster on the stem.
    (a) Protandry   (b) Protogynous
    (c) Cleistogamous   (d) Hercogammy
    (e) None of the above

(143) Tomato flowers anthesis start in the morning at about ————————————.
    (a) 6.0 AM   (b) 10 AM
    (c) 11 AM   (d) 12 noon
    (e) None of the above

(144) The pollen grain of tomato remains viable for a long period ——— at temperature.
    (a) 2-4 days   (b) 4-8 days
    (c) 4-6 days   (d) 8-10 days
    (e) None of the above

(145) For hybridization in tomato, mature buds are emasculated ——— before their opening.
    (a) 12-16 hr   (b) 2-6 hr
    (c) 3-4 hr   (d) 10-12 hr
    (e) None of the above

(146) About ———persons/days required for producing one-kilogram tomato hybrid in India.
    (a) 20-25   (b) 30-40
    (c) 40-45   (d) 60-75
    (e) None of the above

(147) For identification of purity of hybrid seed produced by using male sterile line, the recessive marker is ———.
    (a) Black seed   (b) Brown seed
    (c) Yellowish seed   (d) Whitish
    (e) None of the above

(148) For development of earliness and large fruited character variety of tomato, the ——— breeding method is used.
    (a) Back cross   (b) Pureline
    (c) SSD   (d) Double back cross
    (e) None of the above

(149) In mutation breeding, the tomato seed treated with X rays or gamma rays at ———.
    (a) 10-30 Kr   (b) 10-15 Kr
    (c) 10-20 Kr   (d) 30-40 Kr
    (e) None of the above

(150) Boron deficiency in tomato causes ———.
    (a) Browning   (b) Cat facing
    (c) Puffiness   (d) Sun scalding
    (e) Fruit cracking

(151) Calcium deficiency in tomato causes———.
    (a) Blossom-end rot   (b) Puffiness
    (c) Cat facing   (d) Fruit cracking
    (e) None of the above

(152) The term C: N ratio was first coined by Karaus and Kraywill in——————.
   (a) 1921 (b) 1919
   (c) 1935 (d) 1940
   (e) None of the above

(153) Boron deficiency in tomato can be corrected by soil application of borax —————— kg/ha.
   (a) 20-30 (b) 10-15
   (c) 5-15   (d) 15-20
   (e) None of the above

(154) In normal tomato, plant should have —————— mg/kg of molybdenum.
   (a) 10 (b) 15
   (c) 3   (d) 6
   (e) None of the above

(155) Manganese deficiency is usually found in —————— soil.
   (a) Sandy soil (b) Loamy soil
   (c) Calcarious (d) Sandy loam
   (e) None of the above

(156) Yellow mottling within the interveinal in tomato leaves is deficiency of——————.
   (a) Zinc (b) Iron
   (c) Mo   (d) Ca
   (e) None of the above

(157) In tomato, Mg deficiency could be expected when soil contain less than —————— meq/100g soil exchangable Mg.
   (a) 25 (b) 20
   (c) 15 (d) 30
   (e) None of the above

(158) —————— plant growth regulator is very effective in increasing fruit set, size of the fruit and harvesting of maturity under low and high concentration.
   (a) GA   (b) PCPA
   (c) Cytokinin (d) NAA
   (e) None of the above

(159) Application of —————— controlled plant growth, induced drought resistant and production of early and total yield.
   (a) MH   (b) Cycocel
   (c) NOA (d) IBA
   (e) None of the above

(160) Mature green tomato fruits are stored at 10-15 °C for ———————— days.
- (a) 10
- (b) 15
- (c) 30
- (d) 20
- (e) None of the above

(161) The causes of pocket and puffiness in tomatos are ————————————.
- (a) Environment
- (b) Nutritional
- (c) Poor pollination
- (d) Poor Fertilization
- (e) All of the above

(162) To maintain the genetic purity, the isolation distance for foundation seed ———————— meter and certified seed ———————— meter should kept respectively.
- (a) 20 and 10
- (b) 100 and 200
- (c) 50 and 25
- (d) 400 and 200
- (e) None of the above

(163) In tomato, for seed production, field inspection and roughing should be done at ————————————.
- (a) Before flowering
- (b) Immature stage
- (c) Mature stage
- (d) B and C
- (e) All of the above

(164) Sunscalding physiological disorder in tomato is due to the exposure of fruit to due to ————————————.
- (a) Mg deficiency
- (b) Ca deficiency
- (c) Boron deficiency
- (d) Intense light
- (e) None of the above

(165) Fruit drop and poor fruit setting in tomato are due to ————————————.
- (a) Imbalance supply nutrition
- (b) Incorrect method and time of nutrient application
- (c) High and low temperature
- (d) Only B and C
- (e) All of the above

(166) *Lycopersicon chessmanii* of tomato is resistant to————————————.
- (a) TLCV
- (b) Bacterial wilt
- (c) Salt tolerance
- (d) Nematode
- (e) None of the above

(167) *Lycopersicon pennellii* of tomato is tolerance to————————————.
- (a) Salt tolerance
- (b) Drought tolerance
- (c) TLCV
- (d) Bacterial wilt
- (e) All of the above

(168) Sioux, Marglobe, Keckruth, Keckruth Ageti and Labonita varieties of tomato are introduced from ——————.
    (a) Africa   (b) Asia
    (c) USA   (d) Europe
    (e) None of the above

(169) In tomato, under normal condition, there is more than —————— per cent fruit set after fertilization.
    (a) 20   (b) 30
    (c) 45   (d) 67
    (e) 90

(170) One kg hybrid seed can be produced by ——————persons, using male sterile lines.
    (a) 10-15   (b) 5-10
    (c) 15-20   (d) 25-30
    (e) None of the above

(171) Leaf curl virus in tomato causes —————— per cent yield loss.
    (a) 30   (b) 45
    (c) 55   (d) 67
    (e) 90

(172) Resistance to tomato leaf curl virus in tomato is associated with——————.
    (a) Anthocyanin   (b) Phenol and sugar
    (c) Chlorophyll   (d) Lycopene
    (e) None of the above

(173) Brinjal, croton and cowpea are host plant for rearing ——————.
    (a) Gross hopper   (b) Mite
    (c) Aphid   (d) Whitefly
    (e) None of the above

(174) *Lycopersicon peruvianum* species of tomato is resistant to——————.
    (a) Early blight   (b) Late blight
    (c) TLCV   (d) RKN
    (e) None of the above

(175) Hissar Lalit variety of —————— is resistant to root knot nematode.
    (a) Tomato   (b) Brinjal
    (c) Chilli   (d) Okra
    (e) None of the above

(176) ——————— variety of tomato can set fruit at 8-10 °C.
   (a) Arka Saurabh (b) Pusa Sheetal
   (c) Arka Meghali (d) DVRT-1
   (e) None of the above

(177) ——————— species of tomato which has ability to give adequate flowering at high temperature.
   (a) *Lycopersicon chessmani* (b) *Lycopersicon creasiformae*
   (c) *Lycopersicon esculentum* (d) *Lycopersicon peruvianum*
   (e) *Lycopersicon hirsutum*

(178) In tomato, direct seed sown plants flowered ——— days earlier than transplanted.
   (a) 4-5 (b) 5-6
   (c) 7-8 (d) 9-10
   (e) None of the above

(179) For immediate marketing, the tomato fruit should be picked up at ——— stage.
   (a) Green stage (b) Ripe stage
   (c) Full ripe stage (d) Turning stage
   (e) None of the above

(180) Best cropping intensity in tomato is ——————— per cent.
   (a) 100 (b) 175
   (c) 150 (d) 200
   (e) 300

(181) By drip irrigation, ——— per cent water is saved in tomato.
   (a) 10 (b) 5
   (c) 20-30 (d) 15-44
   (e) None of the above

(182) For processing purpose, pH of juice should be———————————.
   (a) 2.5 (b) 3.5
   (c) 4.5 (d) 6.5
   (e) 7.5

(183) In tomato, ideal variety for processing should have an acid content of ——————— per cent as citric acid.
   (a) 0.10 (b) 0.15
   (c) 0.20 (d) 0.35
   (e) None of the above

(!84) There is a ——————— correlation between fruit size and acidity in tomato.
   (a) Positive (b) Negative
   (c) No relation (d) Positive and negative both
   (e) None of the above

(185) —————————— F_1 tomato hybrid has the highest acidity.
    (a) Vaishali   (b) Karnataka
    (c) KT-4   (d) Rupali
    (e) None of the above

(186) For preparation of tomato ketchup and paste, the prescribed minimum total soluble solids by weight is —————————— per cent.
    (a) 1   (b) 2
    (c) 3   (d) 4
    (e) 5

(187) Arka Saurabh and Arka Ahuti and Arka Ashish varieties of tomato are suitable for —————————— purposes.
    (a) Canning   (b) Dehydration
    (c) Processing   (d) Fresh Marketing
    (e) None of the above

(188) Breaker stage of harvesting in tomato is $1/4^{th}$ of the surface at blossom end shows ——————————.
    (a) Red   (b) Yellow
    (c) Green   (d) Pink   (e) None of the above

(189) Ohio CR-6 is the best-adopted cultivar of tomato for glass house cultivation in the——————————.
    (a) Europe   (b) USA
    (c) Asia   (d) Brazil
    (e) None of the above

(190) Severianin is a —————————— variety of tomato.
    (a) Highly seeded   (b) Parthenocarpic
    (c) Stenospermocarpic   (d) Partehnogensis
    (e) None of the above

(191) —————————— is botanically known as *Cyphomandra betacea*.
    (a) Chow-chow   (b) Tree bean
    (c) Tree tomato   (d) Jack bean
    (e) None of the above

(192) Ostankinsk is a cultivar of —————————— whose fruit set in very low temperature.
    (a) Okra   (b) Tomato
    (c) Pea   (d) Cauliflower
    (e) Brinjal

(193) —————— is botanically known as *Lycopersicon esculentum* var. *cerasiformae*.
    (a) Potato leaf tomato   (b) Upright tomato
    (c) Pear tomato   (d) Cherry tomato
    (e) None of the above

(194) In India the book entitled "Tomato" was written by Dr. G. Kalloo in ——————.
    (a) 1989   (b) 1986
    (c) 1993   (d) 1978
    (e) None of the above

(195) Tomato is the one of the most popular and widely grown vegetable in the world ranking —————— in importance after potato in many countries.
    (a) First   (b) Second
    (c) Third   (d) Four
    (e) None of the above

(196) Pusa Divya hybrid, Pusa Sadabahar, Punjab Upma and Pusa Upkar are varietes of ——————.
    (a) Onion   (b) Tomato
    (c) Capsicum   (d) Okra
    (e) None of the above

(197) Pusa Sadabahar is a variety of —————— suiatble for both stress condition and low temperature fruit setting.
    (a) Chilli   (b) Tomato
    (c) Okra   (d) Capsicum
    (e) None of the above

(198) Pusa Upkar is a variety of tomato tolerant to ——————.
    (a) Jassid   (b) Fruit borer
    (c) White fly   (d) TLCV
    (e) RKN

(199) Arka Alok, Arka Abhijit and Arka Ahreshtha varieties of tomato are resistant to ——————.
    (a) Fusarium wilt   (b) Bacterial wilt
    (c) TLCV   (d) RKN
    (e) None of the above

(200) S-12 and CO-3 varieties of tomato are developed through —————— in breeding.
    (a) Pure line   (b) SSD
    (c) Pedigree   (d) Mutation
    (e) None of the above

(201) Sioux, Manalucie and Crack Proof varieties of tomato are resistant to————.
   (a) Sun scalding (b) Cat facing
   (c) Fruit cracking (d) Winter injury
   (e) None of the above

(202) ———————— is botanically known as *Physalis pruinosa*.
   (a) Tree tomato (b) Husk tomato
   (c) Cherry tomato (d) Potato leaf tomato
   (e) None of the above

(203) The name tomato probably derived from tomatal in the Nahua tongue of ——.
   (a) China (b) Italian
   (c) Mexico (d) None of the above

(204) The first record of tomato in the old world is the description published by Pier Andrea Mattioli of Italy in ————————.
   (a) 1754 (b) 1854
   (c) 1954 (d) None of the above

(205) Thomas Jefferson in ———————— described the planting of tomato in Virginia.
   (a) 1981 (b) 1781
   (c) 1681 (d) 1881

Answer Sheet

1	b	2	d	3	c	4	d	5	d	6	a	7	a	8	c	9	c	10	c
11	b	12	c	13	a	14	c	15	c	16	a	17	a	18	b	19	b	20	b
21	c	22	b	23	b	24	b	25	d	26	d	27	a	28	a	29	c	30	a
31	c	32	b	33	d	34	b	35	a	36	b	37	c	38	b	39	b	40	a
41	c	42	d	43	a	44	b	45	c	46	c	47	b	48	a	49	c	50	b
51	c	52	e	53	a	54	b	55	b	56	c	57	d	58	c	59	c	60	b
61	c	62	b	63	c	64	d	65	a	66	a	67	e	68	b	69	a	70	b
71	b	72	c	73	e	74	b	75	d	76	b	77	b	78	c	79	d	80	c
81	c	82	e	83	a	84	c	85	b	86	d	87	d	88	c	89	c	90	b
91	c	92	a	93	c	94	d	95	b	96	b	97	c	98	c	99	d	100	d
101	a	102	d	103	d	104	b	105	c	106	b	107	b	108	a	109	e	110	a
111	a	112	b	113	a	114	c	115	a	116	b	117	b	118	c	119	a	120	c
121	e	122	c	123	b	124	c	125	a	126	c	127	c	128	a	129	c	130	a
131	b	132	b	133	a	134	a	135	b	136	c	137	b	138	a	139	b	140	a
141	b	142	c	143	a	144	a	145	a	146	c	147	b	148	d	149	a	150	e
151	b	152	b	153	a	154	c	155	c	156	b	157	c	158	b	159	b	160	a
161	e	162	c	163	e	164	d	165	e	166	c	167	b	168	c	169	e	170	d
171	e	172	b	173	d	174	d	175	a	176	b	177	a	178	d	179	d	180	d
181	d	182	c	183	d	184	b	185	c	186	e	187	c	188	d	189	b	190	b
191	c	192	b	193	d	194	b	195	b	196	b	197	b	198	b	199	b	200	d
201	c	202	b	203	c	204	a	205	b										

1.1.2.5. Leafy Vegetables

(1) ———————————— is also known as chauli.
    (a) Palak   (b) Spinach
    (c) Amranths   (d) Basella
    (e) None of the above

(2) ———————— is botanically known as *Amaranthus tricolor* L.
    (a) Amranths   (b) Poi
    (c) Sinach beet   (d) Spinach
    (e) None of the above

(3) *A. hypochondriacus* species of Amranth is of having ———————————— .
    (a) Long day   (b) Short day
    (c) Day netural   (d) Both a and b
    (e) None of the above

(4) Amaranths is rich in ———————————————————— .
    (a) Vitamin K   (b) Vitamin D
    (c) Vitamin A   (d) Vitamin B
    (e) None of the above

(5) Amaranths has ———————————— photosynthetic cycle, which indicates its high productivity.
    (a) C_3 (b) C_4
    (c) Short day   (d) Long day
    (e) Day netural

(6) Amaranths originated from ———————————— .
    (a) Brazil   (b) India
    (c) Europe   (d) Asia
    (e) None of the above

(7) Basic chromosome number in Amaranths is ———————————— .
    (a) 20   (b) 34
    (c) 17   (d) 16
    (e) 10

(8) Amaranth belongs to family ———————————— .
    (a) Amarlydiaceae   (b) Amaranthaceae
    (c) Liliaceae   (d) Araceae
    (e) None of the above

(9) Amaranths are a ———————————— crop.
    (a) Self-pollinated   (b) Cross - pollinated
    (c) Often cross pollinated   (d) Hermaphordite
    (e) None of the above

(10) —————————— is a most serious disease of Amaranths.
   (a) Black spot   (b) White rust
   (c) Sun scarching   (d) Leaf curl
   (e) None of the above

(11) To maintain isolation distance of —————— meters between two cultivars of Amaranths.
   (a) 100   (b) 200
   (c) 400   (d) 800
   (e) 1000

(12) Green yield of Amaranths is about —————— q/ha.
   (a) 40-60   (b) 30-40
   (c) 60-80   (d) 300-400
   (e) None of the above

(13) Per hectare seed rate of bari chauli is —————— kg.
   (a) 10-12   (b) 15-20
   (c) 20-40   (d) 3-4
   (e) None of the above

(14) ——————, a grain type Amaranth is widely grown in Gujarat and Maharashtra.
   (a) Jobner Green   (b) Rajgarh
   (c) Pusa Harit   (d) Pant Haritima
   (e) None of the above

(15) Seeds of Amaranth are dried upto —————— per cent moisture and stored in moisture proof polythene bags.
   (a) 3-4   (b) 5-6
   (c) 1-3   (d) 8-10
   (e) None of the above

(16) Amaranths are excellent as a vegetable for the following reasons ——————
   1. Amaranths are a very fast growing crop with an extremely high yield potential. In warm humid tropic climates the yield of leaves alone amounts to 30 tonnes of fresh green or 4-5 tonnes of dry matter per ha in 4 weeks from the direct sowing.
   2. Soil diseases do not affect it and pests like nematodes, Fusarium sclerotium and bacterial wilt.
   3. Amaranths is easy to grow and very suitable to be fitted in crop rotation in both homestead nutrition gardens and commercial cultivation.
   4. Amaranths grow very well under organic farming conditions. It reacts favourable to fertilizers.
   5. Amaranths are one of the cheapest dark green leafy vegetables because of low production cost and high yield. It is often described as a poor man's vegetable.

(a) 1 and 2 (b) 2 and 3
(c) 3 and 4 (d) 1, 2, 3, 4, and 5
(e) None of the above

(17) ———— is popularly known as Poi, Malabar night shade, or Indian spinach.
(a) Methi (b) Basella
(c) Spinach (d) Spinach beet
(e) None of the above

(18) Basella is commonly grown in————.
(a) North India (b) South India
(c) Westrn India (d) Central India
(e) None of the above

(19) The ———— leaves can be used for preparation of nice pakora.
(a) Basella (b) Spinach
(c) Asparagus (d) Chauli
(e) None of the above

(20) Botanical name of ————is *Basella alba* Linn.
(a) Palak (b) Bari chauli
(c) Choti chauli (d) Basella
(e) None of the above

(21) The origin place of basella is————.
(a) Europe (b) Asia
(c) China (d) India
(e) None of the above

(22) Basic chromosome number in basella is————.
(a) 10 (b) 20
(c) 12 (d) 14
(e) 34

(23) Basella is a ———— plant in nature.
(a) Erect (b) Spreading
(c) Sem-erect (d) Climbing
(e) None of the above

(24) Green leaves cultivars of basella are preferred to grow in ————.
(a) Mahrashta (b) Gujarat
(c) Uttar Pradesh and Punjab (d) Kerala
(e) None of the above

(25) Basella sowing is usually done in March-May in the northern and eastern plains of India whereas sowing time in south India is ———————.
    (a) June - July   (b) August- September
    (c) October- November   (d) December - January
    (e) None of the above

(26) In order to raise one-hectare crop of basella, about ———— kg. seed per hectare will require.
    (a) 4-5   (b) 5-10
    (c) 12-15   (d) 45-60
    (e) None of the above

(27) The basella leaves become ready for harvesting ———————— days after sowing the seeds.
    (a) 30-40   (b) 20-30
    (c) 40-50   (d) 60-75
    (e) None of the above

(28) Basella belongs to family ——————.
    (a) Alliaceae   (b) Amranthaceae
    (c) Basellaceae   (d) Chenopodiaceae
    (e) None of the above

(29) The total yield of basella is ———————— quintal per hectare depending upon the management practices.
    (a) 50-60   (b) 60-70
    (c) 100-150   (d) 150-200
    (e) None of the above

(30) ———————— is important salad crop.
    (a) Bathawa   (b) Amranth
    (c) Celery   (d) Basella
    (e) None of the above

(31) Fleshy ———————— parts of the plant of celery are used as salad.
    (a) Stem   (b) Leaf
    (c) Leaf petiole   (d) Roots
    (e) Infloresence

(32) The temperature ———————— is suitable for celery cultivation.
    (a) 15-21 °C   (b) 5-10 °C
    (c) 1-2 °C   (d) 25-31 °C
    (e) 35-41 °C

(33) High temperature causes ———————————— in leaves.
    (a) Stunted growth   (b) Sun scarching
    (c) Bolting and bitterness   (d) Poor germination
    (e) None of the above

(34) ———————————— seeds are used condiments and medicinal preparation.
    (a) Palak   (b) Poi
    (c) Celery   (d) Amranth
    (e) None of the above

(35) ———————————— is a leafy vegetable.
    (a) Palak   (b) Methi
    (c) Poi   (d) Celery
    (e) None of the above

(36) Wrapping of leaf petioles with dark brown paper around them in celery is known as ————————.
    (a) Bleaching   (b) Packaging
    (c) Blanching   (d) Processing
    (e) None of the above

(37) Per hectare seed yield of celery is ———————————— q.
    (a) 5-6   (b) 6-8
    (c) 8-10   (d) 12-15
    (e) None of the above

(38) Celery is sensitive to ————————————————.
    (a) Frost   (b) Hot
    (c) Cold   (d) Water logging and salinity
    (e) None of the above

(39) From quality point of view, Indian celery is best followed by ————————
    (a) Japanese   (b) Chinese
    (c) English   (d) European
    (e) None of the above

(40) Celery contains ———————————— per cent volatile oil.
    (a) 1-2   (b) 0.5
    (c) 1   (d) 2-3
    (e) None of the above

(41) *Dutch cultivars 'Reskia Wonder, Van Voorfing and Mary King are varieties* ————.
    (a) Basella   (b) Lettuce
    (c) Palak   (d) Methi
    (e) None of the above

(42) 'Great Lakes' is a variety of ——————.
    (a) Lettuce   (b) Spinach
    (c) Methi   (d) Palak
    (e) Amranth

(43) Lettuce belongs to family ——————
    (a) Amranthaceae   (b) Compositae
    (c) Basellaceae   (d) Araceae
    (e) None of the above

(44) Lettuce seeds germinate in —————— days.
    (a) 1-2   (b) 2-3
    (c) 10-15   (d) 4-5
    (e) None of the above

(45) Basic chromosome number in lettuce is——————.
    (a) 10   (b) 11
    (c) 9   (d) 20
    (e) 34

(46) One gram of seed of lettuce contains about —————— seeds.
    (a) 100-200   (b) 200-300
    (c) 900-1000   (d) 400-500
    (e) None of the above

(47) Lettuce seeds are sown in——————.
    (a) June- July   (b) October - November
    (c) September- October   (d) March- April
    (e) None of the above

(48) Average yield of lettuce is —————— q/ha.
    (a) 40-50   (b) 50-60
    (c) 100-140   (d) 400-500
    (e) None of the above

(49) —————— plant growth regulator used in delaying senescence of lettuce.
    (a) GA   (b) BA
    (c) IAA   (d) IBA
    (e) None of the above

(50) Big Vein of lettuce is due to ——————, which transmitted by *Olpidium brassicae* fungus.
    (a) Fungi   (b) Bacteria
    (c) Virus   (d) Algae
    (e) None of the above

(51) ———————— is a physiological disorder of lettuce which is due to light, maturity and duration, an excess of nitrogen, deficiency of Ca, Mn and B, ontogenic age of plant and soil moisture content and high endogenous level of IAA.
    (a) Blindness   (b) Tip burn
    (c) Whiptail   (d) Browning
    (e) None of the above

(52) Lettuce is ———————— crop.
    (a) Cross-pollinated   (b) Self-pollinated
    (c) Often cross-pollinated   (d) All a, b and c
    (e) None of the above

(53) To maintain genetic purity, there should be isolation distance of ——— meters for foundation seed and ——— meters for certified seed between two cultivars.
    (a) 10 and 20   (b) 50 and 25
    (c) 100 and 200   (d) 200 and 100
    (e) None of the above

(54) About ———————— kg seed per hectare is obtained from leaf type and 100-150 kg per hectare from the head type depending upon climatic conditions, soil types, health of the crop etc.
    (a) 100-200   (b) 200-300
    (c) 300-400   (d) 500-600
    (e) None of the above

(55) High temperature ———————— causes bolting or acceleration of seed stalks in lettuce.
    (a) Above 12 ^{0}C   (b) Above 32 ^{0}C
    (c) Above 22 ^{0}C   (d) Above 42 ^{0}C
    (e) None of the above

(56) About ———————— g of seeds is required for raising one-hectare crop.
    (a) 100-200   (b) 200-300
    (c) 400-500   (d) 1-2 kg
    (e) None of the above

(57) Lettuce is originated from————————.
    (a) South America   (b) Mediterranean region
    (c) Brazil   (d) China
    (e) None of the above

(58) The beet leaf is originated from ―――――――.
   (a) India
   (b) Mediterranean region
   (c) South America
   (d) Asia
   (e) Indo- Chinese Region

(59) Parsley (*Petroselinum hortense*, family Umbelliferae), Endive (*Cichorum endivia*, family compositae), Chicory (*Cichorium intybus*, family compositae), Chervil (*Anthriscum cerefolium*, family umbelliferae) Cress (*Lepidium sativum*, family cruciferae) and Water Cress (*Nasturtium officinale*, family cruciferae) are other ――――――――― vegetables.
   (a) Spices
   (b) Salad
   (c) Root
   (d) Tuber
   (e) None of the above

(60) Beet leaf belongs to family ――――――――――――――.
   (a) Amranthaceae
   (b) Chenopodiaceae
   (c) Araceae
   (d) Umbillferae
   (e) None of the above

(61) ――――――――――― is also known as palak.
   (a) Amaranth
   (b) Basella
   (c) Indian Spinach
   (d) Methi
   (e) None of the above

(62) Pusa Harit is a variety of―――――――――――――.
   (a) Amranth
   (b) Basella
   (c) Indian Spinach
   (d) Methi
   (e) None of the above

(63) Jobner Green is a variety of―――――――――――――――.
   (a) Basella
   (b) Tinda
   (c) Methi
   (d) Beet leaf
   (e) None of the above

(64) Poor seed germination of palak is (i) Viability of seeds is low. (ii) Adequate soil moisture is lacking. (iii) Sowing of seeds is done deep i.e. beyond 4 cm. (iv) Formation of soil crust before emergence of seedlings. (v) Attack of pest and/or soil borne/seed borne diseases.
   (a) Only (i)
   (b) Only (i) and (ii)
   (c) Only (iii)and (v)
   (d) Only (v) and (iii)
   (e) All of the above

(65) For genetic purity, maintain isolation distance ———— meters for foundation and ———— meters for certified seed.
   (a) 400 and 800
   (b) 400 and 200
   (c) 1600 and 1000
   (d) 50 and 25
   (e) None of the above

(66) Pusa Jyoti is a giant leaved cultivar of ————.
   (a) Kasuri methi
   (b) Pea
   (c) Bathawa
   (d) Palak
   (e) None of the above

(67) Under favourable growing conditions, about ———— quintals of marketable leaves are obtained from one hectare.
   (a) 40-50
   (b) 50-60
   (c) 30-40
   (d) 80-100
   (e) None of the above

(68) Palak produces ———— flowers.
   (a) Unisexual
   (b) Heterosyle
   (c) Hermaphordite
   (d) Gynomonoecious
   (e) None of the above

(69) Single fruits of palak contain ———— seeds.
   (a) 1
   (b) Rarely 2
   (c) Generally 2-3
   (d) 4-5
   (e) None of the above

(70) The average seed yield of palak is about ———— kg per hectare.
   (a) 400-500
   (b) 500-600
   (c) 700-800
   (d) 1000-1500
   (e) None of the above

(71) ———— are *Spinacia tetrandra* and *S. turkestanica*.
   (a) Bsella
   (b) Spinach beet
   (c) Vilayati Palak
   (d) Amranth
   (e) None of the above

(72) Pyrazone is a ————
   (a) PGR
   (b) Weedicide
   (c) Fungicide
   (d) Insecticide
   (e) None of the above

(73) Leaf yield of palak is governed by factors ———————— (i) Fertility of soil. (ii) Application of nitrogen during cropping. (iii) Availability of soil moisture and (iv) Management practices.

    (a) Only (i)   (b) (ii) and (iii)

    (c) (iii) and (iv)   (d) All of the above

(74) To raise one hectare crop of palak about ———————— kg seeds required.

    (a) 10-20   (b) 15-20

    (c) 25-30   (d) 60-70

    (e) None of the above

(75) Basic chromosome number in beet leaf is ————————.

    (a) 10   (b) 14

    (c) 9   (d) 12

    (e) None of the above

(76) For seed purposes grown palak is grown as ————————.

    (a) Annual   (b) Biennial

    (c) Perennial   (d) None of the above

Answer Sheet

1	c	2	a	3	a	4	c	5	b	6	b	7	c	8	b	9	b	10	b
11	c	12	c	13	d	14	b	15	d	16	d	17	b	18	a	19	a	20	d
21	d	22	c	23	d	24	c	25	c	26	c	27	d	28	c	29	d	30	c
31	c	32	a	33	c	34	c	35	a	36	c	37	d	38	d	39	b	40	b
41	b	42	a	43	b	44	c	45	c	46	c	47	c	48	-	49	b	50	c
51	b	52	b	53	b	54	d	55	c	56	-	57	b	58	e	59	b	60	b
61	c	62	c	63	d	64	e	65	c	66	d	67	d	68	c	69	c	70	d
71	b	72	b	73	d	74	c	75	c	76	b								

1.1.2.6. Legume Vegetables

(1) Broad bean is pollinated by ————————.

    (a) Insects   (b) Winds

    (c) No requirement   (d) Water

    (e) None of the above

(2) ———————— is also known as faba bean, horse bean, bakla bean.

    (a) Cluster bean   (b) Broad bean

    (c) Indian bean   (d) Pea

    (e) None of the above

(3) Sometimes, an illness due to ——————which is known as favism, caused due to allergy of pollen and green pod in some people,
    (a) Cowpea   (b) Rice bean
    (c) Broad bean   (d) Clustered bean
    (e) None of the above

(4) ——————is botanically known as *Vicia faba* L.
    (a) Broad bean   (b) Guar
    (c) Pea   (d) Winged bean
    (e) None of the above

(5) Origin of broad bean is——————.
    (a) India   (b) China
    (c) Europe and Asia   (d) Brazil
    (e) None of the above

(6) ——————are not good for broad bean cultivation.
    (a) Alkali   (b) Saline
    (c) Alkali - saline   (d) Acid soils
    (e) None of the above

(7) Broad bean seeds germinate in ——————days.
    (a) 3-4   (b) 5-8
    (c) 10-15   (d) 20-25
    (e) None of the above

(8) Somatic chromosome numbers in broad bean are ——————.
    (a) 16,18   (b) 20,22
    (c) 12,14   (d) 32,34
    (e) None of the above

(9) About —————— kg of broad bean seeds is needed to sow one hectare.
    (a) 30-40   (b) 50-60
    (c) 70-100   (d) 120-150
    (e) None of the above

(10) About —————— quintals of green pods of broad bean are harvested in one hectare.
    (a) 50-56   (b) 300-400
    (c) 200-300   (d) 70-100
    (e) None of the above

(11) *Vicia galilea* and *V. hyaeniscumus* are relatives of ——————.
    (a) Cluster bean   (b) French bean
    (c) Broad bean   (d) Rice bean
    (e) Jack bean

(12) Aquadule Claudia, Imperial White Long Pod, Masterpiece Green Long Pod, Imperial Green Long Pod, Red Epicure are varieties of ———— having long pod type.
- (a) Indian bean
- (b) Broad bean
- (c) Jack bean
- (d) Tree bean
- (e) None of the above

(13) Imperial White Windsor, Giant Four Seeded Green Windsor and Imperial Green Windsor are Windsor type varieties of ————————————.
- (a) Broad bean
- (b) Tree bean
- (c) Jack bean
- (d) French bean
- (e) None of the above

(14) Cluster bean contains a mucilaginous substance known as————————.
- (a) Galactomanon
- (b) Manon
- (c) Lactomanon
- (d) Triterpines
- (e) All of the above

(15) ———————— is also known as guar.
- (a) Rice bean
- (b) Cluster bean
- (c) Tree bean
- (d) French bean
- (e) None of the above

(16) The plants of cluster bean should be allowed to mature for the animals, as it would change into woody and fibrous substances resulting loss of nutrient and palatability. It may also cause tympanities.
- (a) Human
- (b) Cattle
- (c) Pig
- (d) Both a and c
- (e) None of the above

(17) The guar meal (dry seeds) contains about ———————— per cent protein.
- (a) 20
- (b) 10
- (c) 15
- (d) 33.3
- (e) None of the above

(18) The cluster bean seeds may be sown at a spacing of ———————— cm.
- (a) 45X45
- (b) 30X20
- (c) 60x90
- (d) 45X15
- (e) None of the above

(19) The weight of 100 seeds of cluster bean is ———————— g.
- (a) 20
- (b) 49
- (c) 45
- (d) 6
- (e) 100

(20) ———————— is the most serious disease of cluster bean.
 (a) Fusarium wilt (b) Bacterial wilt
 (c) Leaf spot (d) Rust
 (e) None of the above

(21) To sow one-hectare area, about ———————— kg seeds of cluster bean are required.
 (a) 10-20 (b) 20-30
 (c) 30-40 (d) 100-120
 (e) None of the above

(22) Cluster bean plant required ———————— conditions for induction of flowering.
 (a) Long day (b) Short day
 (c) Day netural (d) Both a and b
 (e) None of the above

(23) Isolation distance of ——— meters for foundation and ——— meters for certified seed, respectively is essential to check contamination.
 (a) 100,200 (b) 50,25
 (c) 400,200 (d) 800,1600
 (e) None of the above

(24) Cluster bean is originated from————————————————.
 (a) China (b) Europe
 (c) Asia (d) West Afica and India
 (e) South Aferica

(25) ———————— is botanically known as *Cyamopsis tetregonolobus* L.
 (a) Sweet pea (b) French bean
 (c) Cluster bean (d) Pea
 (e) None of the above

(26) In ————————————, cluster bean is grown throughout the year.
 (a) North India (b) Western India
 (c) South India (d) Central India
 (e) None of the above

(27) Pusa Mausami, Pusa Sadabahar, Pusa Navbahar and Sharad Bahar are varieties of ————————.
 (a) French bean (b) Cluster bean
 (c) Cowpea (d) Winged bean
 (e) None of the above

(28) *Cyamopsis senegalensis* is relative of——————
  (a) French bean   (b) Rice bean
  (c) Winged bean   (d) Cluster bean
  (e) Jack bean

(29) The protein content in cowpea seeds varies ——————per cent.
  (a) 10-15   (b) 15-20
  (c) 30-40   (d) 23-28
  (e) None of the above

(30) —————— is botanically known as *Vigna sinensis* var. *sesquipedalis*.
  (a) Sem   (b) Cluster bean
  (c) Asparagus or yard long bean (d) Jack bean
  (e) None of the above

(31) —————— is botanically known as *V. sinensis* var. *cylindrical*
  (a) Sweet pea   (b) Pea
  (c) Southern pea   (d) Field pea
  (e) None of the above

(32) —————— is botanically known as *Vigna unguiculata*.
  (a) Catjung cowpea   (b) Tapery bean
  (c) Runner bean   (d) Medagaskar bean
  (e) None of the above

(33) The origin place of cowpea is ——————.
  (a) North America   (b) Central Africa
  (c) North Africa   (d) South Africa
  (e) None of the above

(34) Pusa Rituraj variety of cowpea is a highly photo-thermo-insensitive in nature, can be grown both in summer and rainy seasons.
  (a) Summer season only   (b) Winter season only
  (c) Rainy season only   (d) Summer and rany seasons
  (e) None of the above

(35) Kashi Gauri and Kashi Shyamal are varieties ——————.
  (a) French bean   (b) Indian bean
  (c) Cowpea   (d) Jack bean
  (e) None of the above

(36) The spraying of MH at —————— ppm just before flowering stage increases pod set by about 30-35 per cent.
  (a) 100   (b) 200
  (c) 25   (d) 15
  (e) 50

(37) Basic chromosome number in cowpea is —————————.
    (a) 14   (b) 18
    (c) 16   (d) 20
    (e) 11

(38) Sword bean (*Canavalia gladiala*) and Jack bean (*Canavalia ensiformis*) are the members of family of —————————
    (a) Leguminoaceae   (b) Araceae
    (c) Convoluaceae   (d) Amranthaceae
    (e) None of the above

(39) ————— variety of cowpea is a cross of Pusa Falguni and Philippines Selection.
    (a) Pusa Dophasali   (b) Kashi Gauri
    (c) Kashi Shyamal   (d) Pant Anupma
    (e) None of the above

(40) Philippines Early is a variety of —————————.
    (a) Cowpea   (b) French bean
    (c) Pea   (d) Sem
    (e) None of the above

(41) Cowpea is ready for harvesting after ————— days of sowing.
    (a) 70-80   (b) 40-50
    (c) 60-70   (d) 120-130
    (e) None of the above

(42) ————— of cowpea is scientifically known as *Madhurasia obscurella*.
    (a) Epilenchena beetle   (b) Goss hopper
    (c) Galerucid beetle   (d) Aphid
    (e) None of the above

(43) ————— is chemically mutant male sterile line of cowpea.
    (a) IIHR 61B'   (b) Steppe 287
    (c) Pant Anupma   (d) Kashi Gauri
    (e) None of the above

(44) *Vigna luteola, V. nilotica* and *V. marina* are relatives of —————————
    (a) Cowpea   (b) Broad bean
    (c) Jack bean   (d) Moong bean
    (e) None of the above

(45) 'IIHR 61B' variety of cowpea can be grown —————————.
    (a) Only summer   (b) Only rainy
    (c) Both summer and rainy   (d) Through out the year
    (e) None of the above

(46) Variety 'Goit', 'Dixies-Cream' and 'Alabunch' are resistant to―――――――.
    (a) Leaf spot   (b) Powdery mildew
    (c) Downey mildew   (d) Cowpea mosaic
    (e) None of the above

(47) Varieties 'Grant' resistant to fusarium wilt, 'Chinese Red' resistant to *Phytophthora* stem rot, 'Mississipi57-1' and 'Iron' resistant to ―――――――――――.
    (a) Cowpea mosaic   (b) Downey mildew
    (c) Root-knot nematode   (d) Leaf spot
    (e) None of the above

(48) The presence of anti- nutritional factors such as hydrate, oligosaccharrides and protease inhibitors has also been reported in dry seeds of ―――――――.
    (a) French bean   (b) Cowpea
    (c) Broad bean   (d) Jack bean
    (e) None of the above

(49) ――――――――― is also known as kidney bean and snap bean.
    (a) Snow pea   (b) Pea
    (c) French bean   (d) Broad bean
    (e) None of the above

(50) Arka Komal and Pant Anupama are varieties of―――――――――――.
    (a) Cowpea   (b) French bean
    (c) Cluster bean   (d) Pea
    (e) None of the above

(51) South Mexico and Central America are considered as primary centre of origin of ―――――――.
    (a) Cowpea   (b) Broad bean
    (c) Rice bean   (d) Runner bean
    (e) French bean

(52) ――――――――― is botanically known as *Phaseolus acutifolius* var. *latifolius*.
    (a) Tepary type of bean   (b) Medagaskar bean
    (c) Runner bean   (d) Jack bean
    (e) None of the above

(53) Tender Crop and Cascade are cultivars of――――― suited for processing purpose.
    (a) Indian bean   (b) French bean
    (c) Cowpea   (d) Tree bean
    (e) None of the above

(54) For best growth and yield of French bean, the optimum temperature should be? _____.

    (a) 25-30 °C   (b) 5-10 °C
    (c) 15-20 °C   (d) 35-40 °C
    (e) None of the above

(55) French bean is sown during _____ in south Indian plains.

    (a) July   (b) August
    (c) September   (d) November
    (e) None of the above

(56) About _____ kg/ha seed is required for cultivation of dwarf bean.

    (a) 20-30   (b) 10-20
    (c) 40-50   (d) 50-75
    (e) None of the above

(57) _____ application in French bean enhances nodulation in roots.

    (a) Calcium   (b) Magnesiun
    (c) Phosphorous   (d) Potassium
    (e) None of the above

(58) The French bean is ready for harvest in _____ days.

    (a) 20   (b) 30
    (c) 40   (d) 75
    (e) None of the above

(59) _____ is scientifically known as *Aphis craccivora*.

    (a) Bean aphid   (b) Bean lady beetle
    (c) White fly   (d) Grosshopper
    (e) None of the above

(60) _____ is scientifically known as *Epilachna varivestis*

    (a) Bean aphid   (b) Bean lady beetle
    (c) Grosshopper   (d) White fly
    (e) None of the above

(61) Basic chromosome number in French bean is _____.

    (a) 24   (b) 22
    (c) 12   (d) 14
    (e) 34

(62) _____ is botanically known as *Pachyrrhizus tuberosus*.

    (a) Potato bean   (b) Bean aphid
    (c) Bean lady beetle   (d) White fly
    (e) None of the above

(63) Tender Green and Phenomenal Long Podded are varieties of———————.
- (a) Cowpea
- (b) French bean
- (c) Cluster bean
- (d) Pea
- (e) None of the above

(64) To maintain isolation distance of ————— meters for foundation and ————— meters for certified crop between two cultivars, respectively.
- (a) 10 and 5
- (b) 50 and 25
- (c) 100 and 200
- (d) 400 and 200
- (e) None of the above

(65) About ————— quintals of seed yield per hectare are obtained by French bean cultivation
- (a) 40-50
- (b) 12-18
- (c) 100-120
- (d) 60-70
- (e) None of the above

(66) Ashy stem blight (*Macrophomina phaseolina* Goid), is a serious ————— disease of French bean.
- (a) Bacterial
- (b) Fungal
- (c) Virus
- (d) Mycoplasma
- (e) None of the above

(67) French bean is said to be a —————————————resistant.
- (a) Salanity
- (b) Drought
- (c) Frost
- (d) Cold
- (e) None of the above

(68) Seed corn maggot of French bean attack of—————————————.
- (a) Germinating seeds
- (b) Immature seed
- (c) Mature seed
- (d) Tender pods
- (e) None of the above

(69) The yield of green pods in the bush varieties may vary on an average of 40-50 quintals and the pole varieties of—————quintals per hectare, respectively.
- (a) 30-40
- (b) 70-100
- (c) 40-50
- (d) 60-70
- (e) None of the above

(70) Factors responsible for low production of peas are
- (i) Low yield potential of the varieties.
- (ii) Susceptibility of diseases.
- (iii) Losses due to insect-pests.
- (iv) Late maturity.

(v) Poor responses to inputs like fertilizers and irrigation.
(iv) Cultivation with poor production technology.
(vii) Raising of the crop under marginal and poor soils.
(viii) Physiologically inefficient plant type.
(ix) Instability in the performance of the varieties.
(x) Farmers are reluctant to use costly inputs because of the risk.
    (a) Only (i) and (ii)   (b) Only (iii) and (iv)
    (c) Only (ix) and (x)   (d) All of the above
    (e) None of the above

(71) ——————————wrote the book entitled "Origin of Cultivated Plants".
    (a) DeCandole   (b) Jafferson
    (c) Darwin   (d) Linneaus
    (e) None of the above

(72) ——————————is considered as secondary centre of origin of vegetable pea.
    (a) Asia   (b) India
    (c) Europe   (d) Ethiopia
    (e) None of the above

(73) —————————— is botanically known as *Pisum sativum* var. *arvense*
    (a) Vegetable pea   (b) Field pea
    (c) Arhar   (d) Moong
    (e) None of the above

(74) Alaska cultivar of pea is comparatively tolerant to——————————.
    (a) Wilt   (b) Water logging
    (c) Heat   (d) Salnity
    (e) None of the above

(75) Viability of vegetable pea is ——————————years.
    (a) 3   (b) 5
    (c) 4   (d) 2
    (e) 1

(76) In Peninsular India, pea is sown in ——————————
    (a) August - September   (b) June - July
    (c) October - November   (d) March - April
    (e) None of the above

(77) Phosphorus increases ——————————— of pea.
    (a) Vegetative growth   (b) Only flowering
    (c) Yield and quality   (d) Disease resistance
    (e) None of the above

(78) The high quality of pea is associated with tenderness and ——————— content.
    (a) Fiber   (b) Sugar
    (c) Protein   (d) Fat
    (e) None of the above

(79) Pod maturity of pea is determined by mechanical means such as———————
    (a) Salometer   (b) Hydrometer
    (c) Tendrometer   (d) Lactopmeter
    (e) None of the above

(80) Delwiche Commando is a variety of ——————— resistant to common wilt and near wilt diseases.
    (a) Sword bean   (b) French bean
    (c) Pea   (d) Jack bean
    (e) None of the above

(81) Sylvia is an edible podded variety of————————————————————.
    (a) Snap pea   (b) Pea
    (c) Tree bean   (d) Jack bean
    (e) Non eof the above

(82) Lincoln and Wando are varieties of ——————— having branching in nature.
    (a) Cowpea   (b) French bean
    (c) Guar   (d) Pea
    (e) None of the above

(83) The optimum mean monthly temperature for good growth of pea is———————.
    (a) 10.0-18.3 °C   (b) 5.0-10.3 °C
    (c) 10.0-15.3 °C   (d) 20.0-28.3 °C
    (e) None of the above

(84) The main objectives in breeding pea crop are (i) Yield. (ii) Regional adaptability (iii) Suitable plant type. (iv) Lodging and shattering. (v) Diseases and insect resistance. (vi) Environmental stress resistance. (vii) Quality and effective nitrogen fixation.
    (a) Only (i) and (ii)   (b) Only (iii) and (iv)
    (c) Only Viii   (d) All of them
    (e) None of the above

(85) Feltham First and Meteor varieties of pea are resistance sources of—————.
    (a) Fusarium wilt   (b) Near wilt
    (c) Pea aphid   (d) Pod borer
    (e) None of the above

(86) Arkel, Meteor, Early Badger, Perfection New Line, Alderman, Little Marvel and Super varieties of pea are developed through ——————.
    (a) Pure line   (b) SSD
    (c) Mass selection   (d) Introduction
    (e) None of the above

(87) —————————— is botanically known as *Pisum formosum*.
    (a) Field pea   (b) Vegetable pea
    (c) Ornamental pea   (d) Snap pea
    (e) Non eof the above

(88) —————————— is botanically known as *Pisum fulvum*.
    (a) Ornamental pea   (b) Field pea
    (c) Snap pea   (d) Red yellow pea
    (e) None of the above

(89) —————————— is botanically known as *Pisum elatius*.
    (a) Red yellow pea   (b) Ornamental pea
    (c) Mediterranean pea   (d) Snap pea
    (e) Noneof the above

(90) Isolation distance of —————— m is recommended for production of foundation seed of pea.
    (a) 5   (b) 10
    (c) 20   (d) 100
    (e) 50

(91) In pea, shelling percentage ranges between —————————per cent.
    (a) 35-45   (b) 15-25
    (c) 5-15   (d) 45-65
    (e) 15-20

(92) Khapar Kheda variety of pea is very popular variety for——————.
    (a) Gujarat   (b) Maharashtra
    (c) MP   (d) Bihar
    (e) None of the above

(93) The most important garden pea growing state, which accounts for more than 60 per cent of the total production is ─────────────.
   (a) Meghalaya (b) Assam
   (c) Uttar Pradesh (d) Bihar
   (e) None of the above

(94) In India commercial exploitation of heterosis, crop needs to have certain built in morphological or physiological mechanisms to facilitate hybrid seed production. These are:
   1. Availability of a proven high yielding heterotic combination which could give more profit than the best commercial variety available.
   2. Availability of an efficient and economical system of eliminating or rendering function less male part of the bisexual seed parent.
   3. Economical and efficient transfer of pollen from male to seed parent.
   4. Synchronized flowering of both seed and the pollen parent.
   (a) Only 1 and 2 (b) Only 2 and 3
   (c) Only 3 and 4 (d) All of above
   (e) None of the above

(95) Bacterial blight and powdery mildew of pea are governed by ───────── recessive gene.
   (a) Single (b) Double
   (c) Triple (d) Both single and double
   (e) None of the above

(96) Pea varieties New Line Perfection, Market Prize and Duke of Albany are tolerant to ─────────.
   (a) Water lodging (b) Salinity
   (c) Acidity (d) Near wilt
   (e) None of the abve

(97) Pusa Early Prolific is a variety of ─────────.
   (a) French bean (b) Sem
   (c) Cowpea (d) Guar
   (e) None of the above

(98) About ───────── kg seed is required for raising one-hectare crop of Indian bean.
   (a) 20-30 (b) 50-70
   (c) 120-130 (d) 90-100
   (e) None of the above

(99) About ———— quintals or so seed yield per hectare is obtained for Indian bean cultivation.
   (a) 16-18   (b) 36-48
   (c) 6-8   (d) 76-88
   (e) None of the above

(100) The origin place of sem is ————.
   (a) India   (b) China
   (c) Brazil   (d) South America
   (e) None of the above

(101) For seed production of Dolichos bean, maintain isolation distance of 1 meters for foundation and ———— meters for certified seed between two cultivars.
   (a) 100   (b) 5
   (c) 30   (d) 20
   (e) None of the above

(102) Somatic chromosome number in Indian bean is————.
   (a) 32,34   (b) 22,24
   (c) 12,14   (d) 42,44
   (e) None of the above

(103) Maximum diversity of Indian bean is found in————.
   (a) Andhra Pradesh (b) Gujarat
   (c) Delhi   (d) Madhya Pradesh
   (e) None of the above

(104) Indian bean is sown in the month of————.
   (a) May - June   (b) July- August
   (c) March- April   (d) October- November
   (e) None of the above

(105) ————is also known as hyacinth bean, Indian bean, bonavist bean, lobia bean, Egyptian bean, filed bean, Australian pea, Indian butter bean, sem and salad bean.
   (a) Dolichos bean (b) French bean
   (c) Cluster bean   (d) Tree bean
   (e) None of the above

Answer Sheet

1	a	2	b	3	c	4	a	5	c	6	d	7	c	8	c	9	c	10	-
11	c	12	b	13	a	14	a	15	b	16	b	17	d	18	d	19	d	20	b
21	c	22	b	23	b	24	d	25	c	26	c	27	b	28	d	29	d	30	c
31	c	32	a	33	b	34	d	35	c	36	d	37	e	38	a	39	a	40	a
41	b	42	c	43	b	44	a	45	d	46	d	47	c	48	b	49	c	50	b
51	e	52	a	53	b	54	a	55	c	56	d	57	c	58	c	59	a	60	b
61	b	62	a	63	b	64	a	65	b	66	b	67	b	68	a	69	b	70	d
71	a	72	d	73	a	74	c	75	a	76	b	77	c	78	b	79	c	80	c
81	b	82	d	83	a	84	d	85	c	86	d	87	c	88	d	89	c	90	c
91	a	92	c	93	c	94	d	95	a	96	b	97	b	98	a	99	c	100	a
101	b	102	b	103	d	104	b	105	a										

1.1.2.7. Perennial Vegetables

(1) ——————— is a perennial vegetable crop
    (a) Tomato   (b) Asparagus
    (c) Bottle gourd   (d) Pumpkin
    (e) None of the above

(2) The soft tender shoot of asparagus is known as———————.
    (a) Stem   (b) Spear
    (c) Root   (d) Leaf
    (e) None of the above

(3) Diuretic medicinal properties of asparagus are due to presence of ———
    (a) Asparagines   (b) Tomatine
    (c) Lycopene   (d) Capsaicine
    (e) None of the above

(4) Perfection is the variety of ———————
    (a) Curry leaf   (b) Tree tomato
    (c) Asparagus   (d) Chow-chow
    (e) None of the above

(5) Asparagus is a ——————————in nature.
    (a) Monoecious   (b) Dioecious
    (c) Andromonoecious   (d) Trioecious
    (e) None of the above

(6) Mary Washington, March Washington, Perfection, New Jersey Improved, Book's Special are varieties of———————.
    (a) Asparagus   (b) Curry leaf
    (c) Kundru   (d) Chow - chow
    (e) None of the above

(7) The seed requirement for one-hectare cultivation for asparagus is—————.
   (a) 30-40 kg   (b) 3-4 kg
   (c) 50-60 kg   (d) 23-24 kg
   (e) None of the above

(8) Asparagus is a perennial in nature, economic yield is usually obtained up to ——— years, and afterwards it starts declining.
   (a) 2-3   (b) 5-6
   (c) 8-10   (d) 20-30
   (e) None of the above

(9) ————————— is propagated by seed and crown.
   (a) Curry leaf   (b) Asparagus
   (c) Bottle gourd   (d) Tomato
   (e) None of the above

(10) About ————— quintals of asparagus spear per hectare can be obtained.
   (a) 100-125   (b) 250-400
   (c) 400-800   (d) 50-60
   (e) None of the above

(11) ————————— is scientifically known as *Crioceris asparagi*.
   (a) Curry leaf   (b) Chow-chow
   (c) Asparagus beetle (d) Kudru
   (e) None of the above

(12) Asparagus starts giving marketable crop after a period of ————— years of planting.
   (a) 1   (b) 3
   (c) 10   (d) 15
   (e) None of the above

(13) Before transplanting, seedlings of asparagus raised from seed are allowed to grow in the nursery bed for a period of ————— months.
   (a) 3   (b) 12
   (c) 5   (d) 7
   (e) 9

(14) Wild species of asparagus found throughout—————.
   (a) Asia   (b) Europe and Mideast
   (c) China   (d) India
   (e) None of the above

(15) Edible part of the asparagus 'spear' is the emerging——————— and used as vegetable.
   (a) Root (b) Leaf
   (c) Shoot tip (d) Buds
   (e) None of the above

(16) ——————is known as multivitamin and multimineral packed leafy vegetables.
   (a) Amranth (b) Chekkurmanis
   (c) Palak (d) Curry leaf
   (e) None of the above

(17) The crop is highly cross-pollinated and entomophilous, the reasons for cross pollination being monoecy, alongwith ———————————nature of plant.
   (a) Protogynous (b) Monoecious
   (c) Dioecious (d) Andromonoecious
   (e) None of the above

(18) The other relatives of —————are *Sauropus retroversus* Wight, *S. assimilis* Thw; *S. rigidus* Thw. and *S. quadrangularis* Muell.
   (a) Chekkurmanis (b) Curry leaf
   (c) Chow- chow (d) Tree tomato
   (e) None of the above

(19) ———————————is also known as askas.
   (a) Chow- chow (b) Curry leaf
   (c) Tree tomato (d) Trre bean
   (e) None of the above

(20) ———————————is only single seeded member of cucurbitaceae family.
   (a) Parwal (b) Chow-chow
   (c) Bhar karela (d) Kartoli
   (e) None of the above

(21) Chow-chow is a ———————perennial in habit with annual vine growth.
   (a) Dioecious (b) Hermaphordite
   (c) Monoecious (d) Andromonoecious
   (e) Androgynoecious

(22) ———————————is botanical name of *Sechium edule* Swartz.
   (a) Tree tomato (b) Bhat karela
   (c) Kundru (d) Chow-chow
   (e) Parwal

(23) The ———————— of chow-chow are used as vegetables.
    (a) Only fruit   (b) Only leaves
    (c) Fruits. tender leaves and tuberous roots   (d) Seeds
    (e) None of the above

(24) Chow-chow provides ———————— calories per 100 g of fresh edible portion.
    (a) 56   (b) 26
    (c) 39   (d) 51
    (e) None of the above

(25) Chow- chow can be grown in tropical and subtropical to moderate temperature area up to ———————— m elevation.
    (a) 500   (b) 250
    (c) 1500   (d) 2000
    (e) 1000

(26) Chow-chow plant requires day length of slightly over ———————— hours before flowering.
    (a) 8   (b) 12
    (c) 5   (d) 6
    (e) 10

(27) The origin place of Chow-chow is ————————.
    (a) India   (b) Bangladesh
    (c) Southern America   (d) Italy
    (e) Japan

(28) Basic chromosome number in Chow-chow is ————————
    (a) 24   (b) 14
    (c) 32   (d) 22
    (e) None of the above

(29) Planting of entire fruits generally raises Chow-chow. It can also be propagated by————————.
    (a) Stem   (b) leaf
    (c) Tuberious root   (d) Fruit
    (e) None of the above

(30) Chow-chow vine bears fruits in abundance during————————.
    (a) August-September   (b) June- July
    (c) October - November   (d) May-June
    (e) March- April

(31) If Chow-chow crop is grown under good management, it yields about —— q/ha.
    (a) 120-150   (b) 50-60
    (c) 200-300   (d) 400-500
    (e) None of the above

(32) Chow-chow fruits can be stored for about —————————— without much loss under ordinary conditions.

 (a) One weeks   (b) 15 days

 (c) Three- four weeks   (d) Six months

 (e) None of the above

(33) Curry leaf is a backyard crop in many of the —————————— homesteads.

 (a) Central Indian   (b) Western Indian

 (c) South Indian   (d) North Indian

 (e) None of the above

(34) A volatile oil a crystalline glucoside "koenigin" from the leaves and a glucoside " murrayin" from the flowers are a few industrial product from the trees of———

 (a) Curry leaf   (b) Tree tomato

 (c) Tree bean   (d) Winged bean

 (e) None of the above

(35) Curry leaf is a self-pollinated crop. Variability in the crop is much limited. Two related species of *Murrya koenigii* and *M. paniculata* are indigenous to ——————and M. exotica, an ornamental shrub indigenous to India, Ceylon and China.

 (a) India   (b) Burma

 (c) Bangladesh   (d) Srilanka

 (e) None of the above

(36) *Murrya paniculata* (L), *Murraya exotica* L, *Marsana buxifolia* Sonnerat) and *Murraya koenigii* are relatives of ——————————————

 (a) Drumstick   (b) Curry leaf

 (c) Tree tomato   (d) Tree bean

 (e) None of the above

(37) —————— crop derives from the shape of pod resembling the slender and curved stick used for beating the drum.

 (a) Tree tomato   (b) Tree bean

 (c) Winged bean   (d) Curry leaf

 (e) Drumstick

(38) The crop is also known as horse radish tree or radish tree and west Indian bean. Probably the name 'radish tree' derives from the pendulous slender and thin shape of immature fruits of the tree resembling very much the siliqua of radish is known as——————————.

 (a) Drumstick   (b) Curry leaf

 (c) Winged bean   (d) Tree bean

 (e) Tree tomato

(39) Flowers are bisexual, oblique, stalked, axillary and heteromorphic. ———
    (a) Drumstick   (b) radish
    (c) sponge gourd   (d) Ivy gourd
    (e) None of the above

(40) 'Chavakacherri Murunga', Chemmurunga', 'Palmurungai', 'Punamurangai', 'KM-1' and 'PKM-1' are varieties of———————.
    (a) Broad bean   (b) Curry leaf
    (c) Winged bean   (d) Drumstick.
    (e) Basella

(41) Drumstick belongs to the family ———————————.
    (a) Euphorbiaceae   (b) Araceae
    (c) Moringaceae   (d) Leguminosae
    (e) Fabaceae

(42) ——————————————— is also known as little gourd or kundru or tondali.
    (a) Kakrol   (b) Ivy gourd
    (c) Chow-chow   (d) Pointed gourd
    (e) Snake gourd

(43) Sex form of *Cocconia grandis* is———————————.
    (a) Andromonocieous   (b) Dioceous
    (c) Monoceious   (d) Hermaphrodite
    (e) Gynomonocieous

(44) Little gourd is propagated by———————————.
    (a) Cutting   (b) Seed
    (c) Runners   (d) Suckers
    (e) Tuber

(45) While planting the crop, care is taken that 5-10 per cent male plant should be planted———————.
    (a) Kale   (b) Little gourd
    (c) Lettuce   (d) Horse radish
    (e) Spinach

(46) Monoceious species of little gourd is found in———————————.
    (a) India   (b) South America
    (c) Ghana   (d) Saudi Arabia
    (e) Mediterranen region

(47) The *Coccinia obyssinnica* species of little gourd formed———————————.
    (a) Edible leaves   (b) Edible tuberous roots
    (c) Non edible tuberous roots   (d) Edible shoots
    (e) Poisonous fruits

(48) Planting of kundru is preferably done during ―――――――――.
    (a) April-June
    (b) August-September
    (c) June-July and February-March
    (d) October-December
    (e) April-June and October-December

(49) ―――――――――― yield of ivy gourd can easily be obtained from one hectare area.
    (a) 275-325 quintals
    (b) 250-275 quintals
    (c) 200-250 quintals
    (d) 150-200 quintals
    (e) 100-125 quintals

(50) In which crop, the male plants are distinct having thin flowered base whereas female plants having swollen flower base ―――――――――.
    (a) Kakrol
    (b) Ivy gourd
    (c) Chow-chow
    (d) Pointed gourd
    (e) Snakegourd

(51) Ideal size of cuttings of ivy gourd for planting should be ―――――――――.
    (a) 20-25 cm long and 1.5-2.5 cm in diameter
    (b) 15-20 cm long and 2-2.5 cm in diameter
    (c) 30-40 cm long and 2-3 cm in diameter
    (d) 25-30 cm long and 1.5-2.0 cm in diameter
    (e) 35-40 cm long and 2.5-3.5 cm in diameter

(52) The ――――――――― are relatives of *Coccinia histella* and *C. sessilifolia*.
    (a) Chow-chow
    (b) Snake gourd
    (c) Kakrol
    (d) Pointed gourd
    (e) Ivy gourd

Answer Sheet

1	b	2	b	3	a	4	c	5	b	6	a	7	b	8	c	9	b	10	b
11	a	12	b	13	b	14	b	15	c	16	b	17	a	18	a	19	a	20	b
21	c	22	d	23	c	24	b	25	d	26	b	27	c	28	b	29	c	30	c
31	d	32	c	33	c	34	a	35	a	36	a	37	e	38	a	39	a	40	d
41	c	42	b	43	b	44	a	45	b	46	c	47	b	48	c	49	e	50	b
51	d	52	e																

1.1.2.8. Root Vegetables

(1) ——————————— is also known as chukandar.
    (a) Rhubarb   (b) Beet root
    (c) Brussels sprouts   (d) Chayote
    (e) Artichoke

(2) Beet root fruit contains——————————.
    (a) 5-6 seeds   (b) 3-4 seeds
    (c) 7-8 seeds   (d) 8-9 seeds
    (e) 2-3 seeds

(3) Poor colour development in beet root is due to ——————————.
    (a) High temperature   (b) Low temperature
    (c) High humidity   (d) Long dry spell
    (e) High dose of fertilizers

(4) Beet root fruits are botanically known as ——————————.
    (a) Fleshy thalmus   (b) Pome
    (c) Berry   (d) Siliqua
    (e) Capsule

(5) ——————————— elevation, beet root seed production is taken up in hills.
    (a) < 1200 m   (b) > 1200 m
    (c) > 1500 m   (d) > 1000 m
    (e) < 15000 m

(6) Beetroot belongs to family ——————————.
    (a) Convolvulaceae   (b) Euphorbiaceae
    (c) Dioscoraceae   (d) Araceae
    (e) Chenopodiaceae

(7) Basic chromosome number in beet root is ——————————.
    (a) 9   (b) 10
    (c) 11   (d) 12
    (e) 13

(8) The optimum temperature for beet root growth is——————————.
    (a) 20-23.5°C   (b) 16.3-18°C
    (c) 25-26.2°C   (d) 18.3-21°C
    (e) 20-24.5°C

(9) Beet root is sensitive to——————————.
    (a) High acidity and high temperature   (b) High acidity and low temperature
    (c) Low acidity and low temperature   (d) Low alkalinity and high temperature
    (e) Low alkalinity and low temperature

(10) Beetroot can be grown in saline and alkaline soils up to pH—————.
  (a) 7-8
  (b) 8-9
  (c) 9-10
  (d) 10-11
  (e) 11-12

(11) In plain, beetroot is sown in the month of —————.
  (a) Feburary-March
  (b) March-May
  (c) June-August
  (d) Sept-October
  (e) October-December

(12) Beet root plants have leaves with yellow blotches between the veins and leaves tend to curl up, usually in mid-summer. This situation occurs in beetroot is due to deficiency of —————.
  (a) Manganese
  (b) Copper
  (c) Boron
  (d) Molybdenum
  (e) Sulpher

(13) The central leaves of beetroot dieback and become blackened and the roots may turn black on the inside and the cankered on the outside. This phenomenon in beet root is known as heart rot and it is due to deficiency of —————.
  (a) Molybdenum
  (b) Copper
  (c) Boron
  (d) Manganese
  (e) Sulpher

(14) Beet is cross pollinated crop, therefore, to maintain genetic purity, the isolation distance between two varieties should be for foundation ————— and ————— for certified seed, respectively.
  (a) 1600 m, 1000 m
  (b) 1200 m, 1000 m
  (c) 600 m, 800 m
  (d) 500 m, 1000 m
  (e) 1200 m, 2000 m

(15) Crimson Globe, Detroit Dark Red and Ohio Canrer are varieties of —————.
  (a) Brussels sprouts
  (b) Celery
  (c) Lettuce
  (d) Welsh onion
  (e) Garden beet

(16) The colours red violet pigment, b-cyanins and yellow pigment b-xanthins are in crop —————.
  (a) Beetroot
  (b) Radish
  (c) Carrot
  (d) Cabbage
  (e) Yam

(17) In India, growing of sugar beet was tried during 1959-60 at —————.
  (a) IISR, Lucknow
  (b) IIHR, Bangalore
  (c) IARI, New Delhi
  (d) IISR, Calicut
  (e) CPCRI, Kasargod

(18) Kanji, an appetizing drink is prepared from——————————.
    (a) Turnip (b) Beet root
    (c) Carrot root (d) Radish root
    (e) Elephant foot yam

(19) Carrot is botanically known as *Daucus carota* and belongs to family ——————.
    (a) Convolvulaceae (b) Euphorbiaceae
    (c) Umbelliferae (d) Araceae
    (e) Chenopodiaceae

(20) Parsnip (*Pastinaca sativa,* 2n= 22), Chervil (*Anthiriscum cerebolium,* 2n= 32), Ceralic *(Apicum graveolens,* 2n= 22) and Skirret *(Sium sisarum,* 2n= 12) are the vegetables of——————————.
    (a) Rhubarb family (b) Carrot family
    (c) Turnip family (d) Beetroot family
    (e) Cinnamon family

(21) Red or yellow or purple colour in carrot is due to——————————.
    (a) Carrotal (b) Carotene
    (c) Catechol (d) Quercetin
    (e) Anthocyanin

(22) —————————— content in the carrot roots may vary from trace amount in pink cultivars to 1750 mg /kg in black carrots.
    (a) Catechol (b) Anthocyanin
    (c) Carrotal (d) Quercetin
    (e) Carotene

(23) Carrot is the richest source of vitamin ——————.
    (a) B6 (b) E
    (c) D (d) B 1
    (e) A

(24) The highest accumulation of carotene occurs in——————————.
    (a) New cell of the phloem (b) Older cell of the xylum
    (c) Older cell of the phloem (d) New cell of the phloem
    (e) Cells of the meristmatic tissue

(25) 'Chantaney' is the variety of——————————.
    (a) Turnip (b) Carrot
    (c) Radish (d) Beetroot
    (e) Parsely

(26) It is suggested that the taste of carrot is mainly due to presence of————.
    (a) Glutamic acid   (b) Carrotol
    (c) Acetaldehyde   (d) Anthocyanin
    (e) Butyric acid

(27) ————compound present in carrot has been reported to reduce the risk of cataract formation.
    (a) Butyric acid   (b) Beta –carotene
    (c) Anthocyanin   (d) Glutamic acid
    (e) Carrotal

(28) Pusa Kesar variety of carrot is the cross of————.
    (a) Royal Chantenay x Nantes Half Long
    (b) Jeno x Nantes
    (c) Early Nantes x Imperator
    (d) Chanteney x Danvers
    (e) Local Red x Nantes Half Long

(29) 'Zeno' is the variety of————.
    (a) Cumin   (b) Turnip
    (c) Radish   (d) Carrot
    (e) Beetroot

(30) For one-hectare area, the seed requirement of carrot is————.
    (a) 2-3 kilogram   (b) 5-6 kilogram
    (c) 1-2 kilogram   (d) 500-600 gram
    (e) 8-10 kilogram

(31) Wire worm and elworm are a serious pest of————.
    (a) Knoll khol   (b) Snake gourd
    (c) Fenugreek   (d) Black pepper
    (e) Carrot

(32) Forking of carrot is due to————.
    (a) Undecaying of manures   (b) High dose of nitrogenous fertilizer
    (c) Lack of moisture   (d) Delay in sowing
    (e) Poor management practices

(33) Tropical carrots are sown————
    (a) July-October   (b) September-December
    (c) October-January.   (d) January-March
    (e) March-July

(34) To maintain genetic purity, the isolation distance between two varieties of ——————— for foundation and for certified seed, respectively.
   (a) 1200 m, 1800 m   (b) 1000 m, 800 m
   (c) 500 m, 800 m   (d) 1200 m, 1000 m
   (e) 800 m, 1200 m

(35) The origin place of carrot is———————.
   (a) Mexico   (b) Europe and south western Asia
   (c) South Africa   (d) India
   (e) Mediterrenean region

(36) Carrot is grown at the distance of ——————— between two rows and between plant to plant, respectively.
   (a) 15-20 cm x 10-15 cm   (b) 20-25 cm x 15-20 cm
   (c) 25-30 cm x 20-25 cm   (d) 20-30 cm x 10-25 cm
   (e) 30-35 cm x 20-25 cm

(37) On the basis of mode of pollination, carrot is———————.
   (a) Cross-pollinated crop   (b) Often cross-pollinated crop
   (c) Self-pollinated crop   (d) Partially cross-pollinated crop
   (e) Oftern cross and often self pollinated crop

(38) Although, in carrot there is no incompatibility but main reason for cross-pollination is due to———————.
   (a) Stenospermocarpy   (b) Chasmogamy
   (c) Cleistogamy   (d) Protogyny
   (e) Protandry

(39) For seed production of carrot, minimum temperature ——————— requires for 40-60 days.
   (a) 5.8 °C   (b) 3.8 °C
   (c) 2.8 °C   (d) 6.8 °C
   (e) 4.8 °C

(40) Pusa Yamdagni variety of carrot is cross of———————.
   (a) Royal Chantenay x Nantes Half Long   (b) EC 9981 x Nantes
   (c) Early Nantes x Imperator   (d) Chanteney x Danvers
   (e) Local Red x Nantes Half Long

(41) 'Imperator' is the variety of ———————.
   (a) Cumin   (b) Turnip
   (c) Radish   (d) Carrot
   (e) Beetroot

(42) Basic chromosome number of carrot is ─────────.
    (a) 9   (b) 10
    (c) 11   (d) 12
    (e) 13

(43) For better carrot growth, optimum temperature is ─────────.
    (a) 12.7 to 16.4 °C   (b) 16.3 to 18.7 °C
    (c) 22.1 to 26.3 °C   (d) 20.5 to 24.9 °C
    (e) 18.3 to 23.9 °C

(44) Best temperature for root colour development of carrot is ─────────.
    (a) 15.6-21.1 °C   (b) 12.6-16.1 °C
    (c) 10.6-14.1 °C   (d) 18.6-22.1 °C
    (e) 22.6-24.1 °C

(45) Seeds required for sowing of one-hectare area of carrot is ─────────.
    (a) 6-8 kg   (b) 8-10 kg
    (c) 10-12 kg   (d) 500-600 gm
    (e) 5-6 kg

(46) Orange colour in carrot is due to ─────────.
    (a) Catechol   (b) Anthocyanin
    (c) Carrotal   (d) Quercetin
    (e) Carotene

(47) Nutritive quality of carrot roots has been found to be influenced by ─────────.
    (a) Nitrogen levels   (b) Irrigation interval
    (c) Sowing methods   (d) Nitrogen levels and sowing methods
    (e) Sowing time

(48) In general, the carotene content in carrot is more affected by stage of maturity and ─────────.
    (a) Sowing methods   (b) Time of sowing
    (c) Spacing   (d) Varieties
    (e) Fertilization

(49) Chantenay is an excellent cultivar of ───── for fresh market as well as processing.
    (a) Carrot   (b) Radish
    (c) Turnip   (d) Knoll khol
    (e) Parsely

(50) Carrot seeds contain a germination inhibitor called as ─────────.
    (a) Catechol   (b) Glycosinolates
    (c) Carrotal   (d) Quercetin
    (e) Carotene

(51) Plant growth regulator ———— extended storage life of carrot roots.
   (a) BA (b) GA3
   (c) IAA (d) NAA
   (e) CCC

(52) Cavity spot in carrot is due to increased accumulation of potassium and a decreased accumulation of ————.
   (a) S (b) Mg
   (c) B (d) P
   (e) Ca

(53) Splitting and cavity spot are physiological disorders of ————.
   (a) Turnip (b) Radish
   (c) Carrot (d) Knolkhol
   (e) Onion

(54) Carrot plant bears compound ————.
   (a) Seeds (b) Umbels
   (c) Berry (d) Flowers
   (e) Capsule

(55) The carrot flower is protandrous, hence it requires ————.
   (a) Cross-pollination (b) Self-pollnation
   (c) Male sterility (d) Self-incompatibility
   (e) Often cross pollination

(56) Pusa Meghali is a variety of ————.
   (a) Turmeric (b) Broccoli
   (c) Turnip (d) Carrot
   (e) Radish

(57) Lygus Bug is a serious pest for seed crops of ————.
   (a) Parsely (b) Carrot
   (c) Radish (d) Ginger
   (e) Brussels sprouts

(58) The fleshy root radish is modified form of root is known as ————.
   (a) Moniliform (b) Setaceous
   (c) Fusiform (d) Stolon
   (e) Tap root

(59) Pungency in ———— is due to isothiocynates (tran-4-methyl–thiobutenyl isothiocynate).
   (a) Radish (b) Onion
   (c) Bell pepper (d) Garlic
   (e) Black pepper

(60) Pink colour in radish is due to presence of a pigment is known as ———————.
   (a) Catechol
   (b) Anthocyanin
   (c) Carrotal
   (d) Quercetin
   (e) Carotene

(61) Rat-tail radish is botanically known as *Raphanus sativus* var. *caudatus,* pods are used as ———————.
   (a) Vegetable
   (b) Oil extration
   (c) Medicinal insustry
   (d) Fodder
   (e) Making rodenticides

(62) Arka Nishant is a variety of ———————.
   (a) Tomato
   (b) Cucumber
   (c) Bottle gourd
   (d) Radish
   (e) Cluster bean

(63) Radish root causes forking of roots ———————.
   (a) Delay in sowing
   (b) Poor management practices
   (c) Lack of moisture
   (d) Undecomposed organic matter
   (e) High dose of nitrogenous fertilizer

(64) Radish belongs to the family ———————.
   (a) Cruciferae
   (b) Chenopodiaceae
   (c) Alliaceae
   (d) Umbelliferae
   (e) Compositae

(65) Basic chromosome number in radish is ———————.
   (a) 7
   (b) 6
   (c) 8
   (d) 10
   (e) 9

(66) Suitable temperature for radish cultivation is ———————.
   (a) 10-15 °C
   (b) 15-19 °C
   (c) 18-21 °C
   (d) 20.1-23.6 °C
   (e) 22.3 -24.4 °C

(67) In order to sow one-hectare area, about healthy seed of tropical types ——————— and ——————— seed of temperate type is required, respectively.
   (a) 6-8 kg, 8-10 kg
   (b) 4-6 kg, 6-8 kg
   (c) 8-10 kg, 10-12 kg
   (d) 10-12 kg, 12-14 kg
   (e) 12-14 kg, 14-16 kg

(68) Tropical types of radish are ready in ——————— after sowing.
   (a) 20-25 days
   (b) 25-30 days
   (c) 30-35 days
   (d) 35-40 days
   (e) 15-20 days

(69) Delayed harvesting of radish root causes———————.
  (a) Internal browining (b) Bolting
  (c) Decaying of roots (d) Hollowness of roots
  (e) Pithiness of roots

(70) Radish seesd remain viable under favourable storage conditions ———————.
  (a) 7-8 years (b) 5-6 years
  (c) 6-7 years (d) 4-5 years
  (e) 3-4 years

(71) To maintain genetic purity, the isolation distance is for foundation ——————— and ——————— for certified seed between two cultivars of radish, respectively.
  (a) 1200 m, 800m (b) 1600 m, 1000m
  (c) 800 m, 1200m (d) 1000 m, 1600 m
  (e) 1200 m, 1600 m

(72) Seed yield of radish per hectare is———————.
  (a) 4-6 quintals (b) 10-12 quintals
  (c) 2-4 quintals (d) 8-10 quintals
  (e) 6-8 quintals

(73) White rust of radish is caused by———————.
  (a) *Colletotrichium* spp. (b) *Pernospora parasitica*
  (c) *Alternaria solani* (d) *Fusarium oxysporum*
  (e) *Albugo candida*

(74) Radish is originated from———————.
  (a) South Africa (b) India
  (e) Mexico (d) Europe and Asia
  (e) Mediterrenean region

(75) Cultivated radish is originated from———————.
  (a) *Raphanus strumpnum* (b) *Raphanus fetjee*
  (c) *Raphanus hertsunum* (d) *Raphanus sepfeate*
  (e) *Raphanus rophanistrum*

(76) Fodder type of radish is botanically known as ———————.
  (a) *Raphanus strumpnum* (b) *Raphanus sativus* var. *mongri*
  (c) *Raphanus hertsunum* (d) *Raphanus sativum var. trumpate*
  (e) *Raphanus rophanistrum*

(77) Punjab Safed is the variety of———————.
  (a) Onion (b) Turnip
  (c) White brinjal (d) Radish
  (e) Snake gourd

(78) Pusa Himani variety of radish is cross of——————————.
    (a) Pusa Desi x White Icicle
    (b) Black x Japanese White
    (c) Japanese White x Pusa Reshmi
    (d) Pusa Chetki x scarlet long
    (e) Rapid red white tipped x local white

(79) Radish can be stored for two to three days at room temperature of————— and————— for two months in the cold storage, respectively.
    (a) 4 °C and 80-85 % RH
    (b) 3 °C and 90 % RH
    (c) 7 °C and 90-95 % RH
    (d) 10 °C and 90-95 % RH
    (e) 0 °C and 90-95 % RH

(80) The average yield of Asiatic cultivars ranged between—————————————.
    (a) 350-400 q/ha
    (b) 300-350 q/ha
    (c) 150-200 q/ha
    (d) 250-300 q/ha
    (e) 150-250 q/ha

(81) ————————— is variety of radish reaches their an enormous size with a length upto 75-90 cm and a girth of 50-60 cm and may weigh upto 5-15 kg or even more.
    (a) Jaunpuri Giant
    (b) Pusa deshi
    (c) Arka Nishant
    (d) Pusa Chetki
    (e) RRWT

(82) Pusa Chetki variety of radish is a selection made from a material collected from —————.
    (a) Denmark
    (b) Mexico
    (c) South africa
    (d) Srilanka
    (e) India

(83) Rapid Red White Tipped and White Icicle are medium short European cultivars, which matures ————— after sowing of the seed.
    (a) 30-35 days
    (b) 25-30 days
    (c) 35-40 days
    (d) 15-20 days
    (e) 20-25 days

(84) The best sowing time of radish in south India is April-June in the hills and————— in the plains
    (a) February-April
    (b) June-August
    (c) October-December
    (d) December-February
    (e) August-October

(85) Radish seeds count————————— per gram.
    (a) About 175-200 seeds
    (b) About 100-125 seeds
    (c) About 145-175 seeds
    (d) About 75-100 seeds
    (e) About 200-225 seeds

(86) Radish phyllody is a————————.
   (a) Root disease (b) Seed disease
   (c) Leave disease (d) Neck disease
   (e) Post harvest disease

(87) Radish seed crop is harvested in the———————— Kulu Valley.
   (a) End of February to 15 March (b) End of September to 15 October
   (c) End of August to 15 September (d) End of March to 15 April
   (e) End of June to 15 July

(88) Male sterility in radish has been reported in ————————————.
   (a) Pusa Chetki (b) Pusa Desi
   (c) Japanese Radish (d) Pusa Himani
   (e) Arka Nishant

(89) Low temperature in radish is a critical factor causing flowering, which is accelerated by ———————— .
   (a) Long photoperiod (b) Short photoperiod
   (c) 90-90% RH (d) Low light intensity
   (e) Long dry spell

(90) Pungency of radish increased with————————.
   (a) Delay sowing (b) Maturity
   (c) Early sowing (d) Heavy dose of fertilizer
   (e) Long dry condition

(91) Horse radish is botanically known as ————————————.
   (a) Raphanus *sylindricum L* (b) *Septiphalpa trumpnum*
   (c) *Brassica juncea* var. *hertsunum* (d) *Raphanus sativum var. trumpate*
   (e) *Armoracia rusticana*

(92) China Rose is a variety of————————.
   (a) Turnip (b) Radish
   (c) Broccoli (d) Artichoke
   (e) Chive

(93) Turnip is botanically modified form of root is known as————————————.
   (a) Tape root (b) Setaceous
   (c) Moniliform (d) Filiform
   (e) Napiform

(94) Basic chromosome number in turnip is————————.
   (a) 9 (b) 10
   (c) 11 (d) 8
   (e) 7

148 Vegetable Science: Objective Type

(95) In hills, turnip is sown from ─────────────.
    (a) November-Janauary   (b) September-November
    (c) March – May   (d) May-July
    (e) July-September

(96) Turnip crinkle is a disease of ─────────────.
    (a) Viral   (b) MPLO
    (c) Fungal   (d) Bacterial
    (e) Nematode

(97) Turnip (*Brassica rapa* L; Syn. *Brassica campestris* var rapa) belongs to the family ─────────────.
    (a) Cruciferae   (b) Chenopodiaceae
    (c) Compositae   (d) Malvaceae
    (e) Umbelliferae

(98) ─────────── variety of turnip is cross of Asiatic type (Local Red Round) and the European type (Golden Ball). This variety is most susceptible to salinity.
    (a) Pusa Swati   (b) Pusa Chandrima
    (c) Early Milan Top   (d) Pusa Kanchan
    (e) Purple Top White Globe

(99) The most favourable temperature for the development of the root and the ratio / green are ─────────────.
    (a) 13-10 °C air temperature and 15-18 °C root temperature
    (b) 0-3°C air temperature and 8-15 °C root temperature
    (c) 10-3°C air temperature and 18-23 °C root temperature
    (d) 13-16 °C air temperature and 20-23 °C root temperature
    (e) 16-20°C air temperature and 10-16 °C root temperature

(100) Turnip seeds are found to retain viability for ─────────────.
    (a) 4-5 years   (b) 5-6 years
    (c) 3-4 years   (d) 2-3 years
    (e) 6-7 years

(101) ─────────── seeds of turnip are sufficient to sow on area of one hectare.
    (a) 7-8 kg   (b) 6-7 kg
    (c) 5-6 kg   (d) 4-5 kg
    (e) 3-4 kg

(102) An isolation distance ─────────── should be provided for production of stocks (nucleus) and ─────────── for certified seeds, respectively.
    (a) 1600 meter and 1000 meter   (b) 1200 meter and 800 meter
    (c) 1800 meter and 1400 meter   (d) 600 meter and 1000 meter
    (e) 1000 meter and 800 meter

Answer Sheet

1	b	2	a	3	a	4	e	5	b	6	e	7	a	8	d	9	b	10	c
11	b	12	a	13	c	14	a	15	e	16	a	17	a	18	c	19	c	20	b
21	e	22	b	23	e	24	c	25	b	26	a	27	b	28	e	29	d	30	b
31	e	32	a	33	e	34	b	35	b	36	d	37	a	38	e	39	a	40	b
41	d	42	a	43	e	44	a	45	e	46	e	47	d	48	e	49	a	50	c
51	a	52	e	53	c	54	b	55	a	56	d	57	b	58	c	59	a	60	b
61	a	62	d	63	d	64	a	65	e	66	a	67	c	68	b	69	e	70	d
71	d	72	a	73	e	74	d	75	e	76	b	77	d	78	b	79	e	80	e
81	a	82	a	83	b	84	c	85	b	86	b	87	e	88	c	89	a	90	b
91	e	92	b	93	e	94	b	95	c	96	a	97	c	98	d	99	c	100	b
101	e	102	a																

1.1.2.9. Tuber Vegetables

(1) The other relative of cassava is—————.
    (a) *Manihot sativum*   (b) *Manihot palmate*
    (c) *Manihot furijee*   (d) *Manihot diplosia*
    (e) *Manihot palmate*

(2) Cassava provides————— energy per day to people more than 400-500 million in the world especially in the under developed countries.
    (a) 300K calories   (b) 350K calories
    (c) 400K calories   (d) 450K calories
    (e) 250K calories

(3) Cassava is a dicotyledonous plant belonging to the family—————.
    (a) Convolvulaceae   (b) Euphorbiaceae
    (c) Umbelliferae   (d) Araceae
    (e) Chenopodiaceae

(4) Basic chromosome number in cassava is—————.
    (a) 14   (b) 16
    (c) 18   (d) 12
    (e) 22

(5) Photosynthesis in cassava is peculiar in having a ————— combination of C_3 and C_4 characteristics, and there are suggestions that the C_3 and C_4 pathway operates ,respectively.
    (a) High temperatures, low temperature
    (b) Low temperatures, higher humidity
    (c) Low temperatures, higher temperature
    (d) Low intensity, higher temperature
    (e) Low temperatures, higher intensity

(6) In cassava, carbohydrate partioning in favour of ———— the tuber, and flowering is favoured by ———— respectively.
    (a) long day, short day   (b) short day, long day
    (c) low temp, short day   (d) high temp, long day
    (e) low temp, high temp

(7) Cassava is ————————.
    (a) Perennial   (b) Semi- perennial
    (c) Biannual   (d) Both c & e
    (e) Annual

(8) Cassava is plant whose flowers are borne on terminal panicles, with the axis of the branch being continuous with that of the panicle infloresence ————————.
    (a) Dioecious   (b) Monoecious
    (c) Andromonoceous   (d) Hermophrodite
    (e) Gynodioceous

(9) Cassava is ———————— crop.
    (a) Often-self pollinated   (b) Often-cross pollinated
    (c) Self-pollinated   (d) Cross-pollinated
    (e) Both c & d

(10) Most recommended length of cutting for planting of cassava is ————————.
    (a) 15-30 cm   (b) 10-15 cm
    (c) 30-32 cm   (d) 8-14 cm
    (e) 32-36 cm

(11) There is a wide variation in the cyanogenic potential of different cassava cultivars, ranging from 0.2 to 62.4 mg HCN e/100 g. The cyanogenic potential of cassava leaves is higher than ———— that of edible portion of the roots.
    (a) 2-5 times   (b) 35-40 times
    (c) 25-30 times   (d) 15-20 times
    (e) 5-20 times

(12) Cassava is a native of ————————
    (a) South Africa   (b) South east Asia
    (c) India   (d) South America
    (e) Europe

(13) ———————— are the major exporter of cassava starch in international market where it is used as filler material in paints, medicine and health drinks.
    (a) Sri Lanka and Pakistan   (b) Thailand and India
    (c) Sri Lanka and Singapore   (d) Thailand and Sri Lanka
    (e) South America and China

(14) Cassava is———————————.
- (a) Gynodioecious and hypogynous
- (b) Diocecious and hypogynous
- (c) Monoecious and protogynous
- (d) Hemophrodite and protogynous
- (e) Andromonocious

(15) Sree Harsha, Sree Jaya, Sree Vijaya, Sree Rekha and Sree Prabha are the latest released varieties of———————————.
- (a) Cassava
- (b) Sweet potato
- (c) Yam
- (d) Colocasia
- (e) Elephant foot

(16) Patra is prepared from tender leaves of———————————.
- (a) Cassava
- (b) Yams
- (c) Elephant foot
- (d) Colocasia
- (e) Sweet potato

(17) Colocasia is a———————————.
- (a) Leafy vegetables
- (b) Root crop
- (c) Tuber crop
- (d) Bulb crop
- (e) Leguminous crop

(18) The edible part of colocasia is———————————.
- (a) Modified roots
- (b) Modified flower
- (c) Tuber
- (d) Leaves
- (e) Tuber and leaves both

(19) Tender leaves of colocasia are the most suitable for———————————.
- (a) Vegetables
- (b) Medicine
- (c) Insecticides
- (d) Cosmotic industry
- (e) Fodder

(20) During cooking, a pinch of baking soda is added in leaves———————————.
- (a) To improve quality
- (b) To remove acridity
- (c) To minimize cooking time
- (d) To easy dsigestiion
- (e) To preserve long time

(21) Colocaisa is also known as———————————.
- (a) Arvi
- (b) Ghuiyan
- (c) Taro
- (d) Dasheen
- (e) All of the above

(22) Satmukhi and Kovar are varieties of———————————.
- (a) Colocasia
- (b) Yams
- (c) Elephant foot
- (d) Cassava
- (e) Sweet potato

(23) Basically, colocasia is a ———————— crop.
    (a) Cool season   (b) Warm season
    (c) Round the year   (d) Spring
    (e) Moderate type

(24) White flea bettle is scientifically known as ————————.
    (a) *Radiculas sikaminii*   (b) *Drosica signata*
    (c) *Nymphia puncdata*   (d) *Leptocarca flavicarpa*
    (e) *Monolepta signata*

(25) In colocasia, spadix appears ———————— after planting.
    (a) 5 to 6 months   (b) 3 to 4 months
    (c) 6 to 7 months   (d) 7 to 8 months
    (e) 4 to 6 months

(26) In colocasia, seed viability deteriorates after ————————.
    (a) Seven months   (b) One year
    (c) Two year   (d) Four months
    (e) Three year

(27) ———————— crop is popularly known as suran.
    (a) Elephant foot   (b) Cassva
    (c) Yams   (d) Sweet potato
    (e) Colocasia

(28) The economic yield of suran is obtained from ————————.
    (a) Stem   (b) Roots
    (c) Corm and cormels   (d) Leaves
    (e) All of the above

(29) The earliness and number of leaves in elephant foot yam are influenced by ————.
    (a) Climatic conditions   (b) Duration of crop
    (c) Duration of corms   (d) Exposed to atmospheric rest
    (e) All of the above

(30) Acridity in suran is due to ————————.
    (a) Quercetin   (b) Calcium oxalate
    (c) Sulphonin   (d) HCN
    (e) Glycoalkaloids

(31) Gajendra is a variety of ————————.
    (a) Colocasia   (b) Sweet potato
    (c) Cassava   (d) Elephant foot
    (e) Yams

(32) ——————— corm type of suran is not preferred because of having more acridity.
    (a) Smooth   (b) Rough
    (c) Big size   (d) Small size
    (e) Red type

(33) ———————while eating, acridity in corms causes irritation of mouth and throat.
    (a) Chow-chow   (b) Suran
    (c) Turmeric   (d) Ginger
    (e) Knoll khol

(34) Economic crop duration of suran is———————.
    (a) Four years   (b) Three years
    (c) Two years   (d) One years
    (e) Five years

(35) Suran is rich in———————.
    (a) Lipids   (b) Water soluble vitamins
    (c) Minerals   (d) Amino acid
    (e) Carbohydrates

(36) Suran bettle scientifically known as ———————.
    (a) *Radiculas sikaminii*   (b) *Galercicida bicolour*
    (c) *Nymphia puncdata*   (d) *Leptocarca flavicarpa*
    (e) *Monolepta signata*

(37) In elepahant foot tuber dormancy is ———————.
    (a) 1 year   (b) 2 years
    (c) 7-8 months   (d) 3 years
    (e) 5-6 months

(38) The relatives of elephant foot are ———————.
    (a) *Amorphophalus oncophyllus*   (b) *A. variabilis*
    (c) *A. bulbifera*   (d) *A. konjac*
    (e) All of the above

(39) Cultivation of potato was started first in the Nilgiri hills in the year———————.
    (a) 1842   (b) 1822
    (c) 1869   (d) 1900
    (e) 1790

(40) ——————— crop supplies about two and half time's more calories than wheat and rice.
    (a) Potato   (b) Sweet potato
    (c) Cassava   (d) Elephant foot
    (e) Yams

(41) Potato is rich source of———————.
    (a) Lipids   (b) Water soluble vitamins
    (c) Carbohydrates   (d) Amino acid
    (e) Minerals

(42) Potato is a———————.
    (a) Long day plant but cultivated as short day
    (b) Short day plant but cultivated as long day
    (c) Short day plant but cultivated also short day
    (d) Day neutral plant
    (e) Long day plant but cultivated also long day

(43) Optimum temperature require for tuberization in potato is ———————.
    (a) 16.7 °C   (b) 24.4 °C
    (c) 20 °C   (d) 23.1 °C
    (e) 24-26.1 °C

(44) ——————— is a variety of potato, which is resistant to early and late blight and virus diseases, particularly resistant to virus X.
    (a) K. Jawahar   (b) K. Ashoka
    (c) K. Chandramukhi   (d) Kufri Badshah
    (e) K. Pukhraj

(45) Kufri Sheetman variety of potato is cross of———————.
    (a) Craigs Defiance x Phulwa   (b) Majestic x Ekishiraju
    (c) PJ376 x PH/F 1430   (d) EM/C-1020 x Allerfruii Heste Gelbe
    (e) Kufri Red x Ginek

(46) Sprouts on tubers reduce the quality results in poor acceptability and reduced prices in the markets. The intensity of sprouts, time of sprout initiation and their growth are governed by———————.
    (a) Varieties   (b) Storage conditions
    (c) Temperature   (d) Humidity
    (e) All of the above

(47) In the southern hills near Oatcamund and Nilgiris, planting is done of potato in——————— a year.
    (a) February, April and September   (b) February only
    (c) April and September   (d) February and September
    (e) September only

(48) Botanically seed of potato is known as———————.
    (a) Berry   (b) TPS
    (c) Seed potato   (d) Stolon
    (e) Tubers

(49) Greening of potato starts ———— after planting.
   (a) 35 days (b) 25 days
   (c) 20 days (d) 45 days
   (e) 50 days

(50) ———— is major pest of potato in storage.
   (a) Beetles (b) PTM
   (c) Rats (d) Grubs
   (e) Mites

(51) Cloudy weather is conducive for spreading of ————.
   (a) Common scab (b) Wart disease
   (c) Late blight disease (d) Charcoal rot
   (e) Soft rot

(52) The affected stem or tubers cut cross, the browning of the xylum vessels is seen and upon squeezing the whitish ooze is the main symptom of ————.
   (a) Black scurf (b) Charcoal rot
   (c) Wart disease (d) Common scab
   (e) Bacterial brown rot

(53) Bacteria causes ———— disease of potato.
   (a) Black scurf (b) Charcoal rot
   (c) Wart disease (d) Common scab
   (e) Black leg

(54) The relatives of potato are ————.
   (a) *Solanum demissum, S. curtilobum, S. saltense*
   (b) *S. stoloniferum, S. anomalocalyx, S. antipoviczii*
   (c) *S. multidissectum, S. phureja, S. jamessi*
   (d) *S. vernei, S. chacoense*
   (e) All *of above*

(55) Sweet potato is an important tuber crop belongs to the family ———— of Dicotyledonae.
   (a) Convolvulaceae (b) Euphorbiaceae
   (c) Umbelliferae (d) Araceae
   (e) Chenopodiaceae

(56) Sweet potato contains ———— starch
   (a) 14 per cent (b) 16 per cent
   (c) 18 per cent (d) 20 per cent
   (e) 12 per cent

(57) Sweet potato contains carotene as ——————.
    (a) 5-10 mg/100g
    (b) 25-36.8 mg/100g
    (c) 45-76 mg/100g
    (d) 75-90.2 mg/100g
    (e) 5.4-20 mg/100g

(58) —————— is considered as the home of sweet potato.
    (a) South Africa
    (b) South east Asia
    (c) India
    (d) South America
    (e) Europe

(59) Genetically sweet potato is——————.
    (a) Tetraploid
    (b) Haploid
    (c) Hexaploid
    (d) Diploid
    (e) Triploid

(60) ——————day length is ideal for flowering of sweet potato.
    (a) 8 hours
    (b) 11 hours
    (c) 9 hours
    (d) 6 hours
    (e) 7 hours

(61) In sweet potato, the stigma becomes receptive in the bud between ——————.
    (a) 6-8 PM
    (b) 6-8 AM
    (c) 4-6 PM
    (d) 4-6 AM
    (e) 12 noon

(62) Kalmegh is a variety of——————.
    (a) Colocasia
    (b) Sweet potato
    (c) Cassava
    (d) Elephant foot
    (e) Yams

(63) Verda is variety of——————.
    (a) Cassava
    (b) Potato
    (c) Colocasia
    (d) Elephant foot
    (e) Sweet potato

(64) The soil pH —————— is appropriate for sweet potato cultivation.
    (a) Between 6.2 and 7.7
    (b) Between 5.2 and 5.7
    (c) Between 4.5 and 5.5
    (d) Between 5.2 and 6.7
    (e) Between 6 and 7

(65) Pox and scurf of sweet potato is due to ——————.
    (a) High temperaure
    (b) High pH
    (c) Low pH
    (d) Dry spell
    (e) High humidity

(66) For planting one hectare of land, a primary nursery——————— and a secondary nursery——————————— are required, respectively.
   (a) 100 m², 300 m²
   (b) 50 m², 200 m²
   (c) 1000 m², 500 m²
   (d) 100 m², 500 m²
   (e) 200 m², 700 m²

(67) The best time for planting of sweet potato tuber in Kerala is———————.
   (a) September to December
   (b) December-February
   (c) February-April
   (d) April-June
   (e) June-September

(68) The best PGRs used for increasing tuber yield of sweet potato is———————.
   (a) NAA
   (b) IBA
   (c) CCC and SADH
   (d) GA3
   (e) Urea

(69) The time of harvesting of sweet potato differs with cultivar, which may extend from———————.
   (a) 120-180 days
   (b) 100-120 days
   (c) 180-200 days
   (d) 200-220 days
   (e) 90-100 days

(70) Under good management practices, yield of sweet potato in irrigated——————— and——————— under rainfed conditions, respectively.
   (a) 150-200 q/ha, 50-80 q/ha
   (b) 450-500 q/ha, 200-250 q/ha
   (c) 250-300 q/ha, 100-150 q/ha
   (d) 300-350 q/ha, 180-200 q/ha
   (e) 350-400 q/ha, 80-100 q/ha

(71) Vine production in sweet potato ranges from ———————.
   (a) 250-275 quintals
   (b) 50-75 quintals
   (c) 100-250 quintals
   (d) 300-325 quintals
   (e) 80-100 quintals

(72) Sweet potato leaves become yellow along with discolouration of vascular bundles as well as blackening of vascular tissue. Such situation is due to ———————.
   (a) Wilt disease
   (b) Black rot
   (c) Cercospora leaf spot
   (d) Collar rot
   (e) Alternaria leaf spot

(73) Sexual reproduction in sweet potato is limited because of the———————.
   (a) Non blooming habit
   (b) Low fertility of blossoms
   (c) Poor seed setting
   (d) Saprophytic incompatibility
   (e) All of the above

(74) Horned caterpillar of sweet potato is scientifically known as——————.
    (a) *Cylas formicarius* (b) *Brachmia convolvulli*
    (c) *Spilosoma obliqua* (d) *Euchromia polymena*
    (e) *Agrius convolvuli*

(75) Bihar hairy caterpillar of sweet potato is scientifically known as ——————.
    (a) *Brachmia convolvulli* (b) *Cylas formicarius*
    (c) *Spilosoma obliqua* (d) *Agrius convolvuli*
    (e) *Euchromia polymena*

(76) In India, sweet potato is generally cultivated as a——————.
    (a) Irrigated crop (b) Ratoon crop
    (c) Hedge crop (d) Rainfed crop
    (e) Fodder crop

(77) ——————varieties namely 'Triumph', 'Nancy Hall' and 'Nancy' are introduction to India.
    (a) Cassava (b) Potato
    (c) Colocasia (d) Elephant foot
    (e) Sweet potato

(78) ——————varieties 'Pusa Red' and ' Kanghangad Local' are evolved through clonal selection.
    (a) Colocasia (b) Sweet potato
    (c) Cassava (d) Elephant foot
    (e) Yams

(79) ——————hybrid H-620 and the variety 'Kanghangad Local' are least affected by sweet potato weevil.
    (a) Colocasia (b) Elephant foot
    (c) Yams (d) Sweet potato
    (e) Cassava

(80) The relatives of sweet potato are——————..
    (a) *Ipomea muricata* (b) *I. leucantha*
    (c) *I. Littoralis* (d) *I. trifida*
    (e) All of the above

(81) Yam is also known as——————.
    (a) Suran (b) Ratalu
    (c) Dasheen (d) Kochai
    (e) Tapioca

(82) The genus Dioscorea contains ———————— species.
   (a) 200
   (b) 1200
   (c) 1000
   (d) 800
   (e) 600

(83) ———————— alkaloid obtained from yams is used for preparing contraceptic drugs in family planning.
   (a) Diosgenin
   (b) Galactomanin
   (c) Glycosinolates
   (d) Calcium oxalate
   (e) Sulphonin

(84) Yellow yam is botanically known as————————.
   (a) *D. rotundifolia*
   (b) *Dioscorea cayenensis*
   (c) *Dioscorea alata*
   (d) *D. bulbifera*
   (e) *D. dumetorum*

(85) Yams belong to family————————.
   (a) Convolvulaceae
   (b) Euphorbiaceae
   (c) Dioscoreaceae
   (d) Araceae
   (e) Chenopodiaceae

(86) Lasser yam is botanically known as————————.
   (a) *Dioscorea esculenta*
   (b) *D. bulbifera*
   (c) *Dioscorea alata*
   (d) *Dioscorea cayenensis*
   (e) *D. dumetorum*

(87) ———————— plants accommodate per hectare if yams are planted at spacing of 50 cm from hill to hill and 150 cm from row to row.
   (a) 12330
   (b) 13300
   (c) 10330
   (d) 11330
   (e) 13030

(88) Sreelatha is a variety of————————.
   (a) Suran
   (b) Lesser yam
   (c) Dasheen
   (d) Sweet potato
   (e) Tapioca

(89) Ethylene chlorohydrin like plant growth regulator is used for————————.
   (a) Increase corm growth
   (b) Remove acridity
   (c) Early flowering
   (d) Delay sprouting
   (e) Enhancing early sprouting

(90) Air potato is botanically known as ————————.
   (a) *D. dumetorum*
   (b) *D. bulbifera*
   (c) *Dioscorea alata*
   (d) *Dioscorea cayenensis*
   (e) *Dioscorea esculenta*

(91) Under good crop management, per hectare yield of yam is—————.
    (a) 100-250 quintal   (b) 250-300 quintal
    (c) 300-350 quintal   (d) 350-400 quintal
    (e) 50-100 quintal

(92) Kacheo, Birmanica, Suavior and Sativa are varieties of—————.
    (a) Suran   (b) Sweet potato
    (c) Dasheen   (d) Ariel yam
    (e) Tapioca

(93) Suthni is also known as—————.
    (a) Suran   (b) Sweet potato
    (c) Dasheen   (d) Lesser yam
    (e) Tapioca

(94) Soaking of yam tubers in 1 per cent ————— solution, causes delayed sprouting in storage.
    (a) GA3   (b) MH
    (c) NAA   (d) IBA
    (e) CCC

(95) Yam tuber can store at—————.
    (a) 20-25 °C at 80-85 per cent relative humidity
    (b) 15-20 °C at 70-75 per cent relative humidity
    (c) 10-15 °C at 80-85 per cent relative humidity
    (d) 29-30 °C at 90-95 per cent relative humidity
    (e) 5-10 °C at 90-95 per cent relative humidity

(96) Yam tubers become ready for harvesting————— after planting.
    (a) 10-11 months   (b) One year
    (c) 8-9 months   (d) 7-8 months
    (e) 6-8 months

(97) The problems in yam breeding are—————.
    (a) Irregularity of flowering and seed production
    (b) Slow rate of multiplicity of propagating units
    (c) Degradtion in genetic structure due to virus attack
    (d) Above three
    (e) None of the above

(98) 'De 23', ' Lathani', 'De 11' (Sreelatha), 'Chaparia' and 'De 1Y' have been identified as high yielding varieties of—————.
    (a) Colocasia   (b) Elephant foot
    (c) Lesser yam   (d) Sweet potato
    (e) Cassava

Answer Sheet

1	b	2	a	3	b	4	c	5	c	6	b	7	a	8	b	9	d	10	a
11	e	12	d	13	b	14	c	15	a	16	c	17	c	18	e	19	a	20	b
21	e	22	a	23	b	24	e	25	b	26	d	27	a	28	c	29	e	30	b
31	d	32	a	33	b	34	a	35	e	36	b	37	e	38	e	39	b	40	a
41	c	42	a	43	c	44	d	45	a	46	-	47	a	48	b	49	a	50	b
51	a	52	e	53	e	54	e	55	a	56	b	57	e	58	d	59	c	60	b
61	a	62	b	63	e	64	d	65	b	66	d	67	a	68	c	69	a	70	e
71	c	72	a	73	e	74	e	75	c	76	d	77	e	78	b	79	d	80	e
81	b	82	e	83	a	84	b	85	c	86	a	87	b	88	b	89	e	90	b
91	a	92	d	93	d	94	b	95	d	96	c	97	d	98	c				

1.1.2.10. Spice Vegetables

(1) Black pepper is known as ——————.
    (a) Mother of spices   (b) Queen of spices
    (c) King of spices   (d) Poor man spices
    (e) None of the above

(2) Black pepper belongs to the family——————.
    (a) Piperaceae   (b) Myrtaceae
    (c) Lauraceae   (d) Labiateae
    (e) Umbelliferae

(3) Black pepper is native of tropical forest and Western Ghats of——————.
    (a) Nepal   (b) Pakistan
    (c) Sri Lanka   (d) South Africa
    (e) India

(4) Black pepper is commercially propagated through——————.
    (a) Seeds   (b) Berry
    (c) Roots   (d) Cuttings from runner shoots
    (e) Leaves

(5) Consumption period of black pepper is——————.
    (a) 8 to 10 months   (b) 4 to 6 months
    (c) One year   (d) 6 to 8 months
    (e) Two year

(6) Black pepper is——————.
    (a) Dioecious   (b) Monoecious
    (c) Andromonoceous   (d) Hermophrodite
    (e) Gynodioceous

(7) Black pepper starts yielding from——————.
    (a) 3-4 years   (b) 2-3 years
    (c) 4-5 years   (d) 1-2 years
    (e) 6 months

(8) Fruiting branches of black pepper are known as ——————.
    (a) Fragmotropic shoot   (b) Kalimic shoot
    (c) Cosmotrophic shoot   (d) Ceizotropic shoot
    (e) Plagiotropic shoot

(9) First hybridization in black pepper is done in——————.
    (a) 1933   (b) 1902
    (c) 1954   (d) 1888
    (e) 1900

(10) First hybrid of black pepper released in India was——————.
    (a) Sreekara   (b) Shyma
    (c) Panchami   (d) Krishna
    (e) Penniyur-1

(11) Penniyur-1 & Penniyur-3 are the hybrids of——————.
    (a) Cinnamom   (b) Fennel
    (c) Black pepper   (d) Cumin
    (e) Small cardamom

(12) Piperine has been used to impart a pungent taste to brandy. It has also been tried as an insecticide. It is stated to be more toxic to houseflies than pyrethrum and a mixture of 0.05 per cent piperine and 0.01 per cent pyrethrum is more toxic than 0.10 per cent pyrethrum alone. It is obtained from——————.
    (a) Fennel   (b) Cinnamom
    (c) Cumin   (d) Black pepper
    (e) Small cardamom

(13) —————— breeding method plays an important role in black pepper improvement.
    (a) Pedigree method   (b) Recurrent selection
    (c) Clonal selection   (d) Mass selection
    (e) Mutaton

(14) 'Pournami' is the nematode resistant variety of——————.
    (a) Cinnamom   (b) Clove
    (c) Fenugreek   (d) Cumin
    (e) Black pepper

(15) *Piper colubrinum* is a possible doner of foot rot resistant, which is introduced in India from——————.
    (a) Brazil   (b) Ghana
    (c) Sri Lanka   (d) South Africa
    (e) India

(16) ——————trees are used for climbing of pepper.
    (a) Erytherina   (b) *Grevelia robusta*
    (c) *Garuga pinnata*   (d) *Ailanthus*
    (e) All of the above

(17) Piperine is the alkaloid responsible for bitter taste in ——————.
    (a) Coriander   (b) Black pepper
    (c) Fenugreek   (d) Cumin
    (e) Clove

(18) Piperine is found in——————.
    (a) Roots   (b) Only leaves
    (c) Vine   (d) Only berries
    (e) All parts

(19) India accounts 54 per cent of area and 26.6 per cent of production of the world——————.
    (a) Black pepper   (b) Coriander
    (c) Fenugreek   (d) Ginger
    (e) Turmeric

(20) —————— in Kerala which alone contributes about 96 per cent of the total production in India and Karnataka (3.5 per cent), Tamil Nadu and Pondichery the rest.
    (a) Fenugreek   (b) Coriander
    (c) Black pepper   (d) Ginger
    (e) Turmeric

(21) In —————— major insect and disease are 'Pollu beetel & foot rot', respectively.
    (a) Fenugreek   (b) Cumin
    (c) Ginger   (d) Black pepper
    (e) Clove

(22) Pepper Research Station is situated at——————.
    (a) Calicut   (b) Kasargod
    (c) Penniyur   (d) Cochin
    (e) Trichy

(23) Annual yield of black pepper per hectare is——————.
    (a) 800-1000 kg   (b) 500-800 kg
    (c) 250-500 kg   (d) 1000-1250 kg
    (e) 1250-1500 kg

(24) —————— position or importance of small cardamom in national level (India)
    (a) First important   (b) Second important
    (c) Third important   (d) Fourth important
    (e) No any position

(25) —————— is the world's leading cardamom supplier accounting for three-quarters of world trade in this commodity.
    (a) India   (b) Brazil
    (c) Guatemala   (d) Africa
    (e) Sri Lanka

(26) —————— is the leading state in area and production of cardamom contributing about 65 per cent.
    (a) Karnataka   (b) Kerala
    (c) Tamil Nadu   (d) Andhra Pradesh
    (e) Rajasthan

(27) —————— is the major producer of cardamom in the world.
    (a) Tanzania   (b) India
    (c) Sri Lanka and Thialand   (d) Guatemala
    (e) All of the above

(28) In Arabian countries, —————— coffee is called 'Gawa' is generally offered to guests at social and religious functions.
    (a) Fennel flavoured   (b) Clove flavoured
    (c) Ginger flavoured   (d) Cardamom-flavoured
    (e) Cinnamon flavoured

(29) It is belived that excessive use of——————causes impotency.
    (a) Black pepper   (b) Fennel
    (c) Cardamom   (d) Clove
    (e) Fenugreek

(30) Cardamom belongs to the family——————.
    (a) Lauraceae   (b) Labiateae
    (c) Myrtaceae   (d) Umbelliferae
    (e) Zingiberaceae

(31) Malabar ———— is a variety of drought resistance and mostly suitable for lower elevations of crop.
- (a) Black pepper
- (b) Fennel
- (c) Cardamom
- (d) Clove
- (e) Fenugreek

(32) Mudigere-1 variety of ———— is tolerant to thrips infestation, while, MCC-61 variety is tolerant to azhukal disease.
- (a) Cardamom
- (b) Fennel
- (c) Black pepper
- (d) Clove
- (e) Fenugreek

(33) Cardamom is propageted through suckers and seedlings. However, sucker method is preferable in order to check the spread of the————.
- (a) Fungal diseases
- (b) Bacterial disease
- (c) Viral disease
- (d) Mycoplasm a disease
- (e) Nematode infection

(34) ———— in cardamom, seedlings are economical and considered better planting materials.
- (a) 10-12 months old
- (b) 20-21 months old
- (c) 12-14 months old
- (d) 22-24 months old
- (e) 16-18 months old

(35) Small cardamom plants start bearing from the third year of planting though an economical yield is usually obtained only from the————.
- (a) Fourth year onwards
- (b) Fifth year onwards
- (c) Six year onwards
- (d) Fifth year, 6 months onwards
- (e) Sixth year, 6 months onwards

(36) The bearing span of cardamom varies from ————.
- (a) 8-10 years
- (b) 10-12 years
- (c) 15-20 years
- (d) 20-24 years
- (e) 24-27 years

(37) In small cardamom flowering starts in ————.
- (a) April-May
- (b) June-July
- (c) August-September
- (d) October-Novemebr
- (e) December-January

(38) The peak harvesting period of cardamom is————.
- (a) December-January
- (b) October-November
- (c) August-September
- (d) June-July
- (e) April-May

(39) The average yield of dry capsules of small cardamom from maintained plantation is———————— per hectare.
   (a) About 50-100 kg
   (b) About 200-250 kg
   (c) About 175-200 kg
   (d) About 150-175 kg
   (e) About 100-150 kg

(40) In small cardamom, a disease is characterized by appearance of water soaked spots, which gradually extended into large areas in which chlorophyll is completely destroyed which is known as ————————.
   (a) Azhukal
   (b) Katte
   (c) Chenthal Disease
   (d) Nursery leaf spot
   (e) Chirke Disease

(41) Azhukal rot disease of small cardamom is due to————————.
   (a) *Corynebacterium* spp
   (b) *Phyllosticta elettariae*
   (c) *Phytophthora meedii*
   (d) *Pythium spp.*
   (e) *Alternaria spp.*

(42) Katte or Marble diseasee of small cardamom is due to———————— virus transmitted through aphid
   (a) *Rhopalosiphum maidis*
   (b) *R. padi*
   (c) *Brachycaudus helichrisi*
   (d) *Silobion aveneae*
   (e) *Pentalonia nigronervosa*

(43) Chenthal is bacterial disease of————————.
   (a) Small cardamom
   (b) Fennel
   (c) Black pepper
   (d) Clove
   (e) Fenugreek

(44) The serious pest of small cardamom is————————.
   (a) Shoot and capsule borer
   (b) Cardamom thrips
   (c) Root grubs
   (d) Capsule borer
   (e) Spotted locust

(45) Botanically cardamom fruit is called————————.
   (a) Pome
   (b) Pod
   (c) Siliqua
   (d) Capsule
   (e) Berry

(46) The fruits of———————— are trilocular capsule containing about 10-15 seed/capsules. The seeds are black when fully ripe and covered with a white mucilagenous coat.
   (a) Small cardamom
   (b) Fennel
   (c) Black pepper
   (d) Clove
   (e) Fenugreek

(47) Large cardamom is botanically known as ―――――――――――.
    (a) *Amomum subulatum*   (b) *Laurus nobilis*
    (c) *Pimenta officinalis*   (d) *Anethum sowa*
    (e) *Eugenia carryophyllus*

(48) Large cardmom is native of ―――――――――――.
    (a) Sri Lanka   (b) Mediterrnean region
    (c) Ghana   (d) Eastern Himalayan region
    (e) Brazil

(49) Sikkim is the leading state in area and production followed by Darjeeling hills of West Bengal in India of ―――――――――――.
    (a) Black pepper   (b) Fennel
    (c) Large cardamom   (d) Clove
    (e) Fenugreek

(50) Bhutan and Nepal are the other producers of ―――――――――――.
    (a) Large cardamom   (b) Fennel
    (c) Black pepper   (d) Clove
    (e) Fenugreek

(51) The species of large cardamom are ―――――――――――
    (a) *Amomum aromaticum*   (b) *A. cardamomum*
    (c) *A. krevanth*   (d) All three
    (e) None of the above

(52) The relatives are Churumpa, Boklok, Belak, Tali and Jaker. These relatives are identified from ICAR Complex, Tadang, Sikkim of which crop is ―――――――――――.
    (a) Clove   (b) Fennel
    (c) Fenugreek   (d) Black pepper
    (e) Large cardamom

(53) ――――――――――― planting of large cardamom is generally done after receiving sufficient rainfall.
    (a) During January-February   (b) During May-June
    (c) During July-August   (d) Durting September-October
    (e) During November-December

(54) Vegetative propogated plants of large cardamom starts fruiting in ―――――――――――.
    (a) Four years of planting   (b) Fifth years of planting
    (c) Six years of planting   (d) Two years of planting
    (e) Three years of planting

(55) Large cardamom is harvested from ———————.
- (a) December- February
- (b) February- April
- (c) April-June
- (d) June-August
- (e) August – November

(56) Bearing span of large cardamom varies between 12-15 years, however, some well-managed plantation can be yield profitably upto———————.
- (a) 20 years
- (b) 16 years
- (c) 18 years
- (d) 22 years
- (e) 24 years

(57) Leaf streak, Foorkey and Chirkey are major diseases of———————.
- (a) Clove
- (b) Large cardamom
- (c) Fenugreek
- (d) Black pepper
- (e) Fennel

(58) The relatives of large cardamom are———————.
- (a) *A. corynostachyum*
- (b) *A. pauciflorum, A. aromatium*
- (c) *A. longuiforme, A. costatum*
- (d) *Amonum dealbatum, A. kingii*
- (e) All of the above

(59) Cinnamon is botanically known as *Cinnamomum zeylanicum* Blume belongs to family ——————— having 2n=24.
- (a) Lauraceae
- (b) Labiateae
- (c) Myristicaceae
- (d) Myrtaceae
- (e) Umbelliferae

(60) Cinnamon is a native——————— of south west tropical.
- (a) West indies
- (b) Iran
- (c) Ghana
- (d) South Africa
- (e) India and Srilanka

(61) ——————— part of cinnamom has delicate fragrance and a warm sweet agreeable taste.
- (a) Flowers
- (b) Bark
- (c) Leaves
- (d) Seeds
- (e) Roots

(62) Cinnamom is a ———————.
- (a) Often self-pollinated crop
- (b) Often cross-pollinated crop
- (c) Self-pollinated crop
- (d) Cross-pollinated crop
- (e) None of the above

(63) It can be grown from——————— sea level in Indian conditions.
  (a) 1000 m altitude (b) 3500 m altitude
  (c) 3000 m altitude (d) 2500 m altitude
  (e) 2000 m altitude

(64) 'Sweet' and Honey, Pat or Mat - Pat and Kurunchi are famous varieties of——————— suitable for growing in Sri Lanka.
  (a) Clove (b) Cinnamon
  (c) Fenugreek (d) Black pepper
  (e) Fennel

(65) The relatives of cinnamon are———————.
  (a) *Cinnamomum cassia* (b) *C. burmanni*
  (c) *C. lourcirri* (d) All of the above
  (e) None of the above

(66) Cinnamon is mostly propagated by———————
  (a) Bark (b) Leaves
  (c) Cutting (d) Suckers
  (e) Seeds

(67) Cinnamon sedlings may be transplanted in the———————
  (a) Before the monsoon sets (b) Beginning of the monsoon
  (c) Summer season (d) Winter season
  (e) A & b both

(68) The first cinnamon cutting is made after 2nd or 3rd years of transplanting when the plant reaches a height of———————.
  (a) 150-200 cm (b) 200-2225 cm
  (c) 225-250 cm (d) 250-275 cm
  (e) 275-300 cm

(69) Economically 'clove' is ———————
  (a) Bark (b) Younger leaves
  (c) Above ground arial roots (d) Unopened flower bud
  (e) Seeds inside berry

(70) In India, clove is being cultivated in———————.
  (a) Tamil Nadu (b) Kerala
  (c) Karnataka (d) North easten region
  (e) a, b & c

(71) The problems connected with cloves, which deserve furtheur investigations are—
   (a) Too low percentage of germination
   (b) Slow germination of seeds
   (c) Slow seedling growth
   (d) All three
   (e) None of the above

(72) Although clove-tree may start yielding from the 7th or 8th year after planting. But economic yields are obtained from the ——————————.
   (a) 9 year only
   (b) 10 years only
   (c) 11 year only
   (d) 12 year only
   (e) 15th year only

(73) Clove root oil is obtained by the steam distillation of the root of clove tree with a yield of about ——————————
   (a) 4 per cent
   (b) 12 per cent
   (c) 6 per cent.
   (d) 8 per cent
   (e) 10 per cent

(74) Coriander is a native of——————————.
   (a) South America
   (b) Mediterranean region
   (c) China
   (d) Sri Lanka
   (e) India

(75) Coriander is the richest source of——————————.
   (a) Vitamin C (250 mg/100g)
   (b) Vitamin A (5200 IU/100g).
   (c) Both a & b
   (d) Protein (7.2 g /100 g edible part)
   (e) Fiber (6.8 g / 100 g edible part).

(76) —————————— is a leading state in area and production of coriander.
   (a) Maharashtra
   (b) Rajasthan
   (c) Andhra Pradesh
   (d) Uttar Pradesh
   (e) Karnataka

(77) Coriander is a member of family ——————————
   (a) Lauraceae
   (b) Labiateae
   (c) Myristicaceae
   (d) Myrtaceae
   (e) Umbelliferae

(78) The spanish people call fresh leaves 'Cilantro' and it is also called 'Chinese Parsley' to——————————
   (a) Bay leave
   (b) Coriander
   (c) Fenugreek
   (d) Curry leaf
   (e) Celery

(79) Karan, Swati, and Sadhana are varieties of———————————————.
    (a) Indian dill   (b) Cumin
    (c) Fenugreek   (d) Coriander
    (e) Celery

(80) The fungus *Protomyces macrosporus* causes———————————————.
    (a) Blight   (b) Coriander tumor
    (c) Anthracnose   (d) Mildew
    (e) Rust pustules

(81) The yield of coriander from seed and leaves per hectare ————————, respectively.
    (a) 5.5 to 7.5 quintals, 9 to 14 quintals
    (b) 6.5 to 8.5 quintals, 14 to 16 quintals
    (c) 8.5 to 9.5 quintals, 16 to 18 quintals
    (d) 9.5 to 10.5 quintals, 18 to 20 quintals
    (e) 11.5 to 12.5 quintals, 22 to 24 quintals

(82) IARI Selection 360, IARI Seethal 36-3, CIMPO (CSIR) Selection S-28 and S 52 are recommended varieties of ——————————————— for green leaf crop.
    (a) Celery   (b) Lettuce
    (c) Fenugreek   (d) Coriander
    (e) Bay leave

(83) Hisar Sugandh is the latest released variety of———————————————.
    (a) Coriander   (b) Fennel
    (c) Fenugreek   (d) Clove
    (e) Indian dill

(84) A drink prepared by cumin in Germany is known as———————————.
    (a) Dolly   (b) Brinky
    (c) Jummel   (d) Summel
    (e) Kummel

(85) Major importer of cumin seed is ——————————.
    (a) Turkey   (b) Syria
    (c) China   (d) Iran
    (e) All of the above

(86) Cumin belongs to the family ———————————————.
    (a) Myrtaceae   (b) Umbelliferae
    (c) Myristicaceae   (d) Lauraceae
    (e) Labiateae

(87) Cumin seeds have an aromatic fragrance is due to ―――――.
    (a) Cuminol (b) Suminol
    (c) Ferrol (d) Arol
    (e) Cherrol

(88) Topolka is a variety of――――――― introduced from Bulgaria and resistant to *Pseudomonas cumini*.
    (a) Coriander (b) Fennel
    (c) Fenugreek (d) Cumin
    (e) Indian dill

(89) The time of sowing of cumin is ――――――――.
    (a) Last week of December- 1st week of February
    (b) Last week of February- 1st week of April
    (c) Last week of April- 1st week of June
    (d) 1st week of June-last week of August
    (e) Mid November to 1st week of December

(90) The optimum seed rate of cumin sowing is―――――――.
    (a) 5 to 10 kg per hectare (b) 15 to 19 kg per hectare
    (c) 12 to 15 kg per hectare (d) 22 to 25 kg per hectare
    (e) 19 to 22 kg per hectare

(91) Zeeri (*Plantage pumile*) is an important weed with its morphological similarity with ―――――――.
    (a) Coriander (b) Fennel
    (c) Fenugreek (d) Cumin
    (e) Indian dill

(92) Cumin matures in――――― depending upon variety and agroclimatic conditions.
    (a) 120-140 days (b) 80-120 days
    (c) 60-80 days (d) 180-220 days
    (e) 150-200 days

(93) NSUU, NS32 and Kala Jeera are varieties of――――――――.
    (a) Indian dill (b) Cumin black
    (c) Fenugreek (d) Fennel
    (e) Coriander

(94) Cumin black is a――――――― crop.
    (a) Rabi season (b) Kharif
    (c) Summer (d) Round the year
    (e) Spring season

(95) Fennel is cultivated mostly as a garden or homeyard crop throughout India at all altitudes upto ―――――
    (a) 1,825 m   (b) 2,825 m
    (c) 2,325 m   (d) 3,325 m
    (e) 8,025 m

(96) Pant Madhurika, Rajendra Sourabh and Rf-101 are latest released varieties of ―――――
    (a) Indian dill   (b) Cumin
    (c) Fenugreek   (d) Fennel
    (e) Coriander

(97) In fennel, flavour is due to compounds ―――――
    (a) Centhole   (b) Anithole and fenchone
    (c) Fennetol   (d) Fennerol
    (e) None of the above

(98) Fenugreek is also known as ―――――
    (a) Loung   (b) Methi
    (c) Dhania   (d) Kala jeera
    (e) Ajwain

(99) Fenugreek seeds induces ―――――
    (a) Sterlity in human   (b) Mutation in animal
    (c) Baldness in men   (d) Lactation in women
    (e) None of the above

(100) Methi is a crop of ―――――
    (a) Warm season   (b) Cool season
    (c) Hilly crop   (d) Round the year
    (e) Spring crop

(101) Fenugreek is botanically known as ―――――.
    (a) *Apium graveolens* var. *rapaccum*
    (b) *Cynara scalymus*
    (c) *Crocus sativus*
    (d) *Allium schoenoprasum*
    (e) *Triginellafoenum graecum*

(102) Kasuri types (*T. corniculata*) of fenugreek produces flowers ―――――.
    (a) White to yellow in colour   (b) Violet to bluish in colour
    (c) Bright pink in colour   (d) Light red to pink in colour
    (e) Bright orange to yellow in colour

(103) The seed rate per hectare for common methi and for kasuri type is ———— ————, respectively.
    (a) 30 kg, 15 kg
    (b) 25 kg, 20 kg
    (c) 35 kg, 25 kg
    (d) 15 kg, 10 kg
    (e) 10 kg, 5 kg

(104) Fenugreek belongs to family ————————————.
    (a) Leguminosae
    (b) Umbelliferae
    (c) Myristicaceae
    (d) Lauraceae
    (e) Labiateae

(105) Pusa Early Bunching is a variety of ————————————.
    (a) Black pepper
    (b) Cumin
    (c) Fenugreek
    (d) Fennel
    (e) Coriander

(106) Green leaves per hectare are obtained from common methi and kasuri methi are— ————————, respectively.
    (a) 70-80 quintals, 80-100 quintals
    (b) 90-100 quintals, 110-120 quintals
    (c) 100-110 quintals, 120-130 quintals
    (d) 110-120 quintals, 130-140 quintals
    (e) 50-60 quintals, 60-70 quintals

(107) The origin place of fenugreek is ————————————.
    (a) Brazil
    (b) Sri Lanka
    (c) India
    (d) South Eastern Europe and Ethiopia
    (e) China

(108) Basic chromosome number in fenugreek is ————————————.
    (a) 8
    (b) 9
    (c) 10
    (d) 11
    (e) 12

(109) In the plain, sowing is done in the month of ————————————.
    (a) June
    (b) April
    (c) February
    (d) November
    (e) September

(110) The common methi may take 5-6 days to germinate while kasuri methi ————.
    (a) 2-3 days
    (b) 4-5 days
    (c) 5-6 days
    (d) 7-8 days
    (e) 10-12 days

(111) Rajendra Khushba and Pant Ragini are the latest released varietis of————.
   (a) Ginger
   (b) Cumin
   (c) Coriander
   (d) Fennel
   (e) Fenugreek

(112) Ginger is native of————————————————————.
   (a) Southern Asia
   (b) South Africa
   (c) Ghana
   (d) Europe
   (e) Sri Lanka

(113) Contributing one third of ginger production in——————, is the leading state in area and production of ginger in India.
   (a) Sikkim
   (b) Meghalaya
   (c) Andhra Pradesh
   (d) Kerala
   (e) Karnataka

(114) Ginger belongs to the family———————————————.
   (a) Leguminosae
   (b) Zingiberaceae
   (c) Myristicaceae
   (d) Lauraceae
   (e) Labiateae

(115) Banada spices of ginger is botanically known as————————
   (a) *Z. elatum*
   (b) *Z. chrysanthum*
   (c) *Zingiber zerumbe*
   (d) *Zingiber officinale*
   (e) *Zingiber cassumunar*

(116) ————— countries produce the best quality ginger—————.
   (a) Jamaica and India
   (b) China and Nepal
   (c) Pakistan and Iran
   (d) Brazil and USA
   (e) England and Australia

(117) The mode of ginger consumptions are————————————
   (a) 42-43 per cent dry
   (b) 36 per cent green ginger
   (c) 17-18 per cent seed
   (d) 3-4 per cent spoilage
   (e) All of the above

(118) India contributed of world ginger production of————————.
   (a) 40 percent
   (b) 50 per cent
   (c) 30 percent
   (d) 60 per cent
   (e) 70 percent

(119) In ginger, the peak period of flowering is————————————.
   (a) July
   (b) November
   (c) January
   (d) March
   (e) May

(120) Surabhi variety of———————— is a clonal selection from X-ray mutant of local cultivar (V_1K_1-3).
    (a) Cumin   (b) Black pepper
    (c) Clove   (d) Ginger
    (e) Turmeric

(121) Ginger is universally propagated from cuttings of rhizome i.e. known as———
    (a) Clipss   (b) Netts
    (c) Petts   (d) Setts
    (e) Bits

(122) At higher altitudes the ginger seed rate varies from————————.
    (a) 2000-2500 kg/ha   (b) 2500-3000 kg/ha
    (c) 3500-4000 kg/ha   (d) 1500-2000 kg/ha
    (e) 1000-1500 kg/ha

(123) The relatives of ginger are————————.
    (a) *Z. elatum*   (b) *Z. zerumbet*
    (c) *Zingiber mioga*   (d) *Z. cassumunar*
    (e) All of the above

(124) India is the leading supplier of dill seed to————————.
    (a) Sri Lanka   (b) China
    (c) USA   (d) Egypt
    (e) Pakistan

(125) The aroma in dill herb is due to————————.
    (a) Glucosinolate   (b) Hexahydrobenzofuran
    (c) Sinigrin   (d) Oleoresin
    (e) Gingerine

(126) Turmeric is native to————————.
    (a) USA   (b) England
    (c) Pakistan   (d) India or China
    (e) Nepal

(127) The largest producer and exporter of turmeric in the world is————————.
    (a) USA   (b) India
    (c) Pakistan   (d) England
    (e) China

(128) Andhra Pradesh and Tamil Nadu contributes in total production of————————.
    (a) 80 per cent   (b) 35 percent
    (c) 45 percent   (d) 55 percent
    (e) 70 per cent

(129) The constituent of turmeric is ———————, it is orange yellow crystalline odourless powder, insoluble in water but soluble in ethanol and glacial acid.
- (a) Curcumin
- (b) Cucumerin
- (c) Oleoresin
- (d) Turmeralin
- (e) Gingerine

(130) The Indian varieties viz Nazamabad (whole), Madras (fingers), Rajapori Cuddapayh (whole and fingers) Alleppey (fingers) are in world trades circle of————.
- (a) Fennel
- (b) Black pepper
- (c) Clove
- (d) Ginger
- (e) Turmeric

(131) Lakdong variety of turmeric has curcumin content of————————.
- (a) 6.8-7.2 per cent
- (b) 7.8-8.2 per cent
- (c) 5.8-6.2 per cent
- (d) 8.8-9.2 per cent
- (e) 4.8-5.2 per cent

(132) Turmeric belongs to the family————————————.
- (a) Leguminosae
- (b) Lauraceae
- (c) Myristicaceae
- (d) Zingiberaceae
- (e) Labiateae

(133) A specific and serious pest of turmeric is————————.
- (a) Turmeric shoot borer
- (b) Turmeric skipper
- (c) Lace wing bug
- (d) Thrips
- (e) Sap feeders

(134) ———————— healthy disease free rhozomes is sufficinet for turmeric planting.
- (a) 15-20 q/ha
- (b) 35-40 q/ha
- (c) 30-35 q/ha
- (d) 25-30 q/ha
- (e) 20-25 q/ha

(135) Turmeric crop generally becomes ready for harvest———————— after planting.
- (a) 3-4 months
- (b) 7-9 months
- (c) 9-10 months
- (d) 10-12 months
- (e) 5-7 months

(136) The yield of turmeric ———————————— can be obtained.
- (a) 350 to 400 q/ha
- (b) 250 to 300 q/ha
- (c) 200 to 250 q/ha
- (d) 150 to 200 q/ha
- (e) 100 to 150 q/ha

(137) The other relaties of turmeric are ─────────────
(a) *C. caesia* (b) *Curcuma amada*
(c) *C. angustifolia* (d) *C. aromatica*
(e) All of the above

(138) Sobha and Pant Peetabh are the latest varieties of ─────────────.
(a) Cumin (b) Turmeric
(c) Clove (d) Ginger
(e) Black pepper

(139) Amba is a variety of ─────────────.
(a) Small cardamom (b) Turmeric
(c) Clove (d) Mango ginger
(e) Fennel

Answer Sheet

1	c	2	a	3	e	4	d	5	d	6	b	7	a	8	e	9	c	10	e
11	e	12	d	13	c	14	e	15	a	16	e	17	b	18	d	19	a	20	c
21	d	22	c	23	a	24	b	25	c	26	b	27	e	28	d	29	a	30	e
31	c	32	a	33	c	34	b	35	a	36	c	37	a	38	b	39	e	40	d
41	c	42	e	43	a	44	b	45	d	46	a	47	a	48	d	49	c	50	a
51	d	52	e	53	b	54	e	55	e	56	a	57	b	58	e	59	a	60	e
61	b	62	d	63	e	64	b	65	d	66	e	67	e	68	a	69	d	70	e
71	d	72	e	73	c	74	b	75	c	76	b	77	e	78	b	79	d	80	b
81	a	82	d	83	a	84	e	85	e	86	b	87	a	88	d	89	e	90	c
91	d	92	b	93	b	94	a	95	a	96	a	97	b	98	b	99	d	100	b
101	e	102	e	103	b	104	a	105	c	106	a	107	d	108	a	109	e	110	d
111	e	112	a	113	d	114	b	115	e	116	a	117	e	118	b	119	b	120	d
121	e	122	a	123	e	124	c	125	b	126	d	127	b	128	e	129	a	130	e
131	a	132	d	133	b	134	e	135	b	136	d	137	e	138	b	139	d		

1.1.2.11. Underutilized Vegetable Crops

(1) Cork wood tree is known as ─────────────.
(a) Spinach (b) Bathwa
(c) Agati (d) None of the above

(2) *Sesbania grandiflora* is botanical name of ─────────────.
(a) Sunhemp (b) Soo babool
(c) Agati (d) None of the above

(3) ───────────── is also known as west India pea
(a) Pea (b) Cowpea
(c) Snow pea (d) Agati

(4) Agati is native of ——————————————.
    (a) Asia (b) Africa
    (c) Malaysia (d) USA

(5) Agati is extensively grown in ——————————
    (a) Punjab (b) New Delhi
    (c) Malysia (d) USA

(6) —————————————— is known as elephant ear
    (a) Alocassia (b) Colocassia
    (c) Sweet potato (d) Tapoica

(7) *Alocassia macorrrhiza* is botanical name of ——————————.
    (a) Colocassia (b) Alocassia
    (c) Sweet potato (d) Tapoica

(8) —————————————— is primarily grown in
    (a) India and Nepal (b) SriLanka and India
    (c) Eran (d) USA and Italy

(9) West Indian Arrow root, True arrow root and Jamachipeke are synonymous of ——————————.
    (a) Alocassia (b) Agati
    (c) Arrow root (d) Noe of the above

(10) Arrow root belong to family——————————————.
    (a) Solanaceae (b) Fabaceae
    (c) Euphorbiaceae (d) Marantaceae

(11) Arrow root is native of ——————————————.
    (a) Central America (b) Northern – South America
    (c) Asia (d) Indo- Burma

(12) Edible part of arrow root is——————————————.
    (a) Seed (b) Stem
    (c) Flower (d) Rhizome

(13) Sweet leaf bush is known as——————————————.
    (a) Basella (b) Curry leaf
    (c) Chekurmanis (d) None of the above

(14) *Sauropus androgynous* is botanical name of——————————————.
    (a) Drumstick (b) Curry leaf
    (c) Chekurmanis (d) None of the above

(15) Chekurmanis is highly cross- pollinated and——————————————.
    (a) Entomophyllous (b) Hydrophyllous
    (c) Anemophyllous (d) None of the above

(16) Chekurmanis is popular crop of ―――――――――.
   (a) North India (b) Central India
   (c) South India (d) Western India

(17) Chekurmanis is propogated by―――――――――.
   (a) Seed (b) Rhizome
   (c) Bulb (d) Cuttings

(18) Chekurmanis plants can be grown from the sea level to an elevation of ――――――――― M above MSL.
   (a) 1000 (b) 500
   (c) 250 (d) 100

(19) ――――――――― is known as multivitamin green leay vegetables
   (a) Spinach (b) Basella
   (c) Chekurmanis (d) Bathwa

(20) *Anethum graveolens* L. Var. sowo is botanical name of―――――――――.
   (a) Black pepper (b) Dill
   (c) Cinnamon (d) None of the above

(21) Sowa belongs to the family―――――――――.
   (a) Araceae (b) Dioscoraceae
   (c) Apiaceae (d) None of the above

(22) *Solanum verbascifolium* Linn is botanical name of―――――――――.
   (a) Ban Tamaku (b) Bhat Kataia
   (c) Local Garden egg (d) None of the above

(23) Bhat Kataia is botanically known as ―――――――――.
   (a) *Solanum indicum* (b) *Solanum incanum*
   (c) *Solanum gilo* (d) *Solanum macrocarpum*

(24) Bitter tomato is botanically known as ―――――――――.
   (a) *Solanum gilo* (b) *Solanum incanum*
   (c) *S. indicum* (d) *S. macrocarpum*

(25) *Solanum indicum* is probably originated in―――――――――.
   (a) Tropical West Africa (b) Asia
   (c) Europe (d) Indo Burma

(26) Indian night shade, Black night shade, Brihati, Kantakari and Birhatta are synonymous of―――――――――.
   (a) *Solanum gilo* (b) *Solanum nigrum*
   (c) *Solanum indicum* (d) *Solanum macrocarpum*

(27) Black night shade is native of——————————.
    (a) Asia　　　　　　　　(b) Malysia
    (c) Tropical West Africa　　(d) Europe

(28) Ground cherry and Golden berry is synonymous of——————.
    (a) Jilo　　　　　　　　(b) Indian bean
    (c) Cape goose berry　　(d) None of the above

(29) *Physalis peruviana* is known as——————.
    (a) Pepino　　　　　　(b) Cape goose berry
    (c) Naranjillo　　　　　(d) None of the above

(30) —————— is also known as garden egg.
    (a) Jilo　　　　　　　　(b) Naranjillo
    (c) pepino　　　　　　(d) None of the above

(31) Jilo is native of——————.
    (a) Central Africa　　　(b) North Africa
    (c) South Africa　　　　(d) None of the above

(32) *Solanum quitoense* Lamk is botanical name of——————.
    (a) Pepino　　　　　　(b) Naranjillo
    (c) Tomatillo　　　　　(d) None of the above

(33) Naranjillo is native of——————.
    (a) Asia　　　　　　　　(b) Africa
    (c) Peru and Columbia　　(d) None of the above

(34) *Solanum muricatun* is botanical name of——————.
    (a) Pepino　　　　　　(b) Jilo
    (c) Naranjillo　　　　　(d) Tit baigan

(35) In India, Pepino is grown in——————.
    (a) Bihar　　　　　　　(b) Uttar Pradesh
    (c) Nilgiris in TamilNadu　(d) None of the above

(36) *Solanum intrasum* is botanical name of——————.
    (a) Tit baigan　　　　　(b) Garden huckel berry / sun berry
    (c) Tomatillo　　　　　(d) None of the above

(37) Sunberry is widely grown in ——————.
    (a) Pakistan　　　　　　(b) India
    (c) Brajil　　　　　　　(d) Indonesia

(38) Tit Baigan belongs to family——————.
    (a) Chenopodiaceae　　(b) Leguminoaceae
    (c) Solanaceae　　　　(d) Euphobiaceae

(39) Botanically name of Tit Baigan is—————.
    (a) *Solanum nodiflorum*   (b) *S. deraceum*
    (c) *Solanum torvum*   (d) *Solanum ferrigineum*
    (e) C and D

(40) Tomatillo is also known as—————.
    (a) Sunberry   (b) Husk tomato
    (c) Tit baigan   (d) None of the above

(41) *Physalis ixocarpa* is botanical name of—————.
    (a) Husk tomato   (b) Tree tomato
    (c) Indian cape goose berry   (d) None of the above

(42) Tomatillo is native of—————.
    (a) Asia minor   (b) IndoBurma
    (c) Mexico and Central America   (d) Europe

(43) ————— is an important crop plant in Latin America and Hawaiian gardens.
    (a) Tit baigan   (b) Tomatillo
    (c) Sunberry   (d) None of the above

(44) Tree tomato is botanically known as —————.
    (a) *Cyphomandra fetacea*   (b) *Physalis minima*
    (c) *Physalis ixocarpa*   (d) All of the above

(45) Tree tomato is native of—————.
    (a) USA   (b) Asia
    (c) Europe   (d) Andes Mountains of Peru and Chile

(46) Tree tomato is popularly grown in—————.
    (a) Madhya Pradesh   (b) Bihar
    (c) Uttar Pradesh   (d) Uttaranchal
    (e) Nilgiris in Tamilnadu and Khasi Hills in Meghalaya

(47) Tree tomato belongs to the family—————.
    (a) Solanaceae   (b) Cucurbitaceae
    (c) Leguminoceae   (d) None of the above

(48) Tree tomato can be grown at altitude ranging from————— m MSL
    (a) 750-1800   (b) 100-200
    (c) 500-1000   (d) None of the above

(49) *Physalis minima* is botanical name of—————.
    (a) Suberry   (b) Native goose berry
    (c) Cape goose berry   (d) None of the above

(50) *Solenostemon rotundifolia* is botanical name of———————.
    (a) Chinese potato   (b) Potato
    (c) Sweet potato   (d) None of the above

(51) Chinese potato is also known as———————.
    (a) Coleus   (b) Koorka
    (c) a and b both   (d) None of the above

(52) Sree Dhara is a variety of———————.
    (a) Dill   (b) Koorka
    (c) Clove   (d) None of the above

(53) Sweet colocassia, Queens land Arrowroot and Purple arrow root are synonympus of———————.
    (a) Sweet ptotato   (b) Potato
    (c) Edible canna   (d) Chinese potato

(54) Edible canna belongs to the family———————.
    (a) Malvaceae   (b) Solanaceae
    (c) Cannaceae   (d) None of the above

(55) Indian Sorrel is also known as———————.
    (a) Roselle   (b) Water convolvulus
    (c) Water leaf   (d) water chest nut

(56) Roselle is a native of———————.
    (a) India   (b) Malayasia
    (c) China   (d) West Aferica

(57) *Hibiscuss sabdariffa* is botanical name of———————
    (a) Water spinach   (b) Water chestnut
    (c) Roselle   (d) Swamp spinach

(58) Water spinach, Swamp spinach Kang Kong and Weng Cai are synonymous of ———————.
    (a) Water convolvulus   (b) Water chest nut
    (c) Water leaf   (d) Lotus

(59) Water convolvulus belongs to the family ———————.
    (a) Cannaceae   (b) Malvaceae
    (c) Convolvulaceae   (d) Trapaceae

(60) Water chestnut belongs to the family
    (a) Chenopodiaceae   (b) Onagraceae
    (c) Araceae   (d) None of the above

(61) Water chestnut is an ———————— aquatic plant.
   (a) Annual (b) Biennial
   (c) Perennial (d) None of the above

(62) *Trapa natans* L or *Trapa bicornis* Osbeck is botanical name of————.
   (a) Lotus (b) Water leaf
   (c) Water chestnut (d) None of the above

(63) *Surinam spinach* and *Talinum cariru* are synonymous of————.
   (a) Water chest nut (b) Water leaf
   (c) Lotus (d) Roselle

(64) Water leaf is originated from————.
   (a) India (b) Burma
   (c) Malaysia (d) Brazil

(65) Water leaf belongs to the family————.
   (a) Nymphaeceae (b) Portulaceae
   (c) Onagraceae (d) None of the above

(66) East Indian Lotus belongs to the family————.
   (a) Portulaceae (b) Nymphaceae
   (c) Trapaceae (d) Solanaceae

(67) *Nelumo nucifera* is botanical name of————.
   (a) Basella (b) Water leaf
   (c) Spinach (d) Lotus

(68) *Psophocarpus tetragonolobus* is botanical name of————.
   (a) Guar (b) Winged bean
   (c) Hycinth bean (d) Sword bean

(69) Goa bean and Princess bean are synonymous of————.
   (a) Sword bean (b) Jack bean
   (c) Winged bean (d) None of the above

(70) Winged bean is originated from————.
   (a) Asia (b) Africa
   (c) Americ (d) Papua New Guinea and Indonesia

(71) Sword bean is probably native of————.
   (a) Africa (b) Brazil
   (c) China (d) Peru

(72) *Canavalia gladiata* is botanical name of————.
   (a) Pea (b) Indian bean
   (c) Sword bean (d) None of the above

(73) *Canavalia ensuformis* is also known as————————————————.
    (a) Goa bean   (b) Jack bean
    (c) Indian bean   (d) Guar

(74) Jack bean is native of————————————.
    (a) South America   (b) North America
    (c) Central America   (d) None of the above

(75) Kachariya, pathya, Sengha and Goradi are synonymous of————————.
    (a) Ivy gourd   (b) Chow-chow
    (c) Snapmelon   (d) Kachari

(76) *Cucumis callasus* is botanical name of————————————.
    (a) Muskmelon   (b) Snapmelon
    (c) Watermelon   (d) Kachari

(77) AHK-119 and AHK-200 are varieties of————————————————.
    (a) Snapmelon   (b) Kachari
    (c) Muskmelon   (d) Watermelon

(78) *Cucumis melo* var. *momordica* is botanical name of————————————.
    (a) Snapmelon   (b) Muskmelon
    (c) Snakgourd   (d) Bottle gourd

(79) Snapmelon is especially grown in————————————————.
    (a) Rajasthan   (b) Gujarat
    (c) Punjab   (d) All of the above

(80) Kakrol and kartoli are ———————————————— in nature.
    (a) Annual   (b) Biennial
    (c) Perennial   (d) None of the above

Answer Sheet

1	c	2	c	3	d	4	c	5	e	6	a	7	b	8	b	9	c	10	d
11	b	12	d	13	c	14	c	15	a	16	c	17	d	18	b	19	c	20	b
21	c	22	a	23	a	24	b	25	a	26	b	27	c	28	c	29	b	30	a
31	a	32	b	33	c	34	a	35	c	36	b	37	d	38	c	39	e	40	b
41	a	42	c	43	b	44	a	45	d	46	d	47	a	48	a	49	b	50	a
51	c	52	b	53	c	54	c	55	a	56	d	57	c	58	a	59	c	60	b
61	a	62	c	63	b	64	d	65	b	66	b	67	d	68	b	69	c	70	d
71	a	72	c	73	b	74	c	75	d	76	b	77	b	78	a	79	d	80	c

1.2. Differenciate in Between

Kakrol or sweet gourd (*Momordica cochinchinensis* Roxb)	Kartoli or spine gourd (*Momordica dioica* Roxb)
1. Roots develop bigger tuber.	Roots develop small tuber.
2. Leaves are bigger	Leaves are small.
3. Flowers large and white to light yellow in colour.	Flowers small and yellow in colour.
4. There are three small circular dot at the base of petals which are deep blue	No circular dot on the base of petals.
5. Anthesis during early morning (3.30-6.30 hours) and flowers took 72 hours to open.	Anthesis during evening (16.30-18.00 hours) and flowers took 7-22 minutes to open.
6. Fruits are large and oblong.	Fruits are small and round to oval.
7. Individual fruit weight is around 60-80g and can attain upto 500g.	Individual fruit wieght is around 10-15g and can attain upto 30g.
8. Fruit ripening start from periphery to inner.	Fruit ripening start from inner to periphery.
9. Fruit light green to light yellow in colour.	Fruits dark green in colour.
10. Tough spines on fruit.	Smooth and false spines on fruit.
11. It takes 26 days to reach edible maturity from days to bud formation.	It takes 20 days to reach edible matrurity from days to to bud formation.
12. Flowering and fruiting are short period.	Flowering and fruiting continue for longer period.

Ridge gourd	Smooth gourd
1. Six sex form are seen.	Generally monoecious form is seen.
2. Leaves are shallowly lobed and larger.	Leaves are deeply lobed and smaller.
3. No white patches on leaf.	White patches on leaf.
4. Male flower smaller.	Male flower larger
5. Disc shaped and not tomentose ovary.	Cylindrical and tomentose ovary.
6. Angled and 10-ribbed fruit.	Smooth fruits.
7. Seeds are black and pilted.	Seeds are black, white flat and not pitted.
8. Anthesis takes place in the evening.	Anthesis takes place in the early morning.

C. annuum	*C. fruitescens*
1. Annual and herbaceous	Biennial, perennial, shrubby and tall growing
2. Flower solitary pendulous	Flower in cluster, erect
3. Pedicel:Fruit ratio less than one	Pedicel:Fruit ratio more than one
4. Less pungent (0.1 per cent of capsaicin)	Highly pungent (0.22 per cent) capsaicin.
5. Diploids and polyploids are observed	Only diploids are observed

Spinach beet	Spinach
1. Botanical name is *Beta vulgaris* var. *bengalensis*.	Botanical name is *Spinacea oleracea*.
2. Chromosome number 2n=18.	Chromosome number 2n=12.
3. Spinach beet has leaves with entire margin.	Spinach beet has leaves with lobed leaf margin.
4. It produces bisexual (hermaphrodite flowers).	It may produces staminate, pistillate and/or hermaphrodite flowers.
5. It tolerates high temperature and grows well in hot weather.	It is poorly cool season crop and cannot tolerate high temperature. In warm season and long days it quickly tends to flower.

Phaseolus lunnatus	*Vigna savi*
1. Stipules not prolonged.	Stipules prolonged.
2. Keel in a spiral.	Keel erect, in curved.
3. Styles without apical appendage.	Style with distinct beak.
4. Stipules truncate at base.	Stipules cordate.
5. Keel spirally twisted.	Keel usually straight.
6. Fruit not septate.	Fruit septate.
7. Fruit pair of leaves, petioles.	Fruit pair of leaves, sessile.
8. Fine sculpture pollen grains.	Open reticulation of pollen.
9. Type spp. *P. vulgaris* L.	Type spp. *V. luteola* Jacq.

Garden Pea	Field pea
1. Botanical name is *Pisum sativum* var. *arvense*.	Botanical name is *Pisum sativum* var. *hortense*.
2. In this type young, green seeds are used mostly in vegetables and also for canning purpose.	They are also grown as forage or green manuring purpose.
3. Seeds are bold and wrinkled.	Seeds are rounded and little angular.
4. These are generally white flowered.	They have generally coloured flowers.
5. Leaf axils are generally green.	Leaf axils are often pigmented.
6. Seeds are yellowish, whitish or bluish green.	Seeds are greyish green, greyish brown or greyish yellow.

1.3. Fill In The Blanks

(1) _____ temperature, especially _____ bolting in spinach.

(2) _____ and _____ are non-parasitic troubles of chilli that sometimes cause considerable loss.

(3) _____ is an F_1 hybrid of cucumber recommended by I.A.R.I.

(4) _____ is the most ancient type amongst the various types of gardens.

(5) _____ seeded varieties of garden pea are sweeter.

(6) _____ skinned tubers of sweet potato generally store better than_____ skinned ones.

(7) _____ disease of cabbage can be controlled by hot water treatment as in _____.

(8) _____ gardens are situated near some town or city market.

(9) _____ in turnip is due to deficiency of boron.

(10) _____ is a giant-leaved cultivar of *palak*.

(11) _____ is a wrinkle-seeded variety suitable for early season.

(12) _____ is considered to be the place of origin of the watermelon.

(13) _____ is more important than _____ in seed stalk development.

(14) _____ is probably the place of origin of dolichos bean.

(15) _____ is the German name for cabbage-turnip.

(16) _____ is the most common leafy vegetable grown in India.

(17) _____ is the most common storage disease of onion bulb.

(18) _____ is the most commonly grown variety of green sprouting broccoli.

(19) _____ is the most serious virus disease of tomato in India.

(20) _____ is the non-bulb forming member of the onion family and is grown for its branched _____.

(21) _____ is the single typical member of cucurbitaceae family having single seed in fruit.

(22) _____ types of flowers have been described in brinjal.

(23) _____ variety is resistant to YVMV and is recommended for sowing in the rainy season.

(24) _____ variety of tomato is pear shaped and withstands long transportation.

(25) _____, a cucurbitaceous crop, is considered as perennial vegetable.

(26) _____ and training of tomato plants is a common practice in certain parts of USA.

(27) _____ cultivar of beet leaf is suitable for hills.

(28) _____ (*Bari chaulai*) and _____ (*Chhoti chaulai*) are grown as leafy vegetable.
(29) _____ fruits of wax gourd are harvested for vegetable whereas _____ fruits are taken for preparing sweet known as _____.
(30) _____ humidity and _____ temperature are required for the storage of beet leaf.
(31) _____ and _____ cause flower drop and poor fruit set in brinjal.
(32) _____ is an improved variety of round gourd developed by IIHR, Bangalore.
(33) _____ of tomatoes is caused by *Stemphylian solani* Weber and *Alternaria solani* Ellis.
(34) _____ is a seed less variety of watermelon.
(35) _____ and _____ are improved varieties of wax gourd.
(36) _____ is the most important insect-pest of sweet potato.
(37) _____ a winter squash variety, has been recommended by I.I.H.R., Bangalore.
(38) _____ developed fruits of tomato are seedless.
(39) _____ vegetables contain less of vitamin C than fresh ones.
(40) _____ types of turnip are relatively sweeter and more palatable.
(41) _____ is one of the most dangerous insects causing virus disease in potato.
(42) _____ tubers of sweet potato store better than the white-skinned ones.
(43) _____ can tolerate cool climate better than musk melon and snap melon.
(44) _____ is the variety of okra.
(45) A cauliflower crop often shows deficiency symptoms of _____ and _____ when grown either on an alkaline or highly acidic soil.
(46) A daily supply of _____ kg of fresh vegetables can be assured from a kitchen garden of the size of 50 sq.m.
(47) A hermaphrodite variety known as _____ available in ridge gourd.
(48) A pre-harvest spray of potato with _____ controls sprouting.
(49) A relatively _____ temperature as well as _____ photoperiod is essential for bulb formation in most of the commercial varieties of onion grown in India.
(50) *Abelmoschus* genus is distinct from *Hibiscus* in having a deciduous type of calyx whereas it is _____ in the genus *Hibiscus*.
(51) About _____ quintals per hectare of seed yield of Indian bean can easily be obtained.
(52) According to dieticians, an individual should consume about _____ g of leafy and other vegetables and _____ g of root vegetables daily for a balanced diet.

(53) All beans except _____ are susceptible to frost and are grown as a summer crop.
(54) All cole crops have developed from wild cliff cabbage known as _____ from which the name cole is derived.
(55) All the vegetables come under the sub-community _____.
(56) All vitamins are found in small or large quantities in common _____ crops.
(57) Although a ripe tomato is 94 per cent water, it is a good source of vitamin _____, and an excellent source of vitamin C.
(58) Amaranthus has _____ photosynthetic cycle.
(59) Among diseases, _____ is an important disease causing considerable loss to colocasia.
(60) Among the most important states of India, only four viz, _____, _____, _____ and _____ account for three-fourths of the total area under chilli.
(61) Among the pole types of French bean _____ is the most commonly grown variety.
(62) An edible podded variety of pea is _____.
(63) Arka Chandan is the cultivar of _____.
(64) Artichoke, commonly known as globe artichoke, is a perennial crop grown for its.
(65) Ash gourd (*Benincasa hispida* Cong.) is considered to be of great _____ value.
(66) Asparagus is cultivated for its tender shoots commonly known as _____.
(67) Asparagus plants are _____ in sex expression.
(68) Asparagus starts yielding a sizeable crop after about _____ years and gives an economic yield for about _____ years.
(69) At a temperature of about _____, tuber production stops totally.
(70) Based on their temperature requirements, all vegetables are roughly placed in two groups viz,. _____ and _____ vegetables.
(71) Beet leaf is sown _____ times in a year.
(72) Beetroot belongs to the family _____.
(73) Beetroot can be stored at _____ temperature and _____ relative humidity.
(74) Best curing of tubers of sweet potato is done at a temperature of _____ and relative humidity of _____ maintained for ten days.
(75) Best pod set in French bean is obtained with plants kept at _____ °C for four hours after pollination.
(76) Best temperature and humidity conditions for storing potatoes in cold storage are _____ °C and _____, respectively.

(77) Bitter gourd has heating and _____ effect on the body.
(78) Bitter gourd is generally considered to be an old-world species with its native home in the _____ and Asia.
(79) Bitter gourd responds well to training on _____ during rainy season.
(80) Bordeaux mixture (4 : 4: 50 or 4 : 2 : 50), _____ and Copramat check defoliation in tomatoes.
(81) Boron deficiency causes _____ in sugar beet.
(82) Botanical name of long melon is _____ var _____.
(83) Botanical name of round melon is _____ var. _____.
(84) Botanical variety _____ of *Raphanus sativus* is exclusively grown for its long thin pods.
(85) Botanically bitter gourd is known as _____.
(86) Bottle gourd cannot tolerate _____.
(87) Bottle gourd might have originated in _____.
(88) Brinjal crop yielding _____ of produce removes from soil 175 + 40 + 300 kg/ha of N + P_2O_5 + K_2O and 30 + 10 kg/ha of MgO and sulphur, respectively.
(89) Brinjal grows well in _____.
(90) Brinjal is as _____ in nature but is grown as _____.
(91) Brinjal is a native of _____.
(92) Broad bean is an important food crop in _____.
(93) Broccoli and cauliflower are grown for their _____ as edible parts.
(94) Broccoli is an Italian word from the Latin _____ meaning an arm or branch.
(95) Brussels sprout gets its name from the city of Brussels in _____.
(96) By following succession cropping and inter-cropping _____ of land may be made to supply adequate vegetables for an average family of five members.
(97) Cabbage is a _____ pollinated crop.
(98) Cabbage is heavy feeder of nitrogen and _____.
(99) Cabbage, cauliflower, broccoli, *palak*, okra, onion and muskmelon are slightly tolerant to acidic soils with a pH range of _____.
(100) Carrot belongs to the family _____.
(101) Carrot is an excellent source of vitamin _____ and is rich in sugar.
(102) Carrot is cultivated as _____ for its roots and _____ for seeds.
(103) Carrot, bitter gourd, onion and tomato comprise good source of _____.
(104) Cauliflower has a high requirement of boron and _____.
(105) Cauliflower is sensitive to high _____.

(106) Cauliflower with leaves attached can be stored for 30 days _____ at _____ °C and _____ per cent R. H.
(107) Celery is a biennial crop and its seeds are produced in _____.
(108) Celery thrives well in a soil with pH of _____.
(109) Chayote is _____ but its vines are _____.
(110) Chillies are rich in vitamins, especially in vitamins _____ and _____.
(111) Cluster bean is grown at a spacing of _____ cm.
(112) Coccinia, a semi-perennial cucurbit, is propagated by _____ 25-30 cm long and 1.5-2.0 cm thick.
(113) Commercial triploid seeds are available in some advanced countries like.
(114) Cowpea is a _____ season crop.
(115) Cowpea pods, when they are not picked at the right stage become.
(116) Cracking of the skin of tomato fruits has been associated with deficiency of.
(117) Cucumber can tolerate strongly _____.
(118) Cucumber grows best at a temperature between _____ and _____.
(119) Cucumber grows well at a row to row spacing of _____ m and plant to plant spacing of _____ cm.
(120) Cucumber is one of the oldest _____ vegetable crops having its origin probably in India.
(121) Cucumber mosaic is readily transmitted by _____.
(122) Cucurbits are _____.
(123) Cultivated cowpea consists of three main groups namely (i) _____ (ii) _____ and (iii) _____.
(124) Cuttings from _____ portion of sweet potato vines should be preferred for planting.
(125) Deficiency of molybdenum is the main cause of _____.
(126) Dieticians recommend inclusion of _____ grammes of vegetables in our daily diet.
(127) Dormancy of potato tubers can be broken by treating the tubers with _____.
(128) Double bean is a native of _____ and abundantly grown by American Indians.
(129) Drying of the _____ at the stem end of watermelon fruit is taken as a sign of maturity.
(130) During dry weather, round gourd should be irrigated at _____ days' interval.
(131) Each cutting of sweet potato should have at least _____ nodes.
(132) Early sowing in cauliflower is done from _____ to _____.
(133) Early varieties of cauliflower are those which produce curds in the plains of northern India from _____ to _____.

Vegetable Production 193

(134) Edible fruited species of the genus *Lycopersicon* are _____ and _____.
(135) Egg-plant (brinjal) contains vitamins A and B and is an _____ vegetable.
(136) Elephant's foot botanical name is _____.
(137) Elephant's foot is common in _____ in valleys of Tapti, Purna and Ambika.
(138) Every adult should consume at least _____ gm of vegetables per day.
(139) Exclusion of light from stalks while plants are still growing makes them devoid of chlorophyll, and is referred to as _____.
(140) Faba bean is tolerant to water stress due to higher _____ accumulation.
(141) Faba bean reached India probably from _____ and was grown in north.
(142) Fenugreek crop should be protected well from the incidence of _____ mildew.
(143) Fenugreek is a legume crop which fixes _____ from the _____.
(144) Fenugreek is a rich source of _____, _____, _____ and _____.
(145) Fibrous framework of leaves, stems, bulbs, tubers and roots of vegetables yields _____ which satisfies appetite and prevents constipation.
(146) First earthing-up in potato should be done when the plants are about _____ in height.
(147) For better fruit-set in tomato, application of a mixture of _____ urea and 2, 4-D at _____ ppm as whole plant spray when the first few flower clusters appear has been found most effective and economical.
(148) For economic production of hybrid seeds of onion, _____ lines are used as female parent.
(149) For fixing the dates of sowing and harvest, the help of the knowledge of _____ or _____ is taken.
(150) For protecting tomato seedlings in the field from frost, cover them with polythene bags of the size _____ and of _____ gauge thickness.
(151) For seed production of radish in the plains, _____ type of cultivars should be selected.
(152) For spreading varieties of brinjal, row to row distance should be _____ cm and that of plant to plant _____ cm.
(153) Formation of new roots in plants of beans and cucurbits is _____ and there is a tendency for the roots to be _____ in these plants, which make them less effective in absorbing water.
(154) French bean is probably a native of _____.
(155) French bean, an ancient crop, is probably a native of _____.
(156) Fruits of chilli hybrid variety _____ released by P.A.U. are very suitable for salad and drying.
(157) Fruits of pointed gourd should be harvested at when they are _____ and _____.

(158) Fruits of round gourd are harvested at _____ stage.
(159) Fruits of sponge gourd should be harvested when they are _____.
(160) Ganga, Kaveri and Yamuna are _____ of cabbage.
(161) Garlic belongs to the family _____.
(162) Garlic is _____ season crop.
(163) Garlic is propagated by _____ which are detached individually from the bulb.
(164) Garlic is supposed to have originated from Central Asia and _____ regions.
(165) Generally, the yield of fresh green chillies is _____ times higher than that of dry chillies.
(166) Generally, _____ is not harmed by downy mildew but the disease causes much damage to _____.
(167) Germination percentage of peas is reduced if the moisture percentage of seeds increases _____ during storage period.
(168) Gibberellic acid at higher concentrations induces _____ but at lower concentrations of 10-25 ppm increases _____.
(169) Globe artichoke is propagated by means of _____ from the old root-stocks.
(170) Golden Acre is a _____ or _____ type.
(171) Green spears of asparagus have _____ nutritive value than blanched spears.
(172) Growing of vegetables out of their normal season is known as _____.
(173) Harvested brinjal fruits can be kept for seven to ten days in good condition at _____ °F and _____ per cent relative humidity.
(174) Heading type of broccoli is more like _____.
(175) Heavy irrigation to cabbage after development of heads causes _____ within 24 hours.
(176) Henderson Bush is an important variety of _____.
(177) High _____ is favourable for growth and development of plants and fruit formation of snake gourd.
(178) Higher seed yield in radish is expected from _____ method.
(179) Hill potatoes are nearly free from _____ pathogen.
(180) I.A. R.I. has released an F_1 hybrid _____ of summer squash.
(181) If tomato fruits are exposed to intense sunlight, it may cause a disorder known as _____.
(182) Immature tomato blossoms drop rapidly during _____ due to increased transpiration.
(183) In _____, after first shower, when shoots come out from root tubers, it is the best planting time for sweet gourd in Assam.

(184) In _____ method of seed production in cabbage, selected true-to-type fully matured plants are uprooted during November-December and are reset in new beds.
(185) In Bihar, Himachal Pradesh and Nillgiri hills of India, potato is grown as a _____ crop.
(186) In early varieties of tomatoes, concentric cracking, _____ is more common owing to their light, open foliage.
(187) In floating gardening, a floating base is first made from the roots of _____ grass which grow wild in some parts of the Dal Lake.
(188) In floating gardening, all intercultural operations and occasional sprinkling of water are done from _____.
(189) In hills, the best time for sweet potato planting is during _____.
(190) In India, potato has been cultivated since its introduction in the early part of _____ century.
(191) In most of the cucurbits, an ideal variety is one which has _____ female to male flower ratio.
(192) In Nasik, division of Maharashtra, Ootacamund region of Tamil Nadu and parts of Kerala, cabbage is grown during _____ also.
(193) In Nilgiri hills and Tamil Nadu, potato is rotated with _____ in alternate years.
(194) In northern India, vine cuttings of sweet potato are planted during _____.
(195) In *palak* _____ pollination is effective.
(196) In parts of South America, sweet potato is also called _____ or _____.
(197) In radish, higher genetic purity can be maintained when seed production is done by _____ method.
(198) In the plains of India, for the *rabi* crop, seeds are sown from _____ to _____.
(199) In the plains of India, French beans are sown in _____ for the autumn crop and in _____ for the spring-summer crop.
(200) Rainy season vegetables are generally are sown in _____.
(201) Indian bean is grown throughout tropical regions of Asia, _____ and America.
(202) It is believed that small seeded pea originated from _____, and the large seeded ones from Mediterranean.
(203) It is estimated that potato in Nilgiri Hills of Tamil Nadu was introduced in _____.
(204) Ivy gourd is propagated by _____.
(205) Jerusalem artichoke is cultivated for its _____.
(206) June-July sown cauliflower should be transplanted in new beds before their final transplantation into the _____.

(207) Kakrol seed contains oil which is used as an _____.
(208) Kartoli is propagated by _____.
(209) Knol-khol (*Brassica oleracea* var _____) is little known in India.
(210) Late varieties of cauliflower give curds from _____ to _____.
(211) Leaves of amaranths are _____ in nature.
(212) Leaves of Swiss chard are _____ than those of beet leaf.
(213) Leek belongs to the species _____.
(214) Leek is included in the group of _____.
(215) Lettuce belongs family _____.
(216) Lettuce does well in a relatively cool growing season with a monthly average temperature of _____.
(217) Lettuce mosaic is transmitted through _____.
(218) Lettuce seed does not germinate properly when the soil temperature is above _____.
(219) Long melon is usually sown in _____ and over wintered under artificial protection from frost.
(220) Long melon thrives best in _____ and _____ climate.
(221) Long melon variety Selection-3 is free from _____ principles.
(222) Main crop of potato in hills of Himalayan region is planted in _____.
(223) Manjari Gota is a variety of brinjal recommended for _____.
(224) Methi is _____ season crop.
(225) Minimum monthly average temperature requirement of tomato and sweet pepper is _____.
(226) Most commonly occurring disease in nursery is _____.
(227) Most economic utilization of space can be obtained by making use of fence on the periphery for training _____ during the summer and and rainy season and _____ in winter.
(228) Most ideal variety of radish grown in summer and rainy season only is _____.
(229) Most of the cucurbits are _____ in sex expression and a few are _____.
(230) Most of the French bean varieties are _____.
(231) Most of the members of cucurbitaceae family contain _____.
(232) Mostly orange colour in tomato is due to the presence of _____ which is precursor of _____.
(233) Muskmelon is a native of _____ India.
(234) Muskmelon is planted from _____ in plains and _____ in hills.

(235) Muskmelon variety _____ is grown in Afghanistan and is available in India in October-November.
(236) No vegetables belongs to the division _____.
(237) Normally, snake gourd is trained on _____, _____ and _____.
(238) Nursery of lettuce is sown in _____.
(239) Odour is garlic is due to _____.
(240) One of the most important onion growing states in India is _____.
(241) Onion and other bulb crops belong to the family _____.
(242) Onion bulbs and green onion are good source of vitamin _____.
(243) Oniond does not thrive well in areas receiving more than _____ of annual average rainfall.
(244) Onion is propagated by _____ and _____ as well.
(245) Onion seedlings are transplanted at a spacing of _____ cm.
(246) Onion seedlings for *kharif* crop are transplanted in _____ and produce good bulbs in _____ under North Indian conditions.
(247) Onion, okra, asparagus and summer squash supply _____.
(248) Only _____ and _____ styled brinjal flowers marked by a swollen ovary at the base, bear fruits.
(249) Optimum pH range for growth of okra is _____.
(250) Optimum pH range for onion cultivation is between _____ and _____.
(251) Optimum temperature for growth of muskmelon is about _____.
(252) Outbreak of late blight disease in potato is commonly noticed in _____, _____ and _____ weather.
(253) Pepo is a type of berry in which the outer wall, which is the _____, becomes hard as in curbitaceae family.
(254) Per capita estimated consumption of potato in _____ is about 13 kg per annum.
(255) Perennial and annual types of Indian beans are known as Typicus and _____, respectively.
(256) Perkin's Long Green is recommended for _____ areas.
(257) Pink colour of skin of the radish root is due to _____.
(258) Planting root materials are usually collected from old planting of pointed gourd during _____.
(259) Pointed gourd cannot be grown successfully in _____ climate.
(260) Pointed gourd is a _____ cucurbitaceous crop.
(261) Pointed gourd is a _____, therefore, male and female plants are _____.

(262) Pointed gourd is a native of India and _____ is considered to be the primary centre of its origin.
(263) Pointed gourd is easily digestible, diuretic, laxative, envigorates the _____ and is useful in disorders of the circulatory system.
(264) Pointed gourd is mainly propagate by _____ and _____ cuttings.
(265) Pointed gourd is one of the rare cucurbits which is propagated by _____.
(266) Poor _____ application may result in bolting in onion.
(267) Potato belongs to the family _____.
(268) Potato crops remove large amount of potassium followed by _____.
(269) Potato is a native of South America and was cultivated by _____.
(270) Potato is mainly propagated by _____ and _____ techniques.
(271) Potato requires _____ conditions for its tuber development.
(272) Potato stored at -1 to 0°C suffers from internal break-down known as _____.
(273) Potato was introduced in Europe by the early _____ exporters during sixteenth century.
(274) Potato, when grown in alkaline soils, is attacked by _____ disease.
(275) Pre-harvest foliage sprays of _____ may be helpful in controlling the sprouting of onion in storage.
(276) Premature seeding and or failure of the leaves to form a _____ are common defects of early cabbage.
(277) Pungency in radish is due to _____ Punjab Selection, Pusa Ratnar and Arka Kalyan are the varieties of _____ onion.
(278) Pusa Bedana is a seedless variety of watermelon obtained as hybrid between _____ and _____.
(279) Pusa Chetki variety of radish is suitable for sowing from _____ to _____ in the plains.
(280) Pusa Dofasali variety of cowpea is suitable for growing in both _____ and _____ seasons.
(281) Pusa Early Bunching and Kasuri Selection are the cultivars of _____.
(282) Pusa Himani variety of radish is suitable for sowing from _____ to _____ in the plains.
(283) Pusa Lal variety of sweet potato is a selection from a Japanese variety _____.
(284) Pusa Rituraj can be grown both in _____ and _____.
(285) Pusa Sadabahar and Pusa Naubahar are the varieties of cluster bean mainly grown for _____ purposes.
(286) Pusa Sanyog is a _____ of cucumber.
(287) Radish is found growing wild in _____ region.

Vegetable Production   199

(288) Rhubarb grows best in _____ climatic conditions.
(289) Rhubarb is grown for its _____.
(290) Rhubarb originated from the colder parts of Asia, probably _____.
(291) Ripe watermelon fruit when thumped with finger(s) gives out a _____ sound as against metallic and ringing sound by an immatured one.
(292) Ripening of muskmelon fruits begins first from the _____ accompanied by a change in skin colour.
(293) River-bed cultivation of ridge gourd requires _____ irrigations.
(294) River-bed cultivation of wax gourd usually requires _____ watering than upland cultivation.
(295) Round gourd prefer to grow in has _____ soil rich in _____.
(296) Row to row and plant to plant spacing of _____ cm is kept for the early varieties of cauliflower.
(297) S-48, Arka Pragati and Hissar-2 are the varieties of white _____.
(298) Scientific name of potato tuber moth is _____.
(299) Seed is generally raised from the _____ crop.
(300) Seed production of _____ type of turnip cultivars should be done in plains and for _____ types it should be done in the hills.
(301) Seedlings of lettuce are transplanted at a spacing of _____.
(302) Seeds of Brussels sprout are only produced in _____ at an altitude of _____ m and above.
(303) Seeds of leek are produced in India at _____ in the hills.
(304) Seeds of okra will not germinate below _____.
(305) Sex expression in snake gourd is _____ and in pointed gourd is _____.
(306) Short days are beneficial for _____ in potato.
(307) Single fruit of *palak* contains _____ seeds.
(308) Snake gourd is found growing wild in India and the Indian _____ is thought to be its source of origin.
(309) Some of the varieties of _____ bean are used for extraction of gum.
(310) Spacing recommended for dwarf varieties of tomato is _____ cm.
(311) Specialized gardens away from the market but having good means of transport are known as _____.
(312) Spinach is a _____ season crop.
(313) Spinach is cultivated as an _____ for its leaves and as a _____ for obtaining seed.
(314) Sponge gourd is probably indigenous to _____.
(315) Spraying of tomato plants at _____ stage using 25-50 ppm GA_3 improves the quality of fruits.

(316) Sprouts of Brussels sprout resemble _____ borne on the axils of leaves.
(317) Squash melon fruits become ready for harvesting after _____ days of sowing.
(318) Squash melon is believed to have originated in _____.
(319) Summer crop of okra is sown from _____ to _____.
(320) Summer squash is also known as _____.
(321) Summer squash like pumpkin is of _____ origin and was first of the squashes to be introduced in India from Europe.
(322) Sweet potato contains 16 and 4 per cent of starch and sugar, respectively that is, _____ per cent alcohol producing material.
(323) Sweet potato cuttings are planted at a spacing of _____ cm in rows _____ cm apart.
(324) Sweet potato is believed to have originated in _____.
(325) Sweet potato is one of the most _____ resistant vegetable.
(326) Sweet potato is very tolerant to _____ soil.
(327) Tetraploid watermelon is produced by treating the seedlings with _____.
(328) The _____ of pumpkin are more nutritive than _____.
(329) The age of transplanting asparagus is _____ old seedlings or crown.
(330) The botanical name of common *sitaphal* grown throughout India is _____.
(331) The botanical name of watermelon is _____.
(332) The cole crops which are grown for the consumption of leaves or stem are _____ and _____.
(333) The colour of the outer skin of onion bulb is due to _____.
(334) The commercial production of cabbage is done by _____ method.
(335) The critical stages for irrigation in onion are _____ and _____.
(336) The crops which are highly tolerant to acidic soils (pH 6.8 to 5.0) are _____ and _____.
(337) The cultivated carrot probably originated in the hills of Punjab and Kashmir, with a secondary centre of distribution in _____ and North Africa.
(338) The defects noticed in cauliflower are premature heading, _____ of plants and production of undersized head.
(339) The flowers of ridge gourd open in the _____ while those of the sponge gourd do so in the _____.
(340) The fruit of round gourd _____ effect and contain vitamin A.
(341) The fruits of ivy gourd have _____ keeping quality.
(342) The green leaves of fenugreek are used as _____ and also as _____.

(343) The green pods of cluster bean are rich source of _____, C, and _____.
(344) The important viral diseases which attack chilli are _____, _____ and _____.
(345) The leaves or seed stalks of onion fall down from the point of attack by _____.
(346) The mature corms of elephant's foot store well under _____.
(347) The most favourable range of pH for pea cultivation is between _____ and _____.
(348) The most important viral disease of okra is _____.
(349) The most serious disease of cluster bean is _____.
(350) The most serious insect-pest of cucumber is _____.
(351) The most serious virus disease of bean is _____.
(352) The name cauliflower has originated from the Latin words _____ and _____ which means cabbage and flower, respectively.
(353) The netted varieties of muskmelon are commonly known as _____ in the U.S.A.
(354) The onion originated from the region comprising N-W India, Afghanistan, Tajikistan, Uzbekistan (former USSR) and _____.
(355) The origin of potato is _____.
(356) The original home of okra is _____.
(357) The original home of radish is probably _____ and _____.
(358) The original race of pea came from _____.
(359) The place of origin of celery extends from Sweden to Algeria, _____ and in Asia to Caucasus, Beluchistan, Egypt and the mountains of India.
(360) The potato variety Kufri Kuber has been developed at Potato Breeding Station (now CPRI) Shimla in _____.
(361) The pungency in onion is due to the presence of volatile oil _____.
(362) The pungency is garlic is due to _____.
(363) The rat-tail radish is grown for _____ which are used as _____.
(364) The red colour in tomato is due to pigment, _____.
(365) The seed of watermelon does not germinate satisfactorily below _____.
(366) The seed rate of celery is about _____ g per hectare.
(367) The seed rate of garlic varies from 340 to 570 kg/ha of _____.
(368) The seed rate of knol-khol to raise seedlings for one hectare is _____ grammes.
(369) The seed rate of lettuce for nursery and direct sowing is 500-750 and _____ g, respectively.

(370) The seed rate of radish is _____ kg/ha.
(371) The seeds of snake gourd and bitter gourd take _____ days to germinate.
(372) The tomatoes crops supplied with rich nutrition fail to produce _____.
(373) The tubers of potato are borne underground at the _____.
(374) The type of vegetable garden known as _____ is seen on the Dal Lake of the Kashmir Valley.
(375) The usual size of seed piece in potato is _____ gm with _____ mm diameter.
(376) The usual spacing for chillies in northern India is _____ cm.
(377) The usual spacing for khol-rabi is kept about _____ from row to row and _____ from plant to plant.
(378) The vegetables commonly grown under glasshouses or glass frames are _____ and _____.
(379) The word truck in truck gardening has no relationship with a truck but is derived from a French word _____ meeting _____.
(380) The yield of radish is _____ Q/ha.
(381) The young plants of potato grow best at a temperature of _____ and later growth is favoured at a temperature of _____.
(382) There are _____ types or groups of lettuce.
(383) There are at least _____ types of beans.
(384) There are two types of tomatoes. (i) _____ (ii) _____ type.
(385) Thorough curing of onion bulbs for _____ weeks is required before being placed in storage.
(386) To ensure proper fruit-set, about _____ per cent cuttings from male plants are planted at random in between the female plants of pointed gourd.
(387) To sow one hectare area of knol-khol about _____ kg of seed is required for raising seedlings.
(388) Tomato fruits for canning are picked when they are _____.
(389) Tomato seed sufficient to raise one hectare area is _____ gram.
(390) Tomato was perhaps introduced into India by the _____ though there is no definite record of when and how it came to India.
(391) Tomatoes do best in a soil that has a soil reaction with pH from _____ to _____.
(392) Transplanting of leek is done in _____ about 30 to 45 cm deep.
(393) Transplanting of onion seedlings earlier than _____ gives more number of bolters.
(394) Tuber production in potato is maximum at a temperature of _____ and decreases with the rise of temperature.

(395) Turnip greens are good source of minerals such as _____ and iron and vitamin A.
(396) Under ordinary conditions, round gourd fruits can be kept for _____ days only.
(397) Usually, carrot is grown at a distance of _____ cm between two rows and _____ cm between plant to plant.
(398) Usually, radish is grown at a distance of _____ cm between plants and _____ cm between rows.
(399) Vegetable crops occupy only about _____ per cent of the total cultivated area of the country.
(400) Vegetable growing is generally more lucrative than any other type of _____.
(401) Vegetables viz., fenugreek, spinach and mountain spinach are rich in _____.
(402) Vitamin A is abundantly found in _____ vegetables.
(403) Watermelon belongs to the species _____.
(404) Watermelon is grown in USA since _____.
(405) Watermelon is one of the few vegetables that grow well on a soil having a pH even upto _____.
(406) Wax gourd is popularly known as _____.
(407) Welsh onion (*Allium fistulosum*) is probably of _____ origin.
(408) West is Indian bean is known as _____.
(409) Western Asia and area around _____ are onion's secondary centres of development.
(410) When colour of ground spot (white spot) where the watermelon fruit rests on the ground assumes _____, it is indicative of its maturity.
(411) When muskmelon fruit on ripening part completely and easily from the stem leaving a circular depression, it is said to be at _____.
(412) Where possible, a _____ garden is preferred to a square one.
(413) White Iscle, Scarlet Globe and Pusa Himani are _____ of varieties of radish.
(414) White Vienna, Purple Vienna and King of North are the cultivars of _____.
(415) Wild forms of pointed gourd are found throughout _____.
(416) Wild radishes are found in _____.
(417) Wine is also made from _____ besides grapes.

Answer Sheet

(1) High, long day
(2) Blossom-end rot, sunscald
(3) Pusa Sanyog
(4) Home/kitchen garden
(5) Wrinkle
(6) Red, white
(7) Black leg, black rot
(8) Market
(9) Brown heart
(10) Pusa Jyoti
(11) Arkel
(12) Africa
(13) Temperature, day length
(14) India
(15) Khol-rabi
(16) Beet leaf
(17) Black mould
(18) Calabreeze
(19) Leaf curl
(20) Leek, stem and leaves
(21) *Chyote*
(22) Four
(23) Punjab-8
(24) Punjab Chhuhara
(25) Chow-chow
(26) Pruning
(27) Pusa Harit
(28) *Amaranthus tricolour, A. blitum*
(29) Immature, mature, *petha*
(30) Low, cool
(31) High temperature, dry wind
(32) Arka Tinda
(33) Defoliation
(34) Pusa Bedana
(35) Co 1, Co 2
(36) Sweet-potato weevil (*Cylas formicarius*)
(37) Arka Suryamukhi
(38) Parthenocarpically
(39) Wilted
(40) European
(41) *Myzus persicae*
(42) Red-skinned
(43) Long melon
(44) Pusa Savani
(45) boron, molybdenum
(46) 1.5
(47) Satputia
(48) maleic hydrazide
(49) high, long
(50) persistant
(51) 60-80
(52) 115, 70
(53) broad bean
(54) coleworts
(55) Spermatophyta
(56) vegetable
(57) A&B
(58) C_4
(59) Colocasia blight
(60) A.P., Maharashtra, Mysore, T.N.
(61) Kentucy Wonder
(62) Mithi Phali
(63) pumpkin
(64) flower buds
(65) medicinal
(66) spears
(67) dioecious
(68) 3, 10-15
(69) 30°C

(70) winter, summer
(71) several
(72) Chenopodiaceae
(73) 0°C, 90%
(74) 80°C, 30%
(75) 15-25
(76) 2.2, 75
(77) wormicidal
(78) topical
(79) bower
(80) Dithane Z-78
(81) heart rot
(82) *Cuccumis melo, utilissimus*
(83) *Citrullus vulgaris, fistulosus*
(84) *caudatus*
(85) *Momordica charantia*
(86) frost
(87) Africa
(88) 60 t/ha
(89) long warm season
(90) perennial, annual
(91) India
(92) South America
(93) flower parts
(94) Brachium
(95) Belgium
(96) 250 sqm
(97) cross
(98) Potassium
(99) 6.0-6.8
(100) Cruciferae
(101) A
(102) annual, biennial
(103) Iron
(104) molybdenum.
(105) acidity
(106) 0, 85-90
(107) hills
(108) 5.5-6.7
(109) perennial, annual
(110) A, C
(111) 45-60 x 20-30
(112) stem cutting
(113) Japan
(114) warm
(115) puffy
(116) boron
(117) acidic soil
(118) 18-24°C
(119) 1.5-2.5, 60-90
(120) cultivated
(121) aphids
(122) vine crop
(123) yard long bean, catjung, southern pea
(124) upper
(125) whiptail
(126) 285
(127) thiourea
(128) South America
(129) tendril
(130) 8-10
(131) four
(132) mid May, end of June
(133) mid-September, mid-November
(134) *esculentum, pimpinelifolium*
(135) appetizing
(136) *Amorphophallus compannulatus*
(137) Gujrat
(138) 100-150
(139) blanching
(140) protein

(141) east
(142) powdery
(143) nitrogen, atmosphere
(144) mineral, protein, Vit A, Vit C
(145) fibrous mass
(146) 15-25 cm
(147) 1%, 1-2
(148) male sterile
(149) degree days, heat units
(150) 25 x 35 cm, 100
(151) tropical
(152) 75-90, 60-70
(153) slow, suberized or cutinized
(154) South and Central America
(155) South America
(156) CH-1
(157) young, tender
(158) very tender
(159) tender
(160) hybrids
(161) amaryllidaceae
(162) cool
(163) cloves
(164) Mediterranean
(165) 3-4
(166) watermelon, musk melon
(167) above 90
(168) maleness, femaleness
(169) suckers/off shoots
(170) round head, ball head
(171) more
(172) Vegetable forcing
(173) 10-13, 85-90
(174) cauliflower
(175) bursting/splitting
(176) Lima bean
(177) humidity
(178) seed-to-seed
(179) charcol rot
(180) Pusa Alankar
(181) Sun-scalding
(182) heat and drought
(183) April-May
(184) head-to-seed
(185) rainfed
(186) netting and russeting
(187) typha
(188) boats
(189) April-May
(190) 17th century
(191) high
(192) *kharif*
(193) finger millet
(194) June-July
(195) wind
(196) Camote, Kumara
(197) root-to-seed
(198) mid-October, end November
(199) July-September, February-March
(200) June-July
(201) Africa
(202) South Asia
(203) 1822
(204) stem cutting
(205) oblong tubers
(206) field
(207) illuminant
(208) tuberous root
(209) gongylodes
(210) mid-January, April
(211) perishable
(212) bigger

(213) *porrum*
(214) bulb
(215) *compositeae*
(216) 12-15°C
(217) seeds
(218) 30°C
(219) November
(220) tropical, sub-tropical
(221) bitter
(222) March-April
(223) Maharashtra
(224) cool
(225) 20-22°C
(226) damping off
(227) cucurbits, peas
(228) Pusa ehetaki
(229) monoecious, dioecious
(230) day neutral
(231) cucurbitacin
(232) b-carotene, Vit A
(233) north-west
(234) December to March, April to May
(235) Sarda Melon
(236) gymnospermae
(237) stakes, *jhala*, thatch
(238) September to mid-November
(239) allicin
(240) Maharashtra
(241) amaryllidaceae
(242) B
(243) 750-1000 mm
(244) seed, bulb
(245) 15 x 7.5 cm
(246) August, December- January
(247) iodine
(248) long, medium

(249) 6.0-6.8
(250) 5.8, 6.5
(251) 30°C
(252) cool, wet, cloudy
(253) receptacle
(254) India
(255) Lignosus
(256) hilly
(257) anthrocyanin
(258) Dec.-Jan.
(259) coastal
(260) perennial
(261) dioecious, separate
(262) Assam
(263) heart & brain
(264) stem, root
(265) cuttings
(266) nitrogen
(267) Solanaceae
(268) Nitrogen and Phosphoruss
(269) Incas
(270) tuber, true seeds
(271) short day
(272) black heart
(273) Spanish
(274) scab
(275) Maleic hydrazide
(276) solid head
(277) isothiocyanates
(278) red
(279) Tetra-2, Pusa Rassal
(280) March, August
(281) spring, rainy
(282) fenugreek
(283) mid-December, late-February
(284) Norin

(285) summer, rains
(286) vegetable
(287) F_1 hybrid
(288) Mediterranean
(289) cooler
(290) thick leaf stalks
(291) Siberia
(292) muffled dull/dead
(293) blossom end
(294) less
(295) less
(296) sandy loam, organic matter
(297) 45 x 45
(298) onion
(299) *Gnorimoschema operculelia*
(300) rainy season
(301) tropical, temperate
(302) 45 x 30 cm
(303) himalayas, 1200
(304) higher altitudes
(305) 20°C
(306) monoecious, dioecious
(307) tuber production
(308) 2-3
(309) archipelago
(310) cluster
(311) 75 x 30
(312) Truck
(313) cool
(314) annual, biennial
(315) India
(316) flower initiation
(317) miniature cabbage
(318) 45-50
(319) India
(320) February, March

(321) vegetable-marrow
(322) American
(323) 20
(324) 30 cm, 45 cm
(325) Tropical America
(326) drought
(327) acid
(328) Colchicine
(329) flowers, fruits
(330) one year
(331) *Cucurbita moschata*
(332) *Citrullus vulgaris*
(333) cabbage, Knol-khol
(334) quercetin
(335) seed-to-seed
(336) bulb formation, enlargement
(337) potato, watermelon
(338) Ethiopia
(339) buttoning
(340) evening, morning
(341) cooling
(342) good
(343) powdery
(344) Vlt. A, iron
(345) TMV, CMV, curly top
(346) purple blotch
(347) ordinary temperature
(348) 6.0, 7.5
(349) Y.V.M.V.
(350) bacterial blight
(351) red pumpkin beetle
(352) Common bean mosaic
(353) caulis, floris
(354) Cantaloupe
(355) Tien Shan
(356) Peru

(357) Africa
(358) China, India
(359) Ethiopia
(360) Abyssinia
(361) 1943
(362) allyl-propyl-disulphide
(363) dialyl disulphide
(364) pods, vegetable
(365) Iycopine
(366) 21°C
(367) 125
(368) cloves/bulblets
(369) 1250
(370) 1875-2500
(371) 5-7
(372) 7-10
(373) normal flowers
(374) stolon ends
(375) floating garden
(376) 40-50, 40-50
(377) 45
(378) 30 cm, 20 cm
(379) tomato, cucumber
(380) troquer, to barter
(381) 250-300
(382) 24°C, 18°C
(383) 4
(384) 18
(385) indeterminate, determinate
(386) 3-4
(387) two

(388) 1-15
(389) fully ripe
(390) 500-800
(391) Portuguese
(392) 6.0, 7.0
(393) trenches
(394) December and early January
(395) 20°C
(396) calcium
(397) 3-4
(398) 20-30, 10-15
(399) 20-25, 75-90
(400) 2.2
(401) farming
(402) Calcium
(403) leafy
(404) *lanatus*
(405) 1629
(406) 5
(407) *petha*
(408) Chinese
(409) BonavistlHyacinth bean
(410) Mediterranean
(411) yellow tinge
(412) full slip stage
(413) rectangular
(414) European type
(415) Knol-khol
(416) north India
(417) Mediterranean region
(418) Faba bean

1.4. True and False

(1) 1-5 ppm 2, 4-D spray can increase the yield of tomato by ten times.
(2) 2219A male sterile line of cucumber has been exploited for hybrid production.
(3) A large number of improved cultivars of snap melon are available for commercial cultivation in India.
(4) A pre-harvest spray of maleic hydrazide controls sprouting of potato tubers in storage.
(5) A relatively low temperature as well as short photoperiod is essential for bulb formation in most of the commercial varieties.
(6) About 15-20 days old seedlings of Brussels sprouts are most suitable for transplanting.
(7) About 40,000-50,000 cuttings of sweet potato are required to plant one hectare.
(8) Acridity of tubers in some varieties of colocasia is due to calcium oxide.
(9) Alkaline soils create favourable conditions for scab disease.
(10) All beans are susceptible to frost and are grown as a summer crop.
(11) All bulb crops belong to the family Liliaceae.
(12) All cole crops have a common ancestor.
(13) All Green, Jobner Green and Pusa Jyoti are the improved cultivars of fenugreek.
(14) All varieties can do better under forcing structure.
(15) Almost all cucurbits are sensitive to water logging and freezing temperature.
(16) Amaranthus does not do well on heavy, poorly-drained soil or on sandy soils which are poor in water holding capacity and poor in nutrients.
(17) Amaranthus has C_3 photosynthetic cycle which indicates its high productivity.
(18) Amaranthus is often-cross-pollinated crop.
(19) Amaranthus seeds are very small, therefore, some sand is mixed to get uniform distribution of seed.
(20) *Amaranthus tricolour* is *bari chaulai*.
(21) Application of bacterium culture does not improve the yield of peas.
(22) Application of maleic hydrazide (MH) at 25-50 ppm once or twice at the 2-leaf and again at the 4-leaf stage induces a greater number of female flowers.
(23) Arka Harit and Pusa Do Mausami are the cultivars of cowpea.
(24) Arka Jyoti is the hybrid of cucumber evolved by IARI.
(25) Arka Komal is a variety of French bean developed by IIHR, Bangalore.
(26) Arka Nishant, Punjab Safed and Pusa Reshmi are Asiatic or tropical type of radish varieties.
(27) Arka Sheel and Arka Navneet are the promising hybrids of brinjal.
(28) Artichoke is a popular vegetable in India.
(29) Asparagus is propagated by suckers.

(30) Asparagus plants are dioecious in character.
(31) Asparagus starts yielding a sizeable crop after about three years.
(32) At present, the cultivation of broad bean in India is not so popular on large-scale.
(33) Beetroot is a cool-to-warm season crop.
(34) Beetroot is propagated by cormels.
(35) Being a self-pollinated crop, sponge gourd does not need to maintain any isolation distance for its seed production from other cultivar.
(36) Being perennial in nature, chayote requires judicious application of manure and fertilizers.
(37) Best quality spears of asparagus are obtained from the second to fifth year.
(38) Black scurf affected plant sometimes produce aerial tubers in the axils of leaves.
(39) Blossom-end rot and sunscald are non-parasitic troubles of chilli that sometimes cause considerable loss.
(40) Botanical classification is largely used in almost all text books.
(41) Botanical name of cabbage is *Brassica oleracea* var. *capitata* and family cruciferae.
(42) Botanically, there is no difference in sponge gourd and ridge gourd.
(43) Botanically, tomato is known as *Lycopersicon melongena*.
(44) Bottle gourd can withstand cold climate better than muskmelon and watermelon.
(45) Bottle gourd is a monoecious crop.
(46) Broad bean is warm season vegetable.
(47) Broad bean require cool season but for ripening of pods higher temperatures are beneficial.
(48) Brussels sprouts can be grown on very high acidic soils.
(49) Brussels sprouts require hot season.
(50) Cabbage is a cool season crop.
(51) Cabbage is a cross-pollinated crop and does not cross with other members of the cole group.
(52) Cabbage is a deep-rooted crop and is a poor feeder.
(53) Carrot belongs to family Umbelliferae.
(54) Carrot can tolerate slight acidic and alkali soil reactions.
(55) Carrot grown on heavy soils tend to become rough and coarse because the roots fail to penetrate the hard soil.
(56) Caterpillars are the most troublesome insect-pest on lettuce.
(57) Cauliflower is grown for its white tender head or curd.
(58) Cauliflower is sensitive to high acidity.
(59) Cauliflower varieties are not responsive to temperature and photoperiod.

(60) Celery is a water loving plant.
(61) Celery is an annual plant.
(62) Celery plant is moderately sensitive to salinity.
(63) Celery seedlings are transplanted in trenches as they facilitate earthing up and blanching.
(64) Chayote responds well to lower planting.
(65) Chillies can withstand frost up to some extent.
(66) Chow-chow is considered a perennial cucurbit.
(67) Cluster bean is highly tolerant to drought.
(68) Colocasia corms store well for longer period of time.
(69) Colocasia is a bulb crop.
(70) Colocasia is grown only once in a year.
(71) Colocasia matures in 80-90 days.
(72) Colocasia require winter season.
(73) Compared to cauliflower, Brussels sprouts require long growing period.
(74) Cowpea is a native of India.
(75) Cowpea is probably the native of Central Africa.
(76) Cucumber is quick growing dioecious annual cucurbit.
(77) Cucumber is used mostly for salad.
(78) Cucumber mosaic virus is readily transmitted by aphids.
(79) Cucumber requires cool season and low humidity for the formation of pistillate flowers.
(80) Curing of sweet potato tubers is done best at 80ºC and if 30 per cent R.H. is maintained for 10 days.
(81) *Dioscorea batatas* is more commonly grown in the tropical and sub-tropical regions of India.
(82) Direct contact of fertilizers (particularly nitrogen and potash) with seeds affect germination adversely.
(83) During winter season, vines of pointed gourd die but grow again in spring which follows flowering and fruiting immediately.
(84) Dwarf cultivar require support (staking).
(85) Early blight is the most serious fungal disease of potato.
(86) Early maturing potato varieties are favoured by relatively short days while late ones are better adapted to long days.
(87) Early varieties of cauliflower, if sown late, produce button head and late varieties, if sown early, go on giving leafy growth and produce curd very late.
(88) Earthing up of an individual plant of snake gourd does not give any support or prevent plants from direct contact with water.

(89) Elephant's foot is propagated by small pieces of corms.
(90) Elephant's foot is grown in flat beds and on broad ridges.
(91) Elephant's foot is not as rich in minerals and vitamins A and B as potato.
(92) Elephant's foot is propagated by corm.
(93) Elephant's foot species are found in tropical Asia and America, and about 14 species occur in India.
(94) Excessive application of nitrogen in potato crop results in luxuriant and succulent growth such foliage is generally more liable to be affected by diseases.
(95) Extent of cross-pollination is the same in both beans and peas.
(96) Flowers of *palak* are hermaphrodite.
(97) For distant transportation, pink fruits of tomato are picked.
(98) For economic hybrid seed production in cucumber and muskmelon, gynoecious lines are used as female parents.
(99) For getting higher yield of sponge gourd, balanced supply of nutrients and soil moisture are needed.
(100) For improving the quality of leeks, blanching is essentially done.
(101) For propagating sweet potato from vine cuttings, cuttings from the basal portion of the vines should be preferred.
(102) For raising one hectare crop of sponge gourd about 20 to 25 kg seed is required.
(103) For seed production of knol-khol, low temperature and long-day conditions are needed which are found in the hills.
(104) For seed production, there is no need to have any isolation distance between two cultivars of okra.
(105) French bean does well at places 3000-4000 m above mean sea level.
(106) French bean is a good source of protein, calcium, iron and vitamins.
(107) French bean is richer than the hyacinth bean in its nutritive value.
(108) French bean is tolerant to acid soils, but pH should not be lower than 3.5.
(109) Fruit setting in brinjal can be increased by use of growth enhancing substances and nutrition.
(110) Fruits of sponge gourd should not be harvested when they are tender.
(111) Fruits of winter squash are available for harvesting within 200 days after planting.
(112) Fully matured, spongy and fibrous fruits of sponge gourd are most suitable for vegetable purposes.
(113) Galls formed on the roots of beans are often confused with bacterial nodules.
(114) Garden asparagus is closely related to ornamental asparagus.
(115) Garden pea is grouped as *Pisum sativum* sub sp. *arvense*.
(116) Garlic bulbs are cured for four to five weeks in shade before storing.
(117) Garlic is a spice or condiment used for flavouring.

(118) Garlic needs a richer soil than onion.
(119) Great Lakes is a variety of leafy lettuce.
(120) Green leaf stalks of celery are more nutritive than blanched ones.
(121) Green spears of asparagus are more nutritive than blanched (white) spears.
(122) Green tender fruits of winter squash are never harvested.
(123) Green type of broccoli, which is more nutritive, is the most popular.
(124) Hebbal Avare-3, Wal Konkan-1 and Arka Jay are the vegetable varieties of faba bean.
(125) High moisture content and low nitrogen delay maturity.
(126) High quality carrots are those which do not have a relatively large outer core.
(127) High temperature above 12-15°C promotes seed stems and causes a bitter taste in the leaves.
(128) High temperature reduces yield and quality of curds.
(129) Higher seed yield of radish is expected from root-to-seed method.
(130) Hoeing is required in cauliflower during early stage.
(131) Hot weather and inadequate supply of moisture result in deterioration of root quality.
(132) Hybrid varieties have advantages of increased adaptability to adverse environment, more resistance to diseases, better quality, earliness and increased yield.
(133) If picking of okra fruits is delayed, the quality will be improved.
(134) In beetroot, earthing up is usually done to cover the   swollen roots.
(135) In cold season vegetable plants, hardening is done to withstand possible burning due to sunshine, hot winds and dry soil at their planting season.
(136) In cucumber, a small portion of the stem-end is cut cross wise and rubbed together to remove the black substance that comes out.
(137) In green leafy vegetables, tryptophan is the limiting amino acid.
(138) In India, cabbage seeds are raised in the Kashmir Valley.
(139) In kohlrabi, fleshy edible portion is an enlargement of root.
(140) In northern India, sweet potato vine cuttings are planted during April-May.
(141) In *palak*, insect pollination is effective.
(142) In pea, the germination is epigeal.
(143) In plants like *Brassica,* temperature controls flowering.
(144) In radish, for root-to-seed method, sowing is done early when roots have matured.
(145) In radish, harvest when pods are fully ripe.
(146) In radish, higher genetic purity can be maintained when seed production is done by seed-to-seed method.

(147) In seed-to-seed method of seed production of cabbage, the plants are allowed to over-winter in their original position.
(148) In sweet gourd, male and female flowers occur on the same plant.
(149) In the Nilgiri hills, three potato crops are raised in succession.
(150) In vegetable crops which are difficult to transplant, there is a tendency for the roots to be suberized or cutinized.
(151) In winter squash, ripe fruits have no storage life compared with green fruits.
(152) Indeterminate type of tomato plants exhibit 'self topping' growth habit.
(153) India exports chillies and their products to Abu Dhabi, Australia, Canada, Japan, U.K. and USA.
(154) India is the second largest producer of vegetables next to China.
(155) Indian bean Hebbal Avare-1 has been evolved by the segregation of Local Avare x Red Typicus.
(156) Indian bean is also known as French bean.
(157) Indian bean is sown in temperate climate only.
(158) It is better to avoid application of partially decayed organic matter in the current growing season of the carrot crop because many roots will tend to become malformed or forked.
(159) It is easy to store by hanging the bunch of garlic bulbs with tops in well-ventilated sheds.
(160) It is possible to recognize the sex of asparagus plants in the nursery.
(161) It takes three to four years for the corms to be ready for harvesting.
(162) Ivy gourd is a dioecious vine crop.
(163) Ivy gourd is sensitive to water logging conditions.
(164) Jerusalem artichoke is of great value in the diet of diabetic persons because its tubers store insulin carbohydrate.
(165) Kakrol is a bush shrub.
(166) Kakrol is a monoecious.
(167) Kakrol is not dioecious.
(168) Kartoli fruits are small, roundish and covered with soft spines.
(169) Kartoli is a widely cultivated crop in India.
(170) Kartoli is dioecious in nature.
(171) Kartoli requires warm and humid climate.
(172) Knol-khol is a hardy winter vegetable.
(173) Knol-khol is not as popular as cabbage and cauliflower.
(174) Kufri Chandramukhi has better keeping quality than Kufri Red under ordinary conditions.
(175) Leafy type lettuce is also sown directly in the field.'

(176) Leaves are eaten as salad in celery.
(177) Leaves of amaranthus are non-perishable in nature.
(178) Leek is a biennial crop and its seeds are produced in India at higher altitudes in the hills.
(179) Leek is a bulb forming member of the onion family.
(180) Leek is a hardy biennial but grown as an annual.
(181) Leek is grown for its cloves.
(182) Lettuce is a salad-cum-leafy vegetable of temperate climate.
(183) Lettuce is probably a native of Europe and Asia, and has been in cultivation for at least 2,000 years.
(184) Lettuce plant bolts quickly due to insufficient availability of soil moisture.
(185) Lima bean is also known as double bean.
(186) Lima bean is cross-compatible with French bean
(187) Lima bean is cultivated on large-scale in India.
(188) Lima bean neither grows well in spring nor in rainy season.
(189) Lima bean requires fertile soil as compared to French bean.
(190) Little leaf disease of brinjal is caused by virus, spread by aphids.
(191) Long fruited varieties of bitter gourd are generally grown in the summer season.
(192) Long melon can be stored under ordinary conditions for long period without any deterioration in its quality.
(193) Long melon can stand cool climate better than muskmelon.
(194) Longevity of brinjal seed is four years.
(195) Lower fertility, higher temperature and longer light period induce maleness in cucumber.
(196) Mature leaves of colocasia are most suitable for vegetables.
(197) Maturity of onion is indicated by complete drying of leaves.
(198) Maturity of sweet potato tubers is proper if cut surface of tuber exposed to air, dries up soon.
(199) Melchers *et al.* (1978) produced somatic hybrid 'Pomato' by fusion between potato and tomato.
(200) Mild temperate climate is more suitable for French bean.
(201) Molybdenum deficiency symptoms in cauliflower occur in highly alkaline soils.
(202) Most of the cucurbits are monoecious and a few are dioecious.
(203) Most of the French bean varieties are long-day types.
(204) Mucilage, a sticky substance in okra, is generally extracted from flowers and buds.

(205) Mucilaginous material in okra is due to glycoproteins.
(206) Muskmelon fruit should not be picked at full slip stage for home consumption.
(207) Muskmelon is drought resistant and susceptible to frost.
(208) Muskmelon tolerates a slightly cooler weather than cucumber.
(209) No variety of French bean has yet been reported to be resistant to yellow bean mosaic.
(210) Non-ridged okra fruits are easy to thrash.
(211) Occasionally, flowering is also seen in Elephant's foot.
(212) Okra belongs to the genus *Hibiscus*.
(213) Okra is sown in April-May in the hills.
(214) One or more large central corms are termed as *kachalu* while large number of lateral cormels are termed as *arvi*.
(215) Onion bulbs that are to be stored are usually thoroughly cured for 3 or 4 weeks.
(216) Onion can be cultivated as a rainfed crop between April and August even at 1525-2134 m elevations.
(217) Onion crop transplanted earlier than December and January gives higher yield but the number of bolters may be more.
(218) Onion has got medicinal values. Its use is the best remedy against sunstroke.
(219) Onion is insensitive to high acidity.
(220) Onion is one of the most important vegetable for foreign exchange earning.
(221) Onion seed remain viable for more than one year.
(222) Onion varieties which store well produce seed within the same year.
(223) Only lettuce and celery are grown on a commercial scale in India.
(224) Only pseudo-short-styled and true short-styled flowers in brinjal bear fruits.
(225) Optimum temperature and R.H. for storage of sweet potato tubers are 5°C and 70 per cent, respectively.
(226) Pea has high percentage of digestible protein and good content of vitamins and minerals.
(227) Peas are very sensitive to drought.
(228) Pointed gourd is an important summer crop in north Bihar and eastern U.P.
(229) Pointed gourd is propagated by seed for taking a commercial crop.
(230) Pointed gourd propagates by seeds alone.
(231) Pointed gourd ratoon crop flowers 10-15 days earlier than the planted crop.
(232) Potato has wide adaptability for climate and soil requirements, therefore, it is grown throughout the country.
(233) Potato in India was introduced in the seventeenth century.
(234) Potato is a good source of carbohydrates, calories and proteins.

(235) Potato is a quick growing crop.
(236) Potato is native of Ireland.
(237) Potato tubers are borne underground at the root ends.
(238) Potato tubers stored in cold storage should be kept for 12-24 hours at 15°C before exposure to atmosphere.
(239) Potato varieties *viz.*, Phulwa, DRR, Satha and Hellora, are not chance survivals of early introduction.
(240) Premature production of seed stalks in onion is known as bolting.
(241) Pulling of carrot roots becomes easy when bed is watered lightly about 24 hours or so prior to pulling.
(242) Pungency in chilli is due to an alkaloid solasidine.
(243) Pungency in onion is due to a volatile oil allyl propyl disulphide.
(244) Punjab Chhuara has revolutionized tomato cultivation in Punjab state.
(245) Punjab Chhuhara is a determinate variety of tomato.
(246) Punjab-8 variety of okra has been evolved by irradiation of seeds.
(247) Pusa Alankar is the hybrid of summer squash.
(248) Pusa Chikni is not an improved variety of sponge gourd.
(249) Pusa Early Pwlific is a variety of Indian bean developed by IIHR, Bangalore.
(250) Pusa Hybrid is the popular hybrid of bottle gourd.
(251) Pusa Jyoti. is a variety of watermelon.
(252) Pusa Kesar, a cultivar of carrot, is a selection from cross between Local Red and Nantes Half Long.
(253) Pusa Kranti and Azad Kranti are the hybrid varieties of brinjal.
(254) Pusa Makhmali is a YVMV resistant variety of okra.
(255) Pusa Naubahar is a variety of cluster bean suitable for sowing in both summer season and rainy season.
(256) Pusa Purple Long is not a variety of brinjal.
(257) Pusa Sawani variety of okra is tolerant to YVM disease.
(258) Pusa Suyog is the popular hybrid of capsicum.
(259) Radish can be grown throughout the year.
(260) Radish is a quick growing crop.
(261) Radish is suitable for inter-cropping.
(262) Rhubarb is grown for its large thick leaf stalks and is propagated by the division of crown.
(263) Ridge gourd belongs to cruciferae family.
(264) Ridge gourd can never be grown on flat land.
(265) Ridge gourd is affected by waterlogging condition.

(266) Ridge gourd prefers to grow in heavy black cotton soils.
(267) Ridge gourd require cool climate.
(268) Ripe seeds of okra are sometimes used as a substitute for coffee.
(269) S-12 is an indeterminate variety of tomato.
(270) Satputia is a single fruited variety.
(271) Second earthing up of potato crop is done to keep the soil loose and destroy weeds.
(272) Seed crop of okra is generally raised in the summer season in northern plains.
(273) Seed of Brussels sprouts can be easily produced in the plains of North India.
(274) Seed production of temperate types of turnip cultivars should be done in plains.
(275) Seed production of tropical types of turnip cultivars should be done in the hills.
(276) Seed rate of colocasia for one hectare is about 80 to 120 quintals per hectare.
(277) Seed requirement of okra is generally higher in rainy season crop than in spring-summer season crop.
(278) Seedlings of variety Punjab Naroya when transplanted in August produce good bulbs in December-January under North Indian conditions.
(279) Seeds of Chinese cabbage can be produced under North Indian plain conditions.
(280) Seeds of French bean may rot in the ground if soil temperature is lower than 15ºC.
(281) Seeds of okra will not germinate below 20ºC.
(282) Seed-to-seed or *in situ* method of seed production gives higher seed yield of carrot.
(283) Sel-120 is the first root knot resistant variety of tomato.
(284) Short days are beneficial for tuber production.
(285) Single fruit of *palak* contains ten to thirteen seeds.
(286) Single stem training of tomato is most common.
(287) Snake gourd is a dioecious, perennial bush.
(288) Snake gourd is grown only in subtropical climate.
(289) Snap melon is not a winter season crop.
(290) Snap melon is widely cultivated all over the country.
(291) Soil, climate and disease-free conditions are factors influencing location of seed producing areas.
(292) Some of the smooth seeded varieties are comparatively more resistant to a number of rotting organisms active at high soil temperature.
(293) Some people are allergic to the pollen as well as green pods of broad bean.
(294) Spears of asparagus are used for *saag* preparation.
(295) *Spinacea oleracea* belongs to Chenopodiaceae family.
(296) Spinach is a warm-season crop.

(297) Spinach leaves provide high contents of phosphorus and magnesium.
(298) Sponge gourd can be grown under water logging conditions.
(299) Sponge gourd is dioecious.
(300) Sponge gourd is generally propagated by vine cutting.
(301) Sponge gourd is more commonly cultivated in Europe and the Americas.
(302) Sponge gourd is self pollinated.
(303) Succulency and tenderness of the spinach leaves increase under low atmospheric humidity.
(304) Sugarbeet is a long day plant.
(305) Summer squash is one of the most nutritive and wholesome vegetables.
(306) Sun scalding in tomato may be avoided by not staking the plants.
(307) Sweet chillies (*Capsicum annuum*) are cooked as vegetable.
(308) Sweet gourd is mainly grown in Cochin.
(309) Sweet potato belongs to the species *batatas* of the genus *Ipomoea*.
(310) Sweet potato contains 16 per cent starch, 4 per cent sugar, that is 20 per cent alcohol-producing materials.
(311) Sweet potato does best on slightly high acidic soil.
(312) Sweet potato is a native of tropical America.
(313) Sylvia is an edible podded variety of pea.
(314) Tapioca is propagated by cuttings from mature plants.
(315) Tapioca tubers become ready in 3-4 months.
(316) Temperature is more important than day length in seed stalk development of onion.
(317) The botanical name of kakrol is *Momordica cochinchinensis*.
(318) The botanical name of pea is *Allium sativum*.
(319) The critical stage for irrigation in onion is flowering.
(320) The cucumber seed retains viability for about 5 years and good seed gives 80-90 per cent germination.
(321) The cultural operations of the vegetables belonging to the same family are always similar.
(322) The distance of sowing of cowpea is 20-30 cm row to row and 10 to 15 cm plant to plant.
(323) The edible part in colocasia is leaves.
(324) The edible portion of cabbage is 'head' which consists of thick leaves overlapping lightly on growing bud.
(325) The edible portions of Brussels sprouts are leaves.
(326) The fleshy root of radish is modified root and it develops from both the primary root and the hypocotyle.

(327) The fruit of snake gourd are not round in shape.
(328) The fruits of kartoli are not edible.
(329) The leaves and tender shoots of pointed gourd are used to prepare the soup for convalescents.
(330) The lima bean is a bush type plant.
(331) The plant of winter squash may tolerate frost.
(332) The plants of broad bean bear upright pods in the axil of the leaves along the stem.
(333) The recommended seed rate for spinach is about 60 kg/ha.
(334) The ripe fruits of long melon are not consumed.
(335) The seed production of Brussels sprouts is done in the hills.
(336) The seed rate for planting one hectare area of winter squash is about 10 to 12 kg.
(337) The seed rate of onion is 10-12 kg/ha.
(338) The seed rate of tomato is 400-500 g/ha.
(339) The seedlings obtained from double transplanting nursery bed produce higher yield because such seedlings are strong and stout and they establish early and put on good vegetative growth.
(340) The seeds of broad bean are big, therefore, they are planted individually, adopting double row system of planting.
(341) The seeds of Pusa Barsati variety of cowpea have striped spots.
(342) The snap melon fruits never burst.
(343) The soil reaction for the successful cultivation of turnip should be around neutral to slightly alkaline.
(344) The tuber production in potato totally stops at 36°C.
(345) The vegetable prepared from white brinjal fruits is said to be beneficial for persons suffering from diabetes.
(346) The vegetables are not blanched before filling into cans or bottles sterilized under 10 lb pressure and containing 3 per cent sol ution of salt and sugar each.
(347) The vines of watermelon are good source of mulching which guard sandy soils during the summers from wind erosion.
(348) The yield of dry chillies is 2-2.5 tonnes/ha.
(349) The yield of tomato is 20-25 tonnes/ha.
(350) There are more chances of getting high quality carrot seeds when seed crop is raised by root-to-seed method,
(351) Thinning in carrot becomes an essential operation for maintaining proper distance. .
(352) Thinning in turnip is an essential operation.
(353) Thinning in wax gourd is an essential operation.

(354) TMV is the most serious viral disease of tomato in India.
(355) Tomato can only be grown in winter season.
(356) Tomato fruits have been detected to maintain nicotine when grafted on tobacco root stock.
(357) Tomato requires a warm sunny weather for its proper ripening, colour, quality and high yield.
(358) Tomato varieties which are suitable for processing are not suitable for vegetable purposes.
(359) Though Nantes is a European variety of carrot and its seed can be produced in plain.
(360) Triploid watermelons are seedless, sweet, firm and suitable for transportation.
(361) Tuber production in potato is maximum at 30°C.
(362) Turnip does not transplant well, hence direct sowing is adopted.
(363) Turnip is found growing wild in Russia and Siberia.
(364) Turnip may cross with radish, rutabaga and cabbage.
(365) Under long-day condition, tuberisation in potato is affected adversely but vegetative growth but is affected favourably.
(366) Under decomposed organic matter may cause forking or deformed roots.
(367) Vegetables commonly grown under forcing structures are tomato and cucumber.
(368) Vegetables need sterilization as heat resistant bacteria remain unaffected by water boiling at 115.5°C or more.
(369) Vines of sweet potato grow at the expense of tuber formation when soil temperature goes above 40°C.
(370) Wart disease in potato is more serious in tropical climate.
(371) Water logging improves the quality of colocasia corm.
(372) Wax gourd is a monoecious annual climber.
(373) When turnip is allowed to grow under shade, foliage grows at the expense of root development.
(374) Whip-tail is a disease of cauliflower.
(375) White skinned tubers of sweet potato generally store well better than red-skinned tubers.
(376) Winter squash does not require any manures and fertilizers.
(377) Winter squash is a dioecious.
(378) Winter squash is *Cucurbita moschata*.
(379) Yam is propagated by tubers.
(380) Yam or *rata/u* is a rich source of carbohydrates.
(381) Yield of fresh green chillis is 10 times higher than that of dry chilli.
(382) Yield of potato raised for seed purpose is 60-80 q/ha.

(383) YVM disease is transmitted only by an insect vector known as whitefly.
(384) Indian Institute of Vegetable Research was established in 1971.
(385) IIVR is situated at Varanasi
(386) Kashi Sandesh is variety of tomato
(387) Kashi Haritma is variety of brinjal
(388) Crop combination of coffee + carrot is effective for management of diamond back moth.
(389) Kashi Madhu is variety of watermelon.
(390) Kashi Alankar is variety of pointed gourd.
(391) Kashi Pragati is variety of okra.
(392) Kashi Rajhans and Kashi Sampann are varieties of cowpea
(393) For control of DBM in cabbage, Chinese cabbage may be used as trap crop.
(394) Marigold as trap crop may be used for control of aphid in cabbage
(395) Bittergoard + Maize combination may be used for control of fruitfly
(396) Cabbage + Fafabean may be used for control of root fly
(397) Coriander as trap crop for natural enemy may be used for control of leaf minor and in tomato fruit borer
(398) Dr. G. Kalloo is founder Director of Indian Institute of Vegetable Research, Varanasi
(399) A crop combination of cabbage + carrot may be used for control of Diamond back moth in cabbage
(400) Kashi Khushaal and Kashi Sheetal are varieties of Indian bean
(401) Kashi Shivani is a variety of ridge gourd
(402) Kashi Nandani, Koshi Muketi and Koshi Uday are variety of cowpea
(403) Kashi NIdhi and Kashi Kanchan are varieties of French bean.
(404) All Indian Coordinated Research Project-AICRP(VC)). New Delhi established in 1971.
(405) Arka Rakasha is a variety of tomato
(406) National Horticultural Research and Development Foundation (NHRDF) Nasik was established by National Agricultural Co-operative Marketing Federation of Indian Ltd. (NAFED) and its Associates shippers of onion on November 3, 1977.
(407) The garlic variety Yamuna Safed-5 and Yamuna Safed-4 are not sub suitable for dehydration
(408) The hot arid region of the country is spread over nearly 31.7 million hectare area of which 41.5 per cent is arable and 19 per cent is cultural wasteland (T)
(409) Tar Hariparna is a variety of Palak
(410) Thar Harsha is a variety of Drumstick
(411) Kachri and Mateera are popular crop of West Bengal
(412) Thar Sheetal is a variety of tomato
(413) Thar Maghi and Thar Kartiki varieties are of Cauliflower
(414) Seed spices fennel, fenugreek, coriander and cucumin occuby on area of 47.8% of total species and contributes 22.5% of total spices production
(415) Grafting techniques involving two different species/cultivar was first attempted by a Japanese farmers Ukichi Takenaka who grafted watermelon squash to manage Fusarium wilt in 1920.

(416) Water chestnut (*Trapa fispinosa*), lotus (*Nelumbo nucifera*) and water spinach (*Ipomoea quatic*) are grown as aquatic vegetables
(417) Lotus is also known as sacred lotus, Indian lotus. East Indian lotus, oriental lotus, lily of night, Bean of Indian and scared water lily.
(418) Water spinach (Kalmi saag) is originated in Tropical Asia, possibly in India
(419) Vertical Farming in India is still infancy stage
(420) Vertical farming is likely to replace traditional farming system
(421) Kitchen gardening differs with home gardening
(422) Kitchen garden in French called potage
(423) Tomato is a protective food
(424) Pusa Sambandh and Pusa Mukta are varieties of cauliflower
(425) Onion and garlic protect agains cancer and heart diseases
(426) Indian, Bengaladesh, Turkey and China are major countries for spices production
(427) National Research Center for spices Research (NRCS) was estqablished in 1986 and further elevated to the present Indian Institute of spies (IISR) in 1995
(428) Marsh disease of Pea is due to deficiency of Magnese
(429) Pusa Sheetal is a variety of turmeric
(430) Rajasthan ranks first in production of methi fennel and coriander

Answer Sheet

#		#		#		#		#	
1.	F	45.	T	89.	T	133.	F	177.	F
2.	F	46.	F	90.	T	134.	T	178.	T
3.	F	47.	T	91.	F	135.	F	179.	F
4.	T	48.	F	92.	T	136.	F	180.	T
5.	F	49.	F	93.	F	137.	F	181.	F
6.	T	50.	T	94.	T	138.	T	182.	T
7.	T	51.	F	95.	F	139.	F	183.	T
8.	T	52.	F	96.	T	140.	F	184.	T
9.	T	53.	T	97.	F	141.	F	185.	T
10.	T	54.	T	98.	T	142.	F	186.	F
11.	F	55.	T	99.	T	143.	T	187.	F
12.	T	56.	F	100.	T	144.	T	188.	F
13.	F	57.	T	101.	F	145.	T	189.	T
14.	F	58.	T	102.	F	146.	F	190.	F
15.	T	59.	F	103.	T	147.	T	191.	F
16.	T	60.	T	104.	F	148.	F	192.	F
17.	F	61.	F	105.	F	149.	T	193.	T
18.	F	62.	T	106.	T	150.	T	194.	T
19.	T	63.	T	107.	F	151.	F	195.	T
20.	T	64.	T	108.	F	152.	F	196.	F
21.	F	65.	T	109.	T	153.	T	197.	F
22.	F	66.	T	110.	F	154.	T	198.	T
23.	F	67.	F	111.	F	155.	T	199.	T
24.	T	68.	T	112.	F	156.	F	200.	T
25.	T	69.	F	113.	F	157.	F	201.	F
26.	T	70.	F	114.	T	158.	T	202.	T
27.	T	71.	F	115.	F	159.	T	203.	F
28.	F	72.	F	116.	F	160.	F	204.	F
29.	T	73.	F	117.	T	161.	T	205.	T
30.	T	74.	F	118.	T	162.	T	206.	F
31.	T	75.	T	119.	F	163.	T	207.	T
32.	T	76.	F	120.	F	164.	T	208.	F
33.	T	77.	T	121.	T	165.	F	209.	T
34.	F	78.	T	122.	F	166.	F	210.	F
35.	F	79.	F	123.	T	167.	F	211.	T
36.	T	80.	F	124.	F	168.	T	212.	F
37.	F	81.	F	125.	F	169.	F	213.	T
38.	T	82.	T	126.	F	170.	T	214.	T
39.	T	83.	T	127.	T	171.	T	215.	T
40.	T	84.	F	128.	T	172.	T	216.	T
41.	T	85.	F	129.	F	173.	T	217.	F
42.	T	86.	F	130.	T	174.	F	218.	T
43.	F	87.	T	131.	T	175.	T	219.	F
44.	T	88.	F	132.	T	176.	F	220.	T

#		#		#		#		#	
221.	F	266.	T	311.	F	356.	T	401.	T
222.	F	267.	F	312.	T	357.	T	402.	F
223.	T	268.	T	313.	T	358.	F	403.	F
224.	F	269.	T	314.	T	359.	F	404.	T
225.	F	270.	F	315.	F	360.	T	405.	T
226.	T	271.	T	316.	T	361.	F	406.	T
227.	T	272.	F	317.	T	362.	T	407.	F
228.	T	273.	F	318.	F	363.	T	408.	T
229.	F	274.	F	319.	F	364.	T	409.	T
230.	F	275.	F	320.	T	365.	T	410.	T
231.	T	276.	F	321.	F	366.	T	411.	F
232.	T	277.	F	322.	T	367.	T	412.	F
233.	T	278.	F	323.	F	368.	F	413.	F
234.	T	279.	T	324.	T	369.	F	414.	T
235.	T	280.	T	325.	F	370.	F	415.	T
236.	F	281.	T	326.	T	371.	T	416.	T
237.	F	282.	T	327.	T	372.	T	417.	T
238.	T	283.	T	328.	F	373.	T	418.	T
239.	F	284.	T	329.	T	374.	T	419.	T
240.	T	285.	F	330.	F	375.	F	420.	F
241.	T	286.	T	331.	T	376.	F	421.	F
242.	F	287.	F	332.	T	377.	F	422.	T
243.	T	288.	F	333.	F	378.	F	423.	T
244.	T	289.	T	334.	T	379.	T	424.	F
245.	T	290.	F	335.	T	380.	T	425.	T
246.	T	291.	T	336.	F	381.	F	426.	T
247.	T	292.	T	337.	T	382.	F	427.	T
248.	F	293.	T	338.	T	383.	T	428.	T
249.	F	294.	F	339.	T	384.	F	429.	F
250.	T	295.	T	340.	T	385.	T	430.	T
251.	F	296.	F	341.	T	386.	F		
252.	T	297.	F	342.	F	387.	F		
253.	T	298.	F	343.	T	388.	T		
254.	F	299.	F	344.	T	389.	F		
255.	T	300.	F	345.	T	390.	T		
256.	F	301.	T	346.	F	391.	T		
257.	T	302.	F	347.	T	392.	F		
258.	T	303.	F	348.	T	393.	T		
259.	T	304.	T	349.	T	394.	F		
260.	T	305.	F	350.	T	395.	T		
261.	T	306.	T	351.	T	396.	T		
262.	T	307.	T	352.	T	397.	F		
263.	F	308.	F	353.	T	398.	T		
264.	F	309.	T	354.	F	399.	T		
265.	T	310.	T	355.	F	400.	T		

2
Breeding, Biotechnology and Seed Technology

2.1. Multiple Choice Questions

(1) The process of bringing a wild species under human management is referred to as ―――――.

    (a) Selection   (b) Rejection
    (c) Domestication   (d) None of the above

(2) ――――――――was perhaps the first man to use artificial hybridization to develop several new fruit varieties.

    (a) Dawrin   (b) Knight
    (c) Beal   (d) Mendel

(3) Plant breeding aims to ――――――――――

    (a) Improve yield only   (b) Disease resistant only
    (c) Insect resistance only
    (d) Improve the characteristic of plant that are more desirable agronomically and economically.

(4) A population may be simply defined as the――――――――, which mate or can mate freely with each other.

    (a) Single individual   (b) Group of individual
    (c) A and b both   (d) None of the above

(5) Our present day crops are the products of ――――――――.

    (a) Rare artificial selection   (b) Continued artificial selection
    (c) Only natural selection   (d) None of the above

(6) A process which leads to the transfer of same genes from one species into another is known as―――――.

    (a) Interspecific   (b) Intergeneric
    (c) Hybridization   (d) Back crossing

(7) The germplasm of a crop may be defined as the sum total of hereditary material, i.e. all the alleles of various genes present in a crop species and its wild relatives. Therefore, germplasm consists ――――――――.

    (a) Only land races   (b) Only obsolete varieties
    (c) Only breeding lines   (d) Land races, absolete varieties, varieties in cultivation, breeding lines and wild form of wild relatives

(8) The gradual loss of variability from cultivated species and their wild forms and wild relatives is called ─────────────.
    (a) Peliotrophy   (b) Mutation
    (c) Genetic erosion   (d) Genetic variability

(9) The process of obtaining the various germplasm accessions for a germplasm collection is known as collection of germplasm. This can be done in two chief ways: ─────────────
    (a) Only exploration   (b) Only procurement from other agencies
    (c) Both a and b   (d) None of the above

(10) ───────────── is the primary source of all the germplasm present in various germplasm collections.
    (a) Evaluation   (b) Selection
    (c) Procurement   (d) Exploration

(11) The crop plants evolved from wild species in the areas showing great diversity is termed as ───────────── centers of origin.
    (a) Primary   (b) Secondary
    (c) Adapted   (d) None of the above

(12) The law of ───────────── series in variation states that characters found in one species also occurs in other related species.
    (a) Heterozygous   (b) Homologous
    (c) Epistasis   (d) None of the above

(13) Eight main centers of origin were originally proposed by ─────────────. These are China, Hindustan, Central Asia, Asia Minor, Mediterranean, Abyssinia Central and South America. Later in 1935, Vavilov divided the Hindustan centre of origin into two centers viz; Indo-Verma and Siam-Malaya-Java centers of origin. Similarly, the South America Centres was divided into three cenres of origin.
    (a) Darwin   (b) Fairchild
    (c) Mendel   (d) Vavilov

(14) Eight main centers were grouped in ───────────── Centres of Origin
    (a) 15   (b) 12
    (c) 11   (d) 10

(15) The National Bureau of Plant Genetic Resources (NBPGR) was established in —
    (a) 1978   (b) 1976
    (c) 1980   (d) None of the above

(16) For procurement of germplasm of any individual or institution can introduce germplasm in India. But all the institutions must be routed through the ─────.
    (a) ICAR, New Delhi   (b) IARI, New Delhi
    (c) NBPGR, New Delhi   (d) Ministry of Agriculture, New Delhi

(17) The parts of plant used for propagation of a species are known as —————.
    (a) Runners   (b) Propagules
    (c) Grafting   (d) None of the above

(18) ————— means to keep materials in isolation to prevent the spread of diseases etc. present in them to the other materials.
    (a) Seed certification   (b) Prevention
    (c) Quarantine   (d) None of the above

(19) The process that leads to the adaptation of a variety, line or population to a new environment is known as —————.
    (a) Domestication   (b) Acclimatization
    (c) Adaptation   (d) None of the above

(20) The extent of acclimatization is determined by —————.
    (a) Mode of pollination only   (b) Magnitude of genetic variability
    (c) Both a and b   (d) None of the above

(21) ————— is the oldest and very effective breeding approach to create genetic variation.
    (a) Mutation   (b) Polyploidy
    (c) Mass selection   (d) Plant introduction

(22) Wild relatives of crops and the variability present in the crop species is being depleted due to —————.
    (a) Adaptation   (b) Acclimatization
    (c) Introduction   (d) Genetic erosion

(23) In apomixes, seeds are formed but embryos develop —————
    (a) Without pollination   (b) Without fertilization
    (c) With pollinaion   (d) With fertilization

(24) Asexually reproducing crop species are highly heterozygous and show ————— inbreeding depression. Therefore, breeding methods in such species must avoid inbreeding.
    (a) High   (b) Moderate
    (c) Very high   (d) None of the above

(25) Sexual reproduction involves fusion of male and female gametes to form a zygote, which develops into an —————.
    (a) Embryo   (b) Seed
    (c) Zygote   (d) None of the above

(26) In the process of flowering, the first opening of flower is known as —————.
    (a) Fertilization   (b) Pollination
    (c) Anthesis   (d) None of the above

(27) —————— refers to the transfer of pollen grains from anthers to the stigmas.
   (a) Fertilization (b) Seed formation
   (c) Embryogenesis (d) Pollination

(28) Pollen from an anther may fall on to the stigma of the same flowers leading to ——————.
   (a) Allogamy (b) Autogammy
   (c) Gietnogammy (d) None of the above

(29) When pollen grains from flowers of one plant are transmitted to the stigmas of another plant known as ——————.
   (a) Allogammy (b) Autogammy
   (c) Gietnogammy (d) None of the above

(30) A third situation —————— results when pollen from a flower of one plant falls on the stigma of other flowers of the same plant. e.g. in maize.
   (a) Parthenocarpy (b) Parthenogenesis
   (c) Geitonogammy (d) None of the above

(31) In case of ——————, flowers do not open at all. This ensures complete self-pollination since foreign pollen can not reach the stigma of a closed flower.
   (a) Chasmogamy (b) Cleistogamy
   (c) Hercogamy (d) None of the avove

(32) In case of ——————, the flowers open, but only after pollination has taken place.
   (a) Cleistogamy (b) Chasmogamy
   (c) Dioecious (d) None of the above

(33) Self-pollination leads to a very rapid increase in ——————.
   (a) Homozygosity (b) Heterozygosity
   (c) Both homozygosity and heterozygosity (d) None of the above

(34) —————— refers to the failure of pollen from a flower to fertilize the same flower or other flowers on the same plant.
   (a) Sterility (b) Self- incompatability
   (c) Parthenocarpy (d) None of the above

(35) —————— refers to the absence of functional pollen grains otherwise hermophordite flowers.
   (a) Male sterility (b) Self- incompatibility
   (c) Syngammy (d) Double fertilization

(36) Cross-pollination preserves and promotes —————— in a population.
   (a) Homozgosity (b) Heterozygosity
   (c) Both homozygosity and heterozygosity (d) None of the above

(37) In cross pollinating species, the transfer of pollen from a flower to that stigma of the others may be brought about wind is known as—————
- (a) Entomophiliy
- (b) Anemophilly
- (c) Hydrophylly
- (d) None of the above

(38) In often cross-pollinated species cross-pollination often exceeds ————— and reach—————%.
- (a) 5 and 50
- (b) 10 and 50
- (c) 5 and 30
- (d) None of the above

(39) Self-incompatibility was first reported by Koelreuter in the middle of————————century.
- (a) 16
- (b) 17
- (c) 18
- (d) 12

(40) Gametophytic system of incompatibility was first described by East and Mangelsdorf in 1925 in—————————. The incompatibility reaction of pollen is determined by its own genotypes, and not by the genotype of the plant an which it is produced.
- (a) Frenchbean
- (b) Pea
- (c) Tobacco
- (d) Sweet pea

(41) The sporophytic system of incompatibility was first reported by ————————— in 1950 in *Creopis foetida* and Gerstel in *Parthenium argentatum* in 1950. In this case, incompatibility reaction of pollen is governed by the genotype of the plant on which the pollen is produced, and not by the genotype of the pollen.
- (a) Hughes and Babcock and Gerstel
- (b) Mangelsdrof
- (c) Darwin
- (d) East and Shull

(42) ————————— is known to increase as well as decrease the activities of S-alleles both in the gametophytic as well as sporophytic system.
- (a) Polygenes
- (b) Oligogene
- (c) Additive
- (d) Non- additive

(43) In the—————system, stigma surface is plumose having elongated receptive cells and is commonly known as "wet" stigma. Incompatibility pollen grains generally germinate on reaching the stigma, the incompatibility reaction occurs at a later stage.
- (a) Sporophytic
- (b) Gametophytic
- (c) Self-incompatible
- (d) Male sterile

(44) In the —————system, stigma is the site of incompatibility reaction, once the pollen of the tube crosses the stigmatic barriers, there is no further inhibition of pollen tube growth.
- (a) Sporophytic
- (b) Gametophytic
- (c) Self-incompatible
- (d) Male sterile

(45) ——————is characterized by non-functional pollen grains, while female gametes function normally.
    (a) Male sterility   (b) Self- incompatibilty
    (c) Heterosis   (d) None of the above

(46) Male sterility can be induced by application of certain chemicals like GA_3, NAA, MH, ethrel etc. these chemicals are called male gametocides since that lead to pollen abortion and thereby cause male sterility, they are also called ——————.
    (a) Growth promoter   (b) Growth retardant
    (c) Chemical hybridizing agents   (d) None of the above

(47) Autogammy is promoted by ——————.
    (a) Bisexuality   (b) Homogammy
    (c) Cleistogammy   (d) Chasmogammy
    (e) All of the above

(48) Allogammy is promoted by ——————.
    (a) Dicliny   (b) Dichogamy
    (c) Heterostyly   (d) Male sterility
    (e) Herkogamy   (f) Self-incompatability
    (g) All of the above

(49) Development of embryo either from synergids or antipodal cell is referred to as ——————.
    (a) Parthenogenesis   (b) Androgenesis
    (c) Apogamy   (d) Apospory
    (e) Adventive embrtony

(50) Variabilty, heterozygosity, homozygosity, population mean and genetic correlation between relatives remains constant under ——————.
    (a) Random mating   (b) Genetic assertive mating
    (c) Phenotypic assertive mating   (d) Phenotypic dissortive mating

(51) Sexual reproduction includes ——————.
    (a) Autogamy   (b) Allogamy
    (c) Apomixis   (d) Amphimixis
    (e) a and b

(52) The presence of ——————in a character is a must for any improvement in that character.
    (a) Genetic variation   (b) Genetic erosion
    (c) Genetic similarity   (d) None of the above

(53) ─────── involves the identification and isolation of desirable plants from a variable population.
    (a) Selection   (b) Domestication
    (c) Acclimatization   (d) All of the above

(54) The characters produced by polygenes are reffered to as ─────── characters.
    (a) Qualitative   (b) Quantitive
    (c) Both qualitative and quantitative   (d) None of the above

(55) The behavior of genes during gametogenesis and fertilization is identical with that of the chrosomes.
    (a) Gene   (b) Allele
    (c) Chromosome   (d) None of the above

(56) The contrasting character is determined by alternative forms of the same gene, which are referred to as ───────.
    (a) Gene   (b) Alleles
    (c) Chromosome   (d) None of the above

(57) At the time of gamete formation, the two alleles present in the F_1 (Ww) separate and pass into different gametes, this is known as ───────.
    (a) Fertilization   (b) Meiosis
    (c) Segregation   (d) None of the above

(58) The determination of genotype or genotypic value of a plant by studying the progeny produced by it is known as ───────.
    (a) Mass selection   (b) Pedigree method
    (c) Bulk method   (d) Progeny test

(59) The phenomenon of a single major gene affecting more than one character is known as ───────.
    (a) Threshhold   (b) Peneterance
    (c) Pleiotrophy   (d) None of the above

(60) A line which is identical with respect to all other genes, except for the gene under investigation such lines are known as ───────.
    (a) Multiline   (b) RIL
    (c) NIL   (d) Isogenic line

(61) ─────── is the ability of a gene to express itself in the individuals carrying it in the appropriate genotype.
    (a) Threshold   (b) Peneterance
    (c) Pleiotrophy   (d) None of the above

(62) ─────── denotes the ability of a gene to express itself uniformly in all the individuals that carry it in the appropriate genotype.
    (a) Expressivity   (b) Threshold
    (c) Pleiotrophy   (d) None of the above

(63) Characters whose development depends upon a specific environment are known as——————.
    (a) Expressivity   (b) Threshold
    (c) Pleiotrophy   (d) None of the above

(64) Many perhaps most qualitative characters are determined by two or more oligogenes. These genes show various relationships with each other in producing the concerned characters; these relationships are known as——————.
    (a) Gene deviation   (b) Gene mutation
    (c) Gene interaction   (d) None of the above

(65) Complementary gene action refers——————.
    (a) 9: 7   (b) 9:3;4
    (c) 13:3   (d) None of the above

(66) Inhibitory gene action refers——————.
    (a) 13:3   (b) 9:7
    (c) 9:6:1   (d) None of the above

(67) Epistatic gene action refers——————.
    (a) 13:1   (b) 9:7
    (c) 12:3:1   (d) 1:2

(68) Polymeric gene action refers——————.
    (a) 9: 6: 1   (b) 9: 3: 4
    (c) 9: 7   (d) 13: 3

(69) Supplementary gene action refers——————.
    (a) 9:3:4   (b) 9:6:1
    (c) 13:3   (d) None of the above

(70) —————————, selection permits reproduction only in those plants that have the desirable characteristics i.e. the plants that have been selected.
    (a) Cross- pollinated crops   (b) Self – pollinated crops
    (c) Often cross pollinated crops   (d) None of the above

(71) The progeny test was developed by——————.
    (a) Louis de Vilmorin   (b) George Mendel
    (c) Louis Cornvillus   (d) Batesman

(72) The concept of purelines was proposed by —————————— on the basis of his studies with French bean.
    (a) Mandel in 1888   (b) East and Shull in 1908
    (c) Johanson in 1903   (d) None of the above

(73) Each individual plant progeny selected from a self-fertilized population consists of homozygous plants of identical genotype. Such a progeny is known as———.
  (a) Isogenic line (b) Inbred line
  (c) Pure line (d) None of the above

(74) The term ———denotes the frequency of genes in homozygous condition in the population. Linkage between genes does not affect the percentage of ——— in the population.
  (a) Heterozygosity (b) Homozygosity
  (c) Inbreeding depression (d) None of the above

(75) Continuous self-pollination has two main effects on the population———.
  (a) Only All the plants in the population became completely homozygous.
  (b) Only population of a self- pollinated species represents a mixture of several homozygous genotypes.
  (c) Both a and b
  (d) None of the above

(76) The mating or crossing of two plants or lines of dissimilar genotype are known as———.
  (a) Selection (b) Hybridization
  (c) Mutation (d) None of the above

(77) Sex in plants was discovered by Camararious in ———.
  (a) 1764 (b) 1694
  (c) 1717 (d) None of the above

(78) ——— produced the first artificial hybrid
  (a) Camararious (b) Thomas Fairchild
  (c) Vilmorin (d) None of the above

(79) The main aim of combination breeding is the transfer of ——— into a single variety from another variety or other varieties.
  (a) One or more characters (b) Single character
  (c) Specific character (d) None of the above

(80) ——— refers to the appearance of such plants in an F_2 generation that are superior to both the parents for one or more characters.
  (a) Transgrassive segregation (b) Gene interaction
  (c) Supplementry gene action (d) None of the above

(81) Crossing of parents belong to the same species, they may be two strains, varieties or races of the same species is known as ———.
  (a) Intera specific (b) Interspecific
  (c) Combining ability (d) None of the above

(82) —————includes crosses between different species of the same genus or different genera.
    (a) Interspecific hybridization   (b) Distant hybridization
    (c) Combining ability   (d) None of the above

(83) For ————— at least one of the parents involved in a cross should be a well adapted and proven variety in the area for which new variety is being developed.
    (a) Hybridization   (b) Interspecific hybridization
    (c) Intraspecific hybridization   (d) None of the above

(84) The removal of stamens or anthers or the killing of pollen grains of flower without affecting in any way the female reproductive organs is known as —————.
    (a) Anthesis   (b) Dehiscence
    (c) Emasculation   (d) None of the above

(85) The objective of selfing is to avoid ————— and to ensure self-pollination.
    (a) Mutation   (b) Cross-pollination
    (c) Chromosome abberation   (d) None of the above

(86) The Hardy-Weinberg law is the fundamental law of population genetics and provides the basis for studying Mendelian population. This law was independently developed by—————.
    (a) Hardy (1908) in England and Weinberg (1909) in Germany
    (b) Hayman, 1958   (c) Jones 1954
    (d) Hedric and Booth, 1908

(87) The equilibrium in random mating populations are disturbed by—————.
    (a) Mutation only   (b) Migration only
    (c) Selection only   (d) Random drift only
    (e) All of the above

(88) ————— is the movement of individuals into a population from a different population.
    (a) Mutation   (b) Migration
    (c) Selection   (d) Random drift

(89) ————— is a sudden and heritable change in a organism and is generally due to a structural change in a gene.
    (a) Mutation   (b) Selection
    (c) Random drift   (d) Migration

(90) ————— is a random change in gene frequency due to sampling error.
    (a) Selection   (b) Mutation
    (c) Migration   (d) Random drift

(91) Differential reproduction rates of various genotypes are known as ―――.
   (a) Mutation (b) Selection
   (c) Migrtion (d) Radom drift

(92) There are ――― basic mating schemes
   (a) 6 (b) 5
   (c) 4 (d) 3

(93) Random mating is useful in plant breeding for―――.
   (a) Only progeny testing
   (b) Only production and maintenance of synthetic and composite variety
   (c) Only production of polycross progeny
   (d) a, b, c,

(94) Prepotency is effected by―――.
   (a) Homozygosity (b) Domiannce
   (c) Epistasis (d) Linkage
   (e) All of the above

(95) In genetic assortive mating, the mating is between individuals that are ――― by ancestry than random mating. It is useful in the development of inbreds both partial and complete.
   (a) More closely related (b) Less closely related
   (c) Closely related (d) None of the above

(96) In genetic disassortive mating such individuals are mated which are ――― by ancestry.
   (a) Less closely related (b) Closely related
   (c) More closely related (d) None of the above

(97) Pureline selection is also known as―――.
   (a) Mass selection (b) Modified mass selection
   (c) Individual plant selection (d) None of the above

(98) A pureline variety is characterized as (i) all the plant within a pureline have the same genotype as the plant from which the pureline was derived ―――
   (a) Only same genotype
   (b) Only variation present within pureline
   (c) Genetically variable with time
   (d) All of the above

(99) In pedigree method of breeding, individual plants are selected from ――― and subsequent generations and their progenies are tested in―――.
   (a) F_2 (b) F_3
   (c) F_5 (d) F_{6-7}

(100) Bulk method of breeding was first used by —————.
   (a) Nilsson-Ehle in 1908 at Savalof (b) Nilsson-Ehle in 1918 at Savalof
   (c) Nilsson-Ehle in 1928 at Savalof (d) None of the above

(101) The duration of bulking may vary from ————— generations.
   (a) 3-4   (b) 5-8
   (c) 6-7   (d) 6-7 to 30 or more

(102) The objective of single-seed-descent method is to rapidly advance the generations of crosses at the end of the scheme a random sample ————— is obtained.
   (a) Near heterozygous   (b) Near homozygous
   (c) Multiline   (d) None of the above

(103) A cross between a hybrid (F_1 or a segregating generation) and one of its parents is known as —————.
   (a) SSD   (b) Bulk
   (c) Back cross   (d) None of the above

(104) Genetic consequences of repeated backcrossing resulting —————.
   (a) Rapid increase in homozygosity
   (b) Rapid increase in homozygosity
   (c) To develop isogenic lines
   (d) To develop recombinant inbred line

(105) ————— backcross alongwith selection for the recurrent parent plant type in the early backcross generations will be effective in recovering the genotype of recurrent parent.
   (a) Six   (b) Seven
   (c) Eight   (d) Nine

(106) ————— varieties are mixture of several pureline of similar height, flowering and maturity dates, seed colour and agronomic characteristics each of which has a different gene for resistance to the given disease.
   (a) Multilines   (b) Isogenic
   (c) RIL   (d) NIL

(107) The idea of multiline varieties was put forward by ————— for use in cereals.
   (a) Jenson, 1952   (b) Jenson, 1954
   (c) Jension, 1965   (d) None of the above

(108) In —————, Borlauge suggested that several purelines with different resistance genes should be developed through backcross programmes using one recurrent parent.
   (a) 1952   (b) 1954
   (c) 1965   (d) None of the above

(109) Diallel selective mating scheme was originally suggested by ―――――.
   (a) Goulden 1939 (b) Goulden 1942
   (c) Goulden 1949 (d) None of the above

(110) In case of ――――――――, mass selection or its modifications are used to increase the frequently of desirable alleles, thus improving the characteristics of population.
   (a) Population improvement (b) Progeny selection
   (c) Combination breeding (d) None of the above

(111) Stratified mass selection was suggested by ―――――― which is also known as the grid method of masss selection. The field from which selection is to be done is divided into several small plots e.g. having 40-50 plants each.
   (a) Gardner 1969 (b) Gardner 1961
   (c) Gardner 1967 (d) None of the above

(112) Simple recurrent selection cycle constitutes the ――――――― recurrent selection cycle.
   (a) Second (b) Third
   (c) First (d) None of the above

(113) Recurrent selection for general combining ability is a direct outgrowth of early testing suggested by―――――.
   (a) Jenkin in 1945 (b) Jenkin in 1965
   (c) Jenkin in 1935 (d) None of the above

(114) In 1940 Jenkins proposed a scheme for developing systhetic varieties from short term inbreds. This scheme is essentially――――――. it is effective in changing the GCA in the direction of selection. In addition, it is also effective in increasing the yielding ability of the population obtained at the end of selection cycle.
   (a) RSGCA (b) RRS
   (c) RSSCA (d) None of the above

(115) Reciprocal recurrent selection (RRS) was proposed by―――――. The objective of RRS is to improve two different populations in their ability to combine well with each other.
   (a) Jenkin in 1945 (b) Comstock, Robinson and Harvey in 1949
   (c) Hull 1945 (d) None of the above

(116) Recurrent selection for specific combining ability was first proposed by―――. The objective of RSSCA is to isolate from a population such lines that will combine well with a given inbred.
   (a) Hull in 1945 (b) Jenkin in 1945
   (c) Comstock, Robinson and Harvey in 1949 (d) None of the above

(117) Theoretically in almost all practical situations, RRS may be expected to the superior to ———————.
    (a) Only RSGCA   (b) Only RSSCA
    (c) Both RSGCA and RSSCA   (d) None of the above

(118) ——————— variety is produced by crossing in all combinations a number of lines that combine well with each other.
    (a) A synthetic   (b) A composite
    (c) A multiline   (d) None of the above

(119) The lines that make up a synthetic variety may be inbred lines, clones, open-pollinated varieties, short-term inbred lines or other populations tested for ——————— or for combining ability with each other.
    (a) GCA   (b) SCA
    (c) Both SCA and GCA   (d) None of the above

(120) A ——— is a group of proposed from a single plant through asexual reproduction.
    (a) Clone   (b) Apomixis
    (c) Inbred   (d) None of the above

(121) Genetic variation within clones may rise due ———————.
    (a) Only somatic variation   (b) Mechanical variation
    (c) Occasional sexual reproduction   (d) All of the above

(122) The loss in vigour and productivity of clones with time is known as ———————.
    (a) Clonal degeration   (b) Clonal variation
    (c) Both a and b   (d) None of the above

(123) ——————— was the first to describe plant hybrids during eighteenth century.
    (a) Batesman   (b) Mandel
    (c) J. G Kolereuter   (d) Jones

(124) Charles Darwin discussed hybrid vigour in plants in his book the *Effect of Cross and Self-fertilized in the Vegetable Kingdom*, which appeared in ———————.
    (a) 1876   (b) 1987
    (c) 1786   (d) 1922

(125) In vegetable crops, first suggestion to exploit hybrid vigour was made by Hayes and Jones (1916 a) in ———————.
    (a) Tomato   (b) Chilli
    (c) Onion   (d) Cucumber

(126) F_1 hybrid brinjal was commercially used in Japan before ———————.
    (a) 1925   (b) 1945
    (c) 1956   (d) None of the above

(127) —————— in cabbage was first porposed by Pearson 1931.
 (a) Male sterility   (b) Self- incompatability
 (c) Cytoplasmic male sterility   (d) Geneic male sterility

(128) Male sterility mechanism in onion was first proposed by Jones and Clarke ——————.
 (a) 1921   (b) 1943
 (c) 1958   (d) None of the above

(129) In India first report of hybrid vigour was appeared in 1933 in —————— at IARI New Delhi.
 (a) Chilli   (b) Okra
 (c) Cucumber   (d) None of the above

(130) First hybrid of —————— was developed at IARI, Regional Station Katrain
 (a) Muskmelon   (b) Watermelon
 (c) Bottle gourd   (d) None of the above

(131) Indo – American Hybrid Seed Company, Banglore, took the lead in increase of first tomato hybrid Karnataka and First Capsicum hybrid in —————— for commercial cultivation.
 (a) 1933   (b) 1945
 (c) 1965   (d) None of the above

(132) —————— first product of biotechnology application in vegetable was introduced pilot test market.
 (a) 1999   (b) 1994
 (c) 2001   (d) 1988

(133) A World Bank funded mission mode project under National Agriculture Technology was formulated in ——————.
 (a) 2001   (b) 1998
 (c) 1999   (d) 2004

(134) Heterosis under heterozygous genetic situation break down in subsequent generation and thus is not fixable, such heterosis is termed as ——————.
 (a) Mutational   (b) Relative
 (c) Pseudoheterosis   (d) Liable

(135) The term useful heterosis was used by Meredith and Bridge in ——————.
 (a) 1972   (b) 1998
 (c) 1982   (d) 1975

(136) Dominance hypothesis of heterosis was proposed by Devenport in ——————.
 (a) 1911   (b) 1909
 (c) 1908   (d) None of the above

(137) Over dominance hypothesis of heterosis was first proposed by East and Shull in ─────.

    (a) 1934   (b) 1935

    (c) 1912   (d) 1908

(138) ───────── proposed an alternative hypothesis for over-dominance.

    (a) Milbrow in 1998   (b) East and Shull, 1908

    (c) Mereth and Bridge in 1972   (d) None of the above

(139) The additive dominance model, based on the biometrical procedures developed by Mather and Jinks in ───── and ─────.

    (a) 1982 and 1984   (b) 1971 and 1982

    (c) 1999 and 2000   (d) None of the above

(140) The phenomenon of hybrid vigour in tomato to was first observed as early as in the beginning of ───── century.

    (a) 18th   (b) 19th

    (c) 20th   (d) 21th

(141) Wellington (1922) proposed the large scale use of F_1 hybrids of ───── for the first time.

    (a) Cucumber   (b) Tomato

    (c) Chilli   (d) Brinjal

(142) In tomato, reported range of heterosis for various yield attributes is ─────.

    (a) 10-15%   (b) 20-25%

    (c) 40-60%   (d) None of the above

(143) ───── were probably the first to observe hybrid vigour in brinjal crosses among some Japanese varieties.

    (a) Nagi and Kida (1926)   (b) Wellington (1922)

    (c) Muller (1943)   (d) None of the above

(144) Reported range of heterosis in chilli is ─────.

    (a) 60-70%   (b) 20-30%

    (c) 10-20%   (d) None of the above

(145) In India, first report on heterosis in chilli came from Deshpande in ─────.

    (a) 1943   (b) 1933

    (c) 1956   (d) None of the above

(146) In bell pepper, the average range of heterosis for yield and its contributing characters has been reported as ─────.

    (a) 10-20%   (b) 30-50%

    (c) 20-25%   (d) None of the above

(147) Hayes and Jones were the first to observe heterosis in cucumber in ———.
 (a) 1916 (b) 1930
 (c) 1925 (d) None of the above

(148) ——— reported first time an outstanding hybrid of watermelon by crossing AsahiYamato X Miyako.
 (a) Deshpande, 1933 (b) Wellington, 1922
 (c) Yanagisawa and Hosono, 1951 (d) None of the above

(149) ——— is one of the pioneer crop in which heterosis has been commercially exploited after the discovery of gene- cytoplasmic male sterility in 1944.
 (a) Tomato (b) Cucumber
 (c) Onion (d) None of the above

(150) ——— was perhaps the first to report crossing of two inbred lines of carrot exhibited heterosis.
 (a) Wellington, 1922 (b) Deshpande, 1933
 (c) Poole, 1937 (d) None of the above

(151) ——— are well known for their effect on sexual determination and floral development.
 (a) Auxins (b) Cytokinins
 (c) Gibberellins (d) None of the above

(152) ——— suggested first time gamete method of selection.
 (a) Richey, 1927 (b) Stadler, 1944
 (c) Dhawan, 1965 (d) Wellington, 1922

(153) ——— first suggested convergent method of breeding
 (a) Richey, 1927 (b) Dhawan, 1965
 (c) Poole, 1937 (d) Yanagisawa and Hosono, 1951

(154) ——— first time suggested Step Lader method of breeding
 (a) Poole, 1937 (b) Richey, 1927
 (c) Dhawan, 1965 (d) None of the above

(155) In tomato——— at 0.05 % can be applied on the pollinated flowers as it increases hybridization.
 (a) PCPA (b) CCC
 (c) GA (d) NAA

(156) ——— was first reported male sterility in tomato and geneic male sterile lines are more economical.
 (a) Poole, 1937 (b) Wellington, 1922
 (c) Crane, 1915 (d) None of the above

(157) ———— recessive marker seed parent like potato leaf type, anthocyaninless (aa) and brown seeded can be used to detect the purity of hybrid seed, of ————.

    (a) Chilli   (b) Brinjal
    (c) Tomato   (d) Capsicum

(158) In tomato single cross normally produces ———— seeds depending upon parents used.

    (a) 10-20   (b) 20-30
    (c) 40-50   (d) 300-400

(159) The extent of cross – pollination in brinjal has been reported as high as ———— %.

    (a) 50   (b) 29
    (c) 40   (d) 35

(160) In brinjal ———— crossed fruit will yield 210g hybrid seed.

    (a) 110   (b) 125
    (c) 70   (d) 200

(161) The extent of natural crossing in chilli is reported upto ————.

    (a) 50   (b) 25
    (c) 45   (d) 80

(162) GA_3 at ———— mg/l in bell pepper thrice at on set of flowering and once again after 10 days causing complete male sterility, which lasted through out the season.

    (a) 200   (b) 500
    (c) 10,000   (d) 5000

(163) Pollen grains of okra are ————.

    (a) Polysiphonous   (b) Gamosephallus
    (c) Monosiphonous   (d) None of the above

(164) Double cross method was first utilized by ———— for production of stem kale hybrid and later utilized in other cruciferous crops.

    (a) Sheshadri, 1958   (b) UN, 1939
    (c) Thompson, 1959   (d) None of the above

(165) Cytoplasmic male sterility in cabbage was given by ————.

    (a) Thompson, 1959   (b) Ogura, 1968
    (c) Poole, 1937   (d) None of the above

(166) In bottle gourd, F_1 seeds can be produced by application ethrel at the rate of ———— at two true leaf and four true leaf stages

    (a) 100-200   (b) 200-300
    (c) 500   (d) None of the above

(167) In bottle gourd geneic male sterile line is controlled by ————.
    (a) *ms-1*   (b) *ms-2*
    (c) *ms-3*   (d) *Ms-1*

(168) ———————— advocated the use of gynomonoecious lines for the production of hybrid seed of cucumber.
    (a) Dutta, 1983   (b) Ogura, 1968
    (c) Thompson, 1959   (d) Peterson and Weigle, 1958

(169) Gynoecious lines in muskmelon can be maintained by application of ———— @ 100-200 ppm
    (a) GA_3 (b) NAA
    (c) Ethrel   (d) Silver nitrate

(170) The non-lobing leaf character governed by single recessive gene, which is expressed in the seedling stage, can be used as marker gene for producing hybrid seed in————.
    (a) Muskmelon   (b) Watermelon
    (c) Cucumber   (d) Pumpkin

(171) ———— a variety called Satputia bears, a hermaphrodite flower
    (a) Sponge gourd   (b) Ridge gourd
    (c) Cucumber   (d) Pumpkin

(172) ———————— a brown anther type in which the anthers are shriveled and brown without viable pollen and the petalloid type
    (a) Carrot   (b) Radish
    (c) Tomato   (d) Brinjal

(173) ———————— is prepared by dissolving any substance corresponding to one molecular weight (1 mole in water solvent) so as to obtain a final volume of exactly one litre at 20°C
    (a) Molar solution   (b) Normal solution
    (c) Molal solution   (d) None of the above

(174) ———————— is when one part of a substance is dissolved in one million part of the solvent i.e. if 1 g of a substance is dissolved in 10^6 mg.
    (a) Molar solution   (b) ppm
    (c) Molar solution   (d) Normal solution

(175) Sharkdakov method is used for measurement of————————.
    (a) Water potential   (b) Normality
    (c) Molarity   (d) None of the above

(176) ———————— is used for determination of Relative Vapour Pressure (RVP)
    (a) Psychrometer   (b) Plasmolytic method
    (c) Tensiometer   (d) None of the above

(177) —————— is the sum total of all those properties in seeds which upon planting results in rapid and uniform production of healthy seedling under a wide range of environment both favourable and stress conditions.
    (a) Germination   (b) Seed vigour
    (c) Seed viability   (d) None of the above

(178) —————— is a reliable and quick method for determining the seed viability.
    (a) Sandpaper   (b) Brown paper
    (c) Tetrazolium   (d) None of the above

(179) —————— is the total amount of organic matter assimilated less that lost in respiration.
    (a) Total photosnthesis   (b) Gross primary production
    (c) Net primary production   (d) All of the above

(180) Relative growth rate is the product of ——————.
    (a) NAR X LAR   (b) LAR X LWR
    (c) LWR X SLW   (d) None of the above

(181) —————— is the ratio of dry weight of leaves to whole plant dry weight.
    (a) SLW   (b) AGR
    (c) SLA   (d) LWR

(182) —————— is the function of amount of growing material present and is influenced by the environment.
    (a) SLW   (b) AGR
    (c) LWR   (d) NAR

(183) —————— is the ratio of total assimilatory surface to whole plant dry weight at any instant of time.
    (a) LAR   (b) NAR
    (c) LWR   (d) SLA

(184) —————— is the ratio of leaf area to leaf dry weight
    (a) SLW   (b) AGR
    (c) SLA   (d) LWR

(185) —————— may be defined as the area of leaves (one side only) divided by the ground area over which it is growing.
    (a) CGR   (b) LAI
    (c) LAD   (d) HI

(186) —————— is the ratio of economic yield to the total biological yield expressed in percentage.
    (a) LAD   (b) CGR
    (c) HI   (d) None of the above

(187) In an optimal environment there is no interference by any environment factor with the complete expression of the genotypic potential of an individual / line; such an environment is, therefore, also called ——————.
    (a) Stress –free environment   (b) Stress environment
    (c) Optimal envirnment   (d) None of the above

(188) An ——————is an ideal that is rarely if even achieved and it is difficult to imagine a stress-free area for commercial cultivation of any crop.
    (a) Optimal environment   (b) Stress environment
    (c) Stress –free environment   (d) None of the above

(189) Resistance to an ——————stress may be defined as the ability of a plant / line to produce higher economic yields than other plants/lines subjected to same / comparable levels of the given stress.
    (a) Abiotic   (b) Biotic
    (c) Both a and b   (d) None of the above

(190) —————— may be defined as 'the inadequacy of water availability, including precipitation and soil moisture storage capacity, in quantity and distribution during the life cycle of crop to restrict (the) expression of its full genetic yield potential.
    (a) Drought   (b) Dry period
    (c) Wet period   (d) None of the above

(191) The performance of a crop under water stress will be affected by the integrated effects of water stress at all the levels of plant organization. The level of water stress in reproductive meristems is —————— than that in the transpiring leaves of a plant at any given time.
    (a) Higher   (b) Lower
    (c) Optimum   (d) None of the above

(192) —————— resistance may be defined as the mechanism(s) causing minimum loss of yield in a drought environment relative to the maximum yield in a constraint free, i.e. optimal environment of the crop.
    (a) Drought   (b) Dry period
    (c) Stress period   (d) None of the above

(193) —————— describes the situation where an otherwise drought susceptible variety performs well in a drought environment simply avoiding the period of drought. Early maturity is an important attribute of drought escape and is suitable for environments subjected to late-season drought stress.
    (a) Drought escape   (b) Dehydration avoidance
    (c) Dehydration tolerance   (d) All of the above

(194) ———— is the ability of a plant 'to retain a relatively higher level of hydration under conditions of soil or atmospheric water stress. This results in the various physiological, biochemical and metabolic processes of plants that are involved in growth and not by being internally exposed to stress and thereby, they are protected from water stress.

  (a) Drought escape   (b) Dehydration tolerance
  (c) Dehydration avoidance   (d) None of the above

(195) ———— positively affects growth and yield under stress. It is probably one of the most important and effective components of drought resistance.

  (a) Osmotic turger   (b) Osmotic adjustment
  (c) Osmotic pressure   (d) None of the above

(196) ———— is known as stress hormone as its concentration is incresase in response to stresses, including water stress. ———— is transported via xylem from roots to leaves within minutes to hours, it half-life in leaf being – 30 minutes.

  (a) Ethylene   (b) ABA
  (c) IAA   (d) GA

(197) In stress, expression of several genes are induced by one or more stresses, these genes are called ———— and the protein they encode are known as stress protein.

  (a) Stress gene   (b) Stress-responsive genes
  (c) Stress non-responsive gene   (d) None of the above

(198) Some genes are induced ———— hour after dehydration begins, they are called early responsive to dehydration (ERDS) genes.

  (a) One   (b) Two
  (c) Three   (d) Four

(199) Many of the stress proteins e.g. dehydrin, osmotin, lea proteins etc., are also produced in response to ————; genes.

  (a) ABA   (b) Cytokinin
  (c) GA   (d) None of the above

(200) ———— accumulation are involves in tolerance to water and other stresses.

  (a) Aline   (b) Proline
  (c) GA   (d) None of the above

(201) Genetic control of drought resistance traits are ranges oligogenic to polygenic and generally ———— gene effects are involved.

  (a) Only additive   (b) Only dominance
  (c) Only epistasis   (d) Both additive and dominance

(202) ———— refers to the unintended transfer of undesirable genes due to their tight linkage with the desirable gene being transferred.

  (a) Linkage drag   (b) Hereditary
  (c) Genetic erosion   (d) Receptor

(203) ———————— refers to improved resistance of a genotype to drought as a consequence of a seed / seedling treatment. The various hardening treatments are classfied into two groups (1) pre-sowing and (2) post-sowing treatments.
    (a) Drought succeptibility   (b) Drought hardening
    (c) Drought resistance   (d) None of the above

(204) It is generally accepted that breeding for salinity resistance is most likely to succeed for soils that are upto ————————.
    (a) Saline   (b) Highly saline
    (c) Moderately saline   (d) Low saline

(205) ———————— is that their saturated soil paste extract has an electrical conductivity of more than 4ds/m (deciseimens per metre).
    (a) Alkali soil   (b) Acid soil
    (c) Saline soil   (d) None of the above

(206) The soil solution of wet saline soil may be at water potential of ———————— the value of sea water and higher.
    (a) -30 bar   (b) -35 bar
    (c) -24 bar   (d) None of the above

(207) ———————— adversely affects the activity of enzymes of the glycolytic pathway.
    (a) HCl   (b) NaCl
    (c) H_2SO_4 (d) None of the above

(208) ———————— are adapted to saline conditions and are wild species.
    (a) Halophyte   (b) Bryophyte
    (c) Thalophyte   (d) Hydrophyte

(209) ———————— is an important solute in osmoregulation of halophytes and ———————— may be considered as an ideal osmocutin.
    (a) ABA   (b) Proline
    (c) GA   (d) Cytokinin

(210) ———————— may be defined as a differential effect on various life processes of the same tissue concentration of salt in different genotype of a species
    (a) Salt tolerance   (b) Drought tolerance
    (c) Acid tolerance   (d) None of the above

(211) Osmoregulation may be measured as ———————— accumulation in response to salinity stress.
    (a) Fat   (b) Acid
    (c) Carbohydrate   (d) None of the above

(212) *Lycopersicum cheesmanii, L. peruvianum* and *Solanum pennellii* relatives of tomato are remarkably resistant to ————————.
    (a) Draught   (b) Salinity
    (c) Alkalinity   (d) None of the above

(213) Nutrient absorption by plant roots follows ─────────. Mass flow of ions is usually able to provide the required amount of N, S, Ca and Mg, but not of P, K and some minor elements.
    (a) Nutrient of absorption relationship
    (b) Michaelis-Menten relationship
    (c) Arnon theory
    (d) None of the above

(214) When plant performance is adversely affected by excess of minerals or deficiency of an essential nutrient in soil, it is called ─────────.
    (a) Water stress
    (b) Mineral stress
    (c) Salt stress
    (d) None of the above

(215) The various mechanisms that confer resistance to mineral deficiency are as ─────
    (a) Only mineral redistribution
    (b) Only efficient mineral uptake
    (c) Only increased root / shoot ratio
    (d) Only increased root-hair density / length and increased mineral transfer from root
    (e) All of the above

(216) In several cases, mineral-efficiency / deficiency has been found to govern by a single gene. Boron inefficiency in tomato and celery is controlled by ─────────.
    (a) Single dominant gene
    (b) Single recessive gene
    (c) Double dominant and resessive gene both
    (d) None of the above

(217) In case of tomato ───────── allele is believed to interfere with boron translocation from the roots.
    (a) hpl
    (b) btl
    (c) Nor
    (d) None of the above

(218) The iron efficient genotype of tomato, '─────────' is able to acidify the rhizosphere and reduce F^{3+} to Fe^{2+}.
    (a) Rutgers
    (b) Best of All
    (c) Pusa Rubey
    (d) None of the above

(219) ─────────inefficiency in celery and Fe-inefficiency in maize are conditioned by single recessive genes.
    (a) Ca
    (b) Mn
    (c) Mg
    (d) K

(220) Mineral toxicity stress is generated by the presence of toxic concentration of available minerals in the root zone. Acidic soils may contain toxic concentrations of ───────── and ─────────.
    (a) Ca and Na
    (b) Al and Mn
    (c) Fe and Zn
    (d) None of the above

(221) The most important causes of mineral toxicity as they represent about ——— of the arable land worldwide.
  (a) 10% (b) 40%
  (c) 60% (d) None of the above

(222) ——— toxicity reduces root growth, causes their discoloration and inhibits lateral root formation.
  (a) Mn (b) Mg
  (c) Al (d) None of the above

(223) Generally, seedlings are more sensitive to ——— than older plants.
  (a) Na (b) Fe
  (c) Al (d) All of the above

(224) ——— toxicity is the second most common problem in acid soils.
  (a) Fe (b) Al
  (c) Mn (d) Mg

(225) Mn toxicity may occur at any soil pH below ———.
  (a) 6.5 (b) 7.5
  (c) 5.5 (d) 4.5

(226) ——— decreases the acitivity of various enzymes but increases that of various oxidases, including IAA oxidase, reduces respiration and ATP levels and affects Ca and Fe metabolisms.
  (a) Na and Mn (b) Ca and Fe
  (c) Fe and Al (d) All of the above

(227) ——— toxicity symptom occur mostly in shoots, leaf chlorosis and necrosis, leaf crinking and 'cupping' or 'puckering' are the typical symptoms.
  (a) Fe (b) Mo
  (c) Bo (d) Mn

(228) Al and Mn toxicity is controlled by ———.
  (a) Single gene (b) Double gene
  (c) Polygene (d) None of the above

(229) The adverse effects on plants of temperature higher than the optimal is considered as ———.
  (a) Heat stess (b) Heat tolerant
  (c) Heat sensitive (d) None of the above

(230) ——— is a group of about a dozen or so proteins that normally exist cells, but their synthesis is accelerated by heat. The acceleration phase is about 20-30 min in bacteria and several hours in plants, after this phase, a new steady state is reached. In case of *E.coli*, ——— constitutes about 25% of the total cellular proteins at the upper range of its growth temperature.

(a) HSP (b) HIP
(c) HPL (d) None of the above

(231) ──────── may be defined as the ability of some genotypes to perform better than others when they are subjected to the same level of heat stress.
(a) Heat injury (b) Heat stress resistance
(c) Heat tolerant (d) None of the above

(232) ──────── are a component of heat avoidance.
(a) Root (b) Shoot
(c) Leaf surface (d) None of the above

(233) ──────── may be defined as an improved ability of genotype to withstand a period of high temperature as a consequence of an earlier exposure to high temperatures for a given period of time.
(a) Heat avoidance (b) Heat hardening
(c) Heat tollerance (d) None of the above

(234) The exposure time of heat hardening vary from ────────
(a) 20 sec at 50°C to 4-6 hr <1 hr (b) 30 sec at 50°C to 4-6 hr <3 hr
(c) 40 sec at 50°C to 4-6 hr <4 hr (d) 20 sec at 50°C to 4-6 hr <6 hr

(235) When heat hardened plants are subjected to lower temperatures, they lose their hardening, this is called────────.
(a) Hardening (b) Dehardening
(c) Tolerance (d) Heat accumulation

(236) ──────── are most important component of heat hardening.
(a) Common protein (b) Heat shock proteins
(c) CHO (d) None of the above

(237) Heat tolerance usually improves membrane stability under heat stress. Membrane stability may be determined as ────────.
(a) Only liquid fluidity (b) Only electrolyte leakage
(c) Both a and b (d) None of the above

(238) Callose formation in the sieve tubes is the major cause for inhibition of translocation by heat in ────────.
(a) Brinjal (b) Tomato
(c) Chilli (d) Potato

(239) In general, use of exotic genetic resources is not essential for attaining progress in────────.
(a) Heat tolerance (b) Heat avoidance
(c) Heat resistant (d) None of the above

(240) ──────── (%) = $[1 - \{\frac{1-(T_1/T_2)}{1-(C_1/C_2)}\}]$

where, T_1 and T_2 are mean conductivities of treatment vials for a genotype before and after autoclaving, respectively, while C_1 and C_2 represent the mean conductivities of the control vials of the same genotype before and faster autoclaving, respectively.

 (a) Heat injury   (b) Heat tolerance
 (c) Heat avoidance   (d) None of the above

(241) When temperature remain above freezing i.e. > 0°C, it is called ────────
 (a) Chilling   (b) Freezing
 (c) Cooling   (d) None of the above

(242) ──────── describes temperature below freezing i.e. < 0°C.
 (a) Freezing   (b) Cooling
 (c) Hardening   (d) None of the above

(243) ──────── accumulates in chilli-affected plants and it may be involved in chilli hardening.
 (a) GA   (b) Cytokinin
 (c) ABA   (d) Alar

(244) At chilling temperature, the most affected processes / enzymes are: electron transport, RUBISCO (ribulose – 1, 5- biphosphate carboxylase, particularly in C_4 plants), PEP carboxylase plus NADP malate dehydrogenase in ────────.
 (a) Paddy   (b) Maize
 (c) Tomato   (d) Cowpea

(245) Ability of some genotypes to survive / perform under chilling stress than other genotypes is called ────────. Ordinarily, it is the consequence of chill hardening.
 (a) Chilling avoidance   (b) Chilling tolerance
 (c) Chilling injury   (d) None of the above

(246) Chloroplast and photosynthesis are major sites of ────────.
 (a) Chilling injury   (b) Chilling tolerance
 (c) Chilling resistance   (d) Chilling avoidance

(247) Chilling tolerance in ──────── is related to longer styles, and flower position within a flower cluster.
 (a) Tomato   (b) Chilli
 (c) Brinjal   (d) All of the above

(248) High radiation promotes chlorophyll loss at ────────.
 (a) High temperature   (b) Low temperature
 (c) Medium temperature   (d) None of the above

(249) Chilling injury to photosynthesis is assayed as variable chlorophyll fluorescence at————.
   (a) 985nm   (b) 885nm
   (c) 685nm   (d) 785nm

(250) ———————— is conducive to freezing resistance, while resistance in actively growing tissues is rare.
   (a) Dormant   (b) Growing
   (c) Germination   (d) Growth

(251) The ability of plant tissues / organs (but not the whole plants) to avoid ice formation at subzero temperatures is called freezing avoidance. Supercooling is a mechanism of freezing avoidance. Super cooling may be effective even————.
   (a) -97^0C (b) -67^0C
   (c) -47^0C (d) -110^0C

(252) Ability of plants to survive the stresses generated by extracellular ice formation and to recover and regrow after thawing is known as————————.
   (a) Freezing tolerance   (b) Freezing resistance
   (c) Freezing injury   (d) None of the above

(253) In freezing tolerance osmotic adjustment is effective only to a limited extent and, for this reason, genetic improvement in this characteristic is limited. On the other hand, cytoplasmic water binding and plasma lemma stability under freeze-thaw stress are likely for more effective in inducing freezing tolerance; theoretically they can induce freezing tolerance to upto————————.
   (a) -40^0C (b) -140^0C
   (c) -80^0C (d) -70^0C

(254) Inheritance of winter hardiness is complex and when measured in the field shows poor heritability mainly due to large ————————interaction effects.
   (a) G X G   (b) G X P
   (c) G X E   (d) None of the above

(255) Freezing tolerance is a ———————— trait and that additive component is usually preponderant.
   (a) Monogenic   (b) Diogenic
   (c) Polygenic   (d) None of the above

(256) The term ideotype was introduced by———————— literally means "a form denoting an idea".
   (a) Beaven, 1914   (b) Donald, 1968
   (c) Smith, 1934   (d) None of the above

(257) The ratio of economic yield to biological yield is described as harvest index (HI) by ———— The ideotype therefore must include such morphological and physiological characteristics that result in a high harvest index; this is a critical aspect of plant design.
    (a) Donald, 1962   (b) Donald, 1968
    (c) Donald, 1964   (d) None of the above

(258) Ideotype breeding aims to enhance genetic yield potential by modifying ——. to their predefined optimum levels.
    (a) Complex traits   (b) Double traits
    (c) Individual traits   (d) None of the above

(259) ———— is the ability of a pathogen to attack a host and is synonymous to virulence.
    (a) Expressivity   (b) Pathogenisity
    (c) Threshhold   (d) None of the above

(260) The term ———— refers to the freedom of susceptible host plant varieties from a disease due to environmental factors.
    (a) Disease tolerance   (b) Diseases escape
    (c) Disease succeptible   (d) None of the above

(261) When the host does not show the symptoms of a disease, it is known as immune reaction.
    (a) Desease tolerance   (b) Desease susceptible
    (c) Immune reaction   (d) None of the above

(262) The terms vertical and horizontal resistance were introduced by————.
    (a) Hugo Deveris   (b) Vander Plank.
    (c) Jension   (d) None of the above

(263) The resistance, due to the reaction incited by the attack of the pathogen, may be due to biosynthesis of some chemical substances is known as————.
    (a) Passive resistance   (b) Active resistance
    (c) Negative resistance   (d) None of the above

(264) The resistance is due to the qualities innate in the host prior to the attack of the pathogen is known as passive resistance.
    (a) Passive resistance   (b) Active resistance
    (c) Negative resistance   (d) None of the above

(265) Resistance which influences the epidemics in the field but not immediately under controlled conditions. This relates mainly to horizontal resistance is known as— ————.
    (a) Active resistance   (b) Field resistance
    (c) Passive resistance   (d) None of the above

(266) Oligogenic resistance is synonymous to————————————.
    (a) Active resistance (b) Vertical resistance
    (c) Passive resistance (d) Field resistance

(267) The gene-for gene relationship between a host and its pathogen was postulated by Flor in ———————————— based on his work on linseed (*Linum usitatissimum*) rust. It is the consequence of specific interactions between the products of genes governing host resistance and those conditioning pathogenecity.
    (a) 1956 (b) 1965
    (c) 1967 (d) 1975

(268) ————————————, the term Vertifolia effect derived from the name of a German potato Vertifolia having the late blight resistance genes R_3 and R_4.
    (a) Vander Plank (b) Pristly
    (c) Flor (d) None of the above

(269) 'Boom and bust' cycle was given by Pristly in ————————————.
    (a) 1970 (b) 1980
    (c) 1950 (d) 1960

(270) The terms strong and weak oligogenes were introduced by ————————————. When the pathotype virulent towards a resistance gene has poor survival value in
    (a) Pristly (b) Flor
    (c) Johenson (d) Van der Plank

(271) The varieties having vertical resistance genes usually have low levels of ———————————— resistance.
    (a) Horizontal (b) Vertical
    (c) Both horizonatal and verticle (d) None of the above

(272) ———————————— refers to an adverse effect of feeding on a resistant host plant on the development and or reproduction of insect-pest.
    (a) Antibiosis (b) Pest avoidance
    (c) Pest susceptable (d) None of the above

(273) An insect tolerant variety is attacked by the insect-pest to the same degree as a susceptible variety. But at the same level of infestation, a tolerant variety produce a larger yield than a ———————————— variety.
    (a) Tolerant (b) Resistant
    (c) Suseptable (d) None of the above

(274) ———————————— is the same as disease escape and as such, it is not a case of true insect resistance.
    (a) Pest suseptable (b) Pest resistance
    (c) Pest escape (d) Pest avoidance

(275) A number of morphological factors e.g., hairness, colour, thickness and toughness or tissues etc. are known to confer ——————.
    (a) Insect resistance   (b) Insect avoidence
    (c) Insect tolerance   (d) None of the above

(276) Genetics of insect resistance may be governed by ——————.
    (a) Only oligogenes   (b) Only polygenes
    (c) Only cytoplasmic gene   (d) None of the above

(277) The larger durability of resistance governed by —————— and that by two or more oligogenes stems from the fact that such cases generally involve more than one feature of the host plant.
    (a) Polygenes   (b) Single gene
    (c) Cytoplasmic gene   (d) All of the above

(278) Insect resistance associated with ——————and / or an anatomical character of the host plant is generally more durable than the other kind of resistance.
    (a) Genetical characters
    (b) Morphological charecters
    (c) Both genetical and morphological charecters
    (d) None of the above

(279) In case of tomato, Red fruits are the most preferred. The allele —————— (Crimson colour) increases lycopene (red pigment) concentration at the cost of beta-carotene, while hp gene (for high pigment) increases total carotenoids and produces excellent fruit colour. But hp allele is associated with slow germination and growth and premature defoliation.
    (a) AGC   (b) ogc
    (c) CGO   (d) goc

(280) Gene ogc and hp affect vitamin A content: ogc cause a —————— decline, while hp leads to a —————— increase in the content of beta-carotene.
    (a) 20%   (b) 5%
    (c) 15%   (d) 40%

(281) A variety containing high vitamin C are ——————.
    (a) Extra yielder   (b) Poor yielder
    (c) Yielder   (d) Good yielder

(282) In tomatoes, pH varies from ——————.
    (a) 2.26 to 3.82   (b) 1.26 to 2.82
    (c) 4.26 to 4.82   (d) 7.26 to 8.82

(283) In tomato varieties acidity ranges from ——————.
    (a) 0.4 to 0.91   (b) 0.04 to 0.91
    (c) 1.4 to 1.91   (d) 2.4 to 3.91

(284) TSS is negatively correlated with yield and fruit size and determinate types have lower TSS than ─────── ones.
   (a) Determinate (b) Indeterminate
   (c) Semi-determinate (d) None of the above

(285) ─────── variety of tomato possessing high β-carotene (10 times more β-carotene)
   (a) Balkan (b) Caro Red
   (c) Flora-Dade (d) Punjab Chhuhara

(286) A mutant gene called ─────── in tomato enhances red colour in tomato fruits by lowering β-carotene content and enhancing lycopene content.
   (a) Yellow (b) Blue
   (c) Brown (d) Crimson

(287) In commercial varieties carrots, carotene constitutes about ─────── of root fresh weight.
   (a) 10-20 µg/g (b) 100-120 µg/g
   (c) 50-60 µg/g (d) 5-20 µg/g

(288) Favism is characterized by haemolytic anaemia, haemoglobinusia and jaundice. Favism is caused by consumption of uncooked ───────.
   (a) Sem (b) Cowpea
   (c) Guar (d) Broad bean

(289) Production of first haploid plants by culturing pollen grains was developed by ───────
   (a) JP Nitch and Nitch (b) Guha and Maheshwari
   (c) Skoog and coworker (d) P. R White

(290) Disarmed Ti plasmid contain gene(s)
   (a) iaaM (b) ipt
   (c) iaaH (d) None of the above

(291) ─────── is gusogen
   (a) Glycerol (b) PEG
   (c) Blue vitrole (d) Alum

(292) Tag polymerase does not contain measurable ───────.
   (a) 3' to 5' exonucleaese activity
   (b) 5' to 3 exonuclease activity
   (c) 5' to 3' polymerization activity
   (d) None of the above

(293) In PCR we get annealing programme at ───────
   (a) 55°C (b) 75°C
   (c) 35°C (d) 95°C

(294) In Southern and Nothern blot ————— and ————— sample is transferred to nitrocellulose paper, respectively.
    (a) DNA and protein   (b) RNA and DNA
    (c) DNA and RNA   (d) RNA and protein

(295) Restricted enzyme isolated from—————.
    (a) Bacteria   (b) Virus
    (c) Yeast   (d) Bacteriophase

(296) SSR stand for—————.
    (a) Simple surround Repeats   (b) Sound surface Repeats
    (c) Simple sequence Repeats   (d) Simple satellite Repeats

(297) Ri plasmid found in—————.
    (a) *Agrobacterium rhizocotona*   (b) *A .rhizogenes*
    (c) *Aspergillus indicus*   (d) *A. termefaciens*

(298) Ti plasmid found in —————.
    (a) A . rhizogenes   (b) A .rhizocotona
    (c) A .termifaciens   (d) None of the above

(299) PCR was invented by—————.
    (a) Karry Packers   (b) Karry Mullus
    (c) David Mulford   (d) None of the above

(300) *Bt* gene was isolated from—————.
    (a) *Bacillus thurengensis*   (b) *Bacillo cirus*
    (c) Basidomycetes   (d) *Pseudomonas syrigine*

(301) First transgenic crop for cultivation was released in—————.
    (a) Rice   (b) Tomato
    (c) Potato   (d) Pea

(302) pH range of plant nutrient medium is from—————.
    (a) 5-6   (b) 4.5-7
    (c) 7   (d) 7-8

(303) ————— is not sterilization technique
    (a) Filter sterilization   (b) UV sterilization
    (c) Yellow light sterilization   (d) Heat sterilization

(304) Anther culture is an example of—————.
    (a) Seed culture   (b) Organ culture
    (c) Protoplast culture   (d) All of the above

(305) Which one is false about callus——————.
- (a) It is nonorganised tumer tissue
- (b) Parenchymatus in nature
- (c) Friable and non-friable are two type callus
- (d) Callus tissue from different plant species are alike

(306) Virus free plant could be generated by ——————.
- (a) Callus culture
- (b) Shoot tip culture
- (c) Antibiotic treated callus culture
- (d) Protoplast culture

(307) First haploid culture was successful in ——————.
- (a) Carrot
- (b) Datura
- (c) Okra
- (d) Petunia

(308) Which one is not used as asmoticum during plant protoplast culture——————.
- (a) Mannitol
- (b) $CaCl_2$
- (c) Sorbitol
- (d) Glucose

(309) Which staining method is used in checking the viability of protoplast——————
- (a) Gram stain
- (b) FDA stain
- (c) Both gram and FDA stain
- (d) None of the above

(310) Temperature of solid carbon dioxide is——————.
- (a) $-79°C$
- (b) $-10°C$
- (c) $0°C$
- (d) $-196°C$

(311) —————— is not caryoprotectant.
- (a) DMSO
- (b) Sucrose
- (c) Glycerol
- (d) Protein

(312) —————— is left handed DNA.
- (a) A-DNA
- (b) B-DNA
- (c) c-DNA
- (d) Z-DNA

(313) —————— open reading frame could be occur in a DNA.
- (a) 2
- (b) 3
- (c) 4
- (d) 6

(314) Reverse transcription is——————.
- (a) Formation of RNA
- (b) Formation of protein on RNA
- (c) Formation of DNA on RNA
- (d) Direct formation of protein on DNA

(315) —————— is the controlling sequence on gene.
- (a) Mitron
- (b) Exon
- (c) Peltier box
- (d) Tata box

(316) Modified nitrogenous bases are found in—————————.
  (a) ERNA (b) rRNA
  (c) mRNA (d) RNA

(317) ————————— is not a termination signal in translation.
  (a) VAA (b) VAG
  (c) AGV (d) VGA

(318) ————————— is not a plasmid vector.
  (a) PBR 322 (b) PVC
  (c) Bacteriophase (d) PUN121

(319) YAC stands for—————————————.
  (a) Yeast Agriculture Calender (b) Yeast artificial chromosome
  (c) Yield per artificial crop (d) None of the above

(320) pBhuescript IIks is a—————————————.
  (a) Plasmid (b) Phasemids
  (c) Cosmid (d) Bacteriophase vector

(321) Unknown DNA sequences that lie outside the boundary of known sequences can be amplified by—————————.
  (a) RT-PCR (b) Inverse-PCR
  (c) Simple PCR (d) None of the above

(322) ————————— is not a method of DNA sequencing.
  (a) Sanger technique
  (b) Maximum Gilbert method
  (c) Flourescent Chain terminating dihydroxynuclotide method
  (d) None of the the above

(323) Lypase enzyme is used to—————————.
  (a) Cutting of the DNA
  (b) Joining of the DNA
  (c) Modification of the base of the DNA
  (d) Compressing of the DNA

(324) ————————— is a biochemical marker
  (a) Isoenzyme (b) RAPD
  (c) PCR (d) RFLP

(325) Hairy root disease is induced by—————————————.
  (a) Nitrobactor (b) *Agrobacterium temefaciens*
  (c) *A. rhizogenes* (d) *A. radiobacter*

(326) *Agrobacterium* is a —————————————————.
   (a) Gram+ve bacillus
   (b) Gram negative bacillus
   (c) Gram positive cocci
   (d) Gram negative cocci

(327) ————————— is the best suitable media for root culture.
   (a) MS media
   (b) LB media
   (c) Whites media
   (d) B5 media

(328) All Ti and Ri plasmid has a ————————— bp highly conserved direct repeat DNA sequence that flanks the DNA.
   (a) 10
   (b) 15
   (c) 25
   (d) 35

(329) Vector which can replicase in two organism is known as —————————.
   (a) Plasmid vector
   (b) Shuttle vector
   (c) Express vector
   (d) Cosmid

(330) ————————— is not a vectorless or direct DNA transfer method.
   (a) Micriinjection
   (b) Particle bombardment
   (c) Electroporation
   (d) Agrobacterium method gene transfer

(331) *Bacillus thurengensis* accumulates ————————— during sprulation.
   (a) Alpha- exotoxin
   (b) Mytoxin
   (c) r-exotoxin
   (d) None of the above

(332) Production of transgenic Flavr Savr tomato by ————————— technique.
   (a) Antisense RNA gene
   (b) GUS reporter gene
   (c) Lac gene
   (d) cDNA copy

(333) Terminator technology is based on —————————.
   (a) Epistatic gene
   (b) Lethel gene
   (c) Dominant gene
   (d) Suppressor gene

(334) Nucleoside is composed of —————————.
   (a) Base, phosphse and sugar
   (b) Base and phosphase,
   (c) Base and sugar
   (d) None of the above

(335) p^{BR322} vector have ————————— antibiotic resistant gene
   (a) Kanamycin and tetracycline
   (b) Tetracycline and hygromycin
   (c) Chloramphenicol and ampicillin
   (d) Ampicillin and tetracycline

(336) Our hands are good source of ————————— so we use gloves during RNA isolation.
   (a) DNase
   (b) RNase
   (c) Phosphatase
   (d) None of the avove

(337) ———————— is a vector can cloned maximum length of DNA
    (a) Plasmid   (b) Bacteriophase
    (c) Cosmid   (d) YAC

(338) DNA interaction can be judged by————————.
    (a) DNA foot printing   (b) DNA finger printing
    (c) Electroporation   (d) None of the above

(339) mRNA formed with the help of————————.
    (a) RNA polymerase I   (b) RNA polymerase II
    (c) RNA polymerase III   (d) None of the above

(340) Transfer of DNA from electrophoretic gel to nitrocellulose memberane is termed as————————.
    (a) Southern blot   (b) Northern blot
    (c) Western blot   (d) Eastern blot

(341) Automatically replicating multicopy extrac hromosomal DNA generally present in bacteria is known as————————
    (a) Replicon   (b) Vector
    (c) Plasmid   (d) Artificial chromosome

(342) Radioactive sample can be identified by————————.
    (a) Chramatography   (b) Spectrophotometry
    (c) Liquid scintrillatron counter   (d) None of the above

(343) Restrectron map could be generated by the use of ————————.
    (a) Elisa   (b) Ninhydrine
    (c) SSR   (d) Restrictron enzyme

(344) Klenow fragment does not have
    (a) 5'—3' exonuclease activity   (b) 5'——3' polymerase activity
    (c) 3'—5' exonuclease activity   (d) None of the above

(345) Liposome is ————————
    (a) Artificial phospholipids vesicles
    (b) Fragment of endoplasic reticulum
    (c) Aneukaryoase
    (d) A prokaryose

(346) ———————— an eukaryotic organism where the mRNA does not have any intron
    (a) Salamander   (b) *Atropha bellodona*
    (c) *Saccharmyces ceressarae*   (d) *Datura spp*

(347) Poly A tail of RNA occur on the————————.
    (a) 5' end   (b) 3' end
    (c) 2' end   (d) None of the above

(348) TRIPS stands for ─────────────────.
(a) Term Related Intetectual Properity Rights
(b) Theology Related International Project on Standardization
(c) Trade Related Intelectual Properity Rights
(d) Internationally Recognized Protein structure Typing

(349) Haploid plants could be farmed diploid by treating
(a) Gluteraldehyde   (b) Dihydrofolate
(c) Penicillin   (d) Colchicine

(350) Ropfamrcin inhibits synthesis of─────────.
(a) DNA synthesis   (b) RNA synthesis
(c) Protein synthesis (d) None of the above

(351) ───────── restriction enzyme cut blunt end.
(a) Ecori   (b) Hind III
(c) Pst I   (d) Alu I

(352) The most commonly used tomstrain protein electrophoretic gel is─────
(a) Xylene cyanol   (b) Bromophenol blue
(c) Coomassic brilliant blue (d) None of the above

(353) *Et Br* strain the DNA by─────────────.
(a) Spliting   (b) Intercalating
(c) Overlapping   (d) Suppressing

(354) Flow cytometry is used to─────────.
(a) Separate different cell population
(b) Separate DNA from RNA
(c) Isolate protein from cell extract
(d) None of the above

(355) Hardening of is a process to─────────────────.
(a) Mutant organism that will not grow on minimal medium
(b) Those who manufacture their own food
(c) Lowest organism in a food chain
(d) All of the above

(356) Auxotroph organisms are─────────────.
(a) Mutant organism that will not grow on minimal medium
(b) Those who manufacture their own food
(c) Lowest organism in a food chain
(d) All of the above

(357) C value is —————.
   (a) Thickness of cortical layer in monocots
   (b) Climate range of a plant
   (c) Total amount of a DNA in a haploid genome
   (d) Closeness of two bacterial colony

(358) Central dogma is flow of information from—————.
   (a) DNA to RNA to protein
   (b) DNA to protein to RNA
   (c) DNA to protein directly
   (d) All of the above

(359) Copy number is—————.
   (a) Centromere number in a cell
   (b) Number of different karyotype
   (c) Number of molecules of plasmid per genome contained
   (d) Number of cos site in a bacteriophase DNA

(360) Gene library is —————.
   (a) Random collection of all genetic element of genus
   (b) Collection of gene pool
   (c) Random collection of cloned fragment of all genetic material in a vector
   (d) Germplasm collection

(361) SuperBug is a / an—————.
   (a) Prokaryote
   (b) Eukaryote
   (c) GMO
   (d) None of the above

(362) ————— is false statement.
   (a) Tm increase with increase G: C content
   (b) Tm increases with increasing complex folding of DNA
   (c) Tm is characteristic of each DNA species
   (d) None of the above

(363) Sequence of DNA is the same when one strand is read left or the other read right to left —————.
   (a) Inverted repeats
   (b) Plain drome
   (c) Complex DNA
   (d) cDNA

(364) Polysomes are —————.
   (a) Cell organelle present in yeast
   (b) Polyploidy in gymnospermae
   (c) Ribosome bound together by a single mRNA
   (d) Lysosomes fused with phagosome

(365) Cybrid is——————————.
    (a) Mutationally dominant character
    (b) Somatic hybrid in which nucleus is derived from one parent
    (c) A cryoprotectant
    (d) Synonomous to hybrid

(366) Dedifferentiation is——————————.
    (a) Differentiation into root, shoot etc.
    (b) Difference in base pair composition in different organisms
    (c) Reversion of mutation
    (d) Reversion of differentiation to non- differentiated cells

(367) Synthetic seeds are——————————.
    (a) Encapsulated somatic embryo (b) Non capsulated somatic embryo
    (c) Somatic embryo (d) None of the above

(368) Large DNA fragment could be resolved without difficulty in——————————.
    (a) Pulsed field gel electrophorosesis (b) Agrose concentration
    (c) Polyacrylamide gel electrophoresis (d) Tube electrophoresis

(369) —————————— is macronutrient used in synthetic plant tissue culture
    (a) N (b) K
    (c) P (d) Fe

(370) Root and shoot can be generated through callus by ——————————.
    (a) Auxin and gibberellin ratio (b) Auxin and cytokinin ratio
    (c) Cytokinin and gibberellin ratio (d) None of the above

(371) Nucellus culture is useful to study ——————————.
    (a) Root initiation
    (b) Shoot initiation
    (c) Stress protein formation
    (d) Factors responsible for formation of adventive embryo

(372) Synchronization of cells in tissue culture could be achieved by using——————————.
    (a) Colchicine (b) Bromophenol
    (c) Nacl salt treatment (d) All of the above

(373) Cultivation of virus could be performed on ——————————.
    (a) Simple nutrient agar media (b) LB media
    (c) In embryogenated egg (d) Complex media

(374) Prions are ——————————.
    (a) Polysaccharides present *in Xanthomonas spp.* of bacteria
    (b) Ineffective protein particles
    (c) Yeast
    (d) An organism larger than bacteria but smaller than yeast

(375) Protoculture is——————————.
   (a) Culture of bacteria on petroleum product
   (b) Farming of sugarcane for alcohol to mix in petrol
   (c) Culture of any microorganism on petridishes
   (d) None of the above

(376) Ochre cadon is ——————————.
   (a) UAA   (b) UAG
   (c) UGA   (d) UVV

(377) Hotspot is sites ——————————.
   (a) Where natural flora and fauna are diverse
   (b) Hot springs where thermophilic bacteria found
   (c) At which frequency of mutation is very high
   (d) On bacterial cellwall where virus attaches

(378) —————————— is G banding.
   (a) A technique to appear more band on electrophorsed gel
   (b) Is a technique to generate a stratted pattern of chromosome
   (c) Band of gllanine base of DNA in gel
   (d) Band formed in stratosphere because of green house gas

(379) Ecophenes is——————————.
   (a) Plants with similar genetics that exhibit different morphology
   (b) Ecofriendly plastic manufactured by biotechnology
   (c) Site on DNA where protein attaches preferably
   (d) Pheromones

(380) Shotgun experiments are——————————.
   (a) Short sequence repeats
   (b) Making a restriction fragment by many enzymes
   (c) Short sequence repeats targets for mutation
   (d) Cloning of entire genome in the form of randomly generated fragments

(381) Biological oxygen demand is——————————.
   (a) Amount of oxygen taken up by microorganism that decompose waste marker
   (b) Amount of oxygen required for endoparasite
   (c) Oxygen released by transgenic plant in *in vitro* condition
   (d) Oxygen required assimilating carbohydrate by transgenic plant

(382) —————————— is an antiauxin.
   (a) Ascorbic acid   (b) Viral coat protein
   (c) NO_2 (d) None of the above

(383) Resperine is found in ——————.
   (a) *Bauhinna varigata*   (b) *Cassia fistula*
   (c) *Rawolfia serpentina*   (d) Raspberry

(384) FLAVR SAVR transgenic tomato plant has—————— transgene
   (a) Kanamycin resistance gene   (b) Gene encoding polygalacturonase
   (c) Both a and b   (d) None of the above

(385) In autoclave 15lbs is equaivalent to——————.
   (a) 135°C   (b) 125°C
   (c) 121°C   (d) 105°C

(386) DNA transformation in a cell could be possible with —————— salt
   (a) Mercuric chloride   (b) Calcium chloride
   (c) Silver nitrate   (d) Rubidium sulphate

(387) The Government of India launched a National Seed Project in ——————.
   (a) 1976   (b) 1978
   (c) 1979   (d) 1967

(388) A separate NSP on Vegetable s was approved and initiated by Indian Institute of Vegetabl Research, Varanasi in ——————.
   (a) 1976   (b) 1994
   (C) 1984   (d) 2004

(389) There are —————— cooperating centres under the NSP (veg) dealing with the vegetables excluding potato.
   (a) 11   (b) 12
   (C) 10   (d) 14

(390) During the period of 2001-2002, the breeder seed production was——————kg.
   (a) 71303   (b) 47974
   (C) 55799   (d) 37843

(391) —————— is an endospermatous seed.
   (a) Peas   (b) Beans
   (C) Capsella   (d) Maize

(392) —————— is a non- endospermatous seeds
   (a) Pea   (b) Maize
   (C) Wheat   (d) Rice

(393) In soyabean, a non-endospermatous dicot, development to the torpedo stage takes ——————days.
   (a) 20   (b) 30
   (c) 35   (d) 45

(394) In seeds, lipid is stored in——————.
    (a) Vacuoles   (b) Plastids
    (c) Lipid bodies   (d) None of the above

(395) LEA stands——————.
    (a) Lower embryo abundant proteins
    (b) Late embryo abundant proteins
    (c) Lipids abundant proteins
    (d) None of the above

(396) —————— is the first produced in the maternal tissue and transported to the seeds
    (a) ABA   (b) GA
    (c) LEA   (d) None of the above

(397) —————— releases seed dormancy.
    (a) GA3   (b) Kinetin
    (c) Both a and b   (d) Alar

(398) —————— is commonly used for breaking seed dormancy in vegetable crops.
    (a) Prechill   (b) KNO3
    (c) Acetone   (d) None of the above

(399) —————— is the sum total of all the seed attributes which favour stand establishment under favourable conditions
    (a) Seed storage   (b) Seed vigour
    (c) Field emergence   (d) None of the above

(400) The recoginition of performance differences among high germination seed lots was first described by the German Scientist Nobbe in——————.
    (a) 1878   (b) 1876
    (c) 1987   (d) None of the above

(401) The concept of seed vigour was first differentiated by——————.
    (a) Fredrich Nobbe   (b) Frank
    (c) Smith   (d) Caldwell

(402) PEG Mannitol sucrose glucose and NaCl are used for testing——————.
    (a) Respiration   (b) Osmotic stress
    (c) Condictivity test   (d) None of the above

(403) A chemical red compound formed from tetrazolium salt solution is known as —.
    (a) Formazan   (b) Lactic acid
    (c) Phenol   (d) None of the above

(404) In ordinary storage condition for 1-1.5 years seed moisture content of ——— % is adequate.
  (a) 8% (b) 10%
  (c) 12% (d) 15%

(405) For storage in sealed container drying to——— per cent.
  (a) 8% (b) 2-3%
  (c) 5-6% (d) None of the above

(406) For long term cold storage (-20°C) in sealed container seeds dried to moisture content of ——— and stored for 10-50 years or more.
  (a) 3-5% (b) 8-10%
  (c) 5-7% (d) None of the above

(407) Relative humidity——— % is generally appropriate for good storage.
  (a) 10-15% (b) 35-50%
  (c) 50-60% (d) None of the above

(408) ——— are best for sealed packaging of vegetable seeds.
  (a) Jute bag (b) Polythene bag
  (c) Metal cans and polyethylene lined aluminum foil bags
  (d) None of the above

(409) In general, the drying temperature should not be higher than———.
  (a) 43.5°C (b) 33.5°C
  (c) 23.5°C (d) None of the above

(410) If drying temperature is 37°C, the moisture content should be———.
  (a) Over 18 (b) Under 10
  (c) 10-18 (d) None of the above

(411) If air flow rate is 6 m^3/ min per m^3 seed, the moisture content should be———%.
  (a) 22 (b) 25
  (c) 18 (d) 15

(412) ——— screen grader is used for palak and beet root seeds.
  (a) Round hole (b) Slotted hole
  (c) Oblong hole (d) Triangular hole

(413) As the store requires some space between the lot and for loading or unloading, useable volume reduced to ———%.
  (a) 10-20 (b) 20-30
  (c) 30-40 (d) None of the above

(414) In store house, electric fitting should not be of ———.
  (a) Copper (b) Aluminum
  (c) Zinc (d) Iron

(415) Stamping, coating and rolling are the basic steps involved in——————.
    (a) Priming   (b) Pelleting
    (c) Coating   (d) None of the above

(416) The botanic pelleting are recommended @ of——————.
    (a) 200-300 g/ kg of seeds   (b) 300-400 g/ kg of seeds
    (c) 100-150g/ kg of seeds   (d) None of the above

(417) Filter antidote or absorbent coating can be used ——————.
    (a) At germination   (b) Before seed sowing
    (c) After seed sowing   (d) None of the above

(418) —————————— is found as best as seed pelleting.
    (a) Alar   (b) Cytokinin
    (c) 1, 8, napthalic anhydride   (d) None of the above

(419) —————————— refers to the addition of inert materials added to seed to change the size and shape resulting in substantial weight increase for improved platability.
    (a) Seed coating   (b) Seed pelleting
    (c) Seed primimg   (d) None of the above

(420) —————— are powders for dry seed treatments which contain the active ingredient with additives to prevent cohesion of practicable to the seed surface.
    (a) DS   (b) WP
    (c) WS   (d) LS

(421) —————————— is a powder which contains active ingredients filters and wetting agents for use in slurry.
    (a) FS   (b) WS
    (c) LS   (d) WP

(422) —————————— is a powder used in high concentration, so that they can be used in slurry. The size of particle is up to 90cm.
    (a) LS   (b) FS
    (c) WS   (d) DS

(423) Seed treatment with *Trichderma harzianum* @ —————— of seeds in tomato and cabbage.
    (a) 40g/kg   (b) 4g/kg
    (c) 24g/kg   (d) 44g/kg

(424) Seed treatment with —————— @ 3g/kg seed provides protection from whitefly at nursery stage.
    (a) Captan   (b) Thirum
    (c) Imidacloprid   (d) None of the above

(425) ———————— is a most potent germination promoter in a wide range of vegetables.
    (a) Cytokinin   (b) Gibberellic acid
    (c) Alar   (d) CCC

(426) The General Agreement on Tariffs and Trade (GATT) was established and signed by 22 countires (Including India) in ————————.
    (a) 1948   (b) 1964
    (c) 1986   (d) None of the above

(427) ———————— in at Uruguay first time issue related to agricultural trade, textile and Intellectual Properties Right were discussed.
    (a) 1986   (b) 1988
    (c) 1984   (d) 1977

(428) ———————— in Mexico 125 countires (including India) signed an revised GATT agreement
    (a) On 15th April, 1994   (b) On 30th June 1986
    (c) On 15th April, 1999   (d) None of the above

(429) GATT came in force as World Trade Organization in————————.
    (a) 16th June 1992   (b) 20th March, 1998
    (c) 15th April, 2004   (d) 1st January, 1995

(430) The headquarter of WTO is————————.
    (a) Uruguay   (b) Geneva
    (c) Mexico   (d) China

(431) ———————— is headquarter of Indian Patent.
    (a) Kolkata   (b) Chennai
    (c) Mumbai   (d) New Delhi

(432) Indian patent was passed in————————.
    (a) 1980   (b) 1970
    (c) 1990   (d) None of the above

(433) Second Patent Ammendent Bill was passed in————————.
    (a) 2002   (b) 2001
    (c) 1998   (d) 2003

(434) ———————— is a legal term describing rights given to creators for their literary and artistic works
    (a) Trademark   (b) Copyright
    (c) Trade secret   (d) None of the above

(435) The protection of Plant Varieties and Farmers Rights bill was passed in————.
    (a) 2002   (b) 2001
    (c) 2004   (d) 2003

(436) The first biosafety regulation was prepared by the National Institute of Health in the USA in———————
- (a) 1976
- (b) 1978
- (c) 1979
- (d) None of the above

(437) Terminator technology was invented by———————.
- (a) Melvin Oliver and Coworkers
- (b) R .R .white
- (c) Laljee singh
- (d) None of the above

(438) ——————— involves creation of transgenic plant that contains gene whose expression can be controlles by external stimuli
- (a) Gene expression
- (b) Gene interaction
- (c) Terminator
- (d) None of the above

(439) The verminator and traitor technology was given by firm———————.
- (a) Monasanto
- (b) Giant Zeneca
- (c) Delta and Pineland
- (d) None of the above

(440) The Indian Patent Act (1970) grants for———————.
- (a) 2 or 4 years
- (b) 7 or 14 years
- (c) 14 or 24 years
- (d) None of the above

(441) Union International Pour La Protection Des Obtentions Vegetables comprises——————— member countries
- (a) 14
- (b) 19
- (c) 37
- (d) 42

(442) ——————— is member of UPOV.
- (a) Belgium
- (b) Spain
- (c) Sweden
- (d) All of the above

(443) The condition of seeds when it fail to germinate even through the favourable environmental conditions are present, this is known as———————.
- (a) Quiescence
- (b) Dormancy
- (c) Germination
- (d) All of the above

(444) The dormancy due to unfavourable environmental condition is called———————.
- (a) Imposed dormancy
- (b) Innate dormancy
- (c) Dormancy
- (d) All of the above

(445) The dormancy due to condition within the dormant (germination or growth fails to occur even through the external environmetal condition favourable) plant or organ is called ———————.
- (a) Imposed dormancy
- (b) Dormancy
- (c) Innate Dormancy
- (d) None of the above

(446) Imposed dormancy is also called ———————.
    (a) Quiescence   (b) Rest
    (c) Germination   (d) None of the above

(447) Innate dormancy is also called ———————.
    (a) Quiescence   (b) Rest
    (c) Plant organs   (d) All of the above

(448) During the entire process, there may be following phase of innate dormancy that is ———————.
    (a) Predormancy or early rest   (b) Full dormancy or mid rest
    (c) Post Dormancy or after rest   (d) All of the above

(449) Capacity of germination or growth is not completely lost it's grow by various treatment, it is called ———————.
    (a) Predormancy   (b) Innate dormancy
    (c) Post Dormancy   (d) Imposed dormancy

(450) ——————— means, seed or organ become completely dormant and germination or growth can not be induced by changing in invironmental condition.
    (a) Predormancy   (b) Full dormancy
    (c) Imposed Dormancy   (d) All of the above

(451) When the dormant seed or organ gradually emerges from full dormancy and in it the germination or growth can be induced by changing environmental condition. It is called ———————.
    (a) Full dormancy   (b) Pre-dormancy
    (c) Post dormancy   (d) None of the above

(452) If in any seed or organ, the germination or growth can not be induced under any set of environmental condition, and then it is called ———————.
    (a) Post-dormancy   (b) Pre-dormancy
    (c) True dormancy   (d) All of the above

(453) When in a seed the germination can be induced under specific condition even at the time of its deepest dormancy. It is called ———————.
    (a) True dormancy   (b) Relative dormancy
    (c) Post dormancy   (d) Full dormancy

(454) When shed off many seeds fail to germination because of some intrinsic factor related to seeds. This is called ———————.
    (a) Relative dormancy   (b) Primery dormancy
    (c) Post dormancy   (d) None of the above

(455) ——————— defined the dormancy as any phase in life cycle of a plants in which active growth is temporarily suspended.
    (a) Crocker (1916)   (b) Mendal (1908)
    (c) Wareing (1969)   (d) None of the above

(456) When the seed become dormant again after breaking the dormancy, it is called—
   (a) Primery dormancy
   (b) Secondary dormancy
   (c) Post dormancy
   (d) None of the above

(457) —————divided seed dormancy into seed coat induced and embryo induced.
   (a) Wareing (1969)
   (b) Crocker (1916)
   (c) Both a and b
   (d) None of the above

(458) —————become due to extreme hardness of seed coat.
   (a) Seed coat induced dormancy
   (b) Secondary dormancy
   (c) Primery dormancy
   (d) All of the above

(459) —————become due to rudimentary or complete dormant embryo.
   (a) Seed coat induced dormancy
   (b) Primery dormancy
   (c) Embryo induced dormancy
   (d) Secondary dormancy

(460) When the seed germination is inhibited at temperature above 25^0C (77^0F). Heat sensitive is called—————.
   (a) Thermodormancy
   (b) Post dormancy
   (c) Seed dormancy
   (d) All of the above

(461) The dormant seed of the many species that have temperature sensitive also have light sensitive also known as—————.
   (a) Thermodormancy
   (b) Seed dormancy
   (c) Photo dormancy
   (d) Post dormancy

(462) —————is charecterised by having both seed coat dormancy (lack of water permeability) and a dormant embryo.
   (a) Thermodormancy
   (b) Photodormancy
   (c) Seed dormancy
   (d) Double dormancy

(463) —————that act as seed germination inhibitors have been extracted from various plant parts and indentified.
   (a) Physical dormancy
   (b) Mechanical dormancy
   (c) Chemical dormancy
   (d) None of the above

(464) —————found in plant families that characteristically have seed in which the embryo is not fully developed the time of ripening.
   (a) Chemical dormancy
   (b) Physical dormancy
   (c) Morphological dormancy
   (d) Internal dormancy

(465) —————is characterized principally by the requirement of a period of moist chilling for germination and the inability of the exised embriyo to germinate normally.
   (a) Physical dormancy
   (b) Embryo dormancy
   (c) Epicotyl dormancy
   (d) All of the above

(466) The oxygen requirement in upper seeds is greater than lower seeds. These types of dormancy are found in many gasses and some members of——————.
  (a) Compositae (b) Brassicaceae
  (c) Liliacae (d) All of the above

(467) Many seeds show the primery dormancy or undeveloped because the embryo are——————.
  (a) Mature (b) Immature
  (c) Open (d) None of the above

(468) Many seeds have hard and thick seed-coat but they are compeletly impermeable to ——————.
  (a) Sugar (b) Water and O_2 Immature
  (c) CO_2 and O_2 (d) Both b and c

(469) The light sensitive seeds are called ——————.
  (a) Photosynthesis (b) Photoblostic
  (c) Photochromic (d) All of the above

(470) The seed requiring single exopure of light for germination are called——————.
  (a) Positive-photoblostic (b) Negative- photoblostic
  (c) Non- photoblostic (d) All of the above

(471) The seed requiring single exposure of light for germination are called ——————.
  (a) Positive photoblostic (b) Non photoblostic
  (c) Negative photoblostic (d) All of the above

(472) The seed requiring either light or darkness for germination are called ——————.
  (a) Positive photoblostic (b) Non photoblostic
  (c) Negative photoblostic (d) Both a and c

(473) Differences between the alternating temperature should not be more than ——.
  (a) 4°C-5°C (b) 10°C-20°C
  (c) 50C-200C (d) 20°C-40°C

(474) Alternating temperature of —————— is useful in breaking the dormancy of photoblostic seeds like *Rumex crispus*.
  (a) 10°C-20°C (b) 20°C-40°C
  (c) 15°C-25°C (d) None of the above

(475) Dormancy due to general impermeability of the seed coat can be over come by partial or total removal of the seed coat. Breaking the seed coat barrier is called——————.
  (a) Sacrification (b) Stratification
  (c) Impaction (d) None of the above

(476) When the mechanical breaking of seed coat is done at one or more place, it is called —————————.
    (a) Sacrification   (b) Machanical scarification
    (c) Chemical scarification   (d) All of the above

(477) The treatment of seed coat with strong mineral acid or other chemical is called —————————.
    (a) Impaction   (b) Machanical scarification
    (c) Chemical sacrification   (d) Sacarification

(478) During the —————————, the entry of water and O_2 insides the seeds and shaking of seed some time removes the suberized plug and allows germination of seed.
    (a) Sacrification   (b) Impaction
    (c) Stratification   (d) None of the above

(479) The dormancy due to a special termperature requirements, is the treatment of seed at low temperature (0^0C-10^0C) under moist condition. This treatment is called —————————.
    (a) Impaction
    (b) Sacrification
    (c) Stratification or after ripening
    (d) None of the above

(480) Bud of many types of woody plants also undergo dormancy and they fail to grow and develop even under favourable environmental conditions. This type of dormancy is called—————————.
    (a) Seed dormancy   (b) Bud dormancy
    (c) Embryo induced dormancy (d) All of the above

(481) During the stratification, ————————— changes take place inside the seed.
    (a) Anatomical and biochemical changes
    (b) Temperature
    (c) Size and shape
    (d) None of the above

(482) ————————— Observed that ethylene chlorohydrin treatment of dormant potato buds caused a rapid synthesis of RNA in the buds.
    (a) Tuan and Bonner (1964)   (b) Wareing (1969)
    (c) Polja koff- Maber (1989)   (d) None of the above

(483) ————————— may be defined the step beginning with the uptake of water and leading to the rupture of seed coat by the redicle or plumule.
    (a) Germination   (b) Dormancy
    (c) Quiscence   (d) All of the above

(484) —————defined the germination as "Those consequtive events which case a dry quiescent seed in response to water uptake, to show arise in its general metabolic activities and to initiate the formation of a seedling from the embryo".

 (a) Tuan and Bonner (1964)   (b) Wareing (1969)
 (c) Polja koff- Maber (1989)   (d) None of the above

(485) In ————— germination the lengthening the hypocotyle dose not raise the cotyledon above the ground, and only the epicotyl emerges.

 (a) Epigeous   (b) Hypogeous
 (c) Epicotyle   (d) None of the above

(486) In ————— germination the hypocotyle elongates and raises the cotyledons above the ground.

 (a) Epigeous   (b) Hypogeous
 (c) Dicotyledons   (d) All of the above

(487) Mean days = $\dfrac{N_1T_1+N_2T_2+ \text{----------------} N_xT_x}{\text{Total number of}}$

 (a) Dormancy   (b) Quiescence
 (c) Seed germination   (d) Seed

(488) In wheat four type of proteins these are: glutanin, prolamines, globulins, ——.

 (a) Histone   (b) Non-histone
 (c) Albumins   (d) None of the above

(489) A prolamine called ————————— is a quite aboudant.

 (a) Lysin   (b) Proline
 (c) Methionine   (d) Zein

(490) Below ————————— of water in the seed germination dose not occurs.

 (a) 40 to 50 %   (b) 40 to 60 %
 (c) 10 to 30 %   (d) 20 to 50 %

(491) Most pollutant lie $NO_2, SO_2, O_3, NH_3, H_2S, F$ and high concentration of ethylene inhibits germination, it is also known —————.

 (a) Germination   (b) Dormancy
 (c) Influence germination   (d) All of the above

(492) ————————— is a terms in which death of small seedling from the attack of certain fungi, primarily, *Pythium ultimum* and *Rhizoctona solani* etc.

 (a) Seed dormancy   (b) Germination
 (c) Damping-off   (d) None of the above

Answer Sheet

1	c	2	b	3	d	4	a	5	-	6	a	7	d	8	c	9	c	10	d
11	a	12	b	13	d	14	-	15	b	16	c	17	b	18	c	19	b	20	c
21	d	22	d	23	b	24	c	25	c	26	c	27	d	28	b	29	a	30	c
31	b	32	a	33	a	34	b	35	a	36	b	37	b	38	c	39	c	40	c
41	a	42	a	43	b	44	a	45	a	46	c	47	e	48	g	49	c	50	a
51	e	52	a	53	a	54	-	55	c	56	b	57	c	58	d	59	c	60	d
61	b	62	a	63	b	64	c	65	-	66	a	67	c	68	a	69	a	70	b
71	a	72	c	73	c	74	b	75	c	76	b	77	b	78	b	79	a	80	a
81	a	82	b	83	a	84	a	85	b	86	a	87	e	88	b	89	a	90	d
91	b	92	b	93	d	94	e	95	a	96	a	97	c	98	d	99	a	100	a
101	b	102	b	103	c	104	a	105	a	106	a	107	a	108	b	109	a	110	a
111	b	112	c	113	c	114	a	115	b	116	a	117	c	118	a	119	a	120	a
121	d	122	a	123	c	124	a	125	d	126	b	127	b	128	b	129	a	130	c
131	a	132	b	133	c	134	b	135	a	136	c	137	d	138	a	139	a	140	c
141	b	142	c	143	a	144	-	145	b	146	b	147	a	148	c	149	c	150	c
151	c	152	b	153	a	154	c	155	a	156	c	157	c	158	c	159	b	160	c
161	d	162	c	163	a	164	c	165	b	166	b	167	a	168	d	169	d	170	b
171	b	172	a	173	-	174	b	175	a	176	a	177	c	178	c	179	c	180	a
181	d	182	b	183	a	184	c	185	b	186	c	187	b	188	a	189	a	190	a
191	b	192	a	193	a	194	c	195	b	196	b	197	b	198	a	199	a	200	b
201	d	202	a	203	b	204	c	205	c	206	c	207	b	208	a	209	b	210	a
211	c	212	b	213	b	214	b	215	e	216	b	217	b	218	a	219	c	220	b
221	b	222	c	223	c	224	c	225	c	226	a	227	d	228	c	229	a	230	a
231	b	232	c	233	b	234	a	235	b	236	b	237	c	238	b	239	a	240	a
241	a	242	a	243	c	244	b	245	b	246	a	247	c	248	b	249	c	250	a
251	c	252	a	253	a	254	c	255	c	256	b	257	a	258	c	259	b	260	b
261	c	262	-	263	b	264	a	265	b	266	b	267	a	268	a	269	a	270	d
271	a	272	a	273	c	274	d	275	a	276	c	277	a	278	b	279	b	280	d
281	b	282	c	283	a	284	b	285	b	286	d	287	b	288	d	289	b	290	d
291	b	292	a	293	a	294	c	295	a	296	c	297	b	298	c	299	b	300	-
301	b	302	a	303	c	304	b	305	d	306	d	307	b	308	b	309	-	310	-
311	d	312	d	313	d	314	c	315	d	316	a	317	c	318	c	319	b	320	b
321	b	322	-	323	b	324	a	325	b	326	b	327	c	328	c	329	b	330	d
331	d	332	a	333	b	334	-	335	d	336	b	337	d	338	a	339	b	340	a
341	c	342	c	343	d	344	a	345	a	346	c	347	-	348	c	349	-	350	b
351	d	352	c	353	b	354	a	355	a	356	-	357	c	358	a	359	c	360	c
361	c	362	d	363	b	364	c	365	b	366	d	367	-	368	a	369	d	370	d
371	d	372	a	373	c	374	b	375	b	376	a	377	c	378	b	379	a	380	d
381	a	382	a	383	c	384	c	385	c	386	b	387	a	388	b	389	d	390	-
391	d	392	a	393	a	394	c	395	b	396	a	397	c	398	a	399	b	400	b
401	a	402	b	403	a	404	a	405	c	406	b	407	b	408	c	409	a	410	b
411	a	412	d	413	c	414	a	415	b	416	a	417	-	418	c	419	b	420	-
421	d	422	c	423	b	424	c	425	b	426	a	427	a	428	a	429	d	430	b
431	a	432	b	433	a	434	b	435	b	436	b	437	a	438	c	439	b	440	b

441	c	442	d	443	b	444	a	445	c	446	a	447	b	448	d	449	a	450	b
451	c	452	c	453	b	454	b	455	c	456	b	457	b	458	a	459	c	460	a
461	c	462	d	463	c	464	c	465	b	466	a	467	b	468	d	469	b	470	a
471	c	472	b	473	b	474	c	475	a	476	b	477	c	478	b	479	c	480	b
481	a	482	a	483	a	484	c	485	b	486	a	487	c	488	c	489	d	490	b
491	c	492	c																

2.2. Differentiate In Between

Vertical resistance	Horizontal resistance
1. Complete but not permanent.	Incomplete but permanent.
2. Delay the start of epidemic.	Slow down the epidemic.
3. Show differential interaction.	Significant differences may be but absence of differential interaction.
4. Pathogen can change.	No change.
5. Oligogenic inheritance.	Polygenic inheritance.
6. Easily identifiable and incorporated in population.	Difficult.
7. Early breakable when other races come.	Not easily breakable.
8. Discontinuous variation.	Continuous variation for resistance.
9. Most likely a gene-for-gene system.	Minor genes operate additively.
10. No homeostasis because dominant genes are involved therefore less stability.	Genetic homoeostasis operates strongly where more genes are involved to cause stability of resistance.

Polygenic traits	Oliogenic traits
1. Governed by several genes.	Governed by few genes.
2. Effect of each gene is not detectable.	Effect of each gene is detectable.
3. Usually governed by additive genes.	Governed by non-additive genes.
4. Variation is continuous.	Variation is discontinuous.
5. Separation into different classes is not possible.	Separation into different classes is possible.
6. Highly influenced by environmental factors.	Little influenced by environmental factors.
7. Statistical analysis is based on mean, variances and covariances.	Statistical analysis is based on frequencies or ratios.

Dominance variance	Epistatic variance
1. It is due to interaction between genes of the same locus	It is due to interaction between genes of two or more different loci
2. It is of three types, viz., incomplete, complete and over dominance	It is of three types, viz., additive x additive, additive x dominance and dominance x dominance
3. It is unfixable	A x A type is fixable
4. Magnitude is higher than epistatic variance	Magnitude is lower than dominance variance

Additive variance	Dominance variance
1. It refers to difference between homozygotes (AA/aa).	It refers to deviation of Aa from the mean of AA and aa.
2. Genes show lack of dominance.	Genes show incomplete, complete or over dominance.
3. Associated with homozygosity & is more in breeders.	Associated with heterozygosity & is more in out breeders.
4. It is fixable.	It is non-fixable.
5. Selection is very effective as it is fixable.	Selection is ineffective as it is non-fixable.
6. It is the chief cause of transgressive segregation.	It is the chief cause of heterosis or hybrid vigour.

Pureline selection	Mass selection
1. The new variety is a pureline.	The new variety is a mixture of pureline.
2. The new variety is highly uniform. In fact, the variation with in a pureline variety is purerley environmental.	The variety has genetic variation for quantitative characters, although it is relatively uniform in general appearance.
3. The selected plants are subjected to progeny test.	Progeny test is generally not carried out.
4. The variety is generally the best pureline present in the original population. The pureline selection brings about the greatest improvement over the original variety.	The variety is inferior to the best pureline because most of the purelines included in it will be inferior to the best pureline.
5. Generally, a pureline variety is expected to have a narrower adoptation and lower stability in performance than a mixture of purelines.	Usually the variety has a wider adaptation and greater stability than a pureline variety.
6. The plants are selected for their desirability. It is not necessary that they should have a similar phenotype.	The selected plants have to be similar in phenotype since their seeds are mixed to make up the new variety.
7. It is more demanding because careful progeny tests and yield trials have to be conducted.	If a large number of plants are selected, extensive yield trials are not necessary. Thus, it is less demanding on the breeder.
8. generally, 9-10 years are required to develop a new variety.	Generally 5-7 years are required to develop a new variety.
9. Selection within a pureline variety is ineffective unless it has become genetically variable.	Selection within a variety developed by mass selection in effective since it has genetic variation.
10. The procedure of a pureline variety is uniform in quality.	The procedure is generally not uniform since different purelines making up the variety may differ in the quality of their gains etc.
11. This variety is easily identified in seed certification programmes.	The variety is relatively difficult to identify in seed certification programmes.
12. Pureline selection is used in self-pollinated and often-cross pollinated crops.	Mass selection is used in both self and cross-pollinated crops.

Pedigree method	Bulk method
1. Individual plants are selected in F_2 and the subsequent generations and individual plant progenies are grown.	F_2 and subsequent generations are maintained as bulks.
2. Artificial selection, artificial disease epidemic etc. is an integral part of the method.	Artificial selection, artificial disease epiphytotics etc. may be used to assist natural selection. In certain cases, artificial selection may be essential.
3. Natural selection does not play any role in the method	Natural selection determines the composition of the populations at the end of the bulking period.
4. Pedigree records have to be maintained which is often time consuming and laborious.	No pedigree records are maintained.
5. It generally takes 14-15 years to develop a new variety and to release it for cultivation.	It takes much longer for the development and release of a variety. The bulk population has to be maintained for more than 10 years for natural selection to act.
6. Most widely used breeding method.	Used only to a limited extent.
7. It demands close attention from the breeder from F_2 onward as individual plant selections have to be made and pedigree records have to be maintained.	It is simple, convenient and inexpensive and does not require much attention from the breeder during the period of bulking.
8. The segregating generations are space planted to permit individual plant selection.	The bulk populations are generally planted at commercial planting rates.
9. The size of population is usually smaller than that in the case of bulk method.	Large populations are grown. This and natural selection are expected to increase the chances of the recovery of transgressive segregants.

Pedigree method	Backcross method
1. F_1 and the subsequent generations are allowed for self-pollination.	F_1 and the subsequent generations are backcrossed to the recurrent parent.
2. The new variety developed by this method is different from the parents in agronomic and other characteristics.	The new variety is identical with the recurrent parent, except for the character under transfer.
3. The new variety has to be extensively tested before release.	Usually extensive testing is not necessary before release.
4. The method aims at improving the yielding ability and other characteristics of the variety.	The method aims at improving specific defects of a well adapted, popular variety.
5. It is useful in improving both qualitative and quantitative characters.	Useful for the transfer of both quantitative and qualitative characters provided they have high heritability.
6. It is not suitable for gene transfer from related species and for producing substitution or addition lines.	It is the only useful method for gene transfers from related species and for producing addition and substitution lines.

7. Hybridization is limited to the production of the F_1 generation. | Hybridization with the recurrent parent is necessary for producing every backcross generation.
8. The F_2 and the subsequent generations are much larger than those in the backcross method. | The backcross generations are small and usually consist of 20-100 plants in each generation.
9. The procedure is the same for both dominant and recessive genes. | The procedure for the transfer of dominant and recessive genes is different.

Components	Synthetic variety	Composite variety
Base population	Inbred lines	Varieties/ landraces/ other heterozygous sources
Testing of GCA	Tested	Usually tested
Number of lines / varieties	4-20	2-100

Particulars	Clone	Pureline	Inbred
1. Mode of pollination in the crop species where they occur	Cross-pollination	Cross-pollination	Cross-pollination
2. Natural mode of reproduction in such species	Asexual (in most cases)	Sexual	Sexual
3. Genetic make-up of the natural populations of such species	Heterozygous	Homozygous	Heterozygous
4. Obtained through	Asexual reproduction from a single plant	Natural self pollination from a single homozygous plant	Artificial self pollination (or other froms of inbreeding) and selection for several generations
5. Maintained through	Asexual reproduction	Natural self-pollination	Artificial self pollination or close inbreeding
6. All the plants in a single entry are genotypically	Identical	Identical	Almost identical
7. Used directly as a variety	Yes	Yes	No (Used in developing hybrid or synthetic varieties
8. The genetic make-up of plants within a variety	Heterozygous	Homozygous	Almost homozygous
9. Organism where found	Plants	Plants	Plants, animals

	Traditional breeding	Ideotype breeding
1.	The breeder usually has an idea, however vague, of the type of plant he wishes to develop, but he does not describe it formally.	The breeder must define the ideotype to be developed based on trait analysis other considerations.
2.	Selection uses yield per se as a criterion	Yield is not as a basis of selection. Selection is based on the traits constituting in ideotype.
3.	Selection for individual traits (other than yield) is also done for defect elimination. But the goal for the trait is not defined before hand.	The breeding goal for each trait is defined beforehand; this is done while defining the ideotype.
4.	It is usually based on the improved gene pool (of primary genepool) of a crop.	It will usually necessitate introgression of desirable gene/traits from unimproved gene pool (a primary genepool).
5.	It does not encourage systematic thinking and accumulation of information on how yield is achieved in a crop.	Its basic philosophy is to understand how yield is achieved. The information generated from various studies leads to the development of models that are further tested and refined.
6.	It does not deliberately generate genetic diversity may or may not be useful in the future.	It generates such genetic diversity deliberately to test various models and to achieve the defined ideotypes. Some of the variation so generated will also be useful in traditional breeding.
7.	Progress is relatively rapid, but may reach a plateau after a period.	Progress in rather slow due to its very mature but it may be able to break the plateau reached in traditional breeding.
8.	Constitutes the main activity and is likely to remain so.	It may be considered the augment traditional breeding efforts and not to replace it.

2.3. Fill in the Blanks

(1) The whole library of different alleles of a species is referred to as———————.
(2) The germplasm which is collected from foreign countries is called ———————
(3) Plant materials which are meant for medium term storage are known as ———————.
(4) In long term storage , germplam can be stroed upto ———————
(5) Primary gene pool is also known as ———————.
(6) Secondary gene pool is also referred to as ———————.
(7) Tertiary gene pool is also called as ———————.
(8) Original homes of crop plants are known as ———————.
(9) Small areas within the centre of diversity which exhibit tremendous genetic variability of crop plants are known as ———————.
(10) Land race refers ———————.
(11) The gene pool in which intermating is easy and leads to production of fertile hybrids is known as ———————.

(12) The term microcenter was coined by——————————.
(13) The improved varieties of recent past are referred to as ——————.
(14) Law of homologous series of variation is also known as ——————.
(15) Permanent loss of a crop species is referred to as ———————————.
(16) Conservation of germplasm under natural conditions is called ————.
(17) ———————refers to conservation of germplasm in gene bank.
(18) ——————— refers to various organizations where genetic diversity is maintained in living state.
(19) Gene banks are also known as- ———————————.
(20) Those areas of lands in which germplasm of recalcitrant crop species is maintained is known as- ———————.
(21) Seeds which show drastic loss in viability with decrease in moisture content below 13 or 12 % are called———————.
(22) Seeds which can be dried to low moisture content and stored at low temperature without loss in viability are referred to as ———————.
(23) ———————refers to prophylactic measures which are used to prevent the entry of new diseases, insects and weeds from other countries along with introduced material.
(24) Protected areas of great diversity under natural condition are known as———.
(25) The process by which living organism give rise to the off spring of similar kind is known as ———————
(26) Multiplication of plants without the fusion of male and female gametes is referred to as ———————
(27) Multiplication of plants by means of various vegetative plant parts is called—.
(28) Development of embryo without sexual fusion is known as ———————.
(29) Development of embryo from egg cell without fertilization is referred to as —.
(30) Development of embryo either from synergids or antipodal cells is termed as–.
(31) Origin of embryo from the diploid egg cells of another embryosac developed from diploid tissues of acrhesporium, nucellus or integument is called————.
(32) Origin of embryo directly from diploid cells of either nucellus or integument is known as ———————.
(33) Multiplication of plants by fertilized embryos is known as ———————.
(34) Fertilization of ovules by the pollen grains of same flower refers to ————.
(35) Fertilization of ovules by the pollen grains of another plant is called ————.
(36) Reproduction only through apomixis refers to ——————— type of apomixis.
(37) Reproduction by sexual means and also by apomixes refers to ——————— ——————— type of apomixes,
(38) Autogamy refers to———————.
(39) Allogamy refers to ———————.

(40) Genotypically similar population is known as————.
(41) Genotypically dissimilar population is referred to as ————.
(42) A mating system in which each female gamete has equal chance to unite with every male gamete is known as————.
(43) Mating between genetically similar individuals refers to————.
(44) Mating between genetically dissimilar individuals is known as————.
(45) A system of mating in which mating is done between phenotypically similar individuals is called ————.
(46) ————mating refers to mating between phenotypically dissimilar genotypes.
(47) The term self- incompatability was coined by————.
(48) Self- incompatability was first reported in ————.
(49) Inability of a plant with functional pollen to set seeds when sel-pollinated is known as————.
(50) Self- incompatability promotes ————.
(51) In red clover self- incompatability is ————.
(52) ————self- incompatability is controlled by genetic constitution of gametes.
(53) ———— self- incompatability is controlled by the genotype of pollen producing plant.
(54) ———— system of self- incompability permits recovery of male parent only .
(55) Recovery of both the parents is possible in ————of self- incompability .
(56) ———— system of self incompability permits production of some homozygotes in some crosses.
(57) In gametophytic system of self incompability , a cross between $s_1 s_2 \times s_1 s_3$ would be————.
(58) A cross between parents differing in both alleles , i. e., $s_1 s_2 \times s_3 s_4$ would be——.
(59) The pollen sterility which is caused by nuclear genes is refered to as————.
(60) The———— male sterility is caused by cytoplasmic genes or plasma genes .
(61) Pollen sterility which is caused by both nuclear and plasma gene is known as?–————.
(62) Cytoplasmic sterility is maintained by crossing of A line with————.
(63) Chemicals which are used for induction of male sterility called ————.
(64) Mutagen induced male sterility is ————.
(65) Male sterility induced by gametocides is————.
(66) Cytoplasmic genetic male sterility was first discovered by ———— and ———— .in onions .

(67) Transposition of crop plants from the place of their cultivation to such areas where they were never grown earlier is known ─────────.
(68) Introductions which are immediately adapted to the changed environment are called ─────────.
(69) Introduction which require few year for adaptation in new areas are referred to as─────────.
(70) A foreign variety which is directly recommended for commercial cultivation is known as─────────.
(71) Selection in favour of intermediate type is refered to as─────────.
(72) Selection in favour of an extreme phenotype is called─────────.
(73) Selection in favour of two extreme type is known as ─────────.
(74) Selection in one direction in one generation and in opposite direction in another generation is called ─────────.
(75) progeny of self pollinated homozygous plant obtained by selfing is known as ─────────.
(76) Record of ancestry of an individual selected plant for its various generation is known as ─────────.
(77) Crossing of F_1 with either of its parents is refered as ─────────.
(78) Crossing of F_1 with its homozygous ressesive parent is known as ─────────.
(79) The parent which donates desirable genes in backcrossing programme, is termed as ─────────.
(80) Parent which receives desirable character during backcrossing is refered to as ─────────.
(81) The deliberate seed mixture of isogenic lines, closely related lines or unrelated lines is known as ─────────.
(82) The procedure for development of multiline varieties is called ─────────
(83) Genotylpe having single locus difference is refered to as ─────────
(84) A population of genetically similar plants is known as ─────────
(85) A population which is composed of genetically dissimilar plants is called ─────
(86) A true breeding population is known as ─────────.
(87) A population which segregates on selfing is referred to as ─────────.
(88) Pureline theory was developed by ─────────.
(89) Selection of desirable plants from a mixed population and growing of next generation from the bulk seed of selected plants is called ─────────.
(90) Removal of off type plants from a mixed population and allowing rest of the plants to grow further is refered to as ─────────.
(91) Selection of superior plants from a heterogeneous population on the basis of their progeny performance is called ─────────.

(92) Reselection generation after generation with intermating of selected plants to provide for genetic recombination is known as ─────────────.
(93) The procedure of recurrent selection was outlined by ─────────────.
(94) The term recurrent selection was coined by ─────────────.
(95) Recurrent selection for general combining ability is used to improve ───────────────── of a population for a particular character.
(96) Reciprocal recurrent selection was proposed by ───────────── in ─────.
(97) A breeding approach in which both the extreme phenotypes for a character are selected and intermated in a segregation population is known as ─────────.
(98) Concept of disruptive mating and selection was developed by ───────────── in ─────────.
(99) Progeny of a single plant obtained by asexual reproduction is refered to as ─────────.
(100) A procedure of selecting superior clones from the mixed population of asexually propagated crops is known as ─────────────.
(101) Asexually propagated crops are refered to as ─────────────.
(102) A variety which is developed by intermating in all possible combinations a number of in bred lines with good gca and mixing the seed of F_1 crosses in equal quantity is called ─────────────.
(103) A variety in cross pollinated species, which is developed by mixing the seed of various genotype which are similar in height, maturity, seed colour, and seed size is called ─────────────.
(104) Convergent improvement was proposed by───────── in ─────────────.
(105) Gamete selection was proposed by─────────────.
(106) Convergent improvement is a modified form of ─────────────.
(107) Recurrent selection for sca was proposed by ─────────────.
(108) In Synthetic varieties, heterosis is ───────────── exploited.
(109) Superiority of F_1 in fitness and vigour over its parents is called ─────────────.
(110) The term heterosis was coined by ───────────── in ─────────────.
(111) Superiority of F_1 over its parent in vegetative growth is known as ─────────.
(112) Luxuriance is also referred to as ─────────────.
(113) Masking effect of one allelel over the other on the same locus is called─────.
(114) Dominance hypothesis of heterosis was first proposed by ─────────────.
(115) Superiority of heterozygote over both the homozygotes is refered to as ─────.
(116) Overdominance hypothesis of heterosis was independently proposed by──────.
(117) Term over-dominance was coined by─────────────.
(118) Overdominance hypothesis of heterosis was supported by ─────────────.
(119) Superiority of F_1 over the mean value of both the parents is known as─────.

(120) Hetrosis over the better parent is refered to as ―――――――.
(121) Heterosis over the best commercial variety is called ―――――――.
(122) Economic heterosis is also known as ―――――――.
(123) Mating between closely related individual is known as ―――――――.
(124) Mating between distantly related individuals is called ―――――――.
(125) Loss or decrease in fitness and vigour as a result of inbreeding is known as ―――.
(126) The F_1 population which is used for commercial cultivation is known as ―――.
(127) Hybrid between two varieties or genotypes of the same species is refered to as ―――――――.
(128) A hybrid between two species of the same genus is called ―――――――.
(129) A hybrid involving two parent is called ―――――――.
(130) Hybrid progeny of a cross between a single cross and an inbred is known as ―――――――.
(131) A cross involving four different parents is refered to as ―――――――.
(132) A true breeding line obtained by continuous inbreeding is known as ―――――――.
(133) A cross between an inbred line and open pollinated variety is called ―――――――.
(134) Mating between two different species of the same genus or two different genera of the same family is referred to as ―――――――.
(135) Wide crossing is also known as ―――――――.
(136) Crossing between two different species of the same genus is called ―――――――.
(137) Crossing between two different genera of the same genus is called .
(138) The first interspesific cross between Carnation and Sweet William species of *Dianthus* was made by ―――――――.
(139) The first cross between bread wheat and rye was made by ―――――――.
(140) Distant crosses is also known as ―――――――.
(141) Inability of the functional pollen of one species or genus to effect the fertilization of the female gamets of related species or genus is refered to as ―――――――.
(142) Inability of a hybrid to produce viable off spring is termed as ―――――――.
(143) A condition in which F_1 plants of an interspecific cross are vigorous and fertile but their F_2 progeny is week and sterile is referred to as ―――――――.
(144) Addition of one chromosome of wild species to the normal complements of a cultivated species is known as ―――――――.
(145) Replacement of one pair of chromosomes of cultivated species with those of wild donor species is called ―――――――.
(146) Transfer of some genes from one species into the genome of another species is termed as ―――――――.
(147) Adverse condition for crop growth and production imposed either by biotic factors or abiotic factor or both is called ―――――――.

(148) Adverse conditions for crop growth and production caused by biotic factor is referred to as ——————.
(149) Ability of some genotypes to give higher yield of good quality than other varieties at the same initial level of disese or insect attack under similar environmental conditions is known as ——————.
(150) Resistance of a host to the particular race of a pathogen is called ——————.
(151) —————— refers to resistance of a host to all the prevalent races of a pathogen.
(152) The term vertical and horizontal resistance was coined by —————— in ——
(153) Long lasting resistance is refered to as ——————.
(154) Resistance exhibited by young seedling is called ——————.
(155) Resistance exhibited by adult plants is referred to as ——————.
(156) Various disease causing organisms, such as fungi, bacteria, viruss and mycoplasm are known as ——————.
(157) The concept of gene for gene hypothesis was developed by—————— in ——.
(158) The gene for gene hypothesis is also known as ——————.
(159) Mechanism of insect resistance was given by ——————.
(160) Various features of host plant which make the host undesirable to insects for food, shelter and reproduction is known as ——————.
(161) Non-freference is also refered to as ——————.
(162) Adverse effect of host plant on the development and reproduction of insect —————— pest feeding on resistant host is known as ——————.
(163) Ability of a variety to produce greater yiled than susceptible variety at the same level of insect attack is called ——————.
(164) A host pathogen interaction which leads to death of infested tissues is called ——————.
(165) Incorporation of two or more major genes in a variety for specific resistance to a pest is known as ——————.
(166) Planned geographical distribution of major genes for specific resistance to pests for in varietal development and production is called ——————.
(167) In India Striga is a parasite weed of ——————.
(168) Condition of soil moisture deficit is referred to as ——————.
(169) Ability of plants to maintain a favourable internal water balance under moisture stress is known as ——————.
(170) Ability of plants to withstand low tissue water content is called ——————.
(171) Ability of plants to grow, develop and reproduce under moisture deficit condition is termed ——————.
(172) Improvement in drought tolerance capacity of a genotype due to various seed and seedling treatments is known as ——————.

(173) Ability of plants to prevent, reduce or overcome the injurious effects of soluble salts present in the root zone is called ——————.
(174) Growing plants in nutrients solution is referred to as ——————.
(175) Cemented microplots of various sizes which are used for the study of roots and salt tolerance capacity of plants are known as ——————.
(176) An instrument which is used for measuring stomatal aperture is called ——.
(177) An instrument which is used to measure tissue water potential in plants is referred to as ——————.
(178) Hisar-2 variery of onion is tolerant to——————.
(179) Pusa Sawani variety of okra is tolerant to ——————.
(180) IGFRI -437 variety of cowpea is tolerant to ——————.
(181) Central Soil Salinity Research Institute is located at ——————.
(182) Stomatal aperature is measured by ——————.
(183) A biological model which is expected to behave in a predictable manner within a defined environment is called ——————
(184) A plant model which is expected to yield greater quantity of grains, fibre, oil or other useful product when developed as a cultivar is referred to as ——.
(185) Concept of crop ideotype in wheat was developed by —————— in ——
(186) The term ideotype was coined by —————— in ——————
(187) Ideotype also known as ——————.
(188) Ideotype breeding is also referred to as ——————.
(189) Concept of plant type was introduced in rice breeding by ——————in ——————
(190) Ideal plant type maize was first proposed by ——————in ——.
(191) Ideal plant type for six rowed barley was suggested by —————— in —— ——————.
(192) Ratio of economic yield to the biological yield is referred to as ——————
(193) Total dry matter production per plant is known as ——————.
(194) Suitability of a plant product for human and animal consumption reflects its —
(195) Genetic improvement of crop plants in relation to their quality is referred to as —
(196) Adverse effect of imbalanced diet on human health is called ——————
(197) Amino acid which can not be synthesized in human body are known as- ——
(198) Aminoacid which can synthesize in human body and need not be supplied through diet are known as ——————.
(199) Sulphur containing aminoacids are referred to ——————.
(200) Vital substances which are required in small quantity for normal growth and good health are called——————.
(201) In cassava, the toxic substance is ——————.

(202) Product of a fertilized ovule which consists of embryo, seed coat and cotyledon(s) is called ―――――――――.
(203) A genotype released for commercial cultivation either by state variety release committee is referred to as ―――――――.
(204) Initial seed of an improved variety, limited in quantity and produced by originating plant breeder is known as ―――――――――.
(205) Progeny of nucleus seed is called ―――――――――.
(206) Progeny of breeder seed is referred to as ―――――――.
(207) Progeny of foundation seed produced under the supervision of seed certification agency is called ―――――――――.
(208) A legal system which ensures production of high quality seeds in terms of purity and germination is called ―――――――
(209) Absence of seeds of other variety of the same crop and also of other crops in a seed lot refers to ―――――――――
(210) Genetic purity is determine by ――――――――――test.
(211) Freedom from inert matter and defective seeds of a seed sample indicates its ―――――――.
(212) In a seed sample, non living materials such as sand pebbles, soil particales, straw, etc. are combinedly known as ―――――――――.
(213) Emergence of normal seedlings from the seeds under ideal condition of light, temperature, Moisture, oxygen, and nutrient is referred to as ―――――――.
(214) Seedlings which are unable to develop in to normal plants are called ―――――.
(215) Seeds which do not absorb water are referred to as ―――――.
(216) Process of evaluation of seeds in term of purity and germination is called-------.
(217) Permanent reduction either in genetic or agronomic value of a released variety is known as ――――――――.
(218) Seperation of the field of a variety to the prescribed distance from that of another variety to avoid contamination is called ―――――――
(219) Process of removal of off type plants from the field of an improved variety is known as ――――――
(220) _____ is able to identify all allelic variants of a DNA region present in an artificial mutant collection
(221) _____ and _____ plant breeding requires genetic variability to the selected in order to increase the frequencies of favourable alleles and genetic combinations..
(222) Targeted genome editing (TGE) Technology as an alternative tool for trait improvement in _____.
(223) _____ Is an indirect process where selection is carried out an the basis of a marker instead of the trait itself.

(224) The possibility to predict the outcome of a set of crosses on the basis of molecular markers information is known as _____.

(225) _____ is based on simultaneous estimation of effect on simultaneous estimation effect on phenotype of all loci, haplotype, and markers available.

(226) _____ _____ and _____ are the most promising approaches for the genome editing.

(227) CRISPRs were initially identified in the *Escherichia coli* genome in _____.

(228) CRISPR system has been introduced into human cells, what is more, customized to remove and fix imperfect DNA, giving the incredible potential for treating hereditary maladies, for example _____ arid _____.

(229) The first utilization of CRISPR or RNA guided casg in plants were depicted in _____

(230) The first CRISPR/Cas- mediated genome altering in vegetable crops was accounted for tomato in.

(231) Phytoene synthase (PSY1) working in carotenoid biosynthesis were transformed by _____

Answer Sheet

(1) Gene pool
(2) Exotic collection
(3) Active collection
(4) 100 years
(5) GP1
(6) GP2
(7) GP3
(8) Primitive cultivars
(9) Primary centres of diversity
(10) to Micro- centres
(11) GP1
(12) Harlan (1948)
(13) Obsolete cultivars
(14) Law of parallel variations
(15) Extinction
(16) In- situ - conservation
(17) Ex- situ conservation
(18) Gene bank
(19) Germplasm banks
(20) Field gene banks
(21) Recalcitrant seeds
(22) Orthodox seeds
(23) Quarantine
(24) Gene sanctuaries
(25) Reproduction
(26) Asexual reproduction
(27) Vegetative reproduction
(28) Apomixis
(29) Parthenogenesis
(30) Apogamy
(31) Apospory
(32) Adventive embryony
(33) Sexual reproduction
(34) Autogamy
(35) Allogamy
(36) Obligate apomixes
(37) Facultative apomixis
(38) Self- pollination
(39) Cross- pollination
(40) Homogenous
(41) Heterogenous
(42) Random mating
(43) Genetic assertive mating
(44) Genetic disassertive mating
(45) Phenotypic assertive mating
(46) Phenotypic disssortive mating
(47) Stout in (1917)
(48) Varbascum phoeniceum
(49) Self incompability
(50) Allogamy
(51) Gametophytic monoallelic
(52) Gametophytic
(53) Sporophytic
(54) Gametophytic
(55) Sporophytic
(56) Sporophytic
(57) Partially fertile
(58) Fully fertile
(59) Genetic male sterility
(60) Cytoplasmic
(61) Cytoplasmic genic male sterility
(62) B line
(63) Male gametocides
(64) Heritable
(65) Non-heritable
(66) Jones and Davis in 1944
(67) Plant introduction
(68) Direct introduction
(69) Indirect introduction
(70) Exotic variety

(71) Stabilizing selection
(72) Directional selection
(73) Disruptive selection
(74) Cyclic selection
(75) Pure line
(76) Pedigree
(77) Backcross
(78) Test cross
(79) Donor parent
(80) Recipient parent
(81) Multilines
(82) Multiline breeding
(83) Isogenic line
(84) Homogeous population
(85) Heterogenous population
(86) Homozygous population
(87) Heterozygous population
(88) Johannsen (1903)
(89) Positive mass selection
(90) Negative mass selection
(91) Recurrent selection
(92) Recurrent selection
(93) Jenkins in (1940)
(94) Hull in 1945
(95) gca
(96) Comstock et al. in 1949
(97) Disruptive selection
(98) Mather.in (1953)
(99) Clone
(100) Clonal selection
(101) Clonal crops
(102) Synthetic variety
(103) Composite variety
(104) Richey in (1927)
(105) Stadler in (1944)
(106) Back- cross

(107) Hull, in (1950)
(108) Partially
(109) Heterosis
(110) Shull in (1914)
(111) Luxuriance
(112) Pesudoheterosis
(113) Dominance
(114) Davenport in 1908
(115) Overdominance
(116) Shull (1908) and East (1908)
(117) Hull in 1945
(118) East (1936) and Hull (1945)
(119) Mean heterosis
(120) Heterobeltiosis
(121) Useful heterosis
(122) Useful heterosis
(123) Inbreeding
(124) Out breeding
(125) Inbreeding depression
(126) Hybrid Variety
(127) Intraspecific hybrid
(128) Interspecific hybrid
(129) Single cross hybrid
(130) Three way cross hybrid
(131) Double cross hybrid
(132) Inbred
(133) Top cross
(134) distant hybridization
(135) distant hybridization
(136) Interspecific hybridization
(137) Intergeneric hybridization
(138) Thomas Fairchild in 1717
(139) Rimpu in 1890
(140) Wide crosses
(141) Cross incompatability
(142) Hybrid inability

(143) Hybrid sterility
(144) Hybrid breakdown
(145) Alien substitution
(146) Introgression
(147) Stress
(148) Biotic stress
(149) Genetic resistance
(150) Vertical resistance
(151) Vertical resistance
(152) Van der Plank in 1963
(153) Durable resistance
(154) Seedling resistance
(155) Adult resistance
(156) Pathogens
(157) Flor in 1956
(158) Flor Hypothesis
(159) Painter in 1951
(160) Nonpreference
(161) Antixenosis
(162) Antibiosis
(163) Tolerance
(164) Hypersensitive
(165) Genepyrimiding
(166) Gene deployment
(167) Sorghum
(168) Drought
(169) Drought avoidance
(170) Drougt tolerance
(171) Drought resistance
(172) Drought Hardening
(173) Slinity Resistance
(174) Hydroponics
(175) Lysimeter
(176) Porometer
(177) Psychrometer
(178) Salinity
(179) Salinity
(180) Drought
(181) Karnal
(182) Porometer
(183) Ideotype
(184) Crop ideotype
(185) Donald in 1968
(186) Donald. In 1968
(187) Ideal or model plant type
(188) Plant type breeding
(189) Jennings in 1964
(190) Mock and Pence in 1975
(191) Rasmusson in 1987
(192) Harvest index
(193) Biological yield
(194) Nutritional quality
(195) Quality breeding
(196) Malnutrition
(197) Essential amino acid
(198) Nonessential aminoacids
(199) Limiting aminoacid
(200) Vitamins
(201) CN glycosides
(202) Seed
(203) Variety
(204) Nucleus
(205) Breeder seed
(206) Foundation seed
(207) Certified seed
(208) Seed certification
(209) Genetic purity
(210) Grow out
(211) Physical purity
(212) Inert matter
(213) Germination
(214) Abnormal seedling
(215) Hard seeds
(216) Seed testing

(217) Varietal deterioration
(218) Isolation
(219) Rouging
(220) TILLING (Targeting Induced Local Lesions in Genomes)
(221) TILLING Eco TILLING
(222) Horticultural Crops
(223) Marker assisted selection
(224) Breedign by design
(225) Genomic selection
(226) Interspace short palindromic Repeat (CRISPR) Transcription activator like effector nuclease (TALEN) zinc finger nuclease (ZFN)
(227) 1987
(228) Muscular daystrophy and cystic fibrosis
(229) 2013
(230) 2014
(231) CRISPR/Cas

2.4. True and False

(1) First interspecific hybrid in dianthus was developed by Thomas Fairchild (1717).
(2) Artificial hybridization in fruit crops was first used by Andrew Night (1800).
(3) Johannsen (1903) first used the term genotype and phenotype.
(4) Rimpu (1890) first made intergeneric cross between bread wheat and rye.
(5) Shull (1908) and east (1908) independently proposed the hypothesis of overdominance.
(6) Hull first used the term overdominance.
(7) Law of parallel series of variation was given by Vavilov (1936).
(8) Karpenchenko (1948) first made intergeneric cross between radish and cabbage.
(9) The term recurrent selection was coined by Hull in 1945.
(10) Concept of disruptive selection was developed by Mather in 1963.
(11) Concept of crop ideotype was developed by Donald in 1978.
(12) Term vertical resistance and horizontal resistance were coined by Vander Plank in 1973.
(13) International centre for Agriculture Research in Dryland Areas is located in Syria.
(14) International Crop Research Institute for Semi –Arid tropic is located at Hyderabad in India.
(15) Gene pool refers to the whole library of alleles in a species.
(16) Base collections are plant materials which are meant for long term storage.
(17) Active collections are meant for medium term storage.
(18) Germplasm which is meant for short term storage is referred to as a working collection.
(19) Plant material which are stored for 10 to 15 years are called active collections.
(20) Primary gene pool is also known as GP_3.
(21) Secondary gene pool is also referred to as GP_1.

(22) Primary centre of diversity refer to original home of crop plants.
(23) The term micro-centre was coined by J.R. Harlan in 1958.
(24) Law of homologous series of variation was given by N.I. Vavilov (1961).
(25) Law of parallel variation is also referred to as the low of homologous series of variation.
(26) Domestication is refers to the process of bringing wild species under cultivation.
(27) Extinction refers to permanent loss of crop species.
(28) In situ –conservation means conservation of germplasm under natural conditions.
(29) Conservation of germplasm in gene banks is known as ex situ-conservation.
(30) Gene bank refers to various organizations where genetic diversity is maintained in living state.
(31) Field gene banks are those areas of land in which germplasm of recalcitrant crop species is maintained.
(32) Recalcitrant seeds exhibit drastic loss in viability with decrease in moisture content below 20%.
(33) Gene sanctuaries refer to protected areas of great genetic diversity under natural conditions.
(34) Dr. H.B. Singh made remarkable contribution in the field of plant introduction in India.
(35) In sexual reproduction, the sexual process is by passed.
(36) In apomixis embryo develops without sexual fusion.
(37) Development of embryo from egg cell without fertilization refers to parthenogenesis.
(38) Autogamy is also known as self- pollination.
(39) Allogamy refers to cross pollination.
(40) Amphimixis refers to sexual reproduction.
(41) Autogamy is closest form of outbreeding.
(42) Allogamyis common form of inbreeding.
(43) Allogamy leads to homozygosity.
(44) Allogamy permits combination of desirable desirable gene form different genotypes or sources.
(45) Allogamy maintains genetic purity of species / variety.
(46) When reproduction in a species occurs by apomictic means only, it is called obligate apomixis.
(47) When reproduction also takes place by sexual process in addition to apomixis, it is known as facultative apomixes.
(48) Systems of mating were given by Sewall Wright (1931).
(49) In random mating, variability, heterozygosity, population mean and genetic correlation between relatives remains constant.

(50) Random mating is usefull in development of hybrids.
(51) Genetic assortative mating is useful in maintaining the genetic purity of genotypes and in developing inbred lines.
(52) Phenotypic Assortative mating is useful in development of extreme phenotype.
(53) Prepotency increases with increase in heterozygosity.
(54) The term self incompatibility was coind by stout in 1927.
(55) Self incompatibility was first reported in Verboscum phoeniceum by Koelreuter in middle of 20th century.
(56) Self incompatibility prevents inbreeding and promotes inbreeding.
(57) Heteromorphic system of self incompatibility is important in crop plants.
(58) Gametophytic self incompatibility is governed by genetic constitution of gametes.
(59) Sporophytic self incompatibility is controlled by the genotype of pollen producing plant.
(60) In gametophytic system, recovery of only male parent is possible in some crosses.
(61) In sporophytic system, recovery of both male and female parents is possible.
(62) In some crosses, sporophytic system permits production of some homozygotes.
(63) Gametophytic systems of self incompatibility permit production of homozygotes.
(64) In gametophytic system, crosses may be sterile only.
(65) Genetic male sterility is controlled by plasma genes.
(66) Cytoplasmic sterility is governed by nuclear genes.
(67) Cytoplasmic genic male strictly is controlled by both nuclear and plasma genes.
(68) Male gametocides are chemicals which are used for reduction of male sterility.
(69) Genetic male sterility is maintained by sib mating of sterile (msms) and fertile (Msms) plants.
(70) Mutagen induced male sterility is heritabe and male sterility induced by gametocide is non-heritable.
(71) Direct introductions adapt to the new environment immediately.
(72) Indirect introduction require few years for adaptation in the changed environment.
(73) Exotic variety refers to a foreign variety which is directly recommended for commercial cultivation.
(74) Stabilizing selection is also called centripetal selection.
(75) Directional selection favours one extreme phenotype.
(76) Disruptive selection favours both extreme phenotypes.
(77) Pureline is the progeny of a self –pollinated homozygous plant obtained by selfing.
(78) In single seed descent method, plants are advanced by selecting single seed per plant from F_4 generation onwards.

(79) Mass pedigree method was proposed by Harrington in 1957.
(80) Single seed descent method was first suggested by goulden in 1949 for advancing segregating population of self pollinated crops.
(81) Grafius (1995) first used single seed descent methods in oats.
(82) Back cross method is less effective for transfer of polygenic characters.
(83) Use of multiline cultivars was first suggested by Jensen in 1950 in oats.
(84) Bulk methods improve the adaptation of genotypes.
(85) Back cross method offers opportunities for breaking undesirable linkages.
(86) Pure lines are heterozygous and homogeneous.
(87) F_1's and clones are heterogenous but heterozygous.
(88) Multilines are homozygous but heterogenous.
(89) Composites and synthetics are heterogenous and heterozygous.
(90) Donar parent is also known as recurrent parent.
(91) Recipient parent is also called non recurrent parent.
(92) Heterozygous population will segregate on selfing.
(93) Mass selection is more commonly used in allogamous species than autogamous species.
(94) Positive mass selection refers to removal of off type plants from a heterogenous population and allowing rest of the plants to grow further.
(95) Progeny selection refers to selection of inferior plants from a heterogenous population based on progeny test.
(96) The term recurrent selection was coined by hull in 1955.
(97) Recurrent selection for gca is used to improve the general combining ability of a population for a particular character.
(98) In recurrent selection for gca, a heterozygous tester with narrow genetic base is used.
(99) Recurrent selection for gca is used for the improvement of those characters which are governed by additive gene action.
(100) Recurrent selection for sca is used to improve the gca of a population for a particular character.
(101) In recurrent selection for sca, a homozygous tester with narrow genetic base is used.
(102) Recurrent selection for sca is used for the improvement of those characters which are governed by non-additive gene action.
(103) Reciprocal recurrent selection is used to improve both gca and sca of a population for a particular character using two heterozygous testers.
(104) Reciprocal recurrent selection is used for the improvement of those characters which are governed by both additive and non-additive gene action.
(105) Selection for both extreme phenotypes is referred to as disruptive selection .

(106) Clone refers to the progeny of a single plant obtained by sexual reproduction.
(107) Synthetic variety refers to a mixture of all possible F_1 crosses among inbred lines showing good general combining ability.
(108) Composit variety is a mixture of several phenotypically similar genotypes.
(109) Convergent improvement was proposed by Richey in 1937.
(110) Gametic selection was proposed by Stadler in 1964.
(111) Synthetic varieties are maintained by controlled pollination.
(112) Synthetic variety has a hetrogenous population.
(113) Heterosis refers to superiority of F_1 in fitness and vigour over its parents.
(114) The term heterosis was coined by shull in 1924.
(115) Pseudoheterosis refers to superiority of F_1 over its parents in vegetative growth.
(116) Pseudoheterosis is also known as Luxuriance.
(117) Dominance refers to masking effect of one allele over the other on the same locus.
(118) Dominance hypothesis was first proposed by Davenport in 1918.
(119) Over dominance refers to superiority of heterozygote over both the homozygote.
(120) Over dominance hypothesis of heterosis was independently proposed by Shull and East in 1928.
(121) Hull (1965) coined the term overdominance.
(122) Over dominance hypothesis of heterosis was supported by East (1936) and Hull (1945).
(123) Heterosis over the mean value of parents is called mean heterosis.
(124) Hetero beltiosis refers to superiority of F_1 over the better parent.
(125) Economic heterosis refers to superiority of F_1 over the popular variety of a region.
(126) Inbreeding refers to mating between closely unrelated individuals.
(127) Outbreeding refers to mating between closely related individuals.
(128) Loss in vigour and fitness as a result of inbreeding is called inbreeding depression.
(129) Hybrid refers to the F_1 population which is used for commercial cultivation.
(130) Interspecific hybrid refers to a hybrid between two genotype of the different species.
(131) Intraspecific hybrid refers to a hybrid between two species of the same genus.
(132) Single cross hybrid is a hybrid between two single crosses.
(133) Three way cross hybrid refers to a hybrid between a single cross and an inbred.
(134) Double cross hybrid is a hybrid between two single crosses.
(135) Inbred refers to a true breeding line obtained by continuous inbreeding.
(136) A cross between an inbred and open pollinated variety is called top cross.

(137) Multiple cross is also called composite cross.
(138) Over-dominance is also known as superdominance.
(139) Wide crossing refers to mating between two different species of the same genus or two different genera of the same family.
(140) Wide crossing is also known as hybridization.
(141) Interspecific crosses are fully fertile between those species which have complete chromosomal homology.
(142) Interspecific crosses are partially fertile between those species differ in chromosome number (ploidylevel) but have some chromosomes in common.
(143) Interspecific crosses are fully sterile between those species which have no chromosomal homology.
(144) Wide crosses can be obtained through protoplast fusion.
(145) Embryo culture technique helps in obtaining viable interspecific and intergeneric hybrids.
(146) Interspecific hybrids are of three types, viz., fully fertile, partially fertile and fully sterile.
(147) Intergeneric hybrids are always sterile.
(148) Release of hybrids is possible through interspecific hybridization.
(149) Release of hybrids is not possible through intergeneric hybridization.
(150) Main reason of sterility in wide crosses is lack of structural homology between the chromosomes of two species or two genera.
(151) Stress refers to adverse conditions for crop growth and production imposed either by biotic factors or abiotic factors or both.
(152) Adverse conditions for crop growth and production caused by biotic factors is called biotic stress.
(153) Abiotic stress is caused by environmental factors such as high temperature, low moisture, soil salinity, alkalinity etc.
(154) Vertical resistance is also referred to as single gene resistance.
(155) Horizontal resistance is also called as specific resistance.
(156) The term vertical resistance and horizontal resistance were coined by vander plank in 1973.
(157) Durable resistance refers to long lasting resistance.
(158) Vertical resistance is governed by polygene
(159) Horizontal resistance is controlled by oligogenes
(160) Three mechanisms of insect resistance, viz., non- preference, antibiosis and tolerance were given by Painter in 1951.
(161) Non preference is also referred to as antixenosis.
(162) Tolerant cultivars have greater recovery of damaged parts than susceptible ones.

(163) In brassica, sinigrin acts as feeding stimulant to cabbage aphid.
(164) Wild species of crop plants are potential sources of disease and insect resistance.
(165) Concept of gene for gene hypothesis was developed by Flor in 1966.
(166) Gene for gene hypothesis is also called Flor hypothesis.
(167) Gene pyramiding refers to in corporation of two or more major genes for specific resistance to a pest in two genotype.
(168) Gene deployment refers to planned or strategic use of major genes in development of resistant cultivars for various geographical areas.
(169) Draught resistance refers to survival odf plants under water deficit conditions without injury.
(170) Arid and semi arid areas are more prone to draught than humid zones.
(171) Draught damages chloroplasts and lowers output of the the photosynthetic apparatus.
(172) There is an incerse an proline level in the leaves of plants which are subjected to draught.
(173) Draught resistance is genetically controlled physiological properity of species.
(174) Xerophytic plants are more resistance to draught than mesophytic.
(175) Sunken, small and less numbers of stomata are associated with draught resistance.
(176) Stomatal aperature is measured with the help of inhibitor.
(177) Drought resistance genotypes have rapid closing habit of stomata.
(178) Generally, deep rooted plants exhibit greater draught avoidance than shallow rooted ones.
(179) Seedling root growth is an indication of root growth at maturity.
(180) Resistance genotypes maintain high photosynthesis under moisture stress condition by restricting water loss through transpiration.
(181) Tissue water potential is measured with the help of theromocouple psychrometer.
(182) Rate of photosynythesis during and after moisture stress is an important index of draught resistance.
(183) Photosynthesis is used as a criterion to select for draught resistance.
(184) Root length during seedling stage is used as a measure of drought resistance.
(185) Lysimeter is used to study salt tolerance in plants.
(186) Ideotype refers to a biological model which is expected to behave in a predictable manner in a defined environment.
(187) Crop ideotype is a plant model which is expected to yield greater quantity of grains, fibre, oil or other useful product when developed as cultivar.
(188) The term ideotype was coined by Donald in 1978.
(189) Ideal plants or model plants are expected to give higher yield than old cultivars in a defined environment.

(190) Ideotype is a moving goal which changes with advancement in knowledge, national policy, market requirements, etc.
(191) In ideotype breeding emphasis is given on individual morphological and physiological trait which inhance the yield.
(192) In ideotype breeding, the conceptual theoretical model is prepared before initiation of breeding work.
(193) In ideotype breeding value of each trait is defined in advance.
(194) Ideotype breeding is a slow method of cultivars development.
(195) In ideotype breeding, phenotype of new variety to be developed is specified in advance.
(196) Ideotype breeding is a supplement to conventional breeding.
(197) Breeding for nutritional quality refers to genetic improvement of crop plants in relation to quality and quantity of protein, oil and vitamins, and elimination of toxic substances.
(198) In green leafy vegetables, methionine is limiting amino acid.
(199) Essential amino acid can be synthesized by human body and their requirement has to be met through dietary intake.
(200) Essential amino acid can not be synthesized by human body and they need not be supplied through diet.
(201) Multiple cropping refers to cultivation of two or more crops on the same field in a year.
(202) Cultivation of two or more crops on the different field in a year.
(203) Intercropping refers to growing of two or more crops together on the same field in a year.
(204) Mixed intercropping refers to cultivation of two or more crops together on the same field by mixing their seeds.
(205) In relay cropping, there is the same overlapping period of two crops in a field.
(206) Sole cropping also known as solid planting.
(207) Repetitive cultivation of same sole crop on the same field is called monoculture.
(208) Crop included in multiple cropping systems should have different rooting pattern.
(209) Crop selected for multiplying cropping system should not have common diseases and insects.
(210) Seed is a product of fertilized ovule which consists of embryo, seed coat and cotyledon(s).
(211) Variety refers to genotype which has been released for commercial cultivation either by state variety released committee or central variety release committee.
(212) Improved seed refers to the seed of a released and popular variety produced by scientific method.
(213) Initial seed of an improved variety, limited in quantity and produced by originating breeder is called nucleus seed.

(214) Breeder seed is meant for general distribution to the formers for commercial cultivation.
(215) Foundation seed has physical purity of 98 % in majority of crops.
(216) In foundation and certified seeds of carrot, physical purity of 95% is allowed.
(217) National Seed Corporation was established in 1963 to take up the work of quality seed production in India.
(218) In India National Seed Project was launched in 1986.
(219) Notification of released varieties is done by the Government of India.
(220) Plant breeder's rights enable the breeder to multiply and market seeds of his variety and also allow him to authorize other interested persons for its commercial production and marketing.
(221) Plant biotechnology is a combination of tissues culture and genetic engineering.
(222) The plant part which is used for regeneration is known as explant.
(223) Callus refers to a mass of unorganized regenerated cells in culture medium.
(224) Somatic hybridization is fusion of protoplasts of two different genotypes.
(225) Homokaryons are hybrid cells combining protoplasts of two different species.
(226) Somatic hybridization, leads to direct production of allotetraploids.
(227) In somatic hybridization, there is equal contribution of cytoplasm from both the parents.
(228) Segregation dose not occur in somatic hybrids.
(229) The variation observed among the plants which are regenerated from ovules or anthers is called gametoclonal variation.
(230) Protoclonal variation is observed among the plants which are regenerated from the callus of protoplasts.
(231) Plants containing foreign DNA are called transgenic plants.
(232) DNA which contains genes from different sources and can combine with DNA of any organism is known as recombinant DNA.
(233) Plasmids are extra chromosomal genetic elements found within bacterial cell and are capable of self replication.
(234) The technique of detERmining the order of bases of DNA molecules which constitutes a gene is known as gene sequencing.
(235) The process of removal of introns and joining of exons in heterogeneous nuclear RNA (hn RNA) to produce mature functional RNA (mRNA) is called gene splicing.
(236) Small sequences of DNA with known base sequences, origin and function are known as DNA probes.
(237) Tissue culture is defined that the growth of tisues of living organisms in a suitable culture medium (in vitro).

Answer Sheet

1	T	2	T	3	T	4	T	5	T	6	T	7	F	8	F	9	T	10	F
11	F	12	F	13	T	14	T	15	T	16	T	17	T	18	T	19	T	20	F
21	F	22	T	23	F	24	F	25	T	26	T	27	T	28	T	29	T	30	T
31	T	32	F	33	T	34	T	35	T	36	T	37	T	38	T	39	T	40	T
41	F	42	F	43	F	44	T	45	T	46	T	47	T	48	F	49	T	50	F
51	T	52	T	53	F	54	F	55	F	56	F	57	F	58	T	59	T	60	T
61	T	62	T	63	F	64	F	65	F	66	F	67	T	68	F	69	T	70	T
71	T	72	T	73	T	74	T	75	T	76	T	77	T	78	F	79	F	80	F
81	F	82	T	83	F	84	T	85	T	86	F	87	F	88	T	89	T	90	F
91	F	92	T	93	T	94	F	95	F	96	F	97	T	98	F	99	T	100	F
101	F	102	T	103	T	104	T	105	T	106	F	107	T	108	T	109	F	110	F
111	F	112	T	113	T	114	F	115	T	116	T	117	T	118	F	119	T	120	F
121	F	122	F	123	T	124	T	125	T	126	F	127	T	128	T	129	T	130	F
131	T	132	T	133	T	134	T	135	T	136	T	137	T	138	T	139	T	140	F
141	T	142	T	143	T	144	T	145	T	146	T	147	T	148	T	149	T	150	T
151	T	152	T	153	T	154	F	155	F	156	F	157	T	158	F	159	F	160	F
161	T	162	T	163	T	164	T	165	F	166	T	167	F	168	T	169	T	170	T
171	T	172	T	173	T	174	T	175	T	176	F	177	T	178	T	179	T	180	T
181	T	182	T	183	T	184	T	185	T	186	T	187	T	188	F	189	T	190	T
191	T	192	T	193	T	194	T	195	T	196	T	197	T	198	T	199	F	200	F
201	T	202	F	203	T	204	T	205	T	206	T	207	T	208	T	209	T	210	T
211	T	212	T	213	T	214	F	215	T	216	T	217	T	218	F	219	T	220	T
221	T	222	T	223	T	224	T	225	T	226	T	227	T	228	T	229	T	230	T
231	T	232	T	233	T	234	T	235	T	236	T	237	T						

3
Statistics, Biometrics and Application of Computers

3.1. Multiple Choice Questions
(1) The Σ of probability will be ------------------------
  (a) 1.0 (b) 2.0
  (c) 3.0 (d) 0.5
(2) Frequency distribution can be converted in probability by calculating ———— each class which can be consider the probability of that class
  (a) Individual frequency (b) Relative frequency
  (c) Both relative and individual frequency (d) None of the above
(3) In binomial distribution, mean is always greater than ————————.
  (a) Median (b) Mode
  (c) Variance (d) All of the above
(4) ———————— is parameter of Poisson distribution.
  (a) Median (b) Mode
  (c) Mean (d) Variance
(5) The probability of a particular value in Normal distribution is always————.
  (a) One (b) Two
  (c) Zero (d) Three
(6) The normal curve is symmetric about mean which ———— value of curve.
  (a) Right side (b) Left side
  (c) Middle (d) None of the above
(7) A ———————— mean = variance = mode.
  (a) Normal distribution (b) Binomial distribution
  (c) Passion distribution (d) None of the above
(8) Mean and variance are parameter of———————— .
  (a) Bionomial distribution (b) Passion distribution
  (c) Normal distribution (d) None of the above
(9) The mean of standard normal variance is always zero and variance is ————.
  (a) One (b) 0.5
  (c) Two (d) Three

(10) In chi square distribution curve is ——————— skewed curve
   (a) Negatively (b) Positively
   (c) Zero (d) None of the above
(11) λ^2 variable always takes values between ——————— values.
   (a) 5-10 (b) 0-100
   (c) 50-100 (d) 200-300
(12) In λ^2 distribution probability———————.
   (a) Two (b) One
   (c) Three (d) Four
(13) Conclusion for ——————— test
   a. Population Variance are known and sample size may be large or small.
   b. Population variance are unknown but sample size is large (7^{30})
   (a) Z- test (b) t- test
   (c) Paired t test (d) ë² - test
(14) ——————— distribution
   c. Curve is symmetric
   d. Values lies between – á to + á.
   e. In t –d distribution mean is zero and variance n/n-2
   (a) Z (b) t
   (c) λ^2 (d) None of the above
(15) Condition for ——————— test are ———————.
   • Population variance are unknown
   • Sample size is less than 30
   (a) Z- test (b) t- test
   (c) Paired t- test (d) All of the above
(16) Equality of two mean is tested by ——————— test.
   (a) Z- test (b) F- test
   (c) T- test (d) λ^2- test
(17) Testing of equality at two variance is by ———————.
   (a) T- test (b) λ^2- test
   (c) F- test (d) None of the above
(18) Testing of significance at specified variance by ———————.
   (a) Z (b) T
   (c) F (d) λ^2
(19) Testing of goodness of fit and two independent attributes by ———————.
   (a) λ^2- Test (b) Z test
   (c) F- test (d) t- test

(20) Equality of two proportions is tested by —————.
    (a) Z test   (b) F- test
    (c) T- test   (d) Babarttlet- test

(21) Equality of specified population tested by —————.
    (a) F- test   (b) Z test
    (c) t- test   (d) All of the above

(22) Equality of more than two proportions by —————.
    (a) λ^2 test   (b) t- test
    (c) F- test   (d) All of the above

(23) Equality of more than two variance by —————.
    (a) λ^2 test   (b) t- test
    (c) F- test   (d) Fisher t-test

(24) Conditions for ————— are as
    a. Population variance are unknown
    b. Population variance are equal
    c. Sample size is less than 30
    (a)) λ^2 test   (b) Fisher t-test
    (c) Fisher t-test   (d) Coehron t-test

(25) Condition for ————— are as
    a. Population variance are unknown
    b. Population variance are not equal
    c. Sample size is less than 30
    (a) Cochron t-test   (b) Fisher t-test
    (c) λ^2 test   (d) t- test

(26) Equality of more than two mean is tested by ————— technique.
    (a) Paired t- test   (b) Coehron t- test
    (c) ANOVA technique   (d) None of the above

(27) Model for ————— way ANOVA technique is as
    g. Yij = ì + ái + eij
    h. Yij = It is the jth observation belonging ith class of assignable factor A
    i. ì = General mean effect
    j. ái = Effect due to ith assignable factor
    k. eij = Error component
    (a) One   (b) Two
    (c) Three   (d) None of the above

(28) Degrees of freedom of ——————— ANOVA (k-1, N-k).
    (a) One way   (b) Two way
    (c) Three way   (d) All of the above

(29) Model for ——————— classification data.
    Yij = ì + ái + âj + eij
    Assignable factor A
    d.f. F [(r-1), (r-1)(c-1), á]
    Assignable factor B
    F[(c-1), (r-1)(c-1), á]
    (a) One way   (b) Two way
    (c) Three way   (d) None of the above

(30) Assumptions of ——————— technique are as
- Model must be linear
- Various effect must be additive i.e. effect of F and non assignable factor must be additive
- Error component of model must be independently normally distributed with mean o and $ó^2$ ìj ~ N (O, $ó^2$)
- Error component must be uncorrelated

    (a) Diallele   (b) Triallel
    (c) ANOVA   (d) Design

(31) The formula of ——————— is

$$r = \frac{Sxy}{1.\sqrt{Sxx\,Syy}} \text{ or } \frac{Cov(x,y)}{\sqrt{y(x)\,y(y)}}$$

    (a) Correlation coefficient   (b) Regression coefficient
    (c) Angular transformation   (d) None of the above

(32) The values of Correlation Coefficient always lies between ———————.
    (a) -1 and +1   (b) -1 and -1
    (c) +1 and +1   (d) 0 and +1

(33) There is ——————— unit of Correlation Coefficient
    (a) Cm   (b) O_c
    (c) Mm   (d) No unit

(34) The formula of correlation coefficient was given by ———————.
    (a) Karl Pearson   (b) Fisher
    (c) Coehron   (d) None of the above

(35) ———————— of two variable is used for finding (i) what is exact form of relationship between dependent and independent variable. (ii) To estimate the dependent variable for a given value of independent variable, (iii) To know the change in the value of dependent variable for unit change in the value of independent variable
   (a) Correlation coefficient
   (b) Regression analysis
   (c) Both correlation coefficient and regression analysis
   (d) None of the above

(36) The value of bxy and byx can not greater than ————————.
   (a) One   (b) Two
   (c) Three   (d) Four

(37) Testing of significance of Regression Coefficinet is same as testing of significance of Correlation Coefficient i.e. – 1 = 0 is same âyx =0
   (a) Paired t- test   (b) ANOVA technique
   (c) Correlation coefficient (d) All of the above

(38) R^2 is called ————————.
   (a) Regression coefficient
   (b) Correlation coefficient
   (c) Coefficient multiple determination
   (d) None of the above

(39) R^2 = ————————.
   (a) S R^2/Syy S R^2 = bxy Sxy   (b) S R^2/Sxy S R^2 = bxy Sxy
   (c) S R^2/Syy S R^2 = byy Sxy   (d) S R^2/Syy S R^2 = bxx Sxx

(40) The value of R is always ————————.
   (a) Positive   (b) Negative
   (c) Both positive and negative (d) All of the above

(41) The value of R lies between ————————.
   (a) 0-100   (b) 0-5
   (c) -1-+1   (d) 0-1

(42) The values of partial correlation coefficient are ————————.
   (a) –1 and +1   (b)) 0-5
   (c) 0-1   (d) -1-+1

(43) ———————— is used when sample correlation coefficient both variable x and y are not normally distributed then ———————— give good estimate.
   (a) Simple correlation coefficient (b) Partial correlation coefficient
   (c) Ranked correlation coefficient (d) Regression coefficient

(44) The ranked correlation coefficient is given by ——————.
   (a) Spearman   (b) Fisher
   (c) Karl Pearson   (d) Coehron
(45) The value of rank correlation coefficient lies between ——————.
   (a) 0-1   (b) 0-100
   (c) –1 and +1   (d) -1 and -1
(46) Significance of ranked correlation coefficient is tested by——————.
   (a) Z - test   (b) λ^2 test
   (c) F- test   (d) t- test
(47) Intra class correlation coefficient is to know the relationship between the units belonging to ——————.
   (a) Different class   (b) Same class
   (c) Between the class (d) None of the above
(48) Significance of intra class correlation coefficient is used by ANOVA table, which find out by ——————.
   (a) λ^2 test (b) Z - test
   (c) t- test   (d) F- test
(49) The value of inter class correlation coefficient lies ——————.
   (a) –1 and 1.0   (b) +1 and -1
   (c) 0 and 100   (d) 0 and 1
(50) Intra correlation coefficient used in animal science calculated repeatedly and also measure ——————.
   (a) Variability   (b) Heritability
   (c) Uniformity   (d) Compatibility
(51) The estimate value of correlation coefficient is tested by ——————.
   (a) Least square theory (b) Bionomial theory
   (c) Spearman theory   (d) Fisherman theory
(52) Regression coefficient is tested by both ——————
   (a) $ë^2$ test   (b)) Z - test
   (c) F-test and t-test (d) None of the above
(53) D.F. of Fisher t-test is——————.
   (a) $n_1 + n_2 - 2$ (b) $n2 + n_2 - 2$
   (c) n-1   (d) n-1-2
(54) When original data is in —————— then we are square root transformations $y = \sqrt{x}$
   (a) Binomial distribution (b) Poisson distribution
   (c) Normal distribution   (d) Non of the above

(55) The degree to which numerical data tend to spread about the mean value is called ―――――.
   (a) Dispersion   (b) Frequency
   (c) Interclass   (d) Mode

(56) ―――――. is the difference between the lowest and highest values present in the observation of a sample.
   (a) Frequency   (b) Mean
   (c) Range   (d) None of the above

(57) The square root of variance is known as ―――――.
   (a) Mean   (b) Mode
   (c) Standard deviation   (d) Range

(58) ―――――― is based on all the observations of a sample and is capable of further algebric treatment.
   (a) Frequency   (b) Mean
   (c) Mode   (d) Standard deviation

(59) ―――――― measures of an controlled variation present in a sample.
   (a) Mean   (b) Standard error
   (c) Mode   (d) None of the above

(60) The ratio of standard deviation of a sample to its mean expressed in percentage is called―――――― = SD√X x 100
   (a) Coefficient of variation   (b) Standard error
   (c) Standard deviation   (d) None of the above

(61) A hypothesis of no difference is known as null hypothesis (Ho) and is given by ―――――.
   (a) Pearson   (b) Spearman
   (c) Fisher   (d) Coheron

(62) When null hypothesis is rejected, a alternative hypothesis is used, denoted by ―――――.
   (a) H_1 (b) H0
   (c) R   (d) P

(63) When sample size is large ―――――― is used.
   (a) T- test   (b) Z- test
   (c) F test   (d) None of the above

(64) When the sample size is small ―――――― is used.
   (a) F test   (b) Z- test
   (c) t- test   (d) All of the above

(65) When the observations are paired ———————" test is used.
   (a) Student's t   (b) Z- test
   (c) F- test   (d) None of the above

(66) When the observations are not paired ——————— test is used.
   (a) Student's t   (b) Fisher's- t
   (c) Z- test   (d) F- test

(67) The significance of difference among several mean is tested by ———————
   (a) Z- test   (b) t- test
   (c) F- test   (d) None of the above

(68) Regression coefficient was first estimated by ——————— in relationship between the height of father and sons.
   (a) Galton   (b) Fisher
   (c) Spearman   (d) Cohcorn

(69) The ——————— is estimated as
   byx=cov.(XY)/YX, where cov.(XY) = Co-Variance between X and Y, YX=Variance of X
   (a) Partial correlation coefficient   (b) Multiple correlation coefficients
   (c) Regression coefficient   (d) None of the above

(70) Correlation is represented by ——————— and is estimated by
   rxy = cov(X, Y)/[(YX)(XY)]1/2, where, rxy = Correlation between X and Y, Cov(X,Y) = Covariance between X and Y, YX = Variance of X and VY = Variance of Y.
   (a) R   (b) r
   (c) p   (d) q

(71) The significance of correlation is tested with the help of table value at ——————— degrees of freedom.
   (a) n+1   (b) n-2
   (c) n+2   (d) n-1-n-2

(72) The statistical procedure which separates or splits the total variation into different components is known as ———————
   (a) Degree freedom   (b) Test of significance
   (c) Analysis of variance   (d) None of the above

(73) A scientifically planned method is called an experiment and various affects of comparison is known as ———————.
   (a) Unit   (b) Treatment
   (c) Replication   (d) None of the above

(74) The group of material to which a treatment is applied in a single trial of experiment is known as ─────────.
    (a) Replication   (b) Experimental unit
    (c) Randomization   (d) Local control

(75) The repetition of the treatment under investigation is known as ─────────
    (a) Replication   (b) Randomization
    (c) Local control   (d) None of the above

(76) ───────── is calculated as $(2VE/r)1/2$, where VE = error variance and r = number of replication.
    (a) Standard deviation   (b) Standard error
    (c) Coefficient of variance   (d) None of the above

(77) The allocation of the treatments to the different plots by a random process is known as ───────── of treatment. ───────── gives equal chance to all the treatments for being allotted to a more fertile plot as well as to a less fertile plot.
    (a) Replication   (b) Randomization
    (c) Local control   (d) Heritability

(78) The principle of moding use of greater homogeneity in groups experimental units for reducing experimental error is known as ─────────
    (a) Replication   (b) Randomization
    (c) Local control   (d) None of the above

(79) When experimental material is limited and homogenous ───────── is used.
    (a) RBD   (b) LSD
    (c) CRD   (d) Split plot

(80) Randomized block design is appropriate when fertility variation of the field is in ─────────.
    (a) One direction   (b) Two direction
    (c) More than two direction   (d) None of the above

(81) Latin square design is used when fertility variations moves in ───────── directions.
    (a) One   (b) Two
    (c) Three   (d) None of the above

(82) Factorial design may be adopted when several factors are such that same of them require larger plots and may be studied with ─────────.
    (a) One precision   (b) Two precision
    (c) Different precision   (d) None of the above

(83) Lattice design would be appropriate when the number of treatments is sufficiently large and can form a square.
    (a) Scattered   (b) Rectangular
    (c) Square   (d) Circle

(84) When a very large number of germplasm lines each having small quantity of seed is to be tested on a small piece of land ———————— design would be a proper choice.

    (a) RBD   (b) CRD

    (c) Split plot   (d) Augmented

(85) The variation due to environmental factor or uncontrolled factor is called ———.

    (a) Experimental error   (b) Experimental variation

    (c) Experimental repitiation   (d) None of the above

(86) ———————— refers to the square of grand total divided by the number of observations. Thus correction factor = (Grand total)^2 /number of observations.

    (a) Correction factor   (b) MSS

    (c) SSS   (d) Experimental error

(87) The least significant difference, greater than which all the differences are significant is known as ———————— SE difference X t.

    (a) CD   (b) SE

    (c) CV   (d) GT

(88) ———————— is a test of significance which is used for testing the significance of difference among several treatments.

    (a) Z- test   (b) F - test

    (c) t- test   (d) None of the above

(89) F value is the ratio between the treatment ———————— and error variance. It is also known as variance ratio and is estimated as given below F value= treatment variance/error variance.

    (a) Variance   (b) Mean

    (c) Standard error   (d) CD

(90) Fisher in ———————— provided the initial frame work of biometric for the study of quantitative genetics.

    (a) 1928   (b) 1918

    (c) 1938   (d) None of the above

(91) First degree statistic is also known as first order statistics which includes ———————— , which is used for the measurement of all types of parameters.

    (a) Mean   (b) Estimate of variance

    (c) Estimate of co-variance   (d) All of the above

(92) Second degree of statistics is also known as second order statistics. It includes ———————— .

    (a) Mode   (b) Median

    (c) Mean   (d) Estimate of variance and covariance

(93) Third degree of statistics is also known as high order statistics which includes complete interactions like ———————, which are used in fitting frequently curves, surfaces and flatness.

    (a) Median   (b) Estimate of variance and co-variance

    (c) Curtosis and skewness   (d) None of the above

(94) Both major as well as minor genes are located on the chromosome in the ——————.

    (a) Mitochondria   (b) Cytoplasm

    (c) Nucleus   (d) None of the above

(95) The phenomenon of transgression in polygenes can only be explained by ——————.

    (a) Mendialian theory   (b) Batesman theory

    (c) Hedric and Booth theory   (d) None of the above

(96) Phenotypic variation is the total variability which is observable. It includes ——————.

    (a) Only genotypic variation   (b) Only phenotypic variation

    (c) Genotypic and environmental variation   (d) Only environmental variation

(97) Genetic variation is the inherent or genetic variability which remains unaltered by- ——————.

    (a) Phenotypic condition   (b) Genotypic condition

    (c) Both genotypic and phenotypic conditions (d) Environmental conditions

(98) Environmental variation refers to —————— variation which is entirely due to environmental effects varies under different environmental conditions.

    (a) Heritable   (b) Non- heritable

    (c) Epistasis   (d) None of the above

(99) Additive variance refers to that portion of genetic variance which is produced by the deviation due to average effects of the alleles or genes at segregating loci. Thus, it is the component which arises from difference between two homozygotes of a gene, i.e. ——————.

    (a) aa and aa   (b) AA and AA

    (c) AA and Aa   (d) AA and aa

(100) Dominance variance arises due to the deviation from the additive scheme of gene action resulting from intra-allelic interaction i.e. —————— alleles of the same gene or same locus.

    (a) Homozygotes   (b) Heterozygote

    (c) Interallelic interaction   (d) None of the above

(101) Epistatic variance arises due to the deviations as a consequence of ——————.

    (a) Inter- allelic interaction   (b) Intera-allelic interaction

    (c) Both a and b   (d) None of the above

(102) If the value of genotypic coefficient of variation (GCV) is ——————— than phenotypic coefficient of variation (PCV), it indicates that there is little influence of environment on the expression of character. Selection for improvement of such character will be rewarding.

    (a) Higher   (b) Lower

    (c) Average   (d) None of the above

(103) If the value of ——————— variation is higher than GCV, it means that the apparent variation is not only due to genotypes but due to influence of environment. Selection for such traits sometimes may be misleading.

    (a) PCV   (b) GCV

    (c) ECV   (d) None of the above

(104) If the value of ——————— is higher than PCV and GCV, it indicates that environment of such character. Selection for the improvement of such character will be ineffective.

    (a) ECV   (b) GCV

    (c) PCV   (d) All of above

(105) Metroglyph analysis technique was developed by Anderson in ——————— to investigate the pattern of morphological variation in crop species.

    (a) 1977   (b) 1957

    (c) 1967   (d) 1937

(106) The concept of D^2 Statistics was originally given by PC Mahalanobis in ———.

    (a) 1928   (b) 1938

    (c) 1948   (d) 1958

(107) The D^2 technique can not be applied to ——————— data.

    (a) Replicated data   (b) Unreplicated

    (c) Missing data   (d) None of the above

(108) The important properties of correlation coefficient are :

i. It is independent of the unit of measurement

ii. Its value lies between –1 and 1.

iii. It measures the degree and direction of association between two or more variable.

    (a) Only i   (b) Only ii

    (c) Only iii   (d) All of the above

(109) The association between any two variable is termed as simple correlation or total correlation or ——————— correlation coefficient.

    (a) First order   (b) Second order

    (c) Third order   (d) Zero order

(110) The significance of first order of partial correlation is tested ——— degrees of freedom.
   (a) n+1   (b) n-2
   (c) n-3   (d) n+1+2

(111) If the value of 'r' is significant, the association between two characters is ———.
   (a) Low   (b) Medium
   (c) High   (d) None of the above

(112) If the value of genotypic correlation coefficient is higher than———, it means that there is strong association between these two characters genetically, but the phenotypic value is lessened by the significant interaction of environment.
   (a) Genotypic correlation   (b) Phenotypic correlation
   (c) Environmental correlation   (d) None of the above

(113) If the value of phenotypic correlation coefficient is greater than———————, it shows that the apparent association of two characters is not only due to genes but also due to favourable influence of environment.
   (a) Environmental correlation coefficient
   (b) Genotypic correlation coefficient
   (c) Phenotypic correlation coefficient
   (d) None of the above

(114) If the value of ——————— is greater than genotypic and phenotypic correlation, it means that these two characters are showing high association due to favourable influence of particular environment and this association may change in an other locality or with the change in environment.
   (a) Phenotypic correlation   (b) Genotypic correlation
   (c) Environmental correlation   (d) None of the above

(115) If the value of r is ——————— or significant it means that these two characters are independent. But if the values of rg, rph all are insignificant it clearly indicates the independent nature of two characters.
   (a) 1   (b) 2
   (c) +1   (d) 0

(116) If the value of partial correlation coefficient is ———————, it means the simple correlation between X_1 and X_2 is due to the effect of another variable X_3 but after eliminating the effect of X_3 the two variable may be found as uncorrelated.
   (a) +1   (b) -1
   (c) 0   (d) 2

(117) If the value of r is ——— it indicates true relationship between X_1 and X_2 variable.
   (a) 0   (b) Non-significant
   (c) Significant   (d) None of the above

(118) The concept of path analysis was originally developed by Wright in 1921, but the technique was first used for plant selection by Dewey and Lu in ―――――.
    (a) 1929   (b) 1939
    (c) 1949   (d) 1959

(119) Path coefficient analysis is simply a standardized partial ―――――― which splits the correlation coefficient into the measure of direct and indirect effects.
    (a) Simple correlation   (b) Partial correlation
    (c) Regression coefficient   (d) None of the above

(120) The use of discriminate function for plant selection was first proposed by Smith in ―――――.
    (a) 1936   (b) 1939
    (c) 1949   (d) None of the above

(121) Classical selection index was developed by Smith in 1936. This involves several characters simultaneously in selection index and discriminates between desirable and undesirable genotypes on the basis of selection efficiency.
    (a) Smith   (b) Kempthorne
    (c) Nordskong   (d) None of the above

(122) General Selection Index was proposed by ――――――. This is a modification of the Scheme of Smith.
    (a) Hanson and Johnson, 1957   (b) Hanson and Johnson, 1967
    (c) Hanson and Johnson, 1977   (d) Hanson and Johnson, 1947

(123) Restricted selection index was proposed by ――――――.
    (a) Smith in 1936
    (b) Hanson and Johnson, 1947
    (c) Kenpthorne and Nordskog in 1959
    (d) None of the above

(124) ―――――― are important selection parameters.
    (a) Mean
    (b) Variability
    (c) Heritability and genetic advance
    (d) All of the above

(125) The ratio of genotypic variance to the phenotypic variance or total variance is known as ――――――. It is generally expressed in percent. It is a good index of the transmission of characters from parents to their "offspring"
    (a) Genetic advance   (b) Genetic variability
    (c) Heritability   (d) None of the above

(126) ―――――― is the ratio of variance to total or phenotypic variance.
    (a) Narrow sense heritability   (b) Broad sense heritability
    (c) Genetic advance   (d) None of the above

(127) ———— deals with simultaneous inheritance of two characters. It is the ratio of genotypic covariance to the phenotypic covariance X_1 and X_2.
  (a) Narrow sense heritability   (b) Co-heritability
  (c) Genetic advance   (d) None of the above

(128) Improvement in the mean genotypic value of selected plants over the parental population is known as ————. It is the measure of genetic gain under selection.
  (a) Genetic erosion   (b) Genetic advance
  (c) Variability   (d) None of the above

(129) The proportion of plants or families selected for the study is called as————, which plays important role in the success of ————.
  (a) Selection intensity   (b) Genetic advance
  (c) Heritability   (d) None of the above

(130) The difference between the mean phenotypic value of the progeny of selected plants and the base or parental population is known as ————. It is denoted by R.
  (a) Selection intensity   (b) Genetic advance
  (c) Genetic gain   (d) None of the above

(131) The difference between the mean phenotypic value of selected plants and the mean phenotypic value of parental population is known as ————. It is a measure of the intensity of selection and is denoted by K.
  (a) Genetic gain   (b) Genetic advance
  (c) Selection intensity   (d) Selection differential

(132) If the value of heritability in broad sense is high it indicates that through the character is least influenced by the ———— effects, the selection for improvement of such character may not be genetic because which includes both fixable and non-fixable variance.
  (a) Genotype   (b) Phenotype
  (c) Environmental   (d) None of the above

(133) If broad sense heritability is ————, it reveals that the character is highly influenced by environmental effects and genetic improvement through selection will be difficult due to masking effect of the environment on the genotypic effects.
  (a) Low   (b) High
  (c) Medium (d) None of the above

(134) If the heritability estimates are high in narrow sense, it means the character is largely govern by ———— genes and selection for improvement of such character would be rewarding.
  (a) Additive   (b) Non-additive
  (c) Epistatis   (d) None of the above

(135) If the value of genetic advance is —————, it shows that the character is govern by additive genes and selection for improvement of such character would be rewarding.
   (a) High (b) Low
   (c) Medium (d) None of the above

(136) If the value of genetic advance is low, it indicates that the character is govern by ————— genes and heterosis breeding may be useful.
   (a) Additive (b) Non- additive
   (c) Epistasis (d) None of the above

(137) High heritability accompanied with high genetic advance, it indicates that most likely the heritability is due to ————— gene and selection may be effective.
   (a) Additive (b) Non- additive
   (c) Epistasis (d) None of the above

(138) High heritability accompanied with ————— genetic advance is indicative of non-additive gene action. The high heritability is being exhibited due to favourable influence of environment rather than genotype and selection for such traits may not be rewarding.
   (a) High (b) Low
   (c) Medium (d) None of the above

(139) ————— heritability accompanied with genetic advance reveals that the character is governed by additive gene effects. The low heritability is being exhibited due to high environmental effects. Selection may be effective in such cases.
   (a) High (b) Low
   (c) Medium (d) None of the above

(140) The concept of combining ability as a measure of gene action was proposed by ————— working on maize.
   (a) Spargue and Tatum, 1944 (b) Spargue and Tatum, 1949
   (c) Spargue and Tatum, 1945 (d) Spargue and Tatum, 1942

(141) ————— refers to the capacity or ability of a genotype to transmit superior performance to its crosses.
   (a) Combining ability (b) Specific combining ability
   (c) Mixture (d) Multiline varieties

(142) ————— is the performance of a strain or genotype in a series of hybrid combinations.
   (a) Hybrid (b) GCA
   (c) SCA (d) Reciprocal cross

(143) The GCA variance is primarily a function of the additive genetic variance, but if epistasis is present gca will also include ——————— type of non-allelic interaction.
    (a) Additive X additive (b) Additive X non- additive
    (c) Dominance X dominance (d) None of the above

(144) ——————— refers to the deviation of a particular cross from the genetic combining ability.
    (a) GCA (b) SCA
    (c) Reciprocal cross (d) Hybrid

(145) Diallel mating design technique measures the combining ability from single crosses made in a definite fashion. According to this design the total number of single crosses among n parents would equal to ———————, excluding reciprocals.
    (a) n(n-1)/2 (b) (n-1)/2
    (c) n (-1)/2 (d) n (n-2)/2

(146) Partial diallel design also estimates the combining ability from single crosses, but again crosses have to be made in a definite fashion. This design permits inclusion of number of parents (upto 20) for evaluation than diallel design. In this design, the total number of crosses to be made equal to ——————— where n is the number of parents and s is the number of sample crosses.
    (a) n (n-2)/2 (b) n (-1)/2
    (c) (n-1)/2 (d) ns/2

(147) Line x tester design is modified form of ———————.
    (a) Top cross (b) Back cross
    (c) Test cross (d) None of the above

(148) If gca variance is higher than sca variance it means that there is preponderance of ——————— gene action and progeny selection will be effective for the genetic improvement of such traits.
    (a) Additive (b) Non- additive
    (c) Epistasis (d) None of the above

(149) If sca variance is higher than gca variance it indicates that there is prepondence of ——————— gene action (dominance and epistasis) and therefore heterosis breeding may be rewarding.
    (a) Non- additive (b) Additive
    (c) Epistasis (d) None of the above

(150) If both gca and sca variance are of equal magnitude it shows that ——————— genes are equally important in the expression of character. In such situation, reciprocal selection may be resorted to for population improvement.
    (a) Only additive (b) Only non- additive
    (c) Epistasis (d) Additive and non- additive both

(151) The analysis of a diallel cross is based on ———— assumptions, viz., (1) Normal diploid segregation (2) Lack of maternal effects (3) Obscene of multiple alleles (4) Homozygosity of parents (5) Obscene of linkage (6) Obscene of epistasis and (7) Random mating.
    (a) Only (1) and 2   (b) Only (2) and 3
    (c) Only 4 ,5 6 and 7   (d) All of the above

(152) Hayman's Graphical Approach was initially developed by ————.
    (a) Jinks and Hayman in 1953   (b) Jinks and Hayman in 1956
    (c) Jinks and Hayman in 1958   (d) Jinks and Hayman in 1960

(153) The concept of partial diallel mating design was develop by————.
    (a) Kempthorne in 1957   (b) Kempthorne in 1967
    (c) Kempthorne in 1977   (d) Kempthorne in 1978

(154) The concept of line x tester was developed by ———— in 1957. It is a modified form of top cross scheme. In case of top cross only one tester is used, while in case of line x tester cross, several testers are used.
    (a) Kampthorne   (b) Jinks and Hayman
    (c) Hayman   (d) None of the above

(155) The concept of triallel and quadriallel mating designs was developed by ——.
    (a) Rawlings and Cokerham in 1962 (b) Kempthorne in 1967
    (c) Jinks and Hayman in 1956   (d) None of the above

(156) The concept of biparental mating was originally developed by ————.
    (a) Rawlings and Cokerlam in 1962 (b) Comstock and Robinson in 1948
    (c) Jinks and Hayman in 1956   (d) Kempthorne in 1967

(157) The concept of triple test cross (TTC) analysis was proposed by------------------.
    (a) Kearsey and Jinks in 1968   (b) Comstock and Robinson in 1948
    (c) Jinks and Hayman in 1956   (d) Kempthorne in 1967

(158) Six parameter model was first suggested by ———— for the estimation of various genetic component from the generation means. This method is used when non-allelic interactions are present.
    (a) Hayman   (b) Kearsely
    (c) Jinks   (d) Comstock

(159) The three parameter model was proposed by ————.
    (a) Kearsely   (b) Comstock
    (c) Kempthorne   (d) Jinks and Jones

(160) Test of significance for six and three parameter models are tested with the help of ——————. if additive genetic variance is high reliance should be placed on mass selection in self-pollinated species and synthetic breeding in cross-pollinated species.
    (a) Z- value   (b) F- value
    (c) t- value   (d) None of the above

(161) If the —————— is predominant the breeding objective should towards development of hybrids for commercial purpose.
    (a) Additive   (b) Non- additive
    (c) Epistasis   (d) None of the above

(162) If the —————— variance is relatively high, more reliance should be placed on selection between families and line.
    (a) Non- additive   (b) Epistasis
    (c) Additive   (d) None of the above

(163) The significance of heterosis and inbreeding depression is tested with the help of ——————.
    (a) T- test   (b) F - test
    (c) Z- test   (d) CD

(164) Stability analysis can be carried out using of the —————— models
    (a) One   (b) Two
    (c) Three   (d) Four

(165) The regression coefficient of unity indicates —————— stability.
    (a) High   (b) Low
    (c) Medium   (d) None of the above

(166) If the regression coefficient is greater than one, it means the genotype has greater resistance to environmental change ——————.
    (a) Below stability   (b) Average stability
    (c) Medium stability   (d) Above average stability

(167) Regression coefficient of zero would absolute stability.
    (a) 1   (b) +1
    (c) 0   (d) None of the above

Answer Sheet

1	a	2	b	3	c	4	c	5	c	6	c	7	a	8	c	9	a	10	b
11	b	12	b	13	a	14	-	15	b	16	a	17	c	18	d	19	a	20	a
21	c	22	a	23	-	24	b	25	a	26	c	27	a	28	a	29	b	30	c
31	a	32	a	33	d	34	a	35	b	36	a	37	c	38	c	39	a	40	a
41	d	42	a	43	c	44	a	45	c	46	d	47	b	48	d	49	a	50	b
51	a	52	c	53	a	54	b	55	a	56	c	57	c	58	d	59	b	60	a
61	c	62	a	63	b	64	c	65	a	66	b	67	c	68	a	69	c	70	b
71	b	72	c	73	b	74	b	75	a	76	b	77	b	78	c	79	c	80	a
81	b	82	c	83	c	84	d	85	a	86	a	87	a	88	b	89	a	90	b
91	a	92	d	93	c	94	c	95	c	96	c	97	d	98	b	99	d	100	c
101	a	102	a	103	a	104	a	105	b	106	a	107	b	108	d	109	d	110	c
111	c	112	b	113	b	114	c	115	d	116	c	117	c	118	d	119	c	120	a
121	a	122	a	123	c	124	c	125	c	126	b	127	b	128	b	129	a	130	c
131	d	132	c	133	a	134	a	135	a	136	b	137	a	138	-	139	b	140	d
141	a	142	b	143	a	144	b	145	a	146	d	147	a	148	a	149	a	150	d
151	d	152	a	153	a	154	a	155	a	156	a	157	a	158	a	159	d	160	c
161	b	162	b	163	d	164	d	165	a	166	d	167	c						

3.2. Differenciate in Between

D^2 Statistics	Metroglyph Analysis
1. Estimates are based on second order statistics.	Estimates are based on first order statistics
2. Analysis is complicated.	Analysis is very simple.
3. Estimation is possible from replicated data only.	Possible from both replicated and unreplicated data.
4. This is a numerical approach.	This is semi graphic approach.
5. Genetic diversity is depicted by cluster diagram.	Variability is depicted by glyph on the graph.

Correlation	Regression
1. It measures mutual relationship between two or more variables.	It measures functional relationship between two or more variables.
2. Does not involve cause and effect relationship.	It indicates cause and effect relationship.
3. Confined to the study of linear relationship.	It studies both linear and non-linear relationship.
4. Sometime correlations may be nonsense with no practical relevance.	There is no such thing like non-sense regression.
5. It measures degree and direction of relationship between 2 or more variables.	It measures degree of dependence of one variable on the other(s).
6. It is independent of the unit of measurement.	It is in the unit of measurement of variable.

Correlation Analysis	Path Coefficient Analysis
1. It measures the association between two or more variables.	It measures the cause of association between two variables.
2. Analysis is based on variances and covariances.	Analysis is based on all possible simple correlations.
3. Does not provide information about direct and indirect effects of independent variables on the dependent one.	It provides information about direct and indirect effects of independent variables on dependent variable.
4. Does not provide estimate of residual effect.	Provides estimate of residual effect.
5. Based on the assumptions of linearity and additivity.	Also based on the assumptions of linearity and additivity.
6. Helps in determining yield components.	Also helps in determining yield components.

Path Coefficient Analysis	Discriminant Function Analysis
1. It measures the cause of association between two variables.	It measures the efficiency of various trait combinations in selection.
2. Analysis is based on all possible simple correlations.	Analysis involves variances and covariances.
3. Direct, indirect and residual effects are calculated for path analysis.	Weight coefficients, expected genetic. advance and relative efficiency are estimated
4. Helps in determining yield components.	Also helps in determining yield components.
5. Based on the statistical assumptions of linearity and additivity.	Also based on the same statistical. assumptions.

Broad sense heritability	Narrow sense heritability
1. Estimated from total genetic variance.	Estimated from additive genetic variance.
2. Can be estimated from both parental and segregating material.	Requires crossing in a definite fashion.
3. More useful in animal breeding.	Useful in both plant and animal breeding.
4. Useful in selection of elite types from homozygous lines.	Useful in selection of elite types from segregating material.

Heritability	Co heritability
1. It measures inheritance of one trait at a time.	It measures inheritance of two characters at a time.
2. Estimation is based on genotypic and phenotypic variances.	Estimation is based on genotypic and phenotypic covariances.
3. Helps in improvement of single character.	Helps in improvement of two traits simultaneously.

General combining ability	Specific combining ability
1. It is average performance of a strain in a series of crosses.	It refers to the performance of specific cross in relation to gca.
2. gca is due to additive genetic variance and additive x additive epistasis.	sca is due to dominance genetic variance and all the three types of epistasis.
3. It is estimated from half-sib families.	It is estimated from full-sib families.
4. It helps in the selection of suitable parents for hybridization.	It helps in the identification of superior cross combinations.
5. It has relationship with narrow sense heritability.	It has relationship with heterosis.

Combining ability	Heritability
A. Similarities	
1. Free from genetical assumptions	Also free from genetical assumptions
2. It predicts performance of parents in cross combination	It predicts transmission of characters from parents to the offspring
3. It is of two types, viz., general and specific	It is also of two types, i.e., broad sense and narrow sense
4. gca variance gives an idea of additive gene action	In narrow sense it gives an idea of additive variance
5. Its estimation requires specific mating design	Narrow sense estimate requires specific mating scheme
B. Dissimilarities	
1. Its estimation is based on first order statistics	Estimation is based on second order statistics
2. Can be estimated only from the crosses made in a definite fashion	Can be estimated from both parental populations as well as crosses
3. Aids in selection of superior parents and cross combinations	Aids in selection of elite genotypes from diverse genetic populations

Half-sib Crosses	Full-sib Crosses
1. Crosses have one parent in common.	It is mating between two parents.
2. It provides estimates of additive genetic variance.	It provides estimates of both additive and dominance variances.
3. In pure lines in the absence of epistasis Cov. H.S. = 1/2 VA or VA = 2 Cov. H.S.	In such situation Cov. H.S. = VA + VDVD = Cov. F.S. - 2 Cov. H.S.

Hayman's Approaches	Griffing's Approaches
1. It is a graphical approach.	It is a numerical approach.
2. It is based on the estimation of components of variance.	It is based on the estimates of combining ability variances and effects.
3. It provides information about six components, i.e., D, H_1, H_2, E, F and h^2.	It provides information about D and H components through gca and sca variances
4. Various genetic ratios can be worked out from above components.	Calculation of genetic ratios is not possible.
5. Analysis is not possible without parents.	Analysis is possible even when parents are not included.
6. Does not help in the identification of superior cross combinations.	Helps in the identification of superior cross combinations.

Full Diallel	Half Diallel
1. Total number of crosses is equal to p (p-1), where p is number of parents.	Half crosses are required i.e., p(p-1)/2.
2. Used when reciprocal differences are significant.	Used when reciprocal differences are insifginicant.
3. Used when parents do not have male sterility.	Can be used when male sterility or incompatibility is present.
4. Maternal effects can be estimated.	Maternal effects cannot be estimated.
5. Each parent is used as male and female in the mating.	Each parent is used either as male or female in the mating.

Half sib crosses	Full sib crosses
1. Crosses have one parent in common.	It is mating between two parents.
2. It provides estimates of additive genetic variance.	It provides estimates of both additive and dominance variances.
3. In pure lines in the absence of epistais Cov H.S. = ½ VA or VA=2 Cov H.S.	In such situation Cov F.S. = VA + VD VD= Cov F.S. - 2 Cov H.S.

Diallel Cross	Partial Diallel Cross	Line x Tester Cross
1. Includes all possible single crosses among n parents i.e., n(n-1)/2.	1. Involves ns/2 crosses, where n and s = number of parents and sample crosses.	1. Involves mf crosses, where m and f = number of male and female parents.
2. Each cross involves two parents.	2. Same as diallel cross.	2. Same as diallel cross.
3. Measures gca & sca effects.	3. Calculation is difficult.	3. Calculation is simple.
4. Less crosses are to be made say 45 among 10 parents.	4. Can evaluate more parents than diallel (up to 20).	4. Can evaluate large number of parents (up to 50).
5. Requires one crop season for crossing.	5. Results have less precision than diallel	5. Results have high precision.
6. Involves both first and second degree statistics.	6. Provides information about gca, sca variances and gca effects and D and H.	6. Provides information about gca, sca variances and effects and D and H components.
7. Measures D and H components.	7. Same as in case of diallel.	7. Same as in case of diallel.
8. Helps in choice of parents and breeding procedure.		

Diallel Crosses	Triallel Crosses	Quadriallel Crosses
1. Includes all possible single crosses among n parents i.e., n(n-1)/2.	1. Includes all possible three-way crosses among n parents i.e. n(n-1)(n-1)/2.	1. Includes all possible double crosses among n parents i.e. n(n-1)(n-2)(n-3)/8.
2. Each cross involves two parents.	2. Each cross involves three parents.	2. Each cross involves four parents.
3. Measures gca & sca effects.	3. Measures line effect.	3. Measures line effect.
4. Less crosses are to be made say 45 among 10 parents.	4. More crosses than diallel have to be made i.e., 360 among 10 parents.	4. More crosses than diallel and triallel have to be made i.e., 630 among 10 parents.
5. Requires one crop season for crossing.	5. Requires two crop seasons for crossing.	5. Requires two crop seasons for crossing
6. Involves both first and second degree statistics.	6. Involves second degree statistics.	6. Involves second degree statistics.
7. Measures D and H components.	7. Measures D, H and I components.	7. Measure D, H and I components.
8. Helps in choice of parents and breeding procedure.	8. Helps in deciding mating order of parents.	8. Helps in deciding mating order of parents.

North Carolina Design I	North Carolina Design II	North Carolina Design III
1. Each male is mated to a different set of females.	Each male is mated to the same set of females.	Each male is mated to both parents of original cross.
2. Each set consists of mf crosses.	Each set consists of mf crosses.	Each set consists of 2m crosses.
3. Variance between males provides an estimate of D.	Variances due to males & females provide an estimate of D.	Variance due to male X female provides an estimate of H.
4. Variance among females provides estimates of H & D.	Variance due to male x female provides an estimate of H.	Variance due to male x female provides an estimate of H.
5. Influence by the presence of maternal effects.	Same as in Design I.	Not affected by the presence of maternal effects.
6. Requires 10-12 times more area than Design III.	Requires 2-4 times more area than Design III.	Requires less area than Design I and II.
7. This is the least powerful design.	This is an intermediate design.	This is the most powerful design.

Six Parameter Model	Five Parameter Model
1. Analysis is based on P_1, P_2, F_1, F_2, B_1 and B_2	Analysis is based on P_1, P_2, F_1, F_2 and F_3.
2. Six parameter i.e. m, e, d, h, i, j, and I are estimated.	Five parameters i.e. m, d, h, i and I are estimated.
3. Provides information on all three types of non-allelic gene interaction.	Does not provide information about additive x dominance interaction.
4. Testing of model is not possible.	It provides X^2 test for the model.
5. Requires two crop seasons for generation of material.	Requires three crop seasons for generation of material.

Adaptation	Adaptability
1. It refers to fitness of a genotype in a given environment.	Refers to stable performance of a genotype over a wide range of environment.
2. It favours those characters which are advantageous for survival.	It favours those characters which are useful for productivity.
3. Natural selection operates.	Human selection operates.
4. Survival is the main concern.	Productivity is the main objectives.

Stability Analysis	Combining Ability Analysis	Heritability Analysis
1. Requires multilocation or several year data for analysis.	Analysis is possible even with one year data.	Analysis is possible with one year data.
2. Analysis is possible from both parental as well as segregating population.	Analysis is possible only from crosses made in a definite fashion.	Narrow sense heritability. requires crossing in a definite fashion.
3. There are different models for stability analysis for combining ability analysis.	Diallel, partial diallel and line x tester crosses are used.	There are several methods for estimation of heritability.
4. Helps in the identification of adaptable genotypes.	Helps in the choice of parents and breeding procedures.	Helps in the selection of elite genotypes.
5. Predicts varietals response under different environments	Predicts the performance of parents in hybrid combinations.	Predicts inheritance of traits from parents to their offspring.
6. Estimation is based on first order statistics.	Estimation is based on first order statistics.	Estimation is based on second order statistics.

Eberhart and Russell model	Perkins and Jinks model	Freeman and Perkins model
1. Involves 3 parameter i.e. mean, yield regression coefficient and deviation from regression.	Same as in first model.	Same as in first model.
2. Variation is divided into two fraction i.e. genotype and E + G X E.	Variation is divided into three fraction i.e. genotype, environments and G X E.	Variation is divided into three fraction i.e. genotype, environments and G X E.
3. The E + G X E is subdivided into 3 parts, i.e. Env. (linear), G X E linear and pooled deviations.	The G X e is subdivided into heterogeneity due to regression and is due to remainder.	The environment is divided into combined regression and residual 1. The G XE is subdivided into heterogeneity of regression on residual 2
4. Calculation is simple.	Calculation is difficult.	Calculation is very difficult.
5. Does not provide independent estimation for mean performance and environmental index.	Same as in first model.	It provides independent estimation of mean performance and environmental index.
6. Less expensive.	Same as in first model.	More expensive than first and second model
7. The degree of freedom for environment is 1.	The degree of freedom for environment is e-2.	The degree of freedom for environment is e-2.

3.3. Fill in the Blanks

(1) The square root of variance is known as ―――――.
(2) The percentage ratio of standard direction to its mean is referred to as ―――――.
(3) Z- test is used when sample size is ―――――.
(4) t- test is used when the sample size is ―――――.
(5) X^2 test was developed by ―――――.
(6) F- test was developed by ―――――.
(7) The measure of the coefficient of variation was developed by―――――.
(8) The concept of augmented design was developed by ―――――.
(9) ――――― refers to repetition of treatment under investigation.
(10) Allocation of treatments to different plots by a random process is known as ―――――.
(11) Local control is also known as ―――――.
(12) The ratio of treatment variance to error variance is called ―――――.
(13) The least significant difference, greater than which all the differences are significant is known as―――――.
(14) ――――― is used when the experimental material is limited and plots are homogenous.
(15) ――――― control fertility variation in one direction only.
(16) Experimental design which controls fertility variation in two directions is called ―――――.
(17) Experimental design in which experimental plots are divided into main plots, subplots and ultimate plots is referred to as―――――.
(18) Incomplete block design in which the number of treatments from square is termed as ―――――.
(19) Application of statistical concepts and procedures to the study of biological problems is referred to as ―――――.
(20) Initial frame of biometrics for the study of quantitative genetics was provided by―――――.
(21) Biometrical genetics is also known as ―――――.
(22) Concept of metroglyph analysis was developed by ―――――.
(23) Concept D^2 statistics was originally developed by ―――――.
(24) In plant breeding, D^2 statistics was first used by ―――――for assessment of genetic diversity.
(25) Simple correlation is also known as ―――――.
(26) Partial correlation is also referred to as―――――.
(27) Partial correlation is denoted as―――――.

(28) Net correlation refers to partial correlation————.
(29) The inherent correlation between two variables is known as ————.
(30) Association between two variables eliminating the effect of third variable is called————.
(31) Path coeifficient analysis for plant selection was first used by ————.
(32) Concept of path analysis was originally developed by Wright ————.
(33) A line diagram in path analysis used to depict the cause and effect situation is referred to as————.
(34) Use of discriminate function for plant selection was first proposed by ————.
(35) In animal improvement selection index was first used by————.
(36) General selection index was proposed by ————.
(37) ———— proposed restricted selection index.
(38) Classical selection index was proposed by ————.
(39) The term diallel was coined by ————.
(40) Graphical approach of diallel analysis was given by ————.
(41) Numerical approach of diallel analysis was given by————.
(42) A diallel with F_1's and parent is known as————.
(43) A diallel with F_1's and reciprocal is referred to as————.
(44) With P parents, total number of crosses in a full diallel is equal to————.
(45) With P parents, total number of crosses in a half diallel is equal to————.
(46) Graph contracted in diallel cross analysis is referred to as————.
(47) The concept of partial diallel was developed by————.
(48) Partial diallel was further elaborated by————.
(49) Partial diallel is also known as ————.
(50) Partial diallel fail to provide inform about ———— effects.
(51) Concept of line X tester analysis was developed by ————.
(52) Line X tester cross, is a modified form of————.
(53) In line x tester cross, the common male parents are known as ————.
(54) Concept of triallel analysis was developed by————.
(55) In triallele cross, number of all possible crosses among n parents is equal to ————.
(56) Concept of quadriallel analysis was developed by ————.
(57) Number of all possible double crosses among n parents is equal to ————.
(58) Concept of generation mean analysis was developed by ————.
(59) Scaling test was given by ————.
(60) Joint scaling test was given by————.
(61) Analysis of five parameter model is based on ———— generation of a single cross.

(62) Concept of biparental mating was originally developed by ―――――――.
(63) Designs of biparental mating are known as ――――――.
(64) NCD1 is also known as ――――――.
(65) NCD2 is also referred to as――――――.
(66) Concept of triple test cross originally developed by ――――――.
(67) First systematic approach for stability analysis was provided by――――――.
(68) Fitness of a genotype to a specific environment is called ――――――.
(69) The capacity of a genotype for genetic change in adaptation is known as ――――――.
(70) The buffering capacity of a genotype to environmental fluctuation is referred to as ――――――.
(71) The term homeostasis was coined by ――――――.
(72) The term micro and macro environments were coined by――――――.
(73) The ratio of genotypic variance to phenotypic variance is known as ――――.
(74) The ratio of additive variance to phenotypic variance is called ――――――.
(75) The ratio of genotypic covariance to phenotypic covariance is referred to as ――――――.
(76) Improvement in mean genotypic value of selected plants over the parental population is known as――――――.
(77) Differences between the mean phenotype value of the progeny of selected plants and parental population is known as ――――――.
(78) Differences between the mean phenotypic value selected plants and mean phenotypic value of parental population is called ――――――.
(79) Formula D/(D+ H +E) X 100 was used for estimation of heritability by―――.
(80) Formula ½ D/VF2 X 100 was used for estimation of heritability by ――――.
(81) Concept of combining ability as a measure of gene action was first proposed by――――――.
(82) Capacity of genotype to transmit superior performance to its crosses is termed as――――――.
(83) Average performance of a strain in a series of crosses is called ――――――.
(84) Deviation of a particular cross from the general combining ability is referred to as――――――.
(85) gca is primarily a function of ――――――variance and additive X additive epistasis.
(86) sca is primarily a function of―――――― variance and additive X dominance and dominance X dominance types of epistasis.
(87) gca is estimated ―――――― families.
(87) sca is estimated ―――――― families.

Answer Sheet

(1) Standard deviation
(2) Coefficient of variation
(3) Large
(4) Small
(5) Karl Pearson
(6) Fisher
(7) Karl Pearson
(8) Federer (1956)
(9) Replication
(10) Randomization
(11) Error control
(12) F- value
(13) Critical difference
(14) Completely randomized design
(15) Randomized block design
(16) Latin square
(17) Split plot design.
(18) Lattice design.
(19) Biometry
(20) R.A. Fisher (1918)
(21) Statistical genetics
(22) Anderson 1957
(23) P. C. Mahalanobis (1928)
(24) C. R. Rao (1952)
(25) Total correlation
(26) Net correlation
(27) r_{123}
(28) Partial correlation
(29) Genotypic correlation
(30) Partial correlation
(31) Dewey and Lu (1959
(32) Wright (1921
(33) Path diagram
(34) Smith (1936)
(35) Hazel (1943)
(36) Hanson and Johanson, (1957)
(37) Kempthorne and-Nordskog (1959)
(38) Smith (1936)
(39) Yates (1947).
(40) Hayman(1954).Griffing (1956).
(41) Griffing (1956).
(42) Half diallel
(43) Full diallel
(44) n(n-1).
(45) n(n-1)/2.
(46) Vr-Wr graph
(47) Kempthorne in (1957).
(48) Kempthorne and Curnow.
(49) Fractional Diallel.
(50) sca.
(51) Kemphthorne in (1957).
(52) Top cross
(53) Testers.
(54) Rawlings and Cokerham in (1962a).
(55) $n(n-1)(n-2)/2$.
(56) Rawlings and Cokerham in (1962 b).
(57) $n(n-1)(n-2)(n-3)/8$.
(58) Hayman (1958) and Jinks and Hayman (1958).
(59) Mather (1949).
(60) Cavalli (1952).
(61) P_1, P_2, F_1, F_2 & F_3
(62) Comstock and Robinson (1948, 52).
(63) North Carolina Designs.
(64) Nested design
(65) Factorial design
(66) Kearsey and Jinks (1968)
(67) Finaly & Wilkinson (1963).
(68) Adoptation.
(69) Adaptability

(70) Homeostasis.
(71) Lerner(1954)
(72) Comstock and moll(1963)
(73) Heritability
(74) N.S.heritability
(75) Coheritability
(76) Genetic advance
(77) Genetic gain.
(78) Selection differential
(79) Mather(1949)
(80) Warner(1952)
(81) Sprague and Tatum in (1942)
(82) Combining ability
(83) gca.
(84) sca.
(85) Additive
(86) Dominance
(87) Half
(88) Full

3.4. True and False

(1) Microcomputers are meant for use by one person at a time.
(2) Microcomputers are also known as personal computers.
(3) Microcomputers are built with microprocessor.
(4) Minicomputers are designed to serve multiple users at a time.
(5) Computing capacity of super computers is extremely high.
(6) The main component in the second generation computers is vacuum tube.
(7) Large scale integrated circuits are the main components in the first generation computers.
(8) A device that transfers information from outside world to the computer is called output device.
(9) A device that transfers information from computer to the out side world is known as input device.
(10) The input and out put devices are combinedly known as peripherals.
(11) Program refers to a set of instructions written for computer to perform a task.
(12) Software is a group of programmes.
(13) Hardware refers to the physical components of the computers.
(14) A group of bits the computer recognizes and processes at a time is known as word.
(15) Standard deviation is the square root of variance.
(16) When the coefficient of variation is high, the sample is less consistent.
(17) X^2 – test was developed by Karl person.
(18) z – test is used when the sample size is large.
(19) t – test is used when the sample size is small.
(20) Student's t is used when observations are unpaired.
(21) Fisher's t is used when observations are paired.

(22) The measure of coefficient of variation was evolved by Karl Pearson.
(23) Correlation is independent of the unit of measurement.
(24) Regression is expressed in the unit of measurement of variable.
(25) Completely randomized design is used when the experimental when the experimental material is limited and area is homogeneous.
(26) Randomized block design controls fertility variation in one direction only.
(27) Latin square design controls fertility variation in one direction.
(28) Split plot design permits evaluation of several factors simultaneously with different levels of precision.
(29) Lattice design is used when the number of treatment is small and Forms square.
(30) Replication refers to repetition of treatments under investigation.
(31) Randomization refers to allotment of treatments to different plots by a random process.
(32) F – Test measures the significance of difference among several treatments.
(33) Concept of augmented design was developed by Federer (1966).
(34) Ratio of treatment variance to error variance is called F-value.
(35) High values of gca are indicative of additive gene action.
(36) Biometrics is the science which deals with the application of statistical concept and procedures to the study of biological problems.-
(37) Initial frame of biometrics for the study of quantitative genetics was provided by R.A. Fisher (1928).
(38) Biometrical genetics is also known as statistical genetics.
(39) Biometrical genetics includes both quantitative genetics and population genetics.
(40) Analysis of quantitative genetics is based on means, variances and covariances.
(41) Concept of metroglyph analysis was developed by Anderson in 1977.
(42) P. C. Mahalanobis (1928) originally developed the concept of D2 statistics.
(43) Clusters diagram refers to a line diagram which is used in d2 statistics to depict genetic diversity among various groups.
(44) D2 statistics was first used for assessment of genetic diversity in plant breeding by P.C.Mahalanobis(1928).
(45) Simple correlation is also known as total correlation or zero order correlation.
(46) Partial correlation is also called as net correlation.
(47) Partial correlation refers to association two variables eliminating the effect of third variable.
(48) Phenotypic correlation does not refer to inherent correlation.
(49) Genotypic correlation dose not refer to inherent correlation.
(50) Partial correlation is denoted as r 123.
(51) Multiple correlations refer to a joint influence of two or more independent variables on a depended variable.

(52) Multiple correlations are non –positive estimate.
(53) Multiple correlations are denoted as R 123.
(54) Value of multiple correlations is always higher than simple and partial correlations.
(55) Concept of path analysis was first developed by Dewey and Lu in 1959.
(56) Path analysis was first used for plant selection by Wright in 1921.
(57) Path coefficients are of three types, viz. phenotypic, genotypic and environmental.
(58) In path analysis, a line diagram used to depict cause and effect situation is known as path diagram.
(59) In path analysis, residual effect permits precise explanation about the pattern of interaction of other possible variables not included in the study.
(60) Smith 1946 first suggested the use of discriminant function for plant selection.
(61) Selection index for animal improvement was first used by Hazel (1963).
(62) The term diallel was coined by Yates in 1957.
(63) The numerical approach of diallel analysis was proposed by Hayman (1954).
(64) Graphical approach of diallel cross analysis was proposed by Griffing (1956).
(65) In half diallel, total number of crosses among n parents is equal to n (n-1).
(66) In full diallel, total number of crosses among n parent is equal to n(n-1)/2
(67) In half diallel, each parent is used both as male and female in the crossing program.
(68) In full diallel, each parent is used both as male and female in the crossing program.
(69) Numerical approach of diallel analysis is based on the estimates of combining ability effect and variance.
(70) Vr- Wr graph is constructed in graphical approach of diallel analysis.
(71) Diallel cross fails to provide information about epistatic variance.
(72) Concept of partial diallel analysis was developed by kempthorne in 1967.
(73) Partial diallel also known as fractional diallel.
(74) In partial diallel, total number of crosses is equal to ns/2, where n is the number of parents and s is the number of sample crosses per parents or per array.
(75) Partial diallel provide information about sca effect.
(76) In partial diallel, total number of crosses is equal to ns/2, where n is the number of sample crosses per parent or per array.
(77) Partial diallel dose not provide information about sca effect.
(78) In partial diallel, additive genetic variance is equal to twice gca variance and dominance variance is equal to sca variance.
(79) Concept of line x tester cross analysis was developed by kempthorne; in 1967. Line x tester analysis is a modified form of top cross.
(80) In line tester cross, each male is mated with same group of females.
(81) In line x tester analysis, additive variance is equal to gca variance and dominance variance is equal to sca variance.

(82) In line x tester cross, each parent dose not have equal opportunity to mate with every other parent.
(83) Concept of triallel analysis was developed by Rawlings and Cokerham (1972a).
(84) In triallel cross, number of all possible three way crosses among n parents is equal to n(n-1)(n-2)/2.
(85) The variance components in a series of three way crosses are = ¾ VA + ½ VD + 9/16 VAA + 3/8 VAD + 1/4 VDD + ...
(86) Concept of quadriallel analysis was developed by Rawlings and Cokerham (1972b).
(87) High values of sca are indicative of non-additive gene action.
(88) Concept of gca is exploited in synthetic varieties.
(89) Concept of sca is exploited in development of hybrids.
(90) gca is estimated from full sib families.
(91) sca is estimated from half sib families.
(92) sca has relationship with heterosis.
(93) sca is primarily a function of dominance variance, additive x dominance epistasis, and dominance x dominance epistasis.
(94) gca is primarily a function of additive genetics variance and additive x additive epistasis.
(95) Concept of combining ability as measure of gene action was proposed by Sprague and Tatum (1942).
(96) High value of genetic advance is indicative of additive gene action.
(97) Triallel and quadriallel analyses provide information about additive, dominance epistatic variances. .
(98) Concept of generation means analysis was developed by hayman (1958) and Jinks and Jones (1958). .
(99) Scaling test provides information about the presence or absence of epistasis.
(100) Scaling test was given by Mather (1949).
(101) Significance of A and B scale indicates presence of all three types of epistasis.
(102) Significance of C scale indicates the presence of additive x additive type of epistasis.
(103) Significance of D scale reveals dominance x dominance type of epistasis.
(104) Significance of C and D scale reveals dominance x additive and additive x dominance types of epistasis.
(105) Similar sign of h and l indicates duplicate epistasis.
(106) Joint scaling test was devised by cavalla in 1962.
(107) Concept of biparental mating was originally developed by Comstock and Robinson (1948, 1952).
(108) Designs of biparental cross are referred to as North Carolina Design 1, 2, and 3 or NCD1, NCD2 and NCD3.

(109) NCD1 is also known as Nested design.
(110) NCD2 is referred to as factorial design.
(111) In NCD1, each randomly selected f_2 male is crossed with same group of females.
(112) In NCD2, each randomly selected f_2 male is crossed with different group of females.
(113) In NCD3, randomly selected f_2 plants (males) are crossed with both the parents of original cross.
(114) NCD3 involves F_2, P_1 and P_2 generations of a single cross in crossing.
(115) Concept of triple test cross was originally developed by Kearsey and Jinks in 1968.
(116) Triple test involves F_2, F_1, P_1, and P_2 generation of a single cross in mating programme.
(117) Triple test cross analysis provides information about epistatic variance.
(118) Adaptation refers to fitness of a phenotype to a given environment.
(119) Adaptability refers to capacity of a genotype for genetic change in adaptation.
(120) The term homeostasis was coined by Lerner in 1954.
(121) Stability refers to suitability of a variety for general cultivation over a wide range of environmental conditions.
(122) Homozygosity promotes homeostasis.
(123) Homeostasis refers to buffering capacity of a genotype for environmental changes.
(124) Freeman and Perkins model (1971) of stability analysis provides independent estimation of mean performance and environmental index.
(125) The term macro and micro environments were given by Comstock and Moll in 1973..
(126) Low magnitude of G x E interaction indicates high buffering capacity of the population.
(127) Generally heterogeneous population give more stable yields over several environments that pure line.
(128) Broad sense heritability refers to ratio of genotypic variance to phenotypic variance.
(129) Estimation of narrow sense heritability requires an estimate of additive variance.
(130) Co- heritability refers to the ratio of genotypic covariance to the phenotypic covariance.
(131) Estimate of co heritability helps in simultaneous improvements of two characters.

Answer Sheet

1	T	2	T	3	T	4	T	5	T	6	F	7	F	8	F	9	F	10	T
11	T	12	T	13	T	14	T	15	T	16	T	17	T	18	T	19	T	20	F
21	F	22	T	23	T	24	T	25	T	26	T	27	F	28	T	29	F	30	T
31	T	32	T	33	F	34	T	35	T	36	T	37	F	38	T	39	T	40	T
41	F	42	T	43	T	44	F	45	T	46	T	47	T	48	F	49	F	50	T
51	T	52	F	53	T	54	T	55	F	56	F	57	T	58	T	59	T	60	F
61	F	62	F	63	F	64	F	65	F	66	F	67	T	68	T	69	T	70	T
71	T	72	F	73	T	74	T	75	F	76	T	77	T	78	T	79	F	80	T
81	T	82	T	83	F	84	T	85	T	86	F	87	T	88	T	89	T	90	F
91	F	92	T	93	T	94	T	95	T	96	T	97	T	98	T	99	T	100	T
101	T	102	F	103	F	104	F	105	T	106	F	107	T	108	T	109	T	110	T
111	F	112	F	113	T	114	T	115	T	116	T	117	T	118	F	119	T	120	T
121	T	122	F	123	T	124	T	125	F	126	T	127	T	128	T	129	T	130	T
131	T																		

4
General Agriculture (Question Bank)

4.1. Multiple Choice Questions

(1) GoI has indicated ——————— crop will be introduced without evaluating biosafety and socioeconomic desirability.
    (a) GM   (b) Organic
    (c) Rabi   (d) Kharif

(2) ——————— is AGRI UDAAN.
    (a) Food and agribusiness initiative   (b) Food and agribusiness accelerator
    (c) Food and agribusiness incubator (d) Both b and c
    (e) All the above

(3) Union Minister for Agriculture and Farmer's Welfare launched ——————— quality mark logo at New Delhi.
    (a) National Dairy Development Board
    (b) National Dairy Development Authority
    (c) National Dairy Development Authority
    (d) National Dairy Development Programme

(4) ——————— state is leading in the production of litchi.
    (a) Bihar   (b) Odisha
    (c) Jharkahnd   (d) None of the above

(5) ——————— mango received the geographical indication (GI) tag in May 2017.
    (a) Banganpalle   (b) Malda
    (c) Alphonso   (d) None of the above

(6) India is set to produce ——————— million tones (MT) of food grains in 2017-2018 crop year, according to Agriculture Ministry.
    (a) 253   (b) 273
    (c) 263   (d) 233

(7) ——————— purpose Ministry of Agriculture (under Rashtriya Krishi Vikas (RKVY) Yojana and NRSC (ISRO) have signed an MoU in April 2017.
    (a) Geological analysis of soil   (b) Geological analysis of crops
    (c) Geo tagging of agricultural assets   (d) None of the above

(8) The Union Minister for Agriculture and Farmers Welfare has launched an International Centre for Foot and Mouth Disease (ICFM(D).
  (a) Arugul, Odisha
  (b) Bhubaneshwar, Odisha
  (c) Cuttack, Odisha
  (d) None of the above

(9) ICAR and ICRISAT had signed an agreement for ―――――.
  (a) Crop improvement
  (b) Transfer of agricultural technologies between two institutes
  (c) Development of climate smart crops, smart food
  (d) All of the above

(10) ――――――― is Mission Fingerling for ―――――――.
  (a) Holistic development and management of fisheries
  (b) Initiating Blue Revolution
  (c) Both of the above
  (d) None of the above

(11) AISEF stands ――――― in context of spice farming.
  (a) All India Spices Exporters Forum
  (b) All India Spices Education Forum
  (c) All India Spices Employment Forum
  (d) All India Spices Enforcement Forum

(12) A Royal Monarch has released a new digital green app for Indian farmers.
  (a) Queen Elizabeth
  (b) Queen Victoria
  (c) Prince Charles
  (d) Prince William

(13) RubSIS stands for in the context of commerce and industry.
  (a) Rubber Soil Information System
  (b) Rubber Solar Information System
  (c) Rubber Soil Infirmary System
  (d) Rubber Soil Innovation System

(14) The 1st International Agro Biodiversity Congress led to the adoption of ―――.
  (a) Mumbai Declaration on Agro-biodiversity Management
  (b) Delhi Declaration on Agro-biodiversity Management
  (c) Chennai Declaration on Agro-biodiversity Management
  (d) West Bengal Declaration on Agro-biodiversity Management

(15) NITI Aayog on 31st Oct 2016 launched the first agricultural index of its kind, it is called as ―――――.
  (a) Agricultural Marketing and Farm Friendly Reforms Index
  (b) Agricultural Marketing Index
  (c) Agricultural Farm Friendly Reforms Index
  (d) Agricultural Marketing and Eco-Friendly Reforms Index

(16) Field trials of GM Mustard are complete. If approved, it will become the ――― genetically modified food crop to be grown commercially in the country.
  (a) First
  (b) Second
  (c) Third
  (d) Fourth

(17) MS Swaminathan pitched for the preparation and operation of which code of action for flood hit regions ─────────.
    (a) Flood Code   (b) Operation Flood
    (c) Flood Scheme   (d) Flood system

(18) ───────── seed hubs have been approved under the National Food Security Mission with an outlay of INR 13981.08 lakhs ─────────.
    (a) 93   (b) 94
    (c) 95   (d) 99

(19) Ministry of Agriculture and Farmers Welfare has launched the second phase of which agricultural training in collaboration with USAID ─────────.
    (a) Nurture the Future India Total Training Programme
    (b) Feed the Farmer India Triangular Training Programme
    (c) Feed the Future India Triangular Training Programme
    (d) Nurture the Farmer India Triangular Training Programme

(20) Arvind Subramaniam is headed a high-level committee to review MSP and bonus for ───────── crop.
    (a) Wheat   (b) Rice
    (c) Jowar   (d) Pulses

(21) Union Minister of Agriculture and Farmer's Welfare has launched portal for rural districts in the country.
    (a) Krishi Vigyan Kendra   (b) Krishi Vigyan Sangathan
    (c) Krishi Vigyan Sansthan   (d) Krishi Vigyan Sammelan
    (e) All of the above

(22) CMRS stands for in the context of crop and nutrient management.
    (a) Crop Manager for Rice Based Systems   (b) Crop Middleman for River Side Systems
    (c) Crop Machine for Rice Based Systems   (d) Crop Manager for River Side Systems
    (e) All of the above

(23) Mars soil is perfect for growing _____, according to scientists.
    (a) Fruits   (b) Vegetables
    (c) Crops   (d) Both b and c
    e) All of the above

(24) Rice farming originated in China at ───────── time according to new evidence unearthed by archaeologists
    (a) 7000 years ago   (b) 6000 years ago
    (c) 8000 years ago   (d) 9000 years ago

(25) ———— is 9 million-euro Indo-EU project on water scarce agricultural sector
   (a) Water2crops   (b) Water to crops
   (c) Water4crops   (d) Water for crops

(26) India has recorded ———— tea production (million Kg) at what figure in 2015-2016, according to the ———— Tea Boar(d)
   (a) 1233   (b) 1244
   (c) 1033   (d) 1251

(27) Government has capped ———— royalty fee on new GM cotton traits.
   (a) 5%   (b) 10%
   (c) 15%   (d) 18%

(28) ———— university has been given status of central university on May 12, 2016
   (a) Rajendra Agricultural University PUSA Zamrudpur
   (b) Rajendra Agriculture University PUSA Samastipur
   (c) Rajendra Agriculture University, PUSA New Delhi
   (d) Rajendra Agriculture University, PUSA Jaipur

(29) Nomura has estimated that GVA of India is expected to rise by ———— percent in 2015-2016
   (a) 1.1   (b) 2.1
   (c) 1.3   (d) 1.8

(30) Improved varieties of Khesari dal have been released for general cultivation. ————
   (a) Ratan, Prateek, Beejak   (b) Ratan Beejak, Mahateora
   (c) Beejak, Ashtang, Bhairav   (d) Ashatng, Prateek, Mahateora

(31) IARI has developed early maturing genotype of which dal ————
   (a) Tur   (b) Urad
   (c) Mansoor   (d) Arhar

(32) Fish production in Himachal Pradesh increased by ————percent
   (a) 9%   (b) 9.2%
   (c) 9.4%   (d) 9.5%

(33) ICAR has taken on research for developing HYV of coconuts including ————.
   (a) Veppankulam-3   (b) Kalpatharu
   (c) Chandrakalpa   (d) All of the above

(34) Mission for Integrated Development of Horticulture promotes greenhouse technologies ———— component
   (a) Protracted cultivation   (b) Protected cultivation
   (c) Agrarian cultivation   (d) Mixed cultivation

(35) Government and the jute sector have set out a programme to increase the supply of raw jute called _____ project
   (a) Jute WE-CARE project (b) Jute I-CARE project
   (c) Jute HI-CARE project (d) Jute IN-CARE project

36) India's cotton output fell by what percent during October 2015- March 2016, according to CAI/ Cotton Association of Indi(a)
   (a) 11% (b) 12%
   (c) 13% (d) 14%

(37) MoU has been signed by ———— countries for Cooperation in Agriculture on 23rd March 2016
   (a) India and Lithuania (b) India and Denmark
   (c) India and Russia (d) India and Japan

(38) The Vice President, Shri M. Hamid Ansari has said that we need to put in place a ———— programme to transform the socio-economic fabric of our agricultural sector
   (a) Grow in India (b) Lead in India
   (c) Innovate in India (d) Discover in India

(39) a-IDEA has formed an MOA with NRDC. What does a-IDEA stand ————
   (a) Association for Innovation Development of Entrepreneurship in Agrarian affairs
   (b) Agency for Innovation Development of Entrepreneurism in Agriculture
   (c) Association for Innovation Development of Entrepreneurship in Agriculture
   (d) Agency of Invention Development and Entrepreneurism in Agriculture

(40) ———— department and a top consortium of UK research institutions signed an MoU to further development and partnership in the field of crop science
   (a) Department of Biotechnology (b) Department of Agriculture
   (c) Department of Science (d) Department of Atomic Energy

(41) ———— part of the world have peanut's genetic roots been traced to, marking the first time a high-quality sequencing of the crop genome being carried out
   (a) Cambodia (b) Bolivia
   (c) Serbia (d) Latvia

(42) ———— country has signed an MoU with India for numerous agricultural priority sectors such as plant breeding, plant protection and crop seed breeding
   (a) Tanzania (b) Liberia
   (c) Gambia (d) Armenia

(43) Cabinet has provided ex-post facto approval for agricultural MoUs with ———— country on 17th February 2016.
   (a) Liberia (b) Cyprus
   (c) Nepal (d) Both b and c

(44) First Mega Food Park has been opened at ———— district of MP
    (a) Khargone   (b) Dewas
    (c) Chambal   (d) Both a and b

(45) IARI has released 7 new varieties of field crops including ————
    (a) Wheat   (b) Rice
    (c) Lentil   (d) Both a and b

(46) IARI released 11 new varieties of HYV crops. These include ————
    (a) Pusa Sarda   (b) Pusa Madhurima
    (c) Pusa Bahar   (d) All of the above

(47) Intellectual Property Appellate Board (IPAB) has provided GI tag for Basmati rice. What is the GI tag
    (a) Geographical Indicator   (b) Geography Indicator
    (c) Geographical Indication   (d) Geological Indication

(48) Fair Trade Regulatory Authority CCI as approved the proposed merger of ———— seed firm with agrochemical major UPL
    (a) Advanta Ltd   (b) Advaitya Ltd
    (c) Adventa Ltd   (d) Advantia Ltd

(49) India's sugar production has risen by ———— percent since January 2015
    (a) 5%   (b) 6%
    (c) 7%   (d) 8%

(50) Sikkim's is India's ———— fully organic state
    (a) First   (b) Second
    (c) Third   (d) Fourth

(51) India's public research institutions have developed a low uric acid mustard variety called by which name ————.
    (a) Pusa Mustard-30   (b) Pusa Mustard-40
    (c) Pusa Mustard-50   (d) usa Mustard-60

(52) Two mobile apps have been launched for farmers. These are ————.
    (a) Crop Insurance   (b) AgriMarket Mobile
    (c) Krishi Vigyan   (d) Both a and b

(53) What does PMKSY stand for ————.
    (a) Pradhan Mantri Krishna Sanchar Yojana
    (b) Pradhan Mantri Krishi Sanchar Yojana
    (c) Pradhan Mantri Krishi Sinchayee Yojana
    (d) Pradhan Mantri Krishi Samriddhi Yojana

(54) According to the Agriculture Census, total number of operational holdings was estimated at ——————— million
   (a) 134.35 (b) 136.35
   (c) 137.35 (d) 138.35

(55) National Virtual Academy for Indian Agriculture has launched online course on diseases of horticultural crops and their management in collaboration with which educational body———————.
   (a) ICRISAT (b) IIT-Bombay
   (c) ICAR (d) Both a and b

(56) Agro chemicals sector in the country will touch USD ——————— billion by 2018-2019 with 60% of the contribution coming from exports, according to a report by TATA Strategic Management Group ———————.
   (a) 7.3 (b) 7.4
   (c) 7.5 (d) .6

(57) Which country is set to be the largest producer of cotton in the world in 2015-2016 according to a USDA report ———————?
   (a) China (b) India
   (c) Pakistan (d) Bangladesh

(58) Total cotton output has been estimated at which value for 2015-2016 beginning October 1st ———————.
   (a) 370.80 lakh bales (b) 370.70 lakh bales
   (c) 370.60 lakh bales (d) 370.50 lakh bales

(59) Which corporate group signed a pact with India Pulses Grain Association for cost efficient handling of pulses ———————?
   (a) RIL (b) Adani Group
   (c) TATAs (d) Reliance

(60) India has asked China in October 2015 to provide export clearance to a new variety of custard apple called ———————.
   (a) Arkha Sahan (b) Stifle
   (c) Cherimoya (d) None of the above

(61) Black rice or ——————— grown in Manipur has entered the world market through China in October 2015.
   (a) Chakhao amubi (b) Forbidden rice
   (c) Both of the above (d) None of the above

(62) Scientists at BSI have discovered a new species of banana known by which botanical name is ———————.
   (a) *Musa indandamanensis* (b) *Musa acuminate*
   (c) *Musa balbisiana* (d) None of the above

63) First Anaj Bank in the world has been opened in Varunanagar Colony in Hukulgunj area, Kashi under which NGO's pet bharo-campaign ————.
   (a) Vishal Bharat Sansthan   (b) Jagori
   (c) Goonj   (d) Both a and b

64) Which is the term popularized to describe start ups valued at USD 1 billion or more ————.
   (a) Tiger   (b) Unicorn
   (c) Sigma   (d) Centaur

65) India has agreed to intensify agricultural cooperation with which nation on 5th October 2015 ————.
   (a) Slovenia   (b) Lithuania
   (c) Ethiopia   (d) Sierra Leone

66) ———— app was launched for using space technology and Bioinformatics for crop cutting ————.
   (a) KISAN   (b) NCFC
   (c) CCAFS   (d) BHEEM

67) ———— crop has been made ICRISAT's mandate crop, on 4th October 2015.
   (a) Chickpea   (b) Mustard
   (c) Groundnut   (d) Finger Millet

(68) About fraction of total agrochemical products worth Rs 3,475 crore sold in India per annum are not genuine, as per Tata Strategic Management Group ————.
   (a) 1/4   (b) 2/4
   (c) 3/4   (d) 1/3

(69) The food ministry has set a minimum indicative export quota for manufacturers in the next sugar season in October 2015 at which value————.
   (a) 30 lakh tonnes   (b) 40 lakh tonnes
   (c) 50 lakh tonnes   (d) 60 lakh tonnes

(70) As against present day rice cultivation method, a new rice growing method reduces water use and enhances water efficiency through growth of crops in condition where there is plenty of oxygen in the soil. This new method is ————.
   (a) Anaerobic cultivation   (b) Aerobic cultivation
   (c) Oxygen cultivation   (d) Hydrogen cultivation

(71) Organic cash crops are being cultivated through special programmes in which Indian state————.
   (a) Kerala   (b) Karnataka
   (c) Tamil Nadu   (d) Punjab

(72) Exports of ———— number of agri-products were in the negative zone for July 2015.

(a) 8 (b) 9
(c) 10 (d) 11

(73) Domestic cotton production in 2014-2015 was pegged at the following value-----.

(a) 354.75 lakh bales, each weighing 1 70 kgs

(b) 359 lakh bales, each weighing 1 20 kgs

(c) 359 lakh bales, each weighing 1 70 kags

(d) 354.75 lakh bales, each weighing 1 20 kgs

(74) ———— state is becoming the first state in the country to launch 'agro-solar policy' to encourage the farmers tap the solar energy.

(a) Maharashtra (b) Uttar Pradesh
(c) Gujarat (d) Odisha

(75) ———— will be the new name of the Agriculture Ministry as proposed by the Prime Minister Narendra Modi in his Independence Day address to the nation on 15th Aug'15.

(a) Ministry of Indian Agriculture and Farmers' Welfare

(b) Agriculture and Farmers' Welfare Ministry

(c) Ministry of Indian Agriculture and Welfare

(d) Agriculture and Welfare Ministry

(76) ———— state government has launched a full-fledged confidence building exercise to instill confidence among farming community in the state ————.

(a) Uttar Pradesh (b) Karnataka
(c) Madhya Pradesh (d) Himachal Pradesh

(77) In ———— month, the Bengali Edition of Krishi Darshan Channel of Doordarshan was launched

(a) August'15 (b) September'15
(c) October'15 (d) November'15

(78) In collaboration with ———— state the International Crops Research Institute for Semi-Arid Tropics (ICRISAT) launched the Green Phablet for the farmers.

(a) Telangana (b) Andhra Pradesh
(c) Karnataka (d) Kerala

(79) ———— mobile app launched by Digital India With Laado (NDD) on 7th July'15.

(a) Pashu Poshan (b) Gau Seva
(c) Safed Kranti (d) Dairy App

(80) On 16th July'2015 the Reserve Bank of India (RBI) directed banks to ensure that their overall direct lending to farmers should not fall below the average of last ———— years.

    (a) Three   (b) Five

    (c) Seven   (d) Ten

(81) As per the data released by the Agriculture Ministry, ———— is the total area that has been covered till date (18th July'15) as against 346 lakh hectares at this time last year.

    (a) 450 lakh hectare   (b) 563 lakh hectare

    (c) 132 lakh hectare   (d) 245 lakh hectare

(82) With ———— country has India signed the biggest wheat import deals in over a decade.

    (a) Bangladesh   (b) Austria

    (c) Russia   (d) Australia

(83) ———— organization has set a target of providing 30,000 crore rupees as credit to farmers for irrigation over the next three years.

    (a) NABARD   (b) SIDBI

    (c) EPFO   (d) SBI

(84) On 15th July, 2016———— ministry launched the PGS-India, Soil Health Card and FQCS Web Portals.

    (a) Ministry of Comm. and   (b) Ministry of Agriculture
          Information Technology

    (c) Ministry of Earth Science   (d) Ministry of Commerce and Industry

(85) The ———— state laid foundationstone of Indian Agriculture Research Institute (IARI) is set to be laid by Prime Minister on 28th June'2015.

    (a) Assam   (b) Rajasthan

    (c) Jharkhand   (d) Madhya Pradesh

(86) ———— the Insurance Portal and Weather Alert Service launched for farmers on 19th June'15.

    (a) AGROFARM   (b) FARMCAST

    (c) INSUFARM   (d) NOWCAST

(87) ———— farmer centric TV Channel that was launched by the PM Narendra Modi on 26th May'2015.

    (a) Kisan TV   (b) Green TV

    (c) AGRO TV   (d) DD Kisan

(88) ———— percent food grain was output likely to fall in 2014-15 crop year.

    (a) 10%   (b) 5.2%

    (c) 8%   (d) 3.5%

(89) On 29th Apr'2015, the government was decided to raise import duty on sugar from 25 percent to ——————— percent.
   (a) 30% (b) 35%
   (c) 40% (d) 50%

(90) As announced by the Central Government, for ——————— crop farmers will get full Minimum support prices of Rs. 1,450 per quintal.
   (a) Wheat (b) Pulses
   (c) Rice (d) All of these

(91) The National Conference on Agriculture for Kharif Campaign-2015 was organized on 7th- 8th April, 2015 at ——————— .
   (a) New Delhi (b) Pune
   (c) Bhubaneswar (d) Ahmedabad

(92) For ——————— crop, Central Government has asked to relax quality norms for procurement on 7th April, 2015.
   (a) Wheat (b) Maize
   (c) Pulses (d) Rice

(93) ——————— mega food parks have been approved by the government for food processing across the country.
   (a) 12 (b) 10
   (c) 8 (d) 17

(94) ——————— name Prime Minister announced to name Institute of Horticulture
   (a) Subhash Chandra Bose (b) Shaheed Bhagat Singh
   (c) Rajguru (d) Chandrasekhar Azad

(95) In state, the Union Minister M. Venkaiah Naidu laid the foundation stone for National Kamadhenu Animal Breeding Centre on 21st March, 2015.
   (a) Karnataka (b) Uttar Pradesh
   (c) Madhya Pradesh (d) Andhra Pradesh

(96) ——————— state has launched the "Momai Tamuli Barbaruah Krishak Bondhu Scheme" to increase agricultural production in state on 15th March 2015.
   (a) West Bengal (b) Odisha
   (c) Arunachal Pradesh (d) Assam

(97) ——————— state First Kisan Call Centre (KC(C)) of North-East was opened on 9th March'2015.
   (a) Tripura (b) Assam
   (c) Manipur (d) Mizoram

(98) ——————— bank is providing finance to small tea growers in the Barak Valley and other parts of Assam state.
   (a) Bank of Baroda (b) Bank of India
   (c) State Bank of India (d) ICICI

(99) ———— state government plans to launch about one thousand "Agri-Junctions" to provide not only farm inputs and equipments to the farmers, but also business opportunities to rural youth and employment to agriculture graduates.

(a) Uttar Pradesh (b) Gujarat
(c) West Bengal (d) Maharashtra

(100) Name the state government that has planned to introduce Crop Insurance Scheme for farmers ————.

(a) West Bengal (b) Punjab
(c) Delhi (d) Bihar

(101) For the development and management of fisheries sector in India, Mission fingerling is launched under the Blue revolution program with the total expenditure of ————.

(a) 52000 lakhs (b) 75000 lakhs
(c) 4800 lakhs (d) 9200 lakhs
(e) None of the above

(102) In budget 2017-18, allocated for Deen Dayal Upadhyay Gram Jyoti Yojana ————.

(a) Rs.4814 crores (b) Rs.5500 crores
(c) Rs.6814 crores (d) Rs.1900 crores
(e) Rs.1200 crores

(103) PACS, an institute of NABARD works at the gram panchayat and village level. In PACS, A stands for ————.

(a) Accredit (b) Agriculture
(c) Asset (d) Advisory
(e) Ability

(104) Under the new scheme launched by cabinet for the rural household who are not covered under PMAY-G with the interest subsidy, on how much loan amount the beneficiary will get the interest subsidy ————.

(a) 3 lakhs (b) 4 lakhs
(c) 2 lakhs (d) 1.5 lakhs
(e) 5 lakhs

(105) Bharatnet project is related to————.

(a) Free wifi to students in rural area
(b) High speed internet to farmers
(c) Broadband connectivity to gram panchayats
(d) A project connecting rural area with Urban areas
(e) None of the above

106) The period after parturition in which animal produces milk is called as————.
   (a) Dry period (b) Calving period
   (c) Conception (d) Lactation period
   (e) None of the above

(107) Agricultural census is conducted in every ————.
   (a) 2 years (b) 3 years
   (c) 4 years (d) 5 years
   (e) 6 years

(108) To improve the rural road connectivity in left wing extremism affected area which scheme has been approved by cabinet committee on economic affair————.
   (a) Rural Road Connectivity Scheme
   (b) Pradhan Mantri Gram Sadak Yojana
   (c) Left Wing Extremism road Connectivity Scheme
   (d) Rural Road Connectivity Scheme for Left Wing Extremism Affected Areas
   (e) None of the above

(109) ———— is the National Institute of Jute and Allied Fibre Technology located.
   (a) Haryana (b) Telangana
   (c) Andhra Pradesh (d) West Bengal
   (e) Jharkhand

(110) ———— is world soil day celebrated.
   (a) December 5 (b) November 15
   (c) January 5 (d) April 15
   (e) July 6

(111) ———— amount has been allocated for MGNREGA in budget 2017-18.
   (a) 50,000 crores (b) 68,000 crores
   (c) 1000 crores (d) 48,000 crores
   (e) 23,000 crores

(112) ———— is the position of Indian Fisheries in global market.
   (a) 2nd (b) 3rd
   (c) 4th (d) 5th
   (e) 7th

(113) Ministry of rural development and CIRDAP signing an agreement on establishment of CIRDAP centre at National Institute of Rural Development and Panchayati raj which is located in ————.
   (a) Kochi (b) New Delhi
   (c) Hyderabad (d) Ahmedabad
   (e) Kolkata

(114). PMGDISHA Project has been granted approval from Cabinet with an outlay of ──────────.

    (a) 18,980 crores   (b) 30,000 crores
    (c) 4,819.86 crores   (d) 2,351.38 crores
    (e) 55,000 crores

(115) For setting up of weather stations in various districts of the country by 2019. IMD has signed MoU with Ministry of Agriculture to use ────────── for weather forecasting observatories.

    (a) KVK   (b) PMFBY data
    (c) PMGSY data   (d) Soil Health Card data
    (e) None of the above

(116) Which state hosted 13th edition of National Agriculture Science Congress 2017 ──────────.

    (a) Madhya Pradesh   (b) Haryana
    (c) Andhra Pradesh   (d) Karnataka
    (e) West Bengal

(117) ────────── is a branch of agriculture science which deals with principles and practices of soil, water and crop management.

    (a) Crop Production   (b) Agriculture engineering
    (c) Home science   (d) Agronomy
    (e) Horticulture

(118) The cultivation of crops in regions with an annual rainfall of 750mm is called as ──────────.

    (a) Rain fed farming   (b) Dry farming
    (c) Dryland farming   (d) Intensive farming
    (e) None of the above

(119) ────────── crops are sowed in October to December and Harvested in February to April.

    (a) Rabi crops   (b) Zaid Crops
    (c) Zaid Rabi Crops   (d) Zaid Kharif Crops
    (e) None of the above

(120) Central Soil Salinity Research Institute is located at ──────────.

    (a) Gujarat   (b) Maharashtra
    (c) Haryana   (d) Chhattisgarh
    (e) Rajasthan

(121) ────────── is the scientific name of wheat.

    (a) *Zea mays*   (b) *Solanum melongana*
    (c) *Sorghum vulgare*   (d) *Triticum aestivum*
    (e) None of the above

(122) Under Mission Antyodaya Programme, 50,000-gram panchayats will be poverty free by the year —————.
    (a) 2018   (b) 2019
    c. 2020   (d) 2021
    e. 2022

(123) The order in which the crops are cultivated on a piece of land over a fixed period is called as —————.
    (a) Cropping Pattern   (b) Cropping system
    c. Mono-cropping   (d) Intercropping
    e. Mixed farming

(124) ————— is the name of citizen centric mobile app on MGNREGA launched by department of rural development for providing information of the MGNREGA program.
    (a) Meri sadak   (b) Jai MGNREGA
    (c) E-MGNREGA   (d) Jan MGNREGA
    (e) None of the above

(125) National Panchayati Raj day is commemorated at—————.
    (a) New Delhi   (b) Andhra Pradesh
    (c) Pune   (d) Lucknow
    (e) Mumbai

(126) ————— amount have been allocated by National Bee Board for the year 2016-17 for the development of Bee keeping.
    (a) 12 crores   (b) 19 crores
    (c) 18 crores   (d) 17 crores
    (e) 10 crores

(127) India's place in coconut production in the world is —————.
    (a) Fourth   (b) First
    (c) Third   (d) Second
    (e) Fifth

(128) For ensuring the availability of new kinds of seeds, seed hubs are created with the help of—————.
    (a) ICAR and KVKs
    (b) Union Ministry of Agriculture and Farmers Welfare and KVKs
    (c) Union ministry of Agriculture and Farmers and National seed fund
    (d) ICAR and Union Ministry of Agriculture and Farmers

(129) Crop logging' is the method of ───────.
    (a) Soil fertility evaluation
    (b) Plan analyst for assessing requirements for nutrients for crop production
    (c) Assessing crop damage
    (d) Testing suitability of fertilizers
    (e) None of the above

(130) First separate department of agriculture was established in year ───────.
    (a) 1800
    (b) 1878
    (c) 1881
    (d) 1890
    (e) 1875

(131) The process of growing two or more crops simultaneously on same piece of land with definite row pattern is called as ───────.
    (a) Inter cropping
    (b) Mixed cropping
    (c) Multi storyed system
    (d) Sequence cropping
    (e) None of the above

(132) ─────── is a single window e- learning centre, which provides information of the MGNREGA and feedback on the quality of implementation of programme.
    (a) Bharat Nirman centre
    (b) ITC-E-Chopal
    (c) Bharat nirman kendra
    (d) Sandesh Pathak application
    (e) None of the above

(133) When is National Energy Conservation Day ───────?
    (a) April 16
    (b) June 5
    (c) December 16
    (d) July 5
    (e) May 17

(134) The agriculture and farm friendly index is launched by ───────.
    (a) Union ministry of agriculture and famers welfare
    (b) NITI
    (c) FAO
    (d) ICAR
    (e) None of the above

(135) The scheme targeted to enhance the fish production from 107.95 lakh tonnes in 2015-16 to about 150 lakh tonnes by the end of the financial year ───────.
    (a) 2019-20
    (b) 2020-21
    (c) 2021-22
    (d) 2022-23
    (e) 2023-24

(136) Jawar, Bajra, Maize and Cotton are ─────── type of crops.
    (a) Rabi
    (b) Zaid
    (c) Kharif
    (d) Zaid Rabi
    (e) None of the above

(137) GeoMGNREGA is the initiation of————.
- (a) Ministry of Rural Development, ISRO and ICAR
- (b) Ministry of Rural Development, ISRO, NRSC and ICAR
- (c) Ministry of Rural Development, ISRO, NRSC and National Informatic centre
- (d) Ministry of rural Development, ISRO, National Informatic centre and ICAR
- (e) ISRO and ICAR

(138) Yellow revolution is associated with————.
- (a) Potato
- (b) Fertilizers
- (c) Food grains
- (d) Oil seed production
- (e) None of the above

(139) The soil which is found in the areas having heavy rainfall and high humidity, contains large quantity of organic matter and heavy and black in colour————.
- (a) Black soil
- (b) Peat soil
- (c) Forest soil
- (d) Laterite soil
- (e) None of the above

(140) Regur soil is another name of ————.
- (a) Black soil
- (b) Peaty soil
- (c) Laterite soil
- (d) Arid soil
- (e) Forest Soil

(141) Livestock census is conducted in every ————.
- (a) One year
- (b) Five years
- (c) Two years
- (d) Three years
- (e) Four years

(142) The soil which is poor in organic matter, nitrogen, phosphate and calcium is called as ————.
- (a) Black soil
- (b) Laterite soil
- (c) Peaty soil
- (d) Arid soil
- (e) Forest soil

(143) The name of online portal launched by union government to solve the problems of agriculture sector is ————.
- (a) e-NAM
- (b) e-Krishi samvad
- (c) e-agriculture solution
- (d) e-krishi solution
- (e) Kisan call centre

(144) The process in which soil fertility declined, when the nutritional status declines and depth of the soil goes down due to erosion and misuse is called as————.
- (a) Soil erosion
- (b) Soil degradation
- (c) Soil salinity
- (d) Soil conservation
- (e) None of the above

145) National bovine genomic centre for indigenous breeds is being established for —————————.
    (a) To increase milk production
    (b) Identification of disease free high genetic merit bulls
    (c) Attaining faster genetic gain
    (d) All of the above
    (e) None of the above

(146) ————————— is the annual income criteria for saffron ration card.
    (a) More than Rs. 15,000 and less than 1 lakh
    (b) More than Rs. 10,000 and less than 1 lakh.
    (c) More than Rs. 15,000 and less than 1.5lakh.
    (d) More than Rs. 15,000 and less than 2 lakh.
    (e) More than Rs. 20.000 and less than 2 lakh

(147) Central Soil and Material Research Centre is located in —————————.
    (a) Meerut
    (b) New Delhi
    (c) Madhya Pradesh
    (d) Chandigarh
    (e) Mumbai

(148) The process of covering the top layer of the soil with plant materials to prevent the soil from soil erosion is known as —————————.
    (a) Edging
    (b) Mulching
    (c) Matting
    (d) Plant vegetation
    (e) None of the above

(149) According to census 2011, what percentage of population lives in rural area —————————.
    (a) 70%
    (b) 58%
    (c) 69%
    (d) 65%
    (e) 57%

(150) ————————— smart villages has been developed under Sansad Adarsh Gram Yojana by 2019.
    (a) 25,00
    (b) 5,000
    (c) 30,00
    (d) 20,00
    (e) 15,00

(151) Jhum cultivation is another name of—————————.
    (a) Shifting cultivation
    (b) Intensive farming
    (c) Subsistence farming
    (d) Planting cultivation
    (e) Mixed farming

(152) Under Deen Dayal Gram Jyoti Yojana, government has set a target of achieving 100 percent village electrification by ―――――.
    (a) 1st May 2019   (b) 1st May 2018
    (c) 1st May 2020   (d) 1st May 2021
    (e) 1st May 2022

(153) Intensive agriculture programme was launched in year―――――.
    (a) 1972   (b) 1980
    (c) 1961   (d) 1993
    (e) 1987

(154) ――――― is known as father of agriculture.
    (a) Norman Borlaug   (b) Louis Pasteur
    (c) Benjamin franklin   (d) Ronald fisher
    (e) None of the above

(155) Central Marine Fisheries Research Institute is located at ―――――.
    (a) Goa   (b) Kerala
    (c) Dehradun   (d) Mumbai
    (e) Vishakhapatnam

(156) For soil testing, government is planning to set mini labs in ―――――.
    (a) Every district   (b) Villages
    (c) Every States   (d) KVK
    (e) Every capital of the states

(157) ――――― group of plant has maximum water use efficiency.
    (a) C3   (b) C4
    (c) CAM   (d) None of the above
    (e) All of these

(158) ――――― year's data has been taken as reference data for the 19th livestock census.
    (a) 2014   (b) 2010
    (c) 2015   (d) 2012
    (e) 2013

(159) How many APMC has been integrated under eNAM ―――――?
    (a) 253   (b) 700
    (c) 375   (d) 585
    (e) 550

(160) ――――― deadline set by government to provide safe drinking water to over 28000 arsenic and fluoride affected habitation.
    (a) 2019   (b) 2020
    (c) 2021   (d) 2022   (e) 2023

(161) Under the Swacch Bharat Mission (Gramin), sanitation coverage in rural areas has gone up from 42% to ──────.
    (a) 50%　　　　　　　　　　(b) 58%
    (c) 62%　　　　　　　　　　(d) 70%
    (e) 65%

(162) Which of the following protocol is associated to green-house gas emission ──────.
    (a) Bonne convention　　　　(b) INDC
    (c) Montreal protocol　　　　(d) Minamata convention
    (e) None of the above

(163) The first forest survey of India was conducted in ──────.
    (a) 1981　　　　　　　　　　(b) 1977
    (c) 1988　　　　　　　　　　(d) 1987
    (e) 1985

(164) The amount allocated to Integrated Scheme on Agriculture census and statistics for 2017-18 is ──────.
    (a) 150 crores　　　　　　　(b) 200 crores
    (c) 225 crores　　　　　　　(d) 250 crores
    (e) 100 crores

(165) The scheme to provide the remunerative prices to the farmers in case of glut in production and fall in prices is known as ──────.
    (a) MIS　　　　　　　　　　(b) PSS
    (c) ISST　　　　　　　　　　(d) Both a and b
    (e) Both a and c

(166) ────── missions are launched under National Action Plan on Climate Change.
    (a) 5　　　　　　　　　　　(b) 6
    (c) 7　　　　　　　　　　　(d) 8
    (e) 9

(167) Which of the following is correct about the photorespiration ──────.
    (a) Photorespiration is more efficient in synthesizing glucose than photosynthesis.
    (b) C3 plants are best adapted to handle photorespiration.
    (c) Photorespiration is likely to occur when oxygen level is high and CO_2 level is low.
    (d) During photorespiration plants use oxygen to break down carbohydrates and release oxygen.
    (e) All of the above

(168) The central sector scheme which will supplement agriculture, modernize processing of Agricultural products and decrease their wastage is known as ―――――.
   (a) SAMPADA
   (b) APEDA
   (c) PDS
   (d) FPIA
   (e) None of the above

(169) Which Indian state is known as 'Molessis Basin'―――――?
   (a) Kerala
   (b) Gujarat
   (c) Maharashtra
   (d) Mizoram
   (e) Uttar Pradesh

(170) Centrally Sponsored Schemes are the schemes which are ―――――.
   (a) Implemented by centre and funded by union
   (b) Implemented by states and funded by central government
   (c) Implemented by states and centre and funded by central government with definite portion of state government
   (d) Implemented by states and funded by central government with definite portion of state government
   (e) None of the above

(171) Jai kisan- jai vigyan week is being celebrated across the entire country during ―――――.
   (a) 3rd December – 9th December
   (b) 5th June – 11th June
   (c) 23rd December – 29th December
   (d) 9th December – 15th December
   (e) 3rd July – 10th July

(172) COP 22 was held at―――――.
   (a) Paris
   (b) Bali
   (c) Marrakech
   (d) Bonn
   (e) New York

(173) The method of harvesting a crop which leaves the roots and the lower parts of the plant uncut to give the subtle crops is known as―――――.
   (a) Mulching
   (b) Intensive farming
   (c) Ratooning
   (d) Zero tillage farming
   (e) Mixed farming

(174) The mission to disseminate information and knowledge to the farming community in local language in respect of agricultural schemes is known as―――――.
   (a) NMAET
   (b) PMKSY
   (c) ISAM
   (d) MIDH
   (e) None of the above

(175) Environmental protection act is also known as ―――――.
- (a) Central Law of Environment
- (b) The umbrella Legislation
- (c) Law of Environment
- (d) None of the above
- (e) All of the above

(176) Agriculture education Day is celebrated to commemorate the anniversary of ―――――.
- (a) Dr. (B)R. Ambedkar
- (b) Dr. Rajendra Prasad
- (c) Dr. Sarvapalli Radhakrishnan
- (d) R. Venkatraman
- (e) None of the above

(177) The National Fisheries Development Board headquarter situated at ―――――.
- (a) Kochi
- (b) Vishakhapatnam
- (c) Mangalore
- (d) Hyderabad
- (e) Mumbai

(178) ――――― the % growth forecast by CSO for Agriculture, forestry and fishing sector for 2016-17.
- (a) 4%
- (b) 7%
- (c) 4.1%
- (d) 3.8%
- (e) 5%

(179) Under the Housing for all scheme, safe and secure pucca house will be provided to every rural household by ―――――.
- (a) 2019
- (b) 2020
- (c) 2021
- (d) 2022
- (e) 2023

(180) According to livestock census, the cattle contributes ――――― percentage of livestock population.
- (a) 30%
- (b) 47.80%
- (c) 37.28%
- (d) 29.59%
- (e) 35%

(181) ――――― is deadline set by government to skill the 5 lakh rural people for mason training.
- (a) 2019
- (b) 2022
- (c) 2021
- (d) 2020
- (e) 2018

(182) ――――― is the theme for the International Day for the Biodiversity 2017.
- (a) Biodiversity and sustainable tourism
- (b) Biodiversity and pollution
- (c) Biodiversity and Marine life
- (d) Agriculture, Biodiversity and Tourism
- (e) Biodiversity and Marine life

(183) The crops ———— are used to substitute the crops that have failed on account of unfavourable conditions.
    (a) Emergency crops   (b) Cover crops
    (c) Companion crops   (d) Intermediate crops
    (e) None of the above

(184) Under the Rashtriya Gokul Mission, the two national Kamdhenu breeding centre is being established in ———— and ———— states.
    (a) Haryana and Karnataka   (b) Madhya Pradesh and Haryana
    9c) Madhya Pradesh and Andhra Pradesh   (d) New Delhi and Andhra Pradesh
    9e) New Delhi and Maharashtra

(185) The growing of grass or legumes in rotation with grain or tilled crops as a soil conservation measure is termed as ————.
    (a) Crop rotation   (b) Ley farming
    (c) Shifting cultivation   (d) Dairy farming
    (e) Mixed farming

(186) ———— is an Animal Wellness Programme in which the provision of Animal Health Cards along with UID identification of animals in milk and a National Data Base.
    (a) Pashudhan Patra   (b) Pashu Swasthya Card
    (c) Nakul Swasthya Patra   (d) Pashu Sarva Sansadhan Patra
    (e) None of the above

(187) BPL ration card is issued to families having annual income of————.
    (a) Less than 10,000   (b) Less than 20,000
    (c) Less than 17,000   (d) Less than 27,000
    (e) Less than 30,000

(188) PLCN stands ————
    (a) Permanent Location Consumer Number
    (b) Permanent Location Code Name
    (c) Permanent Livelihood Centre Name
    (d) Permanent Location Code Number
    (e) Permanent Livelihood Consumer Number

(189) ———— committee drafted a Comprehensive Review of Deep Sea Fishing Policy and Guidelines 2014.
    (a) Gopa Kumar Committee   (b) Meena Kumari Committee
    (c) Murari Committee   (d) A. Shah Committee
    (e) None of the above

(190) ———— is the amount allocated in budget 2017-18 for setting up of dairy processing and infrastructure development fund.
- (a) 5,000 crores
- (b) 8,000 crores
- (c) 10,000 crores
- (d) 9,000 crores
- (e) 7, crores

(191) ———— state becomes the first state to use the automated weather stations for farmers.
- (a) Maharashtra
- (b) Haryana
- (c) Madhya Pradesh
- (d) Gujarat
- (e) Kerala

(192) Under the National Mission for Sustainable Agriculture Mission ———— – percentage of allocated fund was decided to invest on women farmers.
- (a) 50%
- (b) 33%
- (c) 25%
- (d) 30%
- (e) 23%

(193) ———— cropping pattern is used in Rainfed farming areas.
- (a) Single cropping
- (b) Relay cropping
- (c) Ratoon cropping
- (d) Intercropping
- (e) Mixed farming

(194) The inherent capacity of soil to provide essential chemical elements for the growth of plants is termed as————.
- (a) Soil productivity
- (b) Soil fertility
- (c) Soil capacity
- (d) Soil flexibility
- (e) Soil Utility

(195) "Medh per ped" programme is associated with ————.
- (a) PMKSY
- (b) Agriculture production
- (c) Seed village
- (d) Agroforestry
- (e) None of the above

(196) ———— percentage of the total allocation of National food security mission was allocated for pulses.
- (a) 50%
- (b) 40%
- (c) 70%
- (d) 60%
- (e) 55%

(197) The tillage in which the way of growing crops or pasture from the year to year disturbing the soil through tillage is known as ————.
- (a) Year - round tillage
- (b) Deep tillage
- (c) Zero tillage
- (d) Secondary tillage
- (e) None of the above

(198) Pundia, Kabirya, Wansi are the varieties of————.
    (a) Sugarcane   (b) Maize
    (c) Rice   (d) Oats
    (e) Wheat

(199) National Plant Protection Training Institute is located in ————.
    (a) New Delhi   (b) Madhya Pradesh
    (c) Hyderabad   (d) West Bengal
    (e) Mumbai

(200) The migratory fish that live in both fresh and salt water, independent of breeding is called as ————.
    (a) Catadromous fish   (b) Potamodromous fish
    (c) Amphidromous fish   (d) Anadromous fish
    (e) None of the above

(201) ———— is known as the 'Father of White Revolution' in India ————.
    (a) P Pal   (b) V. Kurien
    (c) M.S Swaminathan   (d) K.N.Bahal
    (e) None of the above

(202) The Principal cereal crop of India is ————
    (a) Barley   (b) Sorghum
    (c) Rice   (d) Wheat
    (e) Oats

(203) Growing of two or more crops simultaneously on the same piece of land is called ————.
    (a) Fanning   (b) Intercropping
    (c) Mixed cropping   (d) Mixed farming
    (e) None of the above

(204) Wet Agriculture is practiced in ———— state of India.
    (a) Kerala   (b) Tamil Nadu
    (c) Karnataka   (d) Orissa
    (e) West Bengal

(205) ———— of these is not a variety of seeds.
    (a) Composite Seed   (b) Breeder Seed
    (c) Mutant Seed   (d) Hybrid Seed
    (e) None of the above

(206) ———— is the reason for reducing soil fertility.
    (a) Over Irrigation   (b) Poor Drainage
    (c) Continuous cropping   (d) Imbalanced use of fertilizers
    (e) Pesticide

(207) Green Revolution in India has been most successful in the case of ―――――.
    (a) Tea & Coffee   (b) Wheat & Potato
    (c) Mustard & Oil Seeds   (d) Rice & Wheat
    (e) None of the above

(208) The major Agricultural land in India is under―――――.
    (a) Oil Seeds   (b) Cash Crops
    (c) Plantation crops   (d) Food Crops
    (e) None of the above

(209) ―――――― is the main source of irrigation of Agricultural land in India.
    (a) Tanks   (b) Rivers
    (c) Streams   (d) Canals
    (e) Wells

(210) India has been divided into ―――――― cotton growing regions.
    (a) Three   (b) Two
    (c) One   (d) Four
    (e) Five

(211) Which crops is sown in the largest area in M.P――――――?
    (a) Rice   (b) Soybean
    (c) Jowar   (d) Wheat
    (e) Oats

(212) ―――――― the sequence of water erosion.
    (a) Splash, sheet, rill, gully   (b) Sheet, gully, rill
    (c) Rill, Splash, sheet   (d) Gully erosion, splash, sheet, rill
    (e) Splash, Gully erosion, rill

(213) Wind erosion will be higher from.
    (a) Clay Soil   (b) Sandy Soil
    (c) A barren sandy soil   (d) Loam Soil
    (e) None of the above

(214) ―――――― the essential growth of and quantities development of fruits.
    (a) Humidity   (b) Wind
    (c) Solar Radiation   (d) Temperature
    (e) Water

(215) Tarai soil occurs in ――――――state.
    (a) Karnataka   (b) Bihar
    (c) Uttar Pradesh   (d) Gujarat
    (e) Punjab

(216) The process of judicious removal of parts like leaf, flower, fruits etc. to obtain good and qualitative yield is called—————.
    (a) Training (b) Pruning
    (c) Fustigation (d) Sedimentation
    (e) None of the above

(217) Tea is a —————.
    (a) Fiber crop (b) Food crop
    (c) Beverage crop (d) Industrial crop
    (e) None of the above

(218) In ————— farming, the land is used for growing food and fodder crops and rearing livestock—————.
    (a) Intensive Farming (b) Plantation Farming
    (c) Primitive Farming (d) Mixed Farming
    (e) None of the above

(219) The land on ————— are grown.
    (a) Arable Land (b) Wet Land
    (c) Dry Land (d) None of the above
    (e) All of the above

(220) Monoculture i.e single crop grown over a large area is also known as —————.
    (a) Commercial Grain Farming (b) Intensive Farming
    (c) Multiple Farming (d) Mixed Farming
    (e) Plantation Farming

(221) ————— is pearl fishing well developed in India.
    (a) Off the Bengal Coast (b) Off Kerala Coast
    (c) Off the Coast of Rameshwaram (d) None of the above
    (e) Off the Vishakhapatnam Coast

(222) Murrah is a breed of —————.
    (a) Pig (b) Buffalo
    (c) Sheep (d) Goat
    (e) Bull

(223) ————— is correctly defines the term 'Transhumance.
    (a) Economy that solely depends on animals
    (b) Farming in which only one crop is cultivated by clearing hill tops
    (c) Practice of growing crops on higher hill slopes in summers and foothills in winters
    (d) Seasonal Migration of people with their animals up and down the mountains.
    (e) None of the above

(224) The important feature of shifting cultivation is ———————.
    (a) Cultivation by transplantation
    (b) Rotation of crops
    (c) Change of cultivation site
    (d) Cultivation of leguminous crops
    (e) All of the above

(225) Inland Fisheries is referred to ———————.
    (a) Extraction of oil from fish
    (b) Deep sea fisheries
    (c) Trapping and capturing fish
    (d) Culturing fish in freshwater
    (e) None of the above

(226) Which branch of agriculture deals rearing of silk worms ———————?
    (a) Sericulture
    (b) Apiculture
    (c) Floriculture
    (d) Mariculture
    (e) Viticulture

(227) For ——————— purpose Ministry of Agriculture (under RKVY Yojan (a) and NRSC (ISRO) have signed a MoU in April 2017.
    (a) Geological analysis of soil
    (b) Geo analysis of crops
    (c) Geo tagging agricultural assets
    (d) None of the above above
    (e) All of the above

(228) ——————— Royal Monarch released a new digital green App for Indian farmers.
    (a) Queen Elizabeth
    (b) Queen Victoria
    (c) Prince Charles
    (d) Prince Williams
    (e) Prince Harry

(229) Forest Survey in India has it's headquarter located in———————.
    (a) New Delhi
    (b) Dehradun
    (c) Shimla
    (d) Guwahati
    (e) Hyderabad

(230) ——————— categories of protected area in India are local people not allowed to collect and use the biomass.
    (a) Biosphere Reserves
    (b) Wetland declared under Ramsar Convention
    (c) Wildlife Sanctuaries
    (d) National Parks
    (e) None of the above

(231) Rural Infrastructure Development Fund (RIDF) was instituted by ———————.
    (a) NABARD
    (b) RBI
    (c) Finance Ministry
    (d) Ministry of Home Affairs
    (e) None of the above

(232) National Rurban Mission was launched by PM Modi in ——— state.
   (a) New Delhi (b) Haryana
   (c) Gujarat (d) Madhya Pradesh
   (e) Chhattisgarh

(233) The Integrated Child Development Services (ICDS) scheme was launched in which year to provide food, pre-school education and primary healthcare to children below 6 years of age and their mothers———.
   (a) 1980 (b) 1975
   (c) 1982 (d) 1993
   (e) 1972

(234) NREGP stands ———.
   (a) National Regional Employment Guarantee Programme
   (b) National Rural Entrepreneurship Guarantee Programme
   (c) National Rural Educational Guarantee Programme
   (d) National Rural Employment Guarantee Programme
   (e) None of the above

(235) What is the full form of MGNREGA———.
   (a) Maharashtra Government National Rural Employment Guarantee Act
   (b) Mahatma Gandhi National Rural Employment Guarantee Act
   (c) Meghalaya Government National Rural Guarantee Act
   (d) Mahatma Gandhi National Rural Empowerment Act
   (e) None of the above

(236) NITI Aayog on 31st October 2016 launched the first agricultural index of ———.
   (a) Agricultural Marketing and Farm Friendly Reforms Index
   (b) Agricultural Marketing Index
   (c) Agricultural Farm Friendly Reforms Index
   (d) Agriculture Farm Friendly Rural Index
   (e) None of the above

(237) Fertility of soil can be improved by ———.
   (a) Adding living earthworms (b) Adding dead earthworms
   (c) Removing dead earthworms (d) Removing dead earthworms and adding living earthworms
   (e) None of the above

(238) The most important item of exports in marine exports from India is ———.
   (a) Crab (b) Lobsters
   (c) Shrimp (d) Prawn
   (e) None of the above

(239) —————— is the home of Alphanso Mango ——————.
    (a) Ratnagiri   (b) Malda
    (c) Vijaywada   (d) Banaras
    (e) None of the above

(240) —————— is the state do not cultivate wheat.
    (a) Karnataka   (b) Maharashtra
    (c) West Bengal   (d) Tamil Nadu
    (e) Jharkhand

(241) —————— is also known as golden fiber.
    (a) Cotton   (b) Wheat
    (c) Silk   (d) Oats
    (e) Jute

(242) Golden Revolution is related to ——————.
    (a) Oilseed Production   (b) Honey/Horticulture
    (c) Jute Production   (d) Egg Production
    (e) Agricultural Production

(243) —————— Crop is grown in the laterite soil.
    (a) Pear   (b) Coconut
    (c) Tea   (d) Litchi
    (e) Coffee

(244) Indian Pulse Research Institute is located in ——————.
    (a) Allahabad   (b) Kanpur
    (c) Faizabad   (d) Lucknow
    (e) New Delhi

(245) Seed Plot Technique is followed in ——————.
    (a) Paddy   (b) Wheat
    (c) Bajara   (d) Potato
    (e) Rice

(246) —————— is of the record of land cultivation.
    (a) Khatouni   (b) Girawari
    (c) Panchnama   (d) Jamabandi
    (e) None of the above

(247) —————— is not a plantation crop.
    (a) Coffee   (b) Sugercane
    (c) Wheat   (d) Rubber
    (e) Rice

(248) ———— requires high temperature, light rainfall, frost free days and bright sunshine.
- (a) Jute
- (b) Cotton
- (c) Tea
- (d) Coffee
- (e) Rice

(249) Indian Green Revolution started from————.
- (a) Bangaluru
- (b) Kanpur
- (c) Delhi
- (d) Mumbai
- (e) Patnanagar

(250) ———— crop rotations, which is good for increasing soil nutrient status.
- (a) Rice- wheat
- (b) Groundnut- wheat
- (c) Pearlmillet- wheat
- (d) Sorgham- wheat
- (e) None of the above

(251) White rust is an important fungal disease of ————.
- (a) Wheat
- (b) Mustard
- (c) Rice
- (d) Bajra
- (e) Tea

(252) HD 2967 is a high yielding variety of————.
- (a) Rice
- (b) Maize
- (c) Mustard
- (d) Wheat
- (e) Bajra

(253) The scientific study of soil is called ————.
- (a) Earth Study
- (b) Soil Study
- (c) Pedology
- (d) Soil Chemistry
- (e) None of the above

(254) Soil factors are otherwise known as ————.
- (a) Climatic factors
- (b) Biotic factors
- (c) Physiographic factors
- (d) Edaphic factors
- (e) None of the above

(255) Fire curing is followed in————.
- (a) Bidi Tobacco
- (b) Hookah Tobacco
- (c) Cheroot Tobacco
- (d) Chewing Type Tobacco
- (e) All of the above

(256) Food of God is ———— .
- (a) Rice
- (b) Barley
- (c) Wheat
- (d) Maize
- (e) Cacao

(257) Water use efficiency is highest in the case of —————.
    (a) Border irrigation   (b) Drip irrigation
    (c) Sprinkler irrigation   (d) Flood irrigation
    (e) None of the above

(258) The animals produced by mating of two different breeds of the same species is known as —————.
    (a) Mixed breed   (b) Cross breed
    (c) Exotic breed   (d) Pure breed
    (e) All of these

(259) The ratio between marketable crop yield and water used in evapotranspiration is —————.
    (a) Field water use efficiency   (b) Economic irrigation efficiency
    (c) Consumptive use efficiency   (d) Water use efficiency
    (e) None of the above

(260) Kisan Mitra is an employee of —————.
    (a) Central Govt.   (b) Corporation
    (c) NGO   (d) State Govt.
    (e) None of the above

(261) Jalpriya is a variety of —————.
    (a) Maize   (b) Jowar
    (c) Paddy   (d) Barley

(262) Sugarcane + Potato is an intercropping system of —————.
    (a) Autumn season   (b) Zaid season
    (c) Spring season   (d) Rainy season

(263) Seed-rate of potato per hectare is —————.
    (a) 25 quintal/hectare   (b) 10 quintal/hectare
    (c) 15 quintal/hectare   (d) 40 quintal/hectare

(264) Deficiency symptoms of calcium on plants first appear at —————.
    (a) Lower leaves   (b) Middle leaves
    (c) Terminal leaves   (d) All of the above

(265) Which weedicide is used to kill broad leaf weeds in wheat —————?
    (a) 2, 4 – (D)   (b) 2, 4, 5 – T
    (c) 2, 4 – DB   (d) None of the above

(266) Maya is the variety of —————.
    (a) Potato   (b) Gram
    (c) Pea   (d) Mustard

(267) The weed that causes Asthma is ————.
    (a) Hirankhuri   (b) Bathua
    (c) Parthenium   (d) Krishna Neel

(268) Which crop requires maximum amount of nitrogen ———— ?
    (a) Potato   (b) Wheat
    (c) Barley   (d) Sugarcane

(269) First dwarf variety of paddy developed in India is ————.
    (a) Jaya   (b) Saket-4
    (c) Govind   (d) Narendra-97

(270) Sprinkler irrigation is suitable, where the soil has ————.
    (a) Clayey texture   (b) Loamy texture
    (c) Undulating topography   (d) All of these

(271) Endosulphan is also known as ————.
    (a) Lindane   (b) Thiodan
    (c) Aldrin   (d) H.C.

(272) Which of the following is systemic poison————?
    (a) Metasystox   (b) Phosphomidan
    (c) Phorate   (d) All of the above

(273) DDVP is known as————.
    (a) Nuvan   (b) Malathion
    (c) Thiodan   (d) Sulfex

(274) Seed treatment with Vitavex is the main controlling method of————.
    (a) Loose smut   (b) Rust
    (c) Downy mildew   (d) All of the above

(275) Covered smut of barley is a disease of ————.
    (a) Externally seed-borne   (b) Internally seed-borne
    (c) Air-borne   (d) None of the above

(276) Which of the following cakes is not edible ————?
    (a) Castor cake   (b) Mustard cake
    (c) Sesame cake   (d) Groundnut cake

(277) In India, about 1 42 million hectare land is under ————.
    (a) Cultivation   (b) Waste land
    (c) Forest   (d) Eroded land

(278) The headquarter of Indian Meteorological Department was established in 1 875 at ————.
    (a) New Delhi   (b) Hyderabad
    (c) Pune   (d) Calcutta

(279) Moisture condensed in small drops upon cool surface is called ―――――.
    (a) Hail   (b) Dew
    (c) Snow   (d) Fog

(280) How many agro-climatic zones (ACZ) are found in India―――――.
    (a) 16   (b) 18
    (c) 15   (d) 20

(281) Tilt angle of a disc plough is generally―――――.
    (a) 10°   (b) 15°
    (c) 20°   (d) 45°

(282) Pudding is done to―――――.
    (a) Reduce percolation of water   (b) Pulverise and levelling soil
    (c) Kill weeds   (d) All of the above

(283) The Community Development Programme (CDP) was started in India on―――
    (a) 2nd October, 1950   (b) 2nd October, 1952
    (c) 2nd October, 1951   (d) None of the above

(284) The main unit of Integrated Rural Development Programme is―――――.
    (a) Family   (b) Village
    (c) Block   (d) District

(285) Element of Communication is―――――.
    (a) Message   (b) Feedback
    (c) Channel   (d) All of the above

(286) The first Kshetriya Gramin Bank (KGB) was opened in India in ―――――.
    (a) 1972   (b) 1980
    (c) 1975   (d) 1969

(287) The main function of NABARD is―――――.
    (a) Farmers' loaning   (b) Agricultural research
    (c) Refinancing to agricultural financing institutions   (d) Development of agriculture

(288) Rent theory of profit was given by―――――.
    (a) Hawley   (b) C.P. Blacker
    (c) Tanssig   (d) F. Walker

(289) In L.D.R., the profit will be maximum when ―――――.
    (a) MC = MP   (b) MC > MP
    (c) MP = TP   (d) MP > TP

(290) The period of 11th Five Year Plan was ―――――.
    (a) 2000-2005   (b) 2002-2007
    (c) 2007-2012   (d) 2008-2012

(291) Acid rain contains mainly ―――――.
    (a) PO4   (b) NO2
    (c) NO3   (d) CH4

(292) Cell Organelle found only in plants are ―――――.
    (a) Mitochondria   (b) Golgi complex
    (c) Ribosomes   (d) Plastids

(293) Proteins are synthesized in ―――――.
    (a) Centrosomes   (b) Ribosomes
    (c) Mitochondria   (d) Golgi bodies

(294) Milk fever is caused due to the deficiency of ―――――.
    (a) P   (b) Ca
    (c) Mg   (d) K

(295) Milk sugar is a type of ―――――.
    (a) Glucose   (b) Sucrose
    (c) Lactose   (d) Fructose

(296) Muriate of potash is.
    (a) K2SO4   (b) KCl
    (c) K2HPO4   (d) KNO3

(297) Azotobacter fixes atmospheric nitrogen in the soil by ―――――.
    (a) Symbiotically   (b) Non-symbiotically
    (c) Both (a) and (b)   (d) None of the above

(298) The chemical formula of iron pyrites is ―――――.
    (a) FeSO4   (b) FeS
    (c) FeS2   (d) Fe2(SO4)3

(299) Rock phosphates are used in ―――――.
    (a) Saline soil   (b) Sodic soil
    (c) Acidic soil   (d) Neutral soil

(300) Intervenous chlorosis is caused due to the deficiency of ―――――.
    (a) N   (b) Mg
    (c) S   (d) Fe

(301) Kinnow is the hybrid variety of ―――――.
    (a) Citrus   (b) Orange
    (c) Mandarin   (d) Lemon

(302) The permanent preservative, which is used for preservation of fruit and vegetables, is ―――――.
    (a) Sodium chloride   (b) Potassium metabisulphate
    (c) Potassium sulphate   (d) Sugar

(303) Whip Tail disease of cauliflower is caused by deficiency of ———————.
    (a) Nitrogen   (b) Boron
    (c) Molybdenum   (d) Zinc

(304) The word 'Agriculture' is derived from ———————.
    (a) Greek   (b) Latin
    (c) Arabic   (d) French

(305) Motha (Grass nut) belongs to the family of ———————.
    (a) Cruciferae   (b) Tiliaceae
    (c) Cyperaceae   (d) Graminaceae

(306) Which of the followings are short day crops ———————.
    (a) Maize, Lobia, Bajra   (b) Wheat, Mustard, Gram
    (c) Moong, Soybean, Bajra   (d) Wheat, Soybean, Bajra

(307) What is the sequence of C4 plants ———————.
    (a) Sudangrass – Sugarcane – Paddy – Bajra
    (b) Sugarcane – Maize – Sudangrass – Bajra
    (c) Sugarcane – Cotton – Paddy – Maize
    (d) Cotton – Maize – Bajra – Sugarcane

(308) Match List-I (crops) with List-II (water requirement) and select your answer from the code given below ———————.
List-I
    (a) Jowar   (b) Soybean
    (c) Cotton   (d) Groundnut
List–II
    1. 140 mm – 300 mm   2. 350 mm – 450 mm
    3. 200 mm – 300 mm   4. 300 mm – 350 mm
Codes :

	(a)	(b)	(c)	(d)
(a)	3	1	2	4
(b)	4	2	3	1
(c)	1	4	2	3
(f)	3	1	4	2

(309) In which state, are there biggest area, highest production and number of Sugar Mills in relation to Sugarcane ———————.
    (a) Maharashtra   (b) Bihar
    (c) Uttar Pradesh   (d) Andhra Pradesh

(310) ——————— is not prepared by potato.
    (a) Acetic Acid   (b) Paper
    (c) Wine   (d) Fanina

(311) Uttar Pradesh is occupying ——— place in India, for Guava production.
    (a) Second　　　　　　　　　　　(b) First
    (c) Third　　　　　　　　　　　　(d) Fifth

(312) ——— is TPS variety of Potato.
    (a) JH 222　　　　　　　　　　　(b) Chipsona-II
    (c) Anand　　　　　　　　　　　(d) HPS-1/113

(313) ——— is VAM.
    (a) Virus　　　　　　　　　　　　(b) Bacteria
    (c) Algae　　　　　　　　　　　　(d) Fungi

(314) ——— is the main function of zinc in the plants.
    (d) Synthesis of nitrogen　　　　　(b) Synthesis of phosphorus
    (c) Required for synthesis of Tryptophos　(d) To increase activity of the boron

(315) ——— is the area in floriculture (in 000 hectare) in India.
    (a) 40 – 50　　　　　　　　　　　(b) 60 – 80
    (c) 100 – 120　　　　　　　　　　(d) None of the above

(316) ——— factors does not affect the nitrification.
    (a) Air　　　　　　　　　　　　　(b) Seed
    (c) Temperature　　　　　　　　　(d) Moisture

(317) ——— is the correct sequence of soil erosion.
    (a) Rill – Sheet – Gulley　　　　　(b) Gulley – Sheet – Rill
    (c) Sheet – Rill – Gulley　　　　　(d) Sheet – Gulley – Rill

(318) Zinc Sulphate ($ZnSO_4$) should not be mixed with ———.
    (a) (D)(A)P.　　　　　　　　　　(b) Compost fertilizer
    (c) Ammonium Chloride　　　　　(d) Urea

(319) Insecticides are specific inhibitors of ———.
    (a) Excretory system　　　　　　　(b) Digestive system
    (c) Nervous system　　　　　　　　(d) Blood circulatory system

(320) The credit for the success of Krishi Vigyan Kendras (KVK) goes to ———.
    (a) Dr. R. S. Paroda　　　　　　　(b) Dr. Chandrika Prasad
    (c) Dr. Mohan Singh Mehta　　　　(d) Dr. Mangla Rai

(321) Cauliflower belongs to the family ———.
    (a) Cruciferae　　　　　　　　　　(b) poacae
    (c) Malvaceae　　　　　　　　　　(d) Leguminaceae

(322) Which type of soil is best for knol-khol ———?
    (a) Loam　　　　　　　　　　　　(b) Clayey loam
    (c) Silty clayey loam　　　　　　　(d) Clay

(323) Which of the following soil type is most suitable for garlic cultivation ——————?
    (a) Loamy sand   (b) Sandy loam
    (c) Loam   (d) Clay

(324) Average planting distance (RP) of guava is ——————.
    (a) 5 m 5 m   (b) 6 m 6 m
    (c) 8 m 8 m   (d) 10 m 10 m

(325) Which of the following soil type has the highest field capacity ——————?
    (a) Loam   (b) Silty loam
    (c) Clayey loam   (d) Clay

(326) The trade name of phorate is ——————.
    (a) Temic   (b) Thiodan
    (c) Phortox   (d) Metasystox

(327) The sprayers are cleaned before use by ——————.
    (a) 1 % chlorine water   (b) 1 % hydrochloric acid
    (c) 1 % ammonia water   (d) 1 % bromine water

(328) The cyanogas pump is a /an ——————.
    (a) Duster   (b) Fumigator
    (c) Sprayer   (d) Emulsifier

(329) The main reason of Irish Famine in Potato was ——————.
    (a) Late blight disease   (b) Bacterial blight disease
    (c) Blast disease   (d) Ear cockle disease

(330) The instrument, which is used for sowing of seed with fertilizer together at a time, is ——————.
    (a) Seed drill   (b) Dibbler
    (c) Seed sowing behind plough   (d) Ferti-cum Seed drill

(331) Seed treatment is done to control ——————..
    (a) Soil-borne diseases   (b) Air-borne diseases
    (c) Seed-borne diseases   (d) None of the above

(332) Salt tolerant crop is ——————..
    (a) Cowpea   (b) Field pea
    (c) Garlic   (d) Longmelon

(333) —————— is not a dairy breed of cattle.
    (a) Sahiwal   (b) Sindhi
    (c) Nagore   (d) All of the above

(334) Stored grains can be saved from insect damage, if the grain moisture content is ––––––––––.

    (a) > 10% (b) Â10%
    (c) 10% (d) None of the above

(335) –––––––––– pesticides has been banned in India.

    (a) Rogor (b) DDT
    (c) Metasystox (d) Dimecron

(336) Pulses fit well in cropping system as they are ––––––––––.

    (a) Short duration crops (b) Disease resistant crops
    (c) Long duration crops (d) Moisture stress resistant crops

(337) Wheat is a ––––––––––.

    (a) Cash crop (b) Cereal crop
    (c) Covered crop (d) None of the above

(338) Autumn sugarcane is planted in month of ––––––––––.

    (a) February-March (b) July
    (c) October (d) December

(339) Seed-rate for timely sown wheat is ––––––––––.

    (a) 75 kg/ha (b) 1 00 kg/ha
    (c) 1 25 kg/ha (d) 1 50 kg/ha

(340) Most critical stage in wheat for irrigation is ––––––––––.

    (a) Crown root initiation (b) Flowering
    (c) Milk (d) Dough

(341) –––––––––– is most popular variety of wheat in Uttar Pradesh.

    (a) PBW – 343 (b) U.P. – 2338
    (c) K – 7903 (d) K – 91 07

(342) KPG – 59 (Udai) is a variety of ––––––––––.

    (a) Field pea (b) Vegetable pea
    (c) Lentil (d) Gram

(343) In plain, Rajma is cultivated during ––––––––––.

    (a) Kharif (b) Rabi
    (c) Zaid (d) None of the above

(344) –––––––––– crop is recommended for Zaid season cultivation in Uttar Pradesh.

    (a) Vegetable pea (b) Groundnut
    (c) Barley (d) Lentil

(345) The most efficient use of potassium is achieved by ―――――.
    (a) Broadcasting at the sowing time
    (b) Top dressing after one month of sowing
    (c) Basal placement at the sowing time
    (d) Foliar spray

(346) The term 'Extension' was first used in ―――――.
    (a) U.K.
    (b) U.S.A
    (c) India
    (d) France

(347) The first K.V.K. (Krishi Vigyan Kendr(a) in India was established in ―――――.
    (a) Bombay
    (b) Port Blair
    (c) Pondicherry
    (d) Madras

(348) ATMA is related to.
    (a) NARP
    (b) NAARM
    (c) NREP
    (d) None of the above

(349) Albert Mayer is the name associated with ―――――.
    (a) Nilokheri Development Project
    (b) Firka Development Project
    (c) Etawah Pilot Project
    (d) Shriniketan Project

(350) Co-operative Credit Societies Act was passed in India in ―――――.
    (a) 1 902
    (b) 1 904
    (c) 1 906
    (d) 1 91 2

(351) Maximum photosynthesis takes place in ―――――.
    (a) Blue light
    (b) Red light
    (c) Violet light
    (d) Green light

(352) Farm Planning means ―――――.
    (a) Farm Budgetting
    (b) Cropping pattern
    (c) Type of enterprises
    (d) None of the above

(353) The first product of photosynthesis in C3 plant is ―――――.
    (a) Pyruvic acid
    (b) Phospho-glyceric acid
    (c) Oxalo-acetic acid
    (d) Succinic acid

(354) Bending of plants towards light is called ―――――.
    (a) Phototropism
    (b) Vernalization
    (c) Photo-respiration
    (d) None of the above

(355) Germination is inhibited by ―――――.
    (a) Red light
    (b) Blue light
    (c) U.V. light
    (d) I.R. light

(356) The best method of milking is ―――――.
    (a) Knuckling method
    (b) Fisting method
    (c) Stripping method
    (d) None of the above

(357) Line breeding is a type of ─────.
    (a) Inbreeding　　(b) Outbreeding
    (c) Natural breeding　　(d) None of the above

(358) Match List-I with List-II and select answer from the codes given below ─────.
    List-I
    (a) White Revolution　　(b) Grey Revolution
    (c) Blue Revolution　　(d) Green Revolution
    List-II
    1. Fertilizer production　　2. Fish production
    3. Cereal production　　4. Milk production
    Codes :
    (a) (b) (c) (d)
    (a) 4 1 2 3　　(b) 1 2 3 4
    (c) 2 4 3 1　　(d) 1 3 4 2

(359) 'Tharparkar' breed of cow is ─────.
    (a) Milch breed　　(b) Working breed
    (c) Dual purpose breed　　(d) None of the above

(360) Cow and buffalo belong to the family ─────.
    (a) Bovidae　　(b) Suidae
    (c) Equidae　　(d) Cammelidae

(361) ───── is the contribution of Animal Husbandry Sector in the agricultural growth.
    (a) 1 0%　　(b) 1 2% – 1 5%
    (c) 7% – 9%　　(d) 5%

(362) ───── labourers are required to run a 30 cows milch herd.
    (a) 8　　(b) 6
    (c) 4　　(d) 1 0

(363) ───── is the availability of per day per capita milk in India presently.
    (a) 229 gram　　(b) 239 gram
    (c) 21 9 gram　　(d) 252 gram

(364) ───── place is occupied by India in egg production.
    (a) First　　(b) Second
    (c) Third　　(d) Fourth

(365) ───── calories (cal) may be obtained from 100 gram chicken egg.
    (a) 175 cal　　(b) 180 cal
    (c) 160 cal　　(d) 130 cal

(366) Main function of biofertilizer is ─────────
    (a) To increase chemical process
    (b) To increase physiological process
    (c) To increase biological process
    (d) To increase photosynthesis process

(367) How much tomato average production (Quintal) may be yield from one hectare ─────────.
    (a) 100
    (b) 105-150
    (c) 250
    (d) 160-275

(368) ───────── soil is found near the canal banks
    (a) Acidic and alkaline
    (b) Acidic
    (c) Alkaline
    (d) None of the above

(369) ───────── is not biofertilizer.
    (a) Multiflex
    (b) PSB
    (c) Vermicompost
    (d) NADEP

(370) ───────── form of nitrogen is absorbed by paddy under waterlogged condition.
    (a) NH_4 ion
    (b) Nitrate ion
    (c) NO_2 ion
    (d) N_2

(371) ───────── do not relate to groundnut.
    (a) Brazil
    (b) $2n = 40$
    (c) Pink disease
    (d) Tikka disease

(372) Which of the following is produced highest in India ─────────?
    (a) Mango
    (b) Banana
    (c) Papaya
    (d) Grapes

(373) The optimum temperature for the Banana crop is─────────.
    (a) 30°C
    (b) 23°C
    (c) 21°C
    (d) 25°C

(374) Which one of the following varieties has been selected to develop Narendra Aonla-6 variety ─────────?
    (a) Chakaiya
    (b) Hathijhool
    (c) Banarasi
    (d) Narendra Aonla-6

(375) Red soil is poor in which of the following nutrients─────────.
    (a) Phosphorus and Sulphur
    (b) Phosphorus and Nitrogen
    (c) Nitrogen and Zinc
    (d) Nitrogen and Potassium

(376) A farming system in which airable crops are grown in alleys formed by trees or shrubs, to establish soil fertility and to enhance soil productivity, is known as ─────────.
    (a) Relay cropping
    (b) Multiple cropping
    (c) Alley cropping
    (d) Mixed cropping

(377) The cropping intensity of Groundnut + Arhar – Sugarcane is————.
    (a) 200% (b) 300%
    (c) 150% (d) 250%

(378) The scented variety of paddy is————.
    (a) Jaya (b) Bala
    (c) Type-3 (d) Type-1

(379) From which language is the word 'Agronomy' taken————.
    (a) Latin (b) Greek
    (c) French (d) German

(380) Tarameera is belonged to which family————.
    (a) Cruciferae (b) Linaceae
    (c) Compositae (d) Graminae

(381) The size of clay particles is————.
    (a) 1.0 mm (b) 0.2 – 0.02 mm
    (c) < 0.02 mm (d) < 0.002 mm

(382) When one plant has both male and female flowers separately, is called ————?
    (a) Monophrodits (b) Monoecious
    (c) Hermaphrodite (d) Apomixis

(383) Amrapali is the cross of ————.
    (a) Neelam x Dashaheri (b) Dashaheri x Langra
    (c) Langra x Dashaheri (d) Dashaheri x Neelam

(384) Seed-plot technique is adopted in ————.
    (a) Onion (b) Potato
    (c) Sugarcane (d) Tomato

(385) The origin of litchi is ————.
    (a) India (b) Philippines
    (c) China (d) Burma

(386) Jute cultivation in India is concentrated in the delta area of which of the following rivers ————.
    (a) Ganga (b) Mahanadi
    (b) Brahamputra (d) Godavari

(387) ———— is the largest irrigation canal in India.
    (a) Buckingham Canal (b) Sirhind Canal
    (c) Indira Gandhi Canal (d) Sutlej Yamuna Link Canal

(388) —————— is the correct sequence in the decreasing order of production (in million tonnes) of the given food grains in India.
    (a) Wheat - Rice - Pulses - Coarse cereals
    (b) Rice - Wheat - Pulses - Coarse cereals
    (c) Wheat - Rice - Coarse cereals - Pulses
    (d) Rice - Wheat - Coarse cereals - Pulses

(389) The eyes of potato are useful for ——————.
    (a) Nutrition
    (b) Respiration
    (c) Vegetative Propagation
    (d) Protection from predators

(390) —————— is Slash and Burn Agriculture.
    (a) Method of sugarcane cultivation
    (b) Process of deforestation
    (c) Agriculture without irrigation
    (d) Jhum cultivation

(391) Rotation of crops means ——————.
    (a) Growing of different crops in the same area in sequential seasons.
    (b) Shifting of area of same crops.
    (c) Growing two or more crops simultaneously to increase productivity.
    (d) Alternating crops with fruits over a period of years.

(392) The nitrogen present in the atmosphere is ——————.
    (a) of no use to plants
    (b) injurious to plants
    (c) directly utilized by plants
    (d) utilized through micro-organisms

(393) Which is the largest cotton growing State in India ——————?
    (a) Maharashtra
    (b) Madhya Pradesh
    (c) Andhra Pradesh
    (d) Gujarat

(394) The variety of coffee largely grown in India is ——————.
    (a) Old Chicks
    (b) Coorgs
    (c) Arabica
    (d) Kents

(395) Under which plan did the Government introduce an agricultural strategy which gave rise to Green Revolution ——————.
    (a) Third Five-Year Plan (FYP)
    (b) Fourth FYP
    (c) Fifth FYP
    (d) Second FYP

(396) Laterite soil develops as a result of ——————.
    (a) Deposits of alluvial
    (b) Deposition of loess
    c. Leaching of sedimentary rocks
    (d) Wind-blown dust

(397) What kind of soil is treated with gypsum to make it suitable for cropping ——?
    (a) Alluvial
    (b) Acidic
    (c) Water-logged
    (d) Soil with excessive clay content

(398) The United Nations General Assembly was declared 2013 as the International Year of ―――――.
   (a) Quinoa   (b) Buckwheat
   (c) Rice   (d) Jatropha

(399) ――――――― is known as a pseudocereal.
   (a) Wheat   (b) Quinoa
   (c) Horse gram   (d) Barley

(400) Besides Rabi and Kharif, ――――――― is the third crop season in India.
   (a) Zaid   (b) Barsati
   (c) Sharad   (d) Jhum

(401) Kharif crops are harvested in ―――――――.
   (a) June - July   (b) October - November
   (c) May - June   (d) March - April

(402) ――――――― has been found to increase the yield of crops.
   (a) Lime   (b) Ash
   (c) Neem oil   (d) Eucalyptus oil

(403) ――――――― agricultural crops/groups of crops may be grown in abundant in lowlands and river deltas of fertile alluvial soil where there is high summer temperature and rainfall varies' from 180 cm to 250cm.
   (a) Wheat and sugarcane (b) Cotton
   (c) Maize and coarse crops (d) Rice, jute and tea

(404) ――――――― crops would be preferred for sowing in order to enrich the soil with nitrogen.
   (a) Wheat   (b) Mustard
   (c) Sunflower   (d) Gram

((405) ――――――― is black soil not very suitable.
   (a) Cotton   (b) Wheat
   (c) Ground nut   (d) Potato

(406) Extensive areas of grape cultivation in France are especially called ―――――.
   (a) Wine cellars   (b) Grape fields
   (c) Grape farms   (d) Vineyards

(407) To which group does the black cotton soil of India belongs ―――――.
   (a) Laterite   (b) Podzol
   (c) Chernozem   (d) Sierozem

(408) ――――――― is a cash crop in India.
   (a) Rice   (b) Wheat
   (c) Sugarcane   (d) Jowar

(409) ——————— is a salt tolerant rice variety grown in water-logged coastal regions of Kerala.
    (a) Navara   (b) Pokkali
    (c) Ponni   (d) Samba

(410) ——————— is grown mainly on mountain slopes.
    (a) Paddy   (b) Tea
    (c) Groundnut   (d) Potato

(411) ——————— is not a nitrogenous fertilizer.
    (a) KCl   (b) CaCN2
    (c) NH4NO3   )(d) Urea

(412) ——————— is green manure crops contains highest amount of nitrogen.
    (a) Dhaincha   (b) Sunhemp
    (c) Cow Pea   (d) Guar

(413) ——————— is the rank of India in the world as a fruit producer.
    (a) Third   (b) Fourth
    (c) First   (d) Second

(414) White rust is an important fungal disease of ———————.
    (a) Wheat   (b) Mustard
    (c) Rice   (d) Bajra

(415) The Indian Council of Agricultural Research (ICAR) has identified three Bt cotton varieties for cultivation in Punjab, Haryana and Rajasthan which were developed by Punjab Agricultural University in Ludhiana recently. Which one of the following is not among those three———————.
    (a) PAU Bt 1   (b) F1 861
    (c) RS201 3   (d) BT1 861

(416) ——————— is not a nitrogenous fertilizer.
    (a) Ammonium sulphate   (b) Urea
    (c) Ammonium nitrate   (d) Super phosphate

(417) HD 2967 is the new high yielding variety of———————.
    (a) Rice   (b) Maize
    (c) Musterd   (d) Wheat

(418) The scientific study of soil is ———————.
    (a) Earth Study   (b) Soil Science
    (c) Pedology   (d) Soil Chemistry

(419) Prabhat is an early short duration variety of ———————.
    (a) Red gram   (b) Wheat
    (c) Maize   (d) Rice

(420) The Minimum Support Price for Foodgrains was introduced in the year———.
    (a) 1954   (b) 1944
    (c) 1964   (d) 1974

(421) Soil factors are otherwise known as ———.
    (a) Climatic factors   (b) Edaphic factors
    (c) Biotic factors   (d) Physiographic factors

(422) ——— is a commercial crop.
    (a) Cotton   (b) Bajra
    (c) Jowar   (d) Paddy

(423) *Triticum aestivum*, the common bread wheat is ———.
    (a) Tetraploid   (b) Hexaploid
    (c) Haploid   (d) Diploid

(424) Monoculture is a typical characteristics of ———.
    (a) Shifting cultivation   (b) Subsitence farming
    (c) Specialized horticulture   (d) Commercial grain farming

(425) Which crop requires water-logging for its cultivation ———?
    (a) Tea   (b) Coffee
    (c) Rice   (d) Mustard

(426) Agronomy is a branch of Agriculture that deals with ———.
    (a) Breeding of crop plants   (b) Principles of field management
    (c) Principles and practice of crop production   (d) Protection of crops from Diseases and Pests

(427) Soils of Western Rajasthan have a high content of ———.
    (a) Aluminium   (b) Calcium
    (c) Nitrogen   (d) Phosphorus

(428) The Black rust of disease of wheat is caused by ———.
    (a) *Xanthomonas graminis*   (b) *Puccinia graminis*
    (c) *Puccinia recondita*   (d) None of the above

(429) A crop grown in zaid season is ———.
    (a) Soyabean   (b) Water melon
    (c) Jute   (d) Maize

(430) The adoption of High Yielding Variety Programme in Indian Agriculture started in ———.
    (a) 1966   (b) 1965
    (c) 1968   (d) 1967

(431) —————— is a food crop.
    (a) Palm   (b) Jute
    (c) Cotton   (d) Maize

(432) —————— is an oilseed.
    (a) Cardamom   (b) Garlic
    (c) Clove   (d) Mustard

(433) —————— makes a case for intensive, modern farming.
    (a) Cropping pattern   (b) Higher output using organic method
    (c) Remunerative price   (d) None of the above

(434) —————— is not an agricultural product.
    (a) Alum   (b) Cotton
    (c) Jute   (d) Rice

(435) Crop rotation helps to —————— .
    (a) Less use of pesticides   (b) Yield more crops
    (c) Produce a greater choice of plant products   (d) Eliminate parasites which have selective hosts

(436) Potassium chloride contains K —————— .
    (a) 18%   (b) 48%
    (c) 44%   (d) 60%

(437) Plant micronutrient is —————— .
    (a) Carbon   (b) Boron
    (c) Magnesium   (d) Sulphur

(438) Which gas is released from paddy fields —————— ?
    (a) CO   (b) H S
    (c) CH   (d) NH

(439) Which two crops of the following are responsible for almost 75% of pulse production in India —————— ?
    (a) Gram and pigeon pea   (b) Gram and moong bean
    (c) Moong bean and lentil   (d) Pigeon pea and moong bean

(440) Fire curing is followed in —————— .
    (a) Bidi tobacco   (b) Hookah tobacco
    (c) Cheroot tobacco   (d) Chewing type tobacco

(441) Groundnut pegs when developed in the soil from —————— .
    (a) Tubers   (b) Fruits
    (c) Stems   (d) Roots

(442) In Jute growing areas the usual alternate crop is———.
    (a) Cotton   (b) Wheat
    (c) Sugarcane   (d) Rice

(443) Neelum is a variety of———.
    (a) Grape   (b) Papaya
    (c) Mango   (d) Apple

(444) The Commission of Agricultural Costs and Prices fixes the———.
    (a) Support price   (b) Wholesale price
    (c) Retail price   (d) None of the above

(445) Vector of phyllody disease is———.
    (a) Thrips   (b) Mite
    (c) White fly   (d) Jassid

(446) Gynodioecious varieties of papaya produce———.
    (a) Only male plants   (b) Female and hermaphrodite
    (c) Male and hermaphrodite plants   (d) Only female plants

(447) Isolation distance for foundation seed of rice is———.
    (a) 30 metre   (b) 3 metre
    (c) 35 metre   (d) 50 metre

(448) Botanically pineapple is a———.
    (a) Pome   (b) Baluster
    (c) Berry   (d) Sorosis

(449) The basic unit of development under the Integrated Rural Development Programme is a ———.
    (a) District   (b) Family
    (c) Village   (d) Community Development Block

(450) Ratna is a variety of———.
    (a) Wheat   (b) Barley
    (c) Maize   (d) Rice

(451) Seed rate of American cotton is———.
    (a) 20 kg/ha   (b) 30 kg/ha
    (c) 35 kg/ha   (d) 1 2 kg/ha

(452) Greening of potato results in———.
    (a) Increase in nutritional quality   (b) Increase in disease resistance
    (c) Decrease in disease resistance   (d) Decrease in nutritional quality

(453) Sugarbeet nematode is———.
    (a) Heterodera avanae   (b) Heterodera cajani
    (c) Heterodera jae   (d) Heterodera schacti

(454) In medium term storage, material can be stored up to ———.
    (a) 1 00 years   (b) 3 to 5 years
    (c) 1 0 to 1 5 years   (d) None of the above

(455) Which is the saturated fatty acid of the following ———?
    (a) Stearic acid   (b) Arachidonic acid
    (c) Oleic acid   (d) Linoleic acid

(456) Iron is an important component of ———.
    (a) Siroheme   (b) Ferredoxin
    (c) Cytochromes   (d) All of these

(457) The chemical, which is used for controlling the mites, is known as———.
    (a) Fungicides   (b) Mematicides
    (c) Acaricides   (d) Insecticides

(458) WP abbreviates as ———.
    (a) Wettable Paste   (b) Wettable Powder
    (c) Water Paste   (d) None of the above

(459) Which of the following elements is not essential element of plants but proves to be beneficial for some plants ———?
    (a) Boron   (b) Sodium
    (c) Iodine   (d) Copper

(460) The pyrite is mostly found in ———.
    (a) Bihar   (b) Rajasthan
    (c) Andhra Pradesh   (d) Maharashtra

(461) Parthenocarpy occurs in ———.
    (a) Peach   (b) Mango
    (c) Jackfruit   (d) Banana

(462) Water use efficiency is the highest is case of———.
    (a) Border irrigation   (b) Drip irrigation
    (c) Sprinkler irrigation   (d) Flood irrigation

(463) Atmosphere is essential for———.
    (a) Winds   (b) Cloud formation
    (c) Weather phenomena   (d) All of the above

(464) Which one of the following plants belongs to family Anacardiaceae ———?
    (a) Papaya   (b) Orange
    (c) Cashewnut   (d) None of the above

(465) In India post-harvest losses of fruit and vegetable is ——— per cent of the total production ———.
    (a) 1 5-20   (b) 50-50
    (c) 40-45   (d) 25-30

(466) The animals produced by mating of two different breeds of the same species is known as —————.
    (a) Mixed breed   (b) Cross breed
    (c) Exotic breed   (d) Pure breed

(467) Recording of milk production in dairy farm is done mainly for —————.
    (a) Selection of good producer   (b) Increase in production of milk
    (c) Ensuring quality of milk   (d) Quick selling of milk

(468) The average gestation period in buffalo is of how many days —————.
    (a) 310   (b) 210
    (c) 400   (d) 345

(469) Citrus tristeza virus is transmitted through —————.
    (a) Aphid   (b) Thrips
    (c) Nematode   (d) Plant hopper

(470) Which one of the pathogens is monocylic —————?
    (a) Ustilago   (b) Alternaria
    (c) Phytophthora   (d) Puccinia

(471) Caenocytic mycellium is found in —————.
    (a) Mastigomycotina   (b) Deuteromycotina
    (c) Ascomycotina   (b) None of the above

(472) Penicillin acts on —————.
    (a) Cell wall   (b) RNA
    (c) Cell membrane   (d) None of the above

(473) Blue colour tag is issued for —————.
    (a) Foundation seed   (b) Certified seed
    (c) Nucleus seed   (d) Foundation seed

(474) Generally, cereals are deficient in —————.
    (a) Protein   (b) Tryptophan
    (c) Methionine   (d) Lysine

(475) The book 'Mutation Research' is written by —————
    (a) Muller   (b) Stadler
    (c) Nilsson-Enle   (d) Auerbach

(476) Each anther has how many pollen sacs —————?
    (a) Four   (b) Two
    (c) One   (d) Three

(477) Rice and wheat has how many stamens —————.
    (a) 3,3   (b) 6,3
    (c) 3,6   (d) 6,6

(478) The chemical, which attracts opposite sex insects of a species, is known as ―――――.

    (a) Hormones   (b) Allomones
    (c) Kairomones   (d) Pheromones

(479) Damaging satge of potato tuber month is ―――――.

    (a) Adult   (b) Pupa
    (c) Larva   (d) All of the above

(480) Piercing and sucking type mouth-parts are present in the insect ―――――.

    (a) Hemiptera   (b) Orthoptera
    (c) Lepidoptera   (d) Isoptera

(481) Classification of insects and rules of their nomenclature comes under the branch

    (a) Morphology   (b) Taxonomy
    (c) Ecology   (d) Physiology

(483) 'Chanchal' is a variety of ―――――.

    (a) Brinjal   (b) Tomato
    (c) Chilli   (d) Capcicum

(484) Central Soil Salinity Research Institute is located at ―――――.

    (a) Karnal   (b) Bihar
    (c) Dehradun   (d) Jodhpur

(485) Seed plant technique is followed in ―――――.

    (a) Wheat   (b) Bajra
    (c) Potato   (d) Paddy

(486) For providing inputs like quality seeds, fertilizers and pesticides, the agency present at the Village Panchayat Samiti Level is ―――――.

    (a) Nationalised Banks   (b) Cooperative Society
    (c) Insurance Companies   (d) NABARD

(487) Where do the female mango leaf hoppers lay their eggs ―――――?

    (a) Inside the tissue of leaf margin   (b) On the dorsal surface of leaves
    (c) On the ventral surface of leaves   (d) Inside the mid-rib of leaves

(488) In H.T.S.T. pasteurization, which one of the following organisms is chosen as index organism for killing ―――――.

    (a) *S.thermophilus*   (b) *S.lactis*
    (c) *(B)subtilis*   (d) *M.tuberculosis*

(489) Thermophilic micro-organism grow well at temperature ―――――.

    (a) 10°C - 20°C   (b) 20°C - 40°C
    (c) 50°C - 60°C   (d) 5°C - 7°C

(490) Pink bollworm is a pest of —————.
    (a) Gram   (b) Okra
    (c) Mustard   (d) Cotton

(491) The croppping intensity of maize-potato-tobacco is —————.
    (a) 200%   (b) 300%
    (c) 100%   (d) None of the above

(492) Indian Institute of Pulse Research is located at —————.
    (a) Kanpur   (b) New Delhi
    (c) Varanasi   (d) Lucknow

(493) Khaira disease of rice can be controlled by spraying —————.
    (a) Copper sulphate   (b) Calcium sulphate
    (c) Zinc sulphate   (d) Borax

(494) ————— can be suitable for cropping as a wheat mixed crop.
    (a) Cabbage   (b) Cotton
    (c) Mustard   (d) Jowar

(495) The optimum cardinal temparature point for germination of rice seeds is —.
    (a) 20°C - 25°C   (b) 18°C - 22°C
    (c) 30°C - 32°C   (d) 37°C - 39°C

(496) —————. culture should be given priority in groundnut cultivation.
    (a) Mycorrhiza   (b) Rhizobium
    (c) Phosphobacteria   (d) Azospirillum

(497) Prabhat is an early short duration variety of —————.
    (a) Black gram   (b) Red gram
    (c) Green gram   (d) Gram

(498) In maize plants —————.
    (a) Silk appear first   (b) Both of these appear at same time
    (c) Tassels appear first   (d) None of the above

(499) Blind hoeing is recommended for —————.
    (a) Wheat   (b) Maize
    (c) Groundnut   (d) Sugarcane

(500) Kisan Mitra is an employee of —————.
    (a) Central Govt.   (b) Corporation
    (c) None of the above   (d) State Govt.

(501) Red Delicious is a variety of —————.
    (a) Mango   (b) Papaya
    (c) Apple   (d) Guava

(502) The number of essential mineral elements of plant is ——————.
    (a) 16   (b) 21
    (c) 13   (d) 20

(503) Sulphur-coated Urea contains N ——————.
    (a) 21%   (b) 33%
    (c) 40%   (d) 26%

(504) Granite is a ——————.
    (a) Metamorphic   (b) Igneous
    (c) Sedimentary   (d) None of the above

(505) Chemical formula of pyrite is ——————.
    (a) FeS   (b) CuS
    (c) MnS   (d) FeS

(506) Soil fertility is reduced due to ——————.
    (a) Over irrigation   (b) Poor drainage
    (c) Imbalanced use of fertilizers   (d) Continuous cropping

(507) Anemometer measures ——————.
    (a) Wind direction   (b) Relative humidity
    (c) Net radiation   (d) Wind velocity

(508) Cutting of branches of a tree is also called ——————.
    (a) Coppicing   (b) Brashing
    (c) Lopping   (d) Prunning

(509) Cytogenetic male sterility is utilized in ——————.
    (a) Pure line selection   (b) Hybrid seed production
    (c) Bulk method   (d) Progeny test

(510) Dahelia, Marigold and Gaillardia belong to the family ——————.
    (a) Iridaceae   (b) Oleaceae
    (c) Malvaceae   (d) Compositeae

(511) Decomposition of organic matter in submerged soil is carried out by——————.
    (a) Bacteria   (b) Actinomycetes
    (c) Fungi   (d) Earthworm

(512) Desuckering is done in ——————.
    (a) Potato   (b) Tomato
    (a) Sugarcane   (d) Tobacco

(513) Diethane M-45 is a local ——————.
    (a) Insecticide   (b) Fungicide
    (c) Nematicide   (d) Weedicide

(514) Difference in the energy of water in a solution and pure water at the same temperature and pressure is termed as ―――――.
    (a) Diffusion
    (b) Osmosis
    (c) Water potential
    (d) Matric potential

(515) Downy mildew disease is caused by ―――――.
    (a) Phytophthora
    (b) Erysiphe
    (c) Peronospora
    (d) Cercospora

(516) Downy mildew of bajra is called by ―――――.
    (a) *Erysiphe graminis*
    (b) *Claviceps purpurea*
    (c) *Claviceps microcephala*
    (d) *Sclerospora graminicola*

(517) Damping-off disease of vegetable nursery can be controlled by ―――――.
    (a) Solarization
    (b) Mixing of fungicides in soil
    (c) Seed treatment
    (d) All of the above

(518) Whiptail is a disorder of cauliflower due to deficiency of ―――――.
    (a) Zinc
    (b) Boron
    (c) Potassium
    (d) Molybdenum

(519) The colour of tomato is due to the presence of ―――――.
    (a) Carotene
    (b) Lycopene
    (c) Anthocyanin
    (d) Xanthophyll

(520) Mastitis is a disease of which organ ―――――.
    (a) Uternus
    (b) Heart
    (c) Udder
    (d) Lung

(521) Colostrum should be fed to newborn calves for ―――――.
    (a) 1/2 day
    (b) 1 day
    (c) 4 day
    (d) 10 day

(522) Toda is breed of ―――――.
    (a) Cattle
    (b) Goat
    (c) Buffalo
    (d) Sheep

(523) Mastitis in animals is due to ―――――.
    (a) Worms
    (b) Virus and Worms
    (c) Fungi and dry hand milking
    (d) Bacteria and Virus

(524) Asexual reproduction includes ―――――.
    (a) Amphimixis
    (b) Apomixis
    (c) Allogamy
    (d) Autogamy

(525) In India, gene bank of wheat is located at ―――――.
    (a) Kanpur
    (b) Ludhiana
    (c) Karnal
    (d) IARI, New Delhi

(526) TPS technique is related to ——————.
    (a) Tomato   (b) Sugarcane
    (c) Potato   (d) All of these

(527) SRI is a technique used in ——————.
    (a) Maize   (b) Groundnut
    (c) Wheat   (d) Rice

(528) Numbers of agro-climate and ecological zones classified by ICARs, respectively are ——————
    (a) 15,131   (b) 131,8
    (c) 8,131   (d) (b) 1,15

(529) The Green Revolution has mainly been successful for ——————.
    (a) Wheat   (b) Rice
    (c) Maize   (d) Gram

(530) Guttation occurs in plants through ——————.
    (a) Stomata   (b) Hydathodes
    (c) None of the above   (d) Both (a) and (b)

(531) Stomata open at night in ——————.
    (a) C4 plants   (b) C3 plants
    (c) CAM plants   (d) None of the above

(532) Living Cells are not essential for ——————.
    (a) Transpiration   (b) Evaporation
    (c) Guttation   (d) All of the above

(533) α & β-tubulins are protein components of ——————.
    (a) Actin filaments   (b) Microtubules
    (c) Intermediate   (d) All of the abve

(534) 'First blight' of sugarcane is due to deficiency of nutrient ——————.
    (a) Mn   (b) Zn
    (c) P   (d) Fe

(535) Percentage of P O in 'Pelofas' fertilizer is ——————.
    (a) 12   (b) 16
    (c) 11   (d) 18

(536) Which one of the following is not a primary nutrient ——————?
    (a) S   (b) N
    (c) K   (d) P

(537) Which group of plant nutrients involves in N fixation ——————?
    (a) P,S,Mo   (b) P,S,Co
    (c) P,Mo,Co   (d) All of these

(538) —————— one of the following fertilizers is known as 'Kisan Khad'.
    (a) Urea   (b) Calcium Ammonium Nitrate
    (c) None of the above   (d) Ammonium Sulphater

(539) Zoological name of white grub is ——————.
    (a) *Holotrichia consanguinea*   (b) *Helicoverpa armigera*
    (c) *Spodoptera litura*   (d) *Bamisia tabaci*

(540) Botanical name of nobel canes is ——————.
    (a) *Saccharum officinarum*   (b) *Saccharum barberi*
    (c) *Saccharum spontaneous*   (d) *Saccharum sinensis*

(541) Nullisomics produce which type of gametes ——————.
    (a) n-1   (b) n+1
    (c) n   (d) n+1 -1

(542) The crossing over occurs during ——————.
    (a) Diplotene   (b) Zygotene
    (c) Leptotene   (d) Pachytene

(543) 'Pusa Jai Kisan' is a somaclone of ——————.
    (a) Indian Mustard   (b) Khesari
    (c) Citronella java   (d) Basmati Rice

(544) Which one of the following countries, the farm gate milk price per liter (RS) is the highest ——————?
    (a) USA   (b) Canada
    (c) Japan   (d) New Zealand

(545) —————— is one of the best 'poultry feed stuff' and having good source of animal protein as it contains 34-55 percent protein.
    (a) Groundnut Cake   (b) Fishmeal
    (c) Soyabean-meal   (d) Linseed-meal

(546) —————— is one of the following Indian breeds of sheep, suggested meat and carpet wool production.
    (a) Deccani   (b) Nellore
    (c) Gaddi   (d) Mandya

(547) Under the head 'Informatics in Agriculture' which institute has developed Grape Expert System, Cabbage Pest Expert System ——————.
    (a) National Institute of Agriculture Extension Management(MANAGE), Hyderbad
    (b) Indian Institute of Horticultural Research (IIHR), Banglore
    (c) Indian Agriculture Research Institute(IARI), New Delhi
    (d) Centre for Informetics Research and Advancement, Kerala(AGREX)

(548) The 'Eri' Silkworm (*Philosamia recini*) which feeds on common castor-plant, is mainly raised in which of the following states, for producing eri-silk ―――――.
    (a) West Bengal   (b) Bihar
    (c) Orissa   (d) Himachal Pradesh

(549) The Jawahar, Amber are considered as important and good varieties of maize used for ―――――.
    (a) Fodder   (b) Quality Protein Maize
    (c) Orisa   (d) Pop-Corn

(550) ――――― one of the Rice variety recommended for growing in saline areas of Haryana State.
    (a) Panvet-1   (b) CSR 30
    (c) CST 7-1   (d) HKR 1 (b)6

(551) Which of the Following states developed 'Uzavaar Sandies'- an innovative model in agricultural marketing which involved direct sale of farm produce to the consumers ―――――?
    (a) Andhra Pradesh   (b) Tamil Nadu
    (c) Karnataka   (d) Uttarakhand

(552) ――――― seeding and fertilizer application equipments is developed by CCSHAU, Hisar centre.
    (a) Ridge Seeder   (b) Two-row bullock-drawn FESPO plough
    (c) Tractor drawn planter   (d) Malviya seed-cum fertilizer drill

(553) The shoot and fruit borers, jassids, spider mites and leaf rollers are the major pests of which of the following vegetable crops ―――――.
    (a) Tomato   (b) Okra or Bhindi
    (c) Cucumber   (d) Peas

(554) The larvae of which of the following pests of sorghum crop, cut growing points and causedead-hearts ti the plant―――――.
    (a) Aphids   (b) Termites
    (c) Cutworms   (d) Stem-Borer

(555) Which is the most widely distributed and commercially reared honeybee species in the world ―――――?
    (a) *Apis florea*   (b) *Apis dorsata*
    (c) *Apis cerana*   (d) *Apis mellifera*

(556) Of the total area under chillies in India, the share of ――――― state is about 27 to 28 percent.
    (a) Andhra Pradesh   (b) Maharashtra
    (c) Karnataka   (d) Tamil Nadu

(557) ———— one of the state has the largest area under kharif maize.
　　(a) Uttar Pradesh　　(b) Rajasthan
　　(c) Madhya Pradesh　　(d) Bihar

(558) Presently, which of the following chemical fertilizers is not produced in India and is being imported ————.
　　(a) Triple Superphosphate(TSP)　　(b) Potassium Sulphate
　　(c) Ammonium Sulphate　　(d) All of the above

(559) ———— is the most widely used green manuring crop in Assam, West Bengal, Bihar, Haryana, Punjab, Tamil Nadu and Uttar Pradesh (This crop does well on alkaline and waterlogged soils.
　　(a) Sunnhemp　　(b) Dhaincha
　　(c) Egyptian Clover　　(d) Senji (*Melilotus parviflor(a)*)

(560) ———— of the methods of surface irrigation considered as water and fertilizer efficient and ideal for several crops ————.
　　(a) Check basin and border strip irrigation　　(b) Furrow irrigation
　　(c) Surge irrigation　　(d) Flood irrigation

(561) Dhaman, Karad and Anjan are the ————.
　　(a) Varieties of Sheep in Rajasthan　　(b) Varieties of caster seed of Gujarat
　　(c) Varieties of Grass in Rajasthan　　(d) Three heros of Gawari dance

(562) Which of the following award associated with Agriculture ————?
　　(a) Bourlog Award　　(b) Shanti Swaroop Bhatnagar
　　(c) Arjun Award　　(d) Vyasa Samman

(563) Which material should be higher in the ratio of the pregnant cow ————?
　　(a) Fat　　(b) Protein
　　(c) Carbohydrates　　(d) Dry matter

(564) As standard how much is the weight in gram of an egg ————.
　　(a) 70 gm　　(b) 58 gm
　　(c) 68 gm　　(d) 48 gm

(565) In how many years does a bull get ready for insemination ————?
　　(a) 2- years　　(b) 3-3 years
　　(c) 1 - years　　(d) None of the above

(566) ———— percentage of income comes from animals, out of India's total income.
　　(a) (b)0%　　(b) 25%
　　(c) 1 5%　　(d) 1 0%

(567) ———— is not a hybrid variety of Mango.
　　(a) Mallika　　(b) Ratna
　　(c) Arka Puneet　　(d) Dashehari

(568)———— is place of India in the World in Tea production.
    (a) First   (b) Second
    (c) Third   (d) Fourth

(569) Which fruit crop has the largest area in India————?
    (a) Banana   (b) Citrus
    (c) Mango   (d) Guava

(570) ———— country is the second highest producer of fruit and vegetables in the world
    (a) Brazil   (b) North America
    (c) China   (d) India

(571) ———— crop is affected by Tobacco caterpillar.
    (a) Jute   (b) Sugarcane
    (c) Lucerne   (d) Sugarbeet

(572)———— optimum temperature is needed for the growth of Potato tuber.
    (a) 17°C to 19°C   (b) 14°C to 16°C3
    (c) 10°C to 12°C   (d) 22°C to 25°C

(573) ———— variety of Groundnut is grown in summer.
    (a) Avtar   (b) Chitra
    (c) Chandra   (d) Amber

(574)———— is not fodder variety of Cowpea.
    (a) Russian Giant   (b) K-585
    (c) Sirsa-10   (d) Azad-1

(575)———— variety of Barley is not for Malt purpose.
    (a) Pragati (K-508)   (b) Geetanjali (K-1 1 49)
    (c) Ritambhara (K-557)   (d) All of the above

(576)———— is not true in case of lentil.
    (a) Chromosome-1 4   (b) I
    (c) Seed rate-35-40 kg   (d) Quantity of proten in lentil seed-1 8%

(577)———— is not a variety of Vegetable Pea.
    (a) Arkel   (b) Azad Pea-3
    (c) Pant Matar-6   (d) Aparna

(578) ———— days old paddy seedling is used for SRI technology.
    (a) 8-12 days   (b) 12-15 days
    (c) 25-30 days   (d) 30-35 days

(579) —————— nursery area is required for transplanting one hectare of paddy crop.
    (a) 1 000 m   (b) 750 m
    (c) 500 m   (d) 250-400 m

(580) —————— variety is not susceptible to insects & disease.
    (a) Pant Dhan-10   (b) Govind
    (c) Narenda Dhan-2   (d) Saket-4

(581) —————— variety of paddy is not suitable for upland situation.
    (a) Narendra Dhan-I   (b) Renu
    (c) Saraju-52   (d) Narendra Dhan-118

(582) —————— is the Headquarters (HQ) of FAO.
    (a) Rome   (b) Washington
    (c) New York   (d) Venezuela

(583) —————— is the highest digestible protein fodder.
    (a) Lucerne   (b) Cowpea
    (c) Clusterbean   (d) Berseem

(584) —————— biomass in quintal per hectare is obtained from green manure of the sunnhemp and dhaincha crops.
    (a) 300-325 quintals   (b) 225-275 quntals
    (c) 200-225 quintals   (d) 325-350 quintals

(585) —————— is not an organic matter in the following.
    (a) Crude protein   (b) Carbohydrate
    (c) Vitamins   (d) Hormones

(586) —————— days is the sunnhemp fibre rotten in normal conditions.
    (a) 10-12 days   (b) 8-1 0 days
    (c) 6-8 days   (d) 5-6 days

(587) —————— herbicide resides for the longest period in the soil.
    (a) Chlorpropham   (b) Linuron
    (c) Propachlor   (d) Simazine

(588) —————— is not a method of knowing seed life.
    (a) Respiration test   (b) Embryo culture method
    (c) Seed dormancy method   (d) Indigocarmine method

(589) —————— is not a variety of seeds.
    (a) Hybrid seed   (b) Composite seed
    (c) Breeder seed   (d) Mutant seed

(590) Opening and closing of stomata depends on——————.
    (a) Sunlight   (b) Water pressure
    (c) Transpiration   (d) Temperature

(591) At ———— temperature does the wilt disease virus grow faster.
   (a) 40°C - 45°C   (b) 35°C - 40°C
   (c) 24°C - 28°C   (d) 22°C - 26°C

(592) Evaporation is measured by ———— instrument.
   (a) Barometer   (b)) Psychrometer
   (c) Lysimeter   (d) Hygrometer

(593) At which height from soil level are the synoptic meteorological observations done ————.
   (a) (b)   (b) 1
   (c) (b)   (d) 1

(594) Success of rural projects depends upon ————.
   (a) Agriculture Extension   (b) Management
   (c) Soil Science   (d) Regular training of staff

(595) 'Pheromone Trap' attracts ————.
   (a) Male moths   (b) Female moths
   (c) Caterpillar   (d) Female bugs

(596) ———— is the average temperature of cow and buffalo.
   (a) -98   (b) 100°F
   (c) 1 01   (d) 102°F

(597) ———— days can the Aonla be kept in salt solution.
   (a) 15 days   (b) 30 days
   (c) 60 days   (d) 75 days

(598) At the time of apple fruit setting what is the minimum temperature required ——.
   (a) 10°C   (b) 4
   (c) 8°C - 10°C   (d) 2°C - 3°C

(599) ———— variety of papaya gives maximum papain.
   (a) Pusa Delicious   (b) Pusa Majesty
   (c) Pusa Gaint   (d) Pusa Dwarf

(600) ———— cloves of garlic is required for one hectare.
   (a) 400 kg   (b) 500 kg
   (c) 600 kg   (d) 300 kg

(601) ———— state of India is the area of Coffee maximum.
   (a) Kerala   (b) Andhra Pradesh
   (c) Karnataka   (d) Tamil Nadu

(602) Potato tuber growth is stopped at what temperature ————.
   (a) 40°C - 4(b)°C   (b) 30°C - 3(b)°C
   (c) 35°C - 37°C   (d) 38° - 40°C

(603) ———— is not prepared from potato.
- (a) Farina and Alcohol
- (b) Paper
- (c) Wine
- (d) Acetic acid

(604) Varieties of Miscavi and Pusa Visal are of which crop ————.
- (a) Cotton
- (b) Jute
- (c) Oat
- (d) Berseem

(605) Seed treatment of which crop is done by Captan or Cerasan @ 5 gm/kg seed —.
- (a) Cotton
- (b) Jute
- (c) Safflower
- (d) Mondua

(606) In which season is the highest yield of Maize obtained ————.
- (a) Zaid
- (b) Kharif
- (c) Rabi
- (d) All of the above

(607) Soldier insect harms which crop more ————.
- (a) Urad
- (b) Arhar
- (c) Moong
- (d) Paddy

(608) How much paddy seed is needed for transplanting one hectare area by SRI technique ————?
- (a) 30-35 kg
- (b) 1-1
- (c) 5-6 kg
- (d) 10-12 kg

(609) From where is Indian Journal of Agriculture Sciences published ————.
- (a) UPCAR
- (b) ICAR
- (c) CISR
- (d) NBRI

(610) Which is the hybrid of Pusa Giant Napier ————?
- (a) Napier x Jowar
- (b) Napier x Bajra
- (c) Jowar x Bajra
- (d) Bajra x Jowar

(611) What is Azofication ————?
- (a) It is also known as composite fixation of nitrogen
- (b) It is also known as free fixation of nitrogen
- (c) It is also known as free fixation of nitrogen fixed by Rhizobium bacteria
- (d) It is also known as nitrogen gain through rains or snow

(612) Which pH range is not suitable to grow the crop ————?
- (a) Oats - 5
- (b) Cotton - 6
- (c) Bean - 6
- (d) Potato - 4

(613) How much minimum temperature is essential for wheat germination and optimum temperature for crop growth in degree Celsius ————?
- (a) 3
- (b) 5
- (c) 8°C-10°C and 30°C-35°C
- (d) 10°C-12°C and 25°C-30°C

(614) The ideal temperature for most of the cultivable crops is ———.
    (a) 30°C - 50°C   (b) 15°C - 40°C
    (c) 35°C - 40°C   (d) 45°C - 55°C

(615) Growth of early emerging leaves becomes faster because of-
    (a) Cell division and cell enlargement   (b) Stunt growth of plants
    (c) Faster photosynthesis   (d) Low pressure of outer atmosphere

(616) How much percentage dry weight of crops should be at physiological maturity ———?
    (a) (b)8%   (b) 20%
    (c) (b)5%   (d) 30%

(617) To reduce the crop-weed competition, at what stage are the herbicides used —.
    (a) At germination   (b) At tillering
    (c) At ear emergence   (d) After first irrigation

(618) How does the supply of nitortgen in grains take place ———?
    (a) From old leaves   (b) From plant roots
    (c) From stem   (d) By photosynthesis

(619) Which of the following points refers to the economic level of output ———?
    (a) MR < MC   (b) MR > MC
    (c) MR = AC   (d) MR = MC

(620) How much is the fat percentage in the milk of Bhadavari buffalo ———?
    (a) 15-18%   (b) 8-13%
    (c) 13-16%   (d) 6-10%

(621) A hen starts egg laying after how many weeks ———.
    (a) 15 weeks   (b) 20 weeks
    (c) 18 weeks   (d) 16 weeks

(622) Which is the Goat breed reared in the city ———?
    (a) Beetle   (b) Surati
    (c) Marwadi   (d) Barbari

(623) Makhdum is a variety for which crop ———.
    (a) Badam (Almond)   (b) Date palm
    (c) Walnut   (d) Coconut

(624) Generally during summer and kharif seasons annual plants produce bloom in how many days ———.
    (a) 60-70 days   (b) 70-80 dyas
    (c) 70-75 days   (d) 50-60 dyas

(625) Inarching method is used for culture of which fruits ———.
    (a) Mango and Guava   (b) Mango and Grapes
    (c) Guava and Litchi   (d) Phalsa and Guava

(626) ——— tomatoes are required for one kg tomato seeds.
    (a) 50-300 kg   (b) 300-350 kg
    (c) 1 60-210 kg   (d) 200-250 kg

(627) ——— is the average annual production of European variety of honeybee colony.
    (a) 14 kg   (b) 16 kg
    (c) 10 kg   (d) 8 kg

(628) ——— seed per hectare is needed for Berseem.
    (a) 1 8-(b)(b) kg   (b) 25-30 kg
    (c) 30-35 kg   (d) 15-20 kg

(629) ——— is a variety of Oat.
    (a) K-1 (b)   (b) Naveen
    (c) LD-491   (d) Kent

(630) ——— kg of sugarcane seed is required for sowing one hectare.
    (a) 7000-7500 kg   (b) 5500-6000 kg
    (c) 8000-8500 kg   (d) 5200-5500 kg

(631) Which of the following is not matche
    (a) Varuna -Irrigated or unirrigated condition
    (b)) Vardan -Unirrigated condition
    (c) Narendra-85 -Saline soil
    (d) K-88 -Black grain

(632) ——— disease is not related to Bajra
    (a) Green Ear   (b) Ergot
    (c) Wilt   (d) Rust

(633) ——— is the suitable Rabi maize variety.
    (a) Sharadmani   (b) Azad Uttam
    (c) Naveen   (d) Ganga -5

(634) ——— is not the symptom of Khaira disease.
    (a) Dark grey colour spots on leaves   (b) Adverse effect on root growth
    (c) More diseases stop the crop growth   (d) Stem of plants turn and fall down

(635) In case of availability of two irrigations for wheat, at which critical stage crop should be irrigate ———.
    (a) Tillering and Flowering   (b) Crown Root Initiation and Milking stage
    (c) Crown Root Initiation and Flowering   (d) Late joint and Dough stage

(636) From where is Krishak Bharati Magazine published ─────.
    (a) ICAR-IIVR, Varanasi   (b) IARI, New Delhi
    (c) Narendra Dev University of Agriculture and Technology, Faizabad
    (d) Sardar Vallabh Bhai Patel University of Agriculture and Technology, Meerut

(637) More than how much percentage of loss of urea should not be during silage fermentation ─────.
    (a) 20% - 22)%   (b) 15% - 18%
    (c) 10% - 12%   (d) 8% - 10%

(638) ───── is the perennial variety of Elephant Grass (Napier Grass).
    (a) Pusa Giant   (b) NB-21
    (c) Pusa Giant Napier   (d) Napier-1

(639) Due to which reason is the Napier Grass mixed with Cluster bean or Cowpea for feeding ─────.
    (a) Due to high crude fibre   (b) Due to oxalic acid
    (c) Due to more carbohydrates   (d) Due to more HCN content

(640) Approximately ───── HCN is harmful for animal feeding.
    (a) 0   (b) 0
    (c) 2   (d) 3

(641) ───── stage of crop, there is no competition for light, moisture and nutrients.
    (a) Node formation stage   (b) Seedling stage
    (c) Grain formation stage   (d) Before maturity stage

(642) ───── stages are more nutrients essentially required in cereal crops.
    (a) Growth stage of plants   (b) Formation of leaves
    (c) Panicle initiation   (d) Maturity of crops

(643) ───── is the order of C_4 plants.
    (a) Sugarcane - Maize - Sudan grass - Bajra
    (b)) Sugarcane - Cotton - Paddy - Maize
    (c) Sudan grass - Sugarcane - Paddy - Bajra
    (d) Cotton - Maize - Bajra - Sugarcane

(644) A
    (a) Photosynthesis   (b) Phosphorylation
    (c) Transpiration   (d) Oxidation

(645) The monsoon airs in the country reaches at which bank first of all─────.
    (a) Coastal region of Orissa   (b) Bay of Bengal region
    (c) Coastal region of Kerala of South Indian region   (d) Kachchh region of Gujarat

(646) How much percentage of soluble salts are present in Alkali soils——————?
   (a) 1%-2% (b) 0
   (c) 1 (d) 2%-3%

(647) Approximately how much area in lakh hectares of saline and alkali soils are there in Uttar Pradesh ——————.
   (a) 1 5 (b) 12
   (c) 1 0 (d) 1 4

(648)—————— is not natural factor affecting soil fertility.
   (a) Topography (b) Soil age
   (c) Air (d) Parent material

(649) The leaching loss of nitrogen (N) is more in the form of——————.
   (a) Ammonia (b) Nitrate
   (c) Nitrogen (d) Water solution

(650) The major fungi that effect food-grains in storage are —————— .
   (a) Mucor (b) Rhizopus
   (c) Candida (d) Aspergillus

(651) Goal of extension education is —————— .
   (a) To promote income of the farmers (b) To promote production of the crops
   (c) To promote new crops (d) To promote scientific outlook

(652) How much area m^2 is required for a Goat ——————?
   (a) 0 (b) 1-1
   (c) 0 (d) 1

(653) For how many years is Sheep able to breed ——————.
   (a) 1 0 years (b) 7 years
   (c) 5 years (d) 12 years

(654) —————— is the cross of Karan Swiss.
   (a) Sahiwal x Brown Swiss (b) Sahiwal x Holstein
   (c) Hariyana x Brown Swiss (d) Hariyana x Jersey

(655) —————— species of honeybee is not Indian.
   (a) *Apis florea* (b) *Apis dorsata*
   (c) *Apis dorsata* (d) *Apis mellifera*

(656) —————— Brinjal seeds may be obtained from one hectare area.
   (a) 150-200kg (b) 100-1 50kg
   (c) 200-300kg (d) 75-1 25kg

(657) ———— elements are useful in energy storage, transfer and bonding.
  (a) NPK
  (b) NSP
  (c) NKS
  (d) None of the above

(658) ———— states are the highest & lowest producers of potato crop.
  (a) Uttar Pradesh and Jharkhand
  (b) Gujarat and Assam
  (c) Uttar Pradesh & Himachal Pradesh
  (d) Bihar & Kashmir

(659) ———— is not correct in the following.
  (a) X-rays and Gamma rays - 9% of energy
  (b) Visible lighting rays - 41 % of energy
  (c) Infrared rays - 50% of energy
  (d) Ultraviolet rays - 1 0% of energy

(660) Effective cause of atmospheric pressure is not.
  (a) temperature
  (b) altitude from the sea level
  (c) rotation of earth
  (d) soil erosion

(661) Where and when was World Meteorological Organization established ————?
  (a) New York - 1 980
  (b)) Washington - 1 978
  (c) Geneva - 1 978
  (d) Rome - 1 976

(662) ———— is not basic principle of Agronomy.
  (a) To select appropriate materials for seed & sowing
  (b) Management of soil and climate
  (c) Appropriate intercropping activities management for corp
  (d) Livestock management for Agriculture

(663) ———— is not matched in relation to sugarcane.
  (a) Seed treatment—Aglol 3%
  (b) For seed—6-7 month old crop
  (c) Sowing—Upper portion is more used
  (d) After flowering—Used for sowing

(664) ———— does not match in relation to cotton.
  (a) C-520 - Diploid
  (b) G-27 - Diploid
  (c) Vikas - American
  (d) Ranivan - American

(665) ———— is not true in relation to Gram.
  (a) Subfamily -Papilionaceae
  (b) Chromosome NO
  (c) Acid-Malic and Oxalic
  (d) Origin-North America

(666) ———— variety of Barley is huskless.
  (a) Pragati (K 508)
  (b) Ritambhara (K 551)
  (c) Geetanjali (K 1 1 49)
  (d) Karan -3

(667) Which disease occurs, when more sorghum is consumed ─────── ?
    (a) Rickets   (b) Scurvy
    (c) Nightblindness   (d) Pellagra

(668) Which is not true in relation to Bajra ─────── .
    (a) *Pennisetum Typhoides*   (b) Graminae Family
    (c) Chromosome No - 20   (d) Origin Africa

(669) ─────── Sorghum variety is not multicut.
    (a) Pusa Chari-2   (b) M
    (c) M   (d) U

(670) Which is the highest digestible protein non-leguminous crop among the following ───────?
    (a) Napier   (b) Maize Silage
    (c) Maize   (d) Jowar

(671) Which is not an inorganic matter in the following ───────?
    (a) Magnesium   (b) Iron
    (c) Fat   (d) Iodine

(672) How much radiation energy percentage radiating on plant is used is photosynthesis ─────── ?
    (a) 10%   (b) 66%
    (c) 30%   (d) 24%

(673) At the vegetative growth stage, flowering is stopped in food-grain crops, known as ───────
    (a) sigmoid growth curve   (b) determinate growth
    (c) indeterminate growth   (d) grand growth period

(674) How does the moisture stress affect the cell ───────?
    (a) Affect cell division   (b) Affect cell expansion
    (c) Cell mortality rate is affected   (d) No effect on cell

(675) How many factors are identified for influencing plant growth till now ───────?
    (a) 55   (b) 60
    (c) 52   (d) 50

(676) Which type of crops are to be examined by the Union Government ─────── ?
    (a) Horticulture   (b) Plant propagation
    (c) Vertical Farming   (d) Genetically Modified

(677) Agriculture and Farmers Welfare Ministry will celebrate Sankalp to Siddhi - New India Manthan from ─────── .
    (a) Aug 15th to 20th   (b) Aug 1 9th to 30th
    (c) Aug 20th to 30th   (d) Aug 22nd to 29th

(678) Which online facility has been launched by the Union Government to provide direct solution to problem of agriculture sector ———— .
    (a) e-Krishi Samasya   (b) e-Krishi Samvad
    (c) e-Krishi Samveda   (d) e-Krishi Solution

(679) Veteran Agriculture Scientist M S Swaminathan received an honorary medallion from this country for his contribution towards the development of agricultural practices ———— .
    (a) Australia   (b) Canada
    (c) New Zealand   (d) Mexico

(680) Soil factors are otherwise known as
    (a) Biotic factors   (b) Edaphic factors
    (c) Climatic factors   (d) Physiographic factors

(681) *Triticum aestivum*, the common bread wheat is ———— .
    (a) Diploid   (b) Haploid
    (c) Hexaploid   (d) Tetraploid

(682) Winnowing it called ———— .
    (a) To thresh   (b) Cutting a crop
    (c) Cutting the fodder   (d) To separate straw et(c) from threshed lank

(683) Chaff-cutter is driven by ———— .
    (a) Hand   (b) Bullocks
    (c) Electric power   (d) All of the above

(684) 'Olpad' thresher is used in ———— .
    (a) Threshing of wheat, barley,   (b) Oil extraction from mustard, toria pea et(c) et(c)
    (c) Extracting juice from cane   (d) None of the above

(685) 'Seed dresser' is used for ———— .
    (a) Making seeds of high grade   (b) Sowing seeds at proper distance
    (c) Mixing/treating seeds with chemicals   (d) Keeping seeds effective upto longer period

(686) A general farmer used deshi plough for the purpose of ———— .
    (a) Land ploughing   (b) Collecting weeds
    (c) Making soil powdery   (d) Above all works

(687) Which of the following is best for driving machine from low power to slow speed ————?
    (a) Spur gear   (b) Belts and pulley
    (c) Toothed wheel and chains   (d) None of the above

(688) Reapers are used for ———.
- (a) Crop cutting
- (b) Seeds sowing
- (c) Fodder cutting
- (d) Threshing of harvested crop (lank) produce

(689) Function of the seed-drill is ———.
- (a) Making furrow
- (b) Dropping seeds
- (c) Covering the seeds in furrow
- (d) All of the above

(690) The mould-board of a tractor drawn soil turning plough is the type of———.
- (a) Stubble
- (b) Sod (breaker)
- (c) High speed
- (d) General purpose

(691) Among the following, ridger is not used in crop———.
- (a) Gram
- (b) Maize
- (c) Potato
- (d) Sugarcane

(692) Which of the following does not affect the draft of ploughs ———.
- (a) Soil moisture
- (b) Depth of furrow
- (c) Width of furrow
- (d) Length of furrow

(693) Swing-basket (Dhenkuli) is used for———.
- (a) Making furrow
- (b) Destroying weeds
- (c) Leveling of land
- (d) Lifting water from wells

(694) Which of the method of ploughing is mostly practised———.
- (a) Inside to outside ploughing
- (b) Outside to inside ploughing
- (c) Ploughing by halai making
- (d) Ploughing by putting furrow from one side of field

(695) The best method of ploughing through deshi plough is———?
- (a) Inside to outside ploughing
- (b) Outside to inside ploughing
- (c) Ploughing by halai making
- (d) Ploughing by putting furrow from one side of field

(696) Harrow is drawn by ———.
- (a) Tractor
- (b) Bullocks
- (c) Bullocks and Tractor both
- (d) Diesel

(697) Which of the following is not a secondary tillage implement ———?
- (a) Hoe
- (b) Harrow
- (c) Cultivators
- (d) Meston plough

(698) Which of the following 'hoe' is bullock drawn ———?
- (a) Akola hoe
- (b) Sharma hoe
- (c) Wheel hoe
- (d) Naini type hoe

(699) The land levelling implement is————.
- (a) Roller
- (b) Patela
- (c) Scrapper
- (d) All of the above

(700) Soil turning plough makes the furrow of which type (shape)————.
- (a) 'V' shape
- (b) 'L' shape
- (c) 'O' shape
- (d) No definite shape

(701) Which of the following is one (single) handed soil turning plough ————?
- (a) Praja plough
- (b) Punjab plough
- (c) Victory plough
- (d) U. P. No.1 plough

(702) The purpose of tillage is/are————.
- (a) Leveling of soil
- (b) Eradication of weeds
- (c) Soil clods breaking and suppressing in soil
- (d) All of the above

(703) The main function of cultivator is————.
- (a) To turn the soil
- (b) To make furrow in soil
- (c) To pulverize the soil
- (d) All above three functions

(704) Which one of the following organisations is responsible for publishing topographical sheets ————?
- (a) Survey of India (S.O.I.)
- (b) Geological Survey of India (G.S.I.)
- (c) Indian Meteorological Department (I.M.(D))
- (d) National Atlas & Thematic Mapping Organisation (N.(A)T.M.O.)

(705) The package technology which brought about green revolution comprised mainly of————.
- (a) Man power, mechanical cultivators and electricity
- (b) Electricity, irrigation and introduction of dry farming
- (c) Changes in crop pattern, industrialisation and chemical fertilizers
- (d) Irrigation, bio-chemical fertilizers and high yield varieties of seeds

(706) Which one of the following is a global biodiversity hotspot in India————?
- (a) Eastern Ghats
- (b) Western Ghats
- (c) Northern Himalayas
- (d) Western Himalayas

(707) The nature of the winter rainfall in north western India is ————.
- (a) Cyclonic
- (b) Orographic
- (c) Monsoonal
- (d) Convectional

(708) Which food crop in India is sown in October-November and reaped in April —?
- (a) Rice
- (b) Wheat
- (c) Coffee
- (d) Coconut

(709) The variety of coffee, largely grown in India, is ———.
- (a) Kents
- (b) Coorgs
- (c) Arabica
- (d) Old chicks

(710) Growth of early emerging leaves becomes faster because of———.
- (a) Faster photosynthesis
- (b) Stunt growth of plants
- (c) Cell division and cell enlargement
- (d) Low pressure of outer atmosphere

(711) To reduce the crop-weed competition, at what stage are the herbicides used——.
- (a) At tillering
- (b) At germination
- (c) At ear emergence
- (d) After first irrigation

(712) How does the supply of nitrogen in grains take place———?
- (a) From stem
- (b) From old leaves
- (c) From plant roots
- (d) By photosynthesis

(713) Makhdum is a variety for which crop ———.
- (a) Walnut
- (b) Coconut
- (c) Date palm
- (d) Badam (Almon(d)

(714) What is Azofication———?
- (a) It is also known as free fixation of nitrogen
- (b) It is also known as composite fixation of nitrogen
- (c) It is also known as nitrogen gain through rains or snow
- (d) It is also known as free fixation of nitrogen fixed by Rhizobium bacteria

(715) Which is not a hybrid variety of Mango———.
- (a) Ratna
- (b) Mallika
- (c) Dashehari
- (d) Arka Puneet

(716) Where is the Headquarters (HQ.) of FAO———?
- (a) Rome
- (b) New York
- (c) Venezuela
- (d) Washington

(717) Importance of fruits and vegetables in human diet is primarily because they are ———.
- (a) Good source of carbohydrates
- (b) Good source of proteins
- (c) Good source of fats
- (d) Good source of vitamins and minerals

(718) Which is the highest digestible protein fodder in the following ———?
- (a) Cowpea
- (b) Lucerne
- (c) Berseem
- (d) Cluster bean

(719) This is one of the best 'poultry feed stuff' and having good source of animal protein as it contains 34-55 percent protein———.
   (a) Fishmeal   (b) Linseed-meal
   (c) Soyabean-meal   (d) Groundnut Cake

(720) Which one of the following is the structure concerned with photorespiration ———?
   (a) Mitochondria   (b) Peroxisome
   (c) Ribosome   (d) Lysosome

(721) Which one of the following is not narrow leavd weeds ———?
   (a) *Cynodon dactylon*   (b) *Cyprus rotundus*
   (c) *Setaria glouca*   (d) *Melilotus indica*

(722) Range' is the measurement between———.
   (a) Highest and lowest   (b) Medium and owest
   (c) Average and highest   (d) None of above

(723) Which of the following is not correctly matched ———?
   (a) Unorganized market organization   (b) Regulated market - Control
   (c) Perfect market - Competition   (d) Terminal market – Area

(724) Fluid mosaic model relates to the structure of———.
   (a) Cell Wall   (b) Protoplasm
   (c) Biomembrane   (d) Nucleic acid

(725) "Hollow-heart" disease of sugar beet is caused due to ———.
   (a) Salt toxicity   (b) Boron deficiency
   (c) Moisture stress   (d) Air pollutant

(726) $2n + 1$ state is refered to as ———.
   (a) Monosomy   (b) Trisomy
   (c) Tetrasomy   (d) Nullisomy

(727) Sugars present in DNA and RNA respectively are ———.
   (a) Glucose and Fructose   (b) Deoxyribose and Ribose
   (c) Galactose and Raffinose   (d) Erythrose and Starchyos

(728) A character determined by a gene present on X-chromosome is called ———.
   (a) Sex linked character   (b) Sex-limited character
   (c) Sex-influenced character   (d) Hollandric character

(729) A chromosome, which differs either in number or in morphology between male and female of a species, is known as ———.
   (a) Autosome   (b) Sex chromosome
   (c) Accessory chromosome   (d) Homologous chromosome

(730) A combination of horticultural crops, field crops and tree species is called ———.
- (a) Agro-forestry
- (b) Silvi-pastoral system
- (c) Multipurpose forest tree plantation system
- (d) None of the above

(731) A combine cultivation of trees and grasses is referred to as ———.
- (a) Aqua silviculture
- (b) Agroforestry
- (c) Silvopastoral
- (d) Silviculture

(732) A conference at which the views of several experts are given, one after another, it seldom provides time for general discussion it refers to ———.
- (a) Symposium
- (b) Panel Discussion
- (c) Seminar
- (d) Forums

(733) A copping system where arable crops are grown in the inter spaces between rows of planted trees ———.
- (a) Relay cropping
- (b) Mixed cropping
- (c) Inter cropping
- (d) Alley cropping

(734) A demonstration, which shows the value or worth of the new practice ———.
- (a) Result Demonstration
- (b) Method Demonstration
- (c) Whole Plot Demonstration
- (d) National Demonstration

(735) A form of low pruning to about 2 m up the stem ———.
- (a) Zero prunning
- (b) Brashing
- (c) Pollarding
- (d) None of the above

(736) A photosynthetic unit consists———.
- (a) 1000 light harvesting chlorophyll molecules
- (b) 800 light harvesting chlorophyll molecules and trapping centre
- (c) 200 light harvesting chlorophyll molecules
- (d) About 400 light harvesting chlorophyll molecules

(737) A physical or chamical agent that frees plant, organ or tissue from infection is a ———.
- (a) Antiseptic
- (b) Deodorant
- (c) Disinfectant
- (d) Germicide

(738) A process of initiating a conscious and purposeful action is called———.
- (a) Motivation
- (b) Education
- (c) Coordination
- (d) Action

(739) A protein produced by living cells of microorganism, plants and annimals that can catalyze specific organic reaction is known as———.
- (a) Hormone
- (b) Enzyme
- (c) Catalyst
- (d) None of the above

(740) A recessive gene for resistance to disease can be transferred from one parent to another using ———.
    (a) Pureline selection   (b) Backcross breeding
    (c) Pedigree method   (d) Bulk method

(741) A short duration crop in between two main seasonal crops is termed as ———.
    (a) Cash crop   (b) Inter crop
    (c) Companion crop   (d) Catch crop

(742) A spherical particle embedded in the phospholipo protein of thylakoid lamellae is known as ———.
    (a) Granule   (b) Quantosome
    (c) Both (a) and (b)   (d) Bone of these

(743) A symbiotic association of fungus with subterranean part of a plant is said to be ———.
    (a) Ectophyte   (b) Endophyte
    (c) Mycorrhiza   (d) Parasite

(744) A useful pruning practice to produce large quantities of foliage close to ground level for fodder species or for leaf extract product is called ———.
    (a) Coppicing   (b) Pollarding
    (c) Lopping   (d) Thinning

(745) A variety produced by crossing in all combination a number of lines that combine well with each other is ———.
    (a) Composite   (b) Hybrid
    (c) Synthetic   (d) Elite variety

(746) A variety produced by mixing seeds of several phenotypically outstanding lines and allowing all combinations is ———.
    (a) Synthetic   (b) Composite
    (c) Hybrid   (d) Germplasm complexes

(747) Abscisic aid is a ———.
    (a) Gibberellins   (b) Auxin
    (c) Retardant   (d) Inhibitor

(748) Which of the following is non-selective herbicide ———?
    (a) Alachlor   (b) Butachlor
    (c) Paraquat   (d) Atrazine

(749) Aeciospores of *Puccinia recondita* infect to ———.
    (a) Barley crop   (b) Maize crop
    (c) Rice crop   (d) Wheat crop

(750) Agricultural extension worker is ─────.
    (a) A messenger boy
    (b) A passive person
    (c) An educator
    (d) Supplier of inputs for farmers

(751) Aim of seed technology is ─────.
    (a) Rapid multiplication of seeds
    (b) Assured high quality seeds
    (c) Timely supply of improved seeds
    (d) All of the above

(752) All fruits are in general _____ in nature.
    (a) Acidic
    (b) Basic
    (c) Neutral
    (d) All of the above

(753) The bacteroid Rhizobium is─────.
    (a) Macroaerobic
    (b) Anaerobic
    (c) Microaerobic
    (d) None of the above

(754) Alternaria blight disease can be effectively controlled by─────.
    (a) BHC
    (b) Aldrin
    (c) Treating seeds with hot water
    (d) All of the above

(755) Ammonium sulphate contain approximately─────.
    (a) 10% nitrogen
    (b) 20% nitrogen
    (c) 28% nitrogen
    (d) 36% nitrogen

(756) Which one of the following is evergreen temperate plant─────?
    (a) Apricot
    (b) Chilgoza
    (c) Plum
    (d) Almond

(757) An acidic buffer solution can be prepared by mixing solution of─────.
    (a) Ammonium acetate and acetic acid
    (b) Ammonium chloride and ammonium hydroxide
    (c) Sulphuric acid and sodium sulphate
    (d) Sodium chloride and sodium hydroxide

(758) An agent that causes gene mutations is known as─────.
    (a) Mutant
    (b) Mutation
    (c) Mutagen
    (d) Mutator

(759) An area in which seeding are raised to be transplanted elsewhere is called─────.
    (a) Brashing
    (b) Shifting cultivation
    (c) Crop rotation
    (d) Nursery

(760) SMART Horticulture includes ─────.
    (a) Precision farming
    (b) Remote pest monitoring
    (c) Robot planting and harvesting
    (d) Crop modeling
    (e) All of the above

(761) *Ty*-1 and *Ty*-2 genes are present in ———.
- (a) Tomato
- (b) Potato
- (c) Brinjal
- (d) Okra

(762) Tomato yellow leaf curl virus belongs to ———.
- (a) Geminivirus
- (b) Begamovirus
- (c) Poxvirus
- (d) Parvoviruses

(763) Natural enemies of polyhouse pest are ———.
- (a) Amblyseicus swirsukii
- (b) Ambyseius cucumeis
- (c) Neoseiulus largispinosus
- (d) Phytoseiulus persimiles
- e. All of these

(764) RASFF stand for ———.
- (a) Rapid Alert System for Food and Feed
- (b) Robust Agricultural System for Food and Feed
- (c) Relay Alert System for Food and Feed
- (d) Rapid Agricultural System for Food and Feed

(765) Protection of Plant Varieties and Farmers Right Act (PPV&FR(A) came into existence in———.
- (a) 1998
- (b) 2000
- (c) 2001
- (d) 2005

(766) The objectives of PPV&FRA are———.
- (a) protection of plant varieties
- (b) Right of farmers
- (c) Right to plant breeders
- (d) All of the above

(767) Genetic Use Restriction Technologies (GURTs) modifies———.
- (a) Plant varieties
- (b) Local cultivars
- (c) Landraces
- (d) All of these

(768) Flavr Savr tomato was developed in ———.
- (a) 1995
- (b) 1996
- (c) 1997
- (d) 1998

(769) Precision farming includes———.
- (a) Genetically modified organism
- (b) Integrated nutrient approach
- (c) Micro-propagation
- (d) All of these

(770) Bio-enhancer (CSR-BIO) improves———.
- (a) Soil mineralization
- (b) Availability of nutrients
- (c) Tolerance to abiotic and biotic stress
- (d) All of these

(771) SAARC integrated program of action includes————.
 (a) Agriculture and Rural Development
 (b) Science and Technology improvement
 (c) Human Resources Development
 (d) All of these

(772) For registering a variety it should be————.
 (a) Novelty (b) Distinctness
 (c) Uniformity (d) Stability
 e. All of these

(773) Highest pesticides consuming crop in India is ————.
 (a) Cotton (b) Rice
 (c) Sugarcane (d) Pulses

(774) FARMER-FIRST is an initiative of————.
 (a) ICAR (b) CSIR
 (c) DBT (d) ICMR

(775) FARMER-FIRST aims at————.
 (a) Enriching farmers-scientists interface for technology development
 (b) production and productivity privilege
 (c) Development of strong linkage of NARS vis-à-vis farmers
 (d) Awareness and capacity building of farmers
 e. All of these

(776) Embryo rescue technique is use to————.
 (a) Overcome post fertilization barriers (b) Make wide crosses
 (c) Transfer useful genes (d) All of these

(777) Protoplast culture technique is used for————.
 (a) Introgression of desirable genes (b) Disease resistance
 (c) Induced dormancy (d) None of the above

(778) Vertical farming includes————.
 (a) Hydroponics (b) Aeroponics
 (c) Terrace farming (d) All of the above

(779) Vertical farming has following advantages over traditional field farming————.
 (a) Control of light (b) Control of environment
 (c) Fertigation (d) All of these

4.2. Fill in the Blanks

(780) Chemical that can maintain insect damage percentage below IMSCS is ———.
(781) Seed priming in pigeonpea seeds can be done by ———
(782) ——— coated seeds show least defoliation in seed quality parameters
(783) Sectioning of seeds through ——— methods is being done for easier evaluation of seed quality parameters.
(784) Coating of maize seed with ——— enhanced its germination and maintain seed quality parameters.
(785) Emerging new seed borne disease are ———
(886) Bacterial panicle blight disease in rice is caused by ———
(787) Dormancy in seed due to unfavorable environmental condition is called ———
(788) Inhibition of seed germination or growth under any set of environmental condition is called ———
(789) Induced germination under specific condition at the time of deepest or true dormancy is called ———
(790) Among cultivated crops hard seed ness is found chiefly in ———
(791) Morphological seed dormancy is due to ———
(792) Seed dormancy characterize by both seed coat and a dormant embryo is called ———
(793) Dormancy due to the general impermeability of seed coat can be overcome by breaking of seed coat barrier is called ———
(794) Vigorous shaking of the seed removes the suberized plug and allow the germination of seed is called ———
(795) Method used to break dormancy by treating the seed at low temperature is called ———
(796) Plant growth regulators that promote germination of seeds are ———
(797) Seed germination starts with the ———of water which increases the ——— rate rapidly.
(798) The increase in respiration depends upon the activation of ——— which in case of dormant seed are little active.
(799) The quantity of the storage products such as carbohydrate, lipids, proteins ——— during seed germination.
(800) Phosphorous is stored in the seed in the form of ———
(801) Ability of water to move by capillary through the pores of soil to the seed is called ———
(802) Damping off is broad term use to describe the death of seedling resulting from the attack by certain fungi ———

(803) ———— are responsible for the production of breeder seeds.
(804) Production of foundation and certified seed are governed by ————
(805) ———— is generally associated with a higher level of plant performance
(806) During early phases of seed detoraition a reduction in ———— occurs prior to a reduction in germination.
(807) ———— is crucial, critical, vital and basic input to agriculture.
(808) Sum total of all seed attributes which favour rapid and uniform stand establishment in field is called ————
(809) The first edible GM crops, if approved, in India will be ————
(810) Barnase gene is isolated from ———— is driven by tapetum specific promoters
(811) Barstar gene is a specific ———— derived from ————
(812) ———— inhibition of flavonoid biosynthesis causes male sterility in plants.
(813) ———— can also be induced by manipulation of endogenous hormone level.
(814) ———— is the process of enclosing a seed with small quantity of coating material to stimulate its germination.
(815) ———— is done by disease controlling agent against bacterial and fungus disease.
(816) Nutrient seed coating is done by ———— to ensure nutrient balance in seed
(817) ———— coating is done to protect seed from 2-4,D and Alachor injury.
(818) ———— coating is done by starch polymers to increase the germination rate.
(819) Progenitors of seeds are ————
(820) In angiosperm ovules are born on————
(821) The process of formation of seeds without sexual reproduction is called ————
(822) In Apomixis female gamete formation occur without ………
(823) ———— is the process of transfer of pollen grains from anther to stigm(a)
(824) Pollination signals led to the increase in the concentration of which hormone —
(825) Self-pollination is also known as ————
(826) Cross-pollination is also known as ————
(827) Anther dehiscence proceed stigma receptivity is called ————
(828) Inability of plant to produce gamete after self-pollination is called ————
(829) The process by which seed germinate precousiously before maturation is called ————
(830) Physiologically vivipary occur due to insensitivity to.
(831) Recalcitrant seeds loses ———— after drying.
(832) Orthodox seeds ———— drying.

(833) ———— is process where seeds germinate on the plant.
(834) Developing endosperm derive its nutrient from ———— tissue.
(835) In endospermic seeds the endospermic tissue is nonliving at ———— stage.
(836) Exposing ———— to favorable condition will induce them to germinate.
(837) ———— fails to germinate even under favorable condition.
(838) Parthenocarpy can also be induced by exposure of stigma to ————
(839) The hormone ———— is effective producing seedless grape.
(840) Outgrowth of the hilum is called ————
(841) Somatic seeds are also known as ————
(842) ———— induces the expression of alpha amylase.
(843) ———— suppress the expression of alpha amylase.
(844) ———— is a condition where seed fail to germinate under favorable condition.
(845) ———— is due to inherent factor.
(846) ———— is imposed by stressful factor.
(847) Traditional method of seed purity testing is through ————
(848) First report on synthetic seeds was reported in ————
(849) Synthetic seed is prepared by coating somatic embryos in ————
(850) Which fruit is known as "Poor man fruit" ————?
(851) King of arid fruit is known as ————
(852) Bael is richest source of ————
(853) Avacado is also known as ————
(854) King of temperate fruit is ————
(855) China's miracle fruit is ————
(856) Kiwi fruit is also known as horticultural wonder of ————
(857) ———— is queen of beverage crop.
(858) Food of god is ————
(859) Kalpavariksh is also known as ————
(860) ———— is called small holder irrigated crop.
(861) Tamarind fruit is botanically a ————
(862) Hesperidium is botanical form of fruit is ————
(863) Hen and chicken a disorder is observed in ————
(864) De-greening in citrus is done with the application of ————
(865) Fruit introduced from Brazil to India is ————
(866) Aroma of overripe banana is due to ————
(867) Theobromin in cacao is extracted from ————
(868) Type of incompatibility in mango is ————

(869) Modified stem of banana is ———
(870) Fruit most suited for jelly making is ———
(871) Growth curve of fruit development in apple is ———
(872) Growth regulator act anti to gibberellin is ———
(873) Red color of mango skin is controlled by ———
(874) Seedlessness in grape is controlled by ———
(875) Litchi fruit is botanically ———
(876) Avacado fruit is rich in ———
(877) Mango fruits are good source of ———
(878) Botanically Jackfruit is ———
(879) ——— parthenocarpy is obtained in apple.
(880) Fruit require heavy ———
(881) $KMNO_4$ ———
(882) ——— is due to excess water.
(883) ——— is commonly observed in Date.
(884) Stratification in apple is done at ———
(885) ——— apomixis is observed in apple.
(886) Amphidiploidy is present in ———
(887) Botanically Quine is a ———
(888) ——— is salt tolerant fruit crop.
(889) ——— is most suitable fruit crop for waterlogging.
(890) ——— is suitable fruit crop for alkaline soil.
(891) Protandry is a problem in ———
(892) Best rootstock for citrus in South India ———
(893) Grapefan leaf virus disease is transmitted by ———
(894) Iron deficiency in plant is characterized by ———
(895) Fruit thinning growth regulator is ———
(896) Organic phosphate solubilizer is ———
(897) Anab-e-Shahi is the most popular grape variety of ———
(898) Vikram is an improved cultivar of ———
(899) Softnose is a disorder of ———

Answer Sheet

1	a	2	d	3	a	4	a	5	a	6	b	7	c	8	a	9	d	10	c
11	a	12	c	13	a	14	b	15	a	16	a	17	a	18	a	19	c	20	d
21	a	22	a	23	d	24	d	25	c	26	a	27	b	28	b	29	a	30	b
31	d	32	b	33	d	34	b	35	b	36	b	37	a	38	a	39	c	40	a
41	b	42	d	43	d	44	a	45	d	46	d	47	c	48	a	49	c	50	a
51	a	52	d	53	c	54	d	55	d	56	c	57	b	58	d	59	b	60	a
61	c	62	a	63	a	64	b	65	b	66	a	67	d	68	a	69	b	70	b
71	a	72	c	73	a	74	c	75	b	76	b	77	b	78	a	79	a	80	a
81	b	82	c	83	a	84	b	85	c	86	d	87	d	88	c	89	c	90	a
91	a	92	a	93	d	94	b	95	d	96	d	97	a	98	c	99	a	100	a
101	a	102	a	103	b	104	c	105	c	106	d	107	d	108	d	109	d	110	a
111	d	112	b	113	c	114	d	115	a	116	d	117	d	118	c	119	a	120	c
121	d	122	b	123	b	124	d	125	d	126	a	127	b	128	a	129	b	130	c
131	a	132	c	133	c	134	b	135	a	136	c	137	c	138	d	139	b	140	a
141	b	142	b	143	b	144	b	145	d	146	a	147	b	148	b	149	c	150	a
151	a	152	b	153	c	154	a	155	b	156	d	157	c	158	d	159	d	160	b
161	c	162	b	163	d	164	c	165	d	166	d	167	c	168	a	169	d	170	d
171	c	172	c	173	c	174	a	175	b	176	b	177	d	178	c	179	d	180	c
181	b	182	a	183	a	184	a	185	b	186	c	187	d	188	d	189	b	190	b
191	a	192	d	193	d	194	b	195	d	196	d	197	c	198	b	199	c	200	c
201	b	202	c	203	c	204	a	205	c	206	d	207	d	208	d	209	e	210	e
211	a	212	a	213	c	214	d	215	b	216	b	217	c	218	d	219	a	220	e
221	c	222	b	223	d	224	c	225	d	226	a	227	c	228	c	229	b	230	d
231	a	232	e	233	b	234	d	235	b	236	a	237	a	238	c	239	a	240	d
241	e	242	b	243	c	244	b	245	a	246	a	247	a	248	b	249	e	250	b
251	b	252	d	253	c	254	d	255	d	256	a	257	b	258	b	259	a	260	d
261	c	262	a	263	d	264	c	265	a	266	d	267	c	268	d	269	c	270	d
271	b	272	c	273	a	274	d	275	b	276	a	277	a	278	d	279	b	280	c
281	d	282	d	283	b	284	b	285	d	286	c	287	c	288	d	289	d	290	c
291	b	292	d	293	b	294	b	295	c	296	b	297	a	298	c	299	c	300	d
301	c	302	b	303	c	304	b	305	c	306	b	307	b	308	c	309	a	310	b
311	a	312	d	313	d	314	c	315	c	316	b	317	c	318	a	319	d	320	d
321	a	322	b	323	b	324	b	325	d	326	c	327	b	328	d	329	a	330	d
331	c	332	a	333	d	334	c	335	b	336	d	337	b	338	c	339	c	340	a
341	b	342	d	343	a	344	b	345	c	346	b	347	c	348	d	349	c	350	d
351	d	352	b	353	b	354	a	355	c	356	d	357	a	358	a	359	c	360	a
361	c	362	b	363	d	364	a	365	c	366	c	367	d	368	c	369	a	370	b
371	c	372	a	373	b	374	d	375	d	376	c	377	c	378	c	379	b	380	a
381	d	382	d	383	d	384	b	385	c	386	a	387	c	388	d	389	c	390	d
391	a	392	d	393	c	394	c	395	b	396	c	397	d	398	a	399	b	400	a
401	b	402	c	403	d	404	d	405	c	406	d	407	c	408	c	409	b	410	b
411	a	412	a	413	d	414	b	415	d	416	d	417	d	418	c	419	a	420	c
421	b	422	a	423	b	424	d	425	c	426	c	427	b	428	b	429	b	430	a

431	d	432	d	433	a	434	a	435	d	436	d	437	b	438	a	439	a	440	d
441	b	442	d	443	c	444	a	445	a	446	d	447	b	448	d	449	c	450	d
451	a	452	d	453	d	454	b	455	a	456	d	457	b	458	b	459	b	460	d
461	c	462	b	463	c	464	c	465	a	466	b	467	a	468	a	469	d	470	a
471	c	472	a	473	b	474	a	475	a	476	b	477	a	478	d	479	c	480	a
481	b	482	c	483	c	484	a	485	d	486	d	487	c	488	a	489	c	490	d
491	b	492	a	493	c	494	c	495	a	496	c	497	b	498	c	499	d	500	d
501	c	502	a	503	b	504	c	505	d	506	c	507	d	508	c	509	b	510	d
511	c	512	d	513	b	514	c	515	c	516	d	517	d	518	d	519	b	520	c
521	d	522	c	523	d	524	c	525	c	526	c	527	d	528	a	529	a	530	d
531	c	532	b	533	d	534	a	535	c	536	a	537	a	538	b	539	a	540	a
541	a	542	d	543	a	544	a	545	b	546	d	547	b	548	a	549	b	550	d
551	b	552	c	553	b	554	c	555	d	556	d	557	a	558	b	559	b	560	b
561	c	562	c	563	a	564	b	565	b	566	a	567	c	568	d	569	a	570	c
571	d	572	a	573	a	574	a	575	d	576	b	577	b	578	d	579	a	580	a
581	d	582	a	583	a	584	d	585	a	586	d	587	a	588	a	589	d	590	c
591	c	592	a	593	c	594	b	595	b	596	d	597	a	598	b	599	d	600	c
601	d	602	b	603	a	604	d	605	d	606	b	607	a	608	c	609	d	610	a
611	d	612	b	613	b	614	d	615	a	616	c	617	a	618	c	619	c	620	b
621	d	622	a	623	d	624	c	625	a	626	a	627	b	628	d	629	c	630	a
631	b	632	d	633	a	634	a	635	b	636	c	637	b	638	b	639	c	640	b
641	a	642	d	643	c	644	a	645	a	646	d	647	d	648	c	649	a	651	a
651	d	652	c	653	d	654	d	655	d	656	d	657	d	658	d	659	d	660	d
661	d	662	c	663	a	664	b	665	b	666	a	667	c	668	d	669	b	670	b
671	b	672	b	673	c	674	d	675	b	676	c	677	c	678	a	679	d	680	a
681	d	682	d	683	d	684	a	685	c	686	d	687	b	688	a	689	d	690	c
691	a	692	d	693	d	694	c	695	c	696	c	697	d	698	a	699	d	700	b
701	a	702	d	703	a	704	d	705	b	706	a	707	b	708	c	709	d	710	d
711	b	712	d	713	b	714	c	715	a	716	d	717	-	718	c	719	a	720	b
721	d	722	a	723	d	724	c	725	b	726	b	727	b	728	a	729	b	730	d
731	c	732	b	733	d	734	a	735	b	736	d	737	c	738	a	739	b	740	b
741	d	742	b	743	c	744	c	745	c	746	b	747	c	748	c	749	d	750	c
751	d	752	a	753	c	754	c	755	c	756	b	757	a	758	c	759	d	760	e
761	a	762	a	763	e	764	a	765	c	766	d	767	d	768	b	769	d	770	d
771	d	772	e	773	a	774	a	775	e	776	d	777	a	778	d	779	d		

Answer Sheet

780. Emaomectin Benzoate
781. GA_3 @100 ppm and KNO_3 @0.2%
782. Flowable thiram
783. Microtomy
784. Vitavex 200
785. Bacterial panicle blight disease of rice, Bean common mosaic of cluster bean, false head smut of Maize
786. Burkhol Deria Glumae
787. Quiescence
788. True dormancy

789. Relative dormancy
790. Legumes
791. Rudimentary embryo
792. Double dormancy
793. Scarification
794. Impactation
795. Stratification
796. Cytokinin, ethylene and gibberellin
797. Imbibition, Respiration
798. Mitochondria
799. Decreases
800. Phytin
801. Matric Potential
802. Pythium and *Rhizoctonia Solani*
803. ICAR and SAU
804. National Seed Corporation
805. Seed vigor
806. Seed vigor
807. Seed
808. Seed vigor
809. Indian Mustard
810. *Bacillus amyloliquifaceans*, TA29
811. Rnase inhibitor, *Bacillus amyloliquifaceans*
812. Antisense
813. Male sterility
814. Seed pelleting
815. Protective seed coating
816. Micro and macro elements
817. 1,8 napthalic anhydride
818. Hydrophillic seed
819. Ovules
820. Placenta
921. Apomixis
822. Meiosis
823. Pollination
824. Ethylene
825. Autogamy
826. Allogamy
827. Protandry
828. Self incompatibility
829. ABA
830. Vivipary
831. Viability
932. Tolerate
933. Vivipary
834. Nucellar
835. Maturity
836. Quiescent seed
837. Dormant seed
838. IAA, NAA
839. Gibberellin
840. Stereophile
841. Syn seeds
842. Gibberellic acid
843. Abscisic acid
844. Dormancy
845. Innate dormancy
846. Imposed dormancy
847. Grow out test
848. Carrot
849. Calcium alginate
850. Ber
851. Ber
852. Riboflavin
853. "Butter fruit"
854. Apple
855. Kiwi fruit
856. New Zealand
857. Tea
858. Cacao
859. Coconut

860. Oilplam
861. Pod
862. Citrus
863. Grape
864. Ethephon
865. Pineapple
866. Isopentanol
867. bark
868. Gametophytic
869. Rhizome
870. Guava
871. Sigmoid
872. Paclobutrazol
873. Dominant gene
874. Recessive polygenic
875. Single seeded nut
876. Fat
877. Vitamin A
878. Sorosis
879. Vegetative
880. Pruning
881. is ethylene absorbent
882. Endoxerosis
883. Metxenia
884. 4°C
885. Recurrent
886. Mango
887. Pome
888. Date
889. Papaya
890. Wood apple
891. Walnut
892. Rangpur lime
893. Nematode
894. Yellowing of new leaves
895. NAA
896. VAM
897. Andhra Pradesh
898. Acid lime
899. Mango

4.3. Current Affairs

- Genetically modified (GM) crop may be introduced in India unless the biosafety and socio-economic desirability is evaluated. The committee has also recommended that the environment ministry should examine the impact of GM crops on environment thoroughly. The remarks come after India's GM crop regulator Genetic Engineering Appraisal Committee (GEA(C) recently recommended the commercial use of genetically modified mustar(d) This was in a submission to the environment ministry.

- ICAR-NAARM Technology Business Incubator (TBI), a-IDEA and Indian Institute of Management Ahmedabad's (IIM-(A) incubator Centre for Innovation, Incubation and Entrepreneurship (CIIE) announces "AGRI UDAAN"- Food and Agribusiness Accelerator 2.0. This programme will help to selected innovative startups who will be mentored in to scale up their operations in agri value chain for effective improvement in agriculture. This is a 6-month program in which shortlisted agri startups with promising innovative business models will be mentored & guided to scale up their operations. Accelerators are 4-8-month program aiming at scaling up innovative startups with a working prototype and initial market traction. Looking at the impact created through NAARM TBI a-IDEA India's first Food & Agribusiness accelerator 201 5 in partnership with IIM-A CIIE, National Science and Technology Entrepreneurship Development Board (NSTED(B), DST has come forward to support AGRI UDAAN.

- Union Minister for Agriculture and Farmers Welfare Minister, Shri Radha Mohan Singh on July 20, 2017 was launched National Dairy Development Board "NDDB's" Quality Mark "Logo" at Krishi Bhawan, New Delhi. NDDB Quality Mark "Logo" is being launched as an umbrella brand identity. This "Logo" signifies safe and quality milk and milk products from dairy cooperatives.

- Bihar is the top litchi producing State in the country. In Bihar, about 300 thousand metric tonnes of litchi is being produced from 32 thousand hectares of area Bihar's contribution in the production of litchi is about 40 percent. Considering the importance of litchi, National Research Centre on Litchi was established on June 6, 2001. The contribution of Muzaffarpur district in litchi's production is impressive, but there is a need to increase the productivity of litchi, which is currently 8.0 tonne per hectare. For this, all the government institutions, cooperatives and farmers will have to come forward Scientists at Bhabha Atomic Research Centre and National Research Centre on Litchi have succeeded in treating litchi and preserving it for 60 days at low temperature.

- The Banganapalle mango in the first week of May 2017 received a Geographical Indication (GI) tag. This makes Andhra Pradesh the proprietor of the variety known for its sweetness. The registration was given following an application from Andhra Pradesh Horticulture Commissioner. The Andhra Pradesh government is the registered proprietor of the GI tag for Banganapalle mangoes, often hailed as "the king of fruits." It is also known as Beneshan, Baneshan, Benishan, Chappatai and Safeda These mangoes are large sized, weighing on an average 350-400 grams. The pulp is fibre less, firm and yellow with sweet taste. These mangoes have been grown for over 100 years in the state.

- Union Agricultural Ministry has partnered the National Remote Sensing Centre, ISRO for geotagging agricultural assets. Agriculture Ministry under the Rashtriya Krishi Vikas Yojana and NRSC/ISRO have signed an MoU to this effect. It will improve governance due to real-time monitoring and effective utilisation of agricultural assets such as ponds, crop area, warehouses and laboratories etc.
- **Geotagging:** It is the process of adding geographical identification like latitude and longitude to various media such as a photo or video. It helps user to find a wide variety of location-specific information from a device. It provides users the pinpoint location of the content of a given picture. Geomapping is a visual representation of the geographical location of geotagged assets layered on top of map or satellite imagery.
- Union minister for agriculture and farmers' welfare Radha Mohan Singh on April 1, 2017 inaugurated International Centre for Foot and Mouth Disease (ICFM(D). It is built at a cost of INR 200 crore at Arugul on the outskirts of Bhubaneswar. The one-of-its-kind research centre in South Asia, will help analyse exotic virus strains in order to develop diagnostics and vaccines to prevent their incursion. As foot and mouth diseases an infectious and sometimes fatal viral disease that affects cloven-hoofed animals, including domestic and wild bovids, such a research centre will be beneficial to find out the remedies for the disease.
- The Indian Council of Agricultural Research/ICAR and the International Crop Research Institute for the SemiArid Tropics (ICRISAT) on March 15, 2017 signed an agreement for crop improvement. Agreement which was signed between the Director General ICAR and DG ICRISAT. The agreement is called the ICAR-ICRISAT collaborative work plan for 201 6-201 8 facilitates the transfer of agricultural technologies between the 2 institutes. Climate smart crops, smart food and digitalization of breeding database are core research areas covered under the agreement.
- The Union Ministry of Agriculture and Farmer's Welfare on 11 March 2017 launched Mission Fingerling. This is a programme aimed at achieving the Blue Revolution by enabling holistic development and management of fisheries. The main motive behind this programme is to enhance the fisheries production from 10.79 mmt (2014-15) to 15 mmt by 2020-21. To spur development in the fisheries sector, the Union Government envisaged a program named Blue Revolution. The Blue Revolution focuses on an enabling environment for an integrated and holistic development and management of fisheries for the socio-economic development of fish farmers.
- British royalty Prince Charles was launched a new Digital Green App. This is for Indian farmers. The app aims to help them sell their crops online. It is part of the prince's charity initiatives for Indi(a) The 68-year-old British royalty provided details of the app part of the Rural Livelihoods Fund created by the British Asian Trust in 2017. The trust is also in the process of setting up a development impact bond for promoting children's education in Rajasthan and Gujarat.

- Commerce & Industry Minister Smt. Nirmala Sitharaman was launched the Rubber Soil Information System (RubSIS). RubSIS is an online system for application of appropriate mix of fertilizers to specific plantations of rubber growers based on the soil nature. It was launched in New Delhi on 23rd Jan 2017. First rubber plantations in India were set up in 1895 on the hill slopes of Kerala
- The 1st International Agro-Biodiversity Congress was held in New Delhi from Nov 6 - Nov 9th 2016. It concluded with the adoption of the Delhi Declaration on Agro-biodiversity Management.
- NITI Aayog on 31 st Oct 2016 wa slaunched the first ever Agricultural Marketing & Farm Friendly Reforms Index. The index was led by Maharashtra Maharashtra has implemented most of the marketing reforms and offers an amazing environment for agribusiness among states and UTs. Maharashtra topped the index with a score of 81 .7 out of 100. Gujarat ranks second with a score of 71 .5. Rajasthan stood third with a score of 70.0 and MP scored 69.5.
- Field trials of GM Mustard was completed and an application from DU's Centre for Genetic Manipulation of Crop Plants is pending before the government. Once approved, GM mustard will be the first genetically modified crop to be grown commercially in the country.
- 93 seed hubs have been approved under the National Food Security Mission against a target of 150 at the ICAR, State Agricultural Universities and Krishi Vigyan Kendras. Government has also launched numerous web and mobile applications for disseminating information on agriculture related activities: Kisan Suvidha, Pusha Krishi, Crop Insurance, Agri Market and India Weather. Major web portals that have been developed include Farmers Portal, mKisan Portal, Crop Insurance Portal and Participatory Guarantee System of India portal.
- Ministry of Agriculture and Farmers Welfare and US Agency for International development launched the second phase of the Feed the Future India Triangular Training Programme. This brings specialised agricultural training to professionals across Africa and Asi(a)
- In the greenhouses at Wageningen University in the Netherlands, scientists have worked on growing crops in Mars and Moon soil simulants since 2013. The first experiment demonstrated that crops could grow on soil simulants with researchers mixing indelible parts of the 201 3 plants into the simulant and succeeded to grow 1 0 different crops.
- Rice farming grew in China over 9000 years ago, as per new evidence found by archeologists. This pushes back earlier estimate of 8200 years made in 2011 . Discovery was made by archeologists from University of Toronto Mississauga in Canada and China's Zhenjiang Provincial Institute of Cultural Relics and Archaeology in Chin(a) Study sheds new light on origins of rice domestication and history of human agricultural practices.
- India recorded the highest ever tea production at 1 233 million kilos during 2015-2016, and exports crossed 230 million kilos during this perio(d) Tea production rose by 3 percent in 2015-2016, from 36 million kilos in 2014-2015, according to a statement from the Tea Boar(d)

- Three improved varieties of Khesari dal namely Ratan, Mahateora and Prateek have been released for general cultivation in MP, Bihar, Chhatisgarh and Jharkhand
- Research for developing HYV of coconut in the country will be undertaken by Indian Council of Agricultural. Research- Central Plantation Crops Research Institute Kasaragod, State Agricultural Universities and Coconut Development Boar(d) Five drought tolerant, tall varieties from ICAR Institutes and one variety from Tamil Nadu Agricultural University, Coimbatore have been released for cultivation. Veppankulam Drought tolerant, High yield, Kalpatharu: Drought tolerant, High yield, Chandrakalpa- Drought tolerant, High Oil yield, Kalpa Pratibha- Drought tolerant, High Oil yield, tender nut, Kalpa Mitra- Drought tolerant High nut, oil yield, tender nut, Kalpa Dhenu- Drought tolerant, high oil yiel(d)
- Ministry of Agriculture and Farmer's Welfare, GoI under the scheme Mission for Integrated Development of Horticulture promotes green house technologies under protected cultivation. Assistance to the degree of 50 percent is provided to farmers for establishing green houses up to 4000 sq mts beneficiary
- Cotton supplies in Indian spot markets fell 12 percent from a year ago to 28 million bales between October and March as adverse weather curtailed production, stated the Cotton Association of India on April 18.
- a-IDEA, technology business incubator of NAARM or National Academy of Agricultural Research Management has entered into a memorandum of agreement with National Research Development Corporation for supporting agribusiness startups. a-IDEA is funded by DST. As part of the agreement between the two, NRDC shall provide services like Technology Management, Mentoring Services, Value Addition Services, Skill Development and Cluster Development as well as Seed Fund support to agri based startup.
- Department of Biotechnology, Ministry of Science and Technology and a group of top UK research institutions signed MOU for establishment of a joint India-UK collaboration programme in crop science. The aim of the agreement is to enhance collaborative research, promote knowledge exchange, and support capacity building to develop resilience in food security.
- India and Armenia signed an MoU for cooperation in the agricultural sector and envisaged numerous priority sectors such as plant breeding, agricultural crop seed breeding, buffalo breeding, plant protection, milk production and processing as well as agrarian education.
- To provide the impetus of the food processing sector in MP, first Mega Food Park within the state was set up at Village Panwa, Tehsil Kasarwad, District Khargone. A 2nd Mega Food Park has been sanctioned in the Dewas district of MP.
- Intellectual Property Appellate Board has registered Basmati Rice for a GI or Geographical Indications Registry 7 month Indian Basmati rice producing states in India will now carry the GI tag. GI tag can be issued for manufactured, natural

or agricultural goods. Products sold with this tag attract premium pricing and gain protection from WTO.
- In Sikkim around 75,000 hectares of agricultural land has bean gradually converted to certified organic land by implementing organic practices and principles as per guidelines laid down in National Programme for Organic Production,
- Indian public research institutions - Indian Council of Agricultural Research and Indian Agricultural Research Institute have developed a low erucic acid mustard variety namely Pusa Mustard-30, PM-30 will not only benefit health but increase productivity.
- Government has launched the Pradhan Mantri Krishi Sinchayee Yojana for providing end to end solutions in irrigation supply chain and it has been formulated keeping the following schemes in mind: Accelerated Irrigation Benefit Programme (AIBP) of Ministry of Water Resources, River Development & Ganga Rejuvenation; Integrated Watershed Management Programme (IWMP) of Department of Land Resources; and On Farm Water Management (OFWM) component of National Mission on Sustainable Agriculture (NMS(A) of Department of Agriculture, Cooperation & Farmers Welfare. Moreover, a Farmer's Portal with Kisan portal to serve as a one stop shop for farmers for accessing information on agricultural activities has also been launched.
- Agro-chemicals sector in the nation will touch USD 7.5 billion by 2018-2019 with majority (60%) of the contribution coming from exports, according to a TATA Strategic Management Group report.
- Country's total cotton output has been pegged at 370.50 lakh bales for 2015-2016 commencing October 1st, according to an industrial body, CAI or Cotton Association of India.
- Adani Group firm on 19th October 2015 formed a pact with IPGA or Indian Pulses Grains Association to ensure commodity handling across its ports for seamless availability of lentils across India. India's largest port developer Adani Ports and Special Economic Zone has signed a MoU with IPGA for handling pulses across ports.
- Black rice is grown in Manipur where it is called Chakhao amubi. It is also known as purple or forbidden rice. Its botanical name is Oryza Sativa L. This little known rice variety has entered world markets through China which cultivates it for consumption and export.
- BSI scientists have found a new species of banana named *Musa indandamanensis* from a remote tropical rain forest on the Andaman Islands. The species was located around 16 km within the Krishna Nalah forest in the Island. This banana species has unique green flowers and fruit bunch axis thrice the size of a regular banana. Fruit pulp of this banana species is orange in colour as against yellow and white of regular bananas.
- The first Anaj bank or Food grain bank has been started to help poor families who find it hard to get two meals in a day. Bank was opened by NGO Vishal

Bharat Sansthan under pet bharo-campaign in Hukugunj area on the first day of Navratri.
- Unicorn, a term proposed by venture investor Aileen Lee for describing startups valued at USD 1 billion or more is becoming a buzzword in tech entrepreneurism.
- Finger millet [Eleusine coracana (L.) Gaertn.], which figured among the six small millets in research portfolio of the International Crops Research Institute for the Semi-Arid Tropics (ICRISAT), has now been formally announced as the mandate crop.
- India is the fourth largest producer of crop protection chemicals globally. The domestic crop protection industry's valuation stood at USD 2.3 billion in 2014 with the sector tipped to reach USD 4.2 billion by FY2018-19.
- Conventional method of rice cultivation uses 5000 litres of water for producing 1 kg of rice from which 2000 litres is lost due to seepage losses and flooding. New approach called aerobic rice cultivation involves growing crops in aerobic condition, where there is plenty of oxygen in the soil.
- Organic cash crops are being cultivated via special programmes in numerous parts of the spice bowl of India, Kerala.
- Exports of close to 10 agricultural products have seen negative growth in the month of July.
- According to Gujarat Energy Research and Management Institute, the farmers and power generation companies will be in a win-win situation with generation of solar energy in agricultural fields. Farmers can use the solar energy for their own purpose and sell the surplus energy to the state run power generation companies.
- The Hyderabad-headquartered ICRISAT launched the Green Phablet, priced about .18,600, for farmers in Telangan(a) The institute would maintain servers to safe keep the information. And even if the users lose their device, they could get back the information in no time.
- National Dairy Development Board launched a mobile application that will recommend a balanced diet for cows and buffaloes. It will help boost dairy farmers' income by raising milk yield and cutting feed cost.
- The state - run National Bank for Agriculture and Rural Development, NABARD has set a target of providing 30,000 crore rupees as credit to farmers for irrigation over the next three years.
- Fertilizer Quality Control System (FQCS) is a web based and configurable workflow application developed by NIC for processing of sample collection, testing and generation of analysis reports.
- The Mission Mode Project in Agriculture under National e-Governance Plan aims to achieve rapid development of agriculture in India through the use of ICT by ensuring timely access to agriculture related information for the farmers of the country.

- The 24-hour channel will air information on farmer-related issues. The government has made it a must carry channel making it mandatory for cable and Direct to Home, DTH operators to provide it to their subscribers.
- Out of 17 Mega Food Parks, 7 parks have been allotted to state agencies whereas 1 0 to private players in 11 states, envisaging an investment of over 6,000 crore rupees.
- Mission fingerling - a programme to enable holistic development and management of fisheries sector in India with a total expenditure of about Rs. 52000 lakh.
- PACS – Primary Agriculture credit society a basic unit and smallest credit co-operative institute of NABARD
- Bharatnet project – It is a new name of National optical fibre network. It aims to provide broadband connectivity to gram Panchayats. Phase – II of the project targets to 1,50,000-gram panchayats. Universal Service obligation fund is funding this project.
- In every 5 years the agricultural census is conducte(d) First agriculture census was conducted in 1970.
- National Institute of Jute and Allied Fibre Technology located in West Bengal.
- CIRDAP - Centre on Integrated Rural Development for Asia and Pacific.
- PMGDISHA – Pradhan Mantri Gramin Digital Saksharta Abhiyan. Aims to digitally literate rural households by March 2019
- National Agriculture Science Congress has been hosted by Karnataka at Science Bangaluru. The theme of the congress is – Climate Smart Agriculture.
- Agronomy deals with the methods which provide favourable environment to crop for higher productivity.
- Cropping System may be defined as the order in which the crops are cultivated on a piece of land over fixed period. The productivity of land is maintained or even increased through proper soil management practices.
- Seed hubs - Seed Hubs are being created through ICAR, State Agriculture Universities and Krishi Vigyan Kendras (KVKs) for ensuring the availability of new kinds of seeds. For this purpose, Rs. 225.31 crore have been approved for establishment of 150 seed centers during 2016-17 to 2017-18.
- Crop logging is a useful tool to keep track of the nutrient status of fruit trees. Crop logging can help extension personnel and farmers diagnose possible nutrient deficiency or toxicity problems and schedule fertilizer practices.
- Department of Revenue and Agriculture and Commerce was set up in June 1871 to deal with all the agricultural matters in India. Until this ministry was established, matters related to agriculture were within the portfolio of the Home Department. In 1881, Department of Revenue & Agriculture was set up to deal with combined portfolios of education, health.
- The India's first Agriculture Index is launched by NITI Ayog. Maharashtra topped the index.

- GeoMGNREGA – For geo tagging of assets created under MGNREGA in each gram panchayat.
- Revolutions related to agriculture sector

    White Revolution (In India: Operation Flood – Milk/Dairy production

    Yellow Revolution – Oil seeds production

    Evergreen Revolution – Overall development of Agriculture

    Golden Revolution – Fruit production

    Blue Revolution – Fish Production

    Green Revolution – Food grains

    Grey Revolution – Fertilizer

    Silver Revolution – Egg/Poultry Production

    Black Revolution – Petroleum Production

    Round Revolution – Potato

    Pink Revolution – Prawn Revolution

    Red Revolution – Meat & Tomato Production
- Peaty soil – It occurs widely in northern parts of Bihar, Southern parts of Uttaranchal and coastal areas of west Bengal, Orissa and Tamil Nadu.
- Livestock Census in our country started in the year 1919 and since this the process has been continuing on quinquennialy basis
- The laterite soils develop in areas with high temperature and high rainfall. These soils are poor in organic matter, nitrogen, phosphate and calcium, while iron oxide and potash are in excess.
- e-Krishi Samvad has been launched by union government to solve the problems of farmers in Ariculture. Through this platform, the farmers will directly connected to the ICAR and get the solutions from experts.
- Soil degradation is the main factor leading to the depleting soil resource base in India. The degree of soil degradation varies from place to place according to the topography, wind velocity and amount of the rainfall.
- In developed dairy countries, genomic selection is used to increase milk production and productivity for attaining faster genetic gain. By using genomic selection indigenous breeds can be made viable within few generations.
- Criteria for saffron ration card: Families having total annual income of more than Rs. 15,000 and less than 1 lakh. None of the members in the family should have four wheeler mechanical vehicle (excluding taxi- driver). The family in all should not posses four hectare or more irrigated land.
- Mulching – It is the process of applying a layer of mulch and fertilizer over the soil. A mulch cover enhances the activity of soil organisms such as earthworms. They help to create a soil structure with plenty of smaller and larger pores through which rainwater can easily infiltrate into the soil, thus reducing surface runoff. As the mulch material decomposes, it increases the content of organic

matter in the soil. Soil organic matter helps to create a good soil with stable crumb structure. Thus, the soil particles will not be easily carried away by water. Therefore, mulching plays a crucial role in preventing soil erosion.
- Shifting cultivation is known by different names in different regions Jhum in Assam, Ponam in Kerala, Podu in Andhra Pradesh and Odisha and bewar masha penda and bera in various parts of Madhya Pradesh
- Crassulacean acid metabolism, also known as CAM photosynthesis, is a carbon fixation pathway that evolved in some plants as an adaptation to arid conditions.
- The nineteenth livestock census was conducted with 15 October 2012 as the reference date.
- NAPCC consist of –

  National Solar Mission

  National Mission for Enhanced Energy Efficiency

  National Mission on Sustainable Habitat

  National Water Mission

  National Mission for Sustaining the Himalayan Ecosystem

  National Mission for a Green India

  National Mission for Sustainable Agriculture

  National Mission on Strategic Knowledge for Climate Change
- Photorespiration - A respiratory process in many higher plants by which they take up oxygen in the light and give out some carbon dioxide, contrary to the general pattern of photosynthesis.
- SAMPADA – Scheme for Agro-Marine processing and development of agro-processing clusters. It will help in providing better prices to farmers and doubling farmer's income.
- Centrally sponsored schemes – MGNREGA, PMGSY etc
- The main benefit of ratooning is that the crop matures earlier in the season. Ratooning can also decrease the cost of preparing the field and planting.
- Agriculture Education Day is celebrated on 3rd December at ICAR.
- The total number of Cattle in country as per census 2012 is 190.90 million in numbers.
- International day for Biological Diversity is observed on 22th May with the theme of Biodiversity and Sustainable Tourism.
- Two National Kamdhenu Breeding Centre, one in northern region-Madhya Pradesh and other in Southern region- Andhra Pradesh, are being established in the country with an allocation of Rs 50 crores.
- In Ley Farming, the field is alternately used for grain or other cash crops for a number of years and "laid down to ley" i.e. left fallow, used for growing hay or used for pasture for another number of years. After that period it is again ploughed and used for cash/field crops.

- BPL Ration Card is issued to the families that live below the poverty line. In India, we will be on BPL list if our annual income is Rs.27,000 or less (It was earlier Rs. 10,000 but revised in 2011 to Rs. 27000)
- Permanent location code number is assigned by office of registrar general of India.
- State's first Automatic Weather Station inaugurated at Nagpur. Mahavedh Portal – Maharashtra's first agriculture weather information network.
- Intercropping- Cultivation of two or more crop simultaneously on the same field. Relay cropping is the type of intercropping. Relay cropping is the growing of two or more crops on the same field with the planting of the second crop after the first one has completed its development.
- Soil fertility refers to the ability of a soil to sustain agricultural plant growth, i.e. to provide plant habitat and result in sustained and consistent yields of high quality.
- Sub-Mission on Agroforestry has been initiated which will accelerate the programme "Medh Par Ped"
- Out of the total allocation of Rs.1700 crores under NFSM for 2016-17, Rs.1,100 crore (central share) was allocated for pulses which amounts to more than 60% of total allocation
- Zero tillage – It is an agriculture technique which increases the amount of water that infiltrates into soil and increases the organic matter retention and cycling of nutrients in the soil.
- Potamodromous fish – migratory fish that move within fresh water only. Catadromous fish – migratory fish that live in fresh water but breed in salt water. Anadromous fish – migratory fish that live in salt water but breed in fresh water.
- Cereal Crops are members of the grass family grown for their edible starchy seeds
- Cultivation of crops along with rearing of animals for meat or milk is called Mixed Farming.For example, the same farm may grow cereal crops, and keep cattle, sheep, pigs or poultry.
- Wet rice agriculture is labor-intensive, meaning that many people are required to do the job (as in the cultivation of silk worms and tea. Labor is particularly important when the fields are prepared, seedlings transplanted, and again when the rice is harvested. It is mostly practiced in Kerala.
- Imbalanced use of fertilizers reduces the soil fertility making it less productive for the crops.
- Green Revolution has got success mostly in case of Rice and Wheat as they are the two major crops in India.
- Wells are the most important source in agriculture as it available in almost all areas in abundance making a major source without which the plantation of crops is not possible.

- Rice is the crop which suits the growth of Rice in Madhya Pradesh.
- Temperature is the basic factor needed for the better development of fruits.
- Tea is basically a crop in India that falls in the beverage category.
- The Murrah breed of Water buffalo is a breed of domestic water buffalo kept for dairy production
- National parks are basically kept without the reach of normal people to conserve the biomass.
- NABARD looks after Rural Development Infrastructure Fund.
- On 22 February 2016, Prime Minister Narendra Modi launched the National Rurban Mission (NRM) from Kurubhat in Rajnandgaon district of Chhattisgarh.
- The Integrated Child Development Services launched in 1975 in 33 community development blocks is now one of the largest scheme and with universalization now envisaged to cover 14 Lakh habitations across the country.
- The NREGA is designed as a safety net to reduce migration by rural poor households in the lean period by providing them at least a hundred days of guaranteed unskilled manual labour on demand at minimum wages
- Adding living earthworms dig the soil and helps in increasing the fertility of the soil.
- Ratnagiri is considered to be the best place for the production of this variety of mangoes.
- Jute is considered to be the golden fibre amongst all the crops.
- Golden revolution is about Honey and Horticulture production.
- Increase germination rates by planting seeds at the correct depth. As a rule, seeds should not be buried any deeper than their diameter.
- PM-KISAN is a scheme of the central government that was introduced in 2019. This scheme was launched to help farmers holding land with their financial needs. This year is the PM- KISAN scheme's third anniversary. Overview: Through Direct Benefit Transfer (DBT), Rs 6000/- is deposited into the bank accounts of farmers across the country*
- The National Statistical Office recently released the first revised GDP estimates for the fiscal year 2021. According to the estimates, the GDP contracted by 6.6%. Earlier, the GDP had contracted by 7.3%. The contraction is mainly due to COVID pandemic and the lockdown imposed.
- The Government of India is to promote the use of drones in agriculture. For this, the Ministry of Agriculture and Farmer Welfare has amended the SMAM guidelines. SMAM means Sub - Mission on Agricultural Mechanisation. Also, the ministry has issued funding guidelines.
- The District Good Governance Index (DGGI) was released by the Home Minister Amit Shah. The index was released for the state of Jammu and Kashmir. The index was prepared by the Department of Administrative Reforms and Public Grievances. The top five districts in the index were Jammu, Doda, Samba, Pulwama and Srinaga.

- The Government of India is working on creating a digital stack of agricultural data. That is storing data related to farmers. Apart from farmer details, the digital agristack also holds all information about which seeds to buy, weather updates, what manures and fertilizers to use, insurance, best agricultural practices to maximize the yield.
- The AgFunder and Omnivore, a capital released the "India AgriFood Startup - Investment Report". According to the report, the investment in agri and food tech start-ups increased by 97% in 2020-21. It increased to 2.1 billion USD. The investments are mainly driven by restaurant funding.
- Zero tillage is an agricultural practice. It maintains permanent soil cover. It increases natural biological processes that occur in the soil. It improves sustained crop production and increases soil nutrition.
- Agro - Climatic Zone is a land that is suitable for growing particular type of crop. It is essential to delineate the land in the country into agi o - climatic zones for sustainable agricultural production. There are 15 agro - climatic zones in India.
- 8th Meeting of Agricultural Experts of BIMSTEC (Bay of Bengal Initiative for Multi- Sectoral Technical and Economic Cooperation) Countries was hosted by India on August 31, 2021. Meeting was attended by Agricultural Ministries from Bangladesh, Bhutan, Nepal, Sri Lanka, India, Myanmar and Thailand.
- India's export of agricultural and allied products in 2020-21 have increased by 17.34 percent. It now equals to USD 41.25 billion.
- World Bee Day is observed annually by the United Nations on 20th May. This day is observed to create awareness about the importance of bees for this planet and nature.
- On May 19, 2021, the Government of India increased fertilizer subsidies to Rs 1,200 per bag. Earlier, it was Rs 500 per bag. Thus, the subsidies have been increased by 140%.
- The Maharashtra Chamber of Commerce Industries and Agriculture and NABARD (National Bank for Agriculture and Rural Development) recently launched the first Agricultural Export Facilitation Centre.
- The National Bank for Agriculture and Rural Development (NABARD) recently announced that it will provide Rs 1,236 crores to Assam in 2020-21 from its Rural Infrastructure Development Fund (RIDF). In 2020-21, Rs 29,848 crores was allocated to RIDF.
- Every year the World Agri-Tourism Day is celebrated on May 16. In 2021, there was fourteenth World Agri-Tourism Day, Government of Maharashtra, Agri-Tourism Development Corporation has been organised International Conference on Agri- Tourism. In 2021, the day is to be celebrated under the following theme Theme: Rural Women Sustainable Entrepreneurship Opportunities through Agri Tourism.
- The Ministry of Agriculture and Farmer Welfare recently announced that the area of summer crops has sharply increased by 21.58% as compared to 2020. The

increase has almost doubled. Summer crops are also called Zaid crops. They are grown between March and June.
- The Telangana Government recently imposed a total ban on Glyphosate. The Glyphosate is a controversial herbicide usually use to kill weeds in cotton farms. The usage of Glyphosate is banned because it is carcinogenic. Also, it is being banned to control the illegal cultivation of HTBt cotton.
- The Beal Seed Viability Experiment was begun by William James Beam. He was an American Botanist who worked at the Michigan State University. In 1879, Beal began one of the longest running experiments in botany
- Italy recently launched the first "Mega Food Park" in India involving food processing facilities. It is the first Italian-Indian Food Park project launched in the country.
- The Department of Science and Technology recently launched the National Climate Vulnerability Report. According to the report, eight Eastern States in the country are highly vulnerable to climate change. They are Mizoram, Bihar, West Bengal, Arunachal Pradesh, Chhattisgarh, Odisha, Jharkhand and Assam.
- The Artificial Intelligence (AI) for Agriculture Hackathon was begun by Google, MyGov and HUL to improve AI solutions in the field of Agriculture.
- The Government of India has launched a new Central Sector Scheme called the "Formation and Promotion of 10,000 Farmer Produce Organizations (FPOs)". Highlights The scheme has been launched with a clear strategy and committed resources in order to form and promote 10,000 new FPOs across India.
- The Uttar Pradesh Government has been launched Kisan Kalyan Mission. The Kisan Kalyan mission aims to double the farmer income covering ail the Assembly constituencies of the state of Uttar Pradesh. This is to be achieved through campaigns that create awareness among farmers.
- The Karnataka State Legislative Council recently passed the Land Reforms (Amendment) Bill 2020. The 2020 amendment bill seeks to amend the Karnataka Land reforms Act of 1961. The 1961 Act brought in restrictions on ownership of agricultural lands in the state.
- The Government of India recently issued a hundred page e-booklet called "Putting Farmers First". The booklet highlights the success stories of farmers who have benefited from contract farming after the legislation of three farm laws.
- The Union Territory of Lakshadweep has been declared as Organic Agricultural Area by the Ministry of Agriculture and Farmer's welfare. The UT is second after Sikkim to achieve the status of 100% organic region. It is first in the Union Territories of India to achieve the status.
- On November 20, 2020, Government of India sanctioned a loan of 3,971 crores of rupees for micro irrigation projects in Tamil Nadu. This is one of the biggest loans being sanctioned for the state of Tamil Nadu. The loan has been sanctioned at subsidised interest rate under micro irrigation fund.

- The India Greenhouse Horticulture market held a market value of USD 190.84 Million in 2021 and is estimated to reach USD 271.25 Million by the year 2030. The market is expected to register a growth rate of 4.19% over the projected period. In 2021, India's greenhouse horticulture production was 27.71 million.
- Fair and Remunerative Price (FRP)Maharashtra Government issued a government resolution which will allow sugar mills to pay the basic Fair and Remunerative Price (FRP) in two tranches. The first instalment would have to be paid within 14 days of delivery of cane, and would be as per the average recovery of the district.'
- Lavender Cultivation' under CSIR-IIIM's Aroma Mission will be started in Ramban district (Jammu Kashmir) as a part of Purple Revolution. Aromatic Plants include lavender, damask rose, mushk bala, etc. Council of Scientific & Industrial Research (CSIR) is a contemporary R&D organization under the Ministry of Science and Technology.
- Crop Diversification: In the Annual Economic Survey, the Department of Economic Affairs said that there is an urgent need for Crop Diversification in view of the severe water stress in areas where paddy, wheat and sugarcane are grown as well as to increase oil seed production and reduce dependency on imports of cooking oil.
- The Ministry of Fisheries, Animal Husbandly and Dailying organised a webinar on "Cage aquaculture in Reservoir: Sleeping Giants" as a part of "Azadi Ka Amrit Mahotsav". Department of Fisheries, GOI earmarked the investment targets for promoting cage aquaculture under flagship scheme Pradhan Mantri Matsya Sampada Yojana (PMMSY).
- NECBDC: NECBDC sponsored a training programme on "Bamboo Shoot Processing and Preservation" which was conducted by the NECBDC empanelled cluster M/s Delicacies food processing Center, Meghalaya the same venue from 13th to 17th, December 2021. North East Cane and Bamboo Development Council (NECBDC) is under the Ministry of DoNER, Govt. of India
- India has emerged as the largest exporter of gherkins in the world. India has crossed the USD 200 million mark of export of agricultural processed product, - pickling cucumber, which is globally referred as gherkins or cornichons.
- Bioenergy Crops: A new study has found that converting annual crops to perennial bioenergy crops can induce a cooling effect on the areas where they are cultivated. The researchers simulated the biophysical climate impact of a range of future bioenergy crop cultivation scenarios. Eucalyptus, poplar, willow, miscanthus and switchgrass are the bioenergy crops.
- What Is The Rythu Bandhu? The total funds disbursed under Rythu Bandhu, Telangana government's direct benefit transfer scheme for farmers, will soon touch Rs 50,000 crore in the coming days. The scheme was launched in 2018. Rythu Bandhu scheme or Farmer's Investment Support Scheme (FISS) is a welfare program to support farmer's investment.

- According to a new study, cropland area across the world increased 9% and cropland Net Primary Production (NPP) by 25% from 2003-2019. The growth was primarily due to agricultural expansion in Africa and South America. Cropland Area: Cropland is defined as 'land used for annual and perennial herbaceous crops for human consumption.
- World Fisheries Day (WFD) is celeberated on the 21 st November every year. Digital Agriculture:The Agriculture Ministry signed agreements with Reliance's Jio Platforms, ITC, Cisco, NCDEX e-Markets and Ninjacart to develop agritech solutions using its National Farmers Database, which includes information of 5.5 crore farmers.This is part of an effort to modernise the agriculture sector by infusing new technologies so that farmers can increase their income,
- Transport and .Marketing Assistance" (TMA) Scheme for Specified Agriculture Products': Centre has revised "Transport and Marketing Assistance" (TMA) scheme for Specified Agriculture Products.' In February 2019, the Department of Commerce had introduced "Transport and Marketing Assistance (TMA) for Specified Agriculture Products Scheme' to provide assistance for the international component of freight.
- Minimum Support Price (MSP): The Cabinet Committee on Economic Affairs (CCEA) has approved the increase in the MSP for all mandated Rabi crops for the Rabi Marketing Season 2022-23. This will ensure maximum remunerative price for farmers and also encourage them to sow a wide variety of crops.
- Assam Cattle Preservation Act, 2021: Various pressure groups in Assam groups recently held a rally against the Assam Cattle Preservation Act, 2021, stating that the law was an assault on the farm economy in the name of religion. It is aimed at regulating slaughter, consumption and transportation of cattle.
- National Farmers Database: The Centre has created a National Fanners' Database with records of 5.5 crore farmers, which it hopes to increase to 8 crore farmers by December by linking it to State land record databases, according to Agriculture Minister Narendra Singh Tomar. Farmers' database is key to advances in digital agriculture.
- Mission on Palm Oil: The Union Cabinet has given its approval to launch a new Mission on Oil palm to be known as the National Mission on Edible Oils - Oil Palm (NMEO-OP) as a new Centrally Sponsored Scheme with a special focus on the North east region and the Andaman and Nicobar Islands.
- Fortification of Rice: Emphasising that malnutrition is a "hurdle" in the development of women and children, Prime Minister Narendra Modi announced fortification of rice distributed under various government schemes including Public Distribution System (PDS) and Mid-Day-Meal scheme by 2024.
- Bhalia Variety of Wheat: In a major boost to wheat exports, the first shipment of Geographical Indication (GI) certified Bhalia variety of wheat was exported today to Kenya and Sri Lanka from Gujarat. The GI certified wheat has high protein content and is sweet in taste.

- Rice Bran Oil: Department of Food and Public Distribution E-launched "NAFED Fortified Rice Bran Oil". Rice bran oil is the oil extracted from the hard outer brown layer of rice called chaff (rice husk). It is known for its high smoke point of 232 °C making it suitable for high-temperature cooking methods
- Anti-Methanogenic Feed Supplement: Harit Dhara: Indian Council of Agricultural Research (ICAR) has developed an anti-methanogenic feed supplement 'Harit Dhara' (HD), which can cut down cattle methane emissions by 17-20% and can also result in higher milk production. HD decreases the population of protozoa microbes in the rumen, responsible for hydrogen production and making it available.
- Crop Insurance Week: Government has launched the Crop Insurance Awareness Campaign for Fasal Bima Yojana during the Crop Insurance Week (being observed from July 1 to 7). Till date, the scheme has insured over 29.16 crore farmer applications (5.5 crore farmer applications on year-on-year
- National Food Security Act (NFSA), 2013: Centre amends Food Security rules to prevent ration leakage, corruption. The government said that this amendment has been made as an attempt to take forward the reform process envisaged under Section 12 of the National Food Security Act (NFSA), 2013 by way of improving the transparency of the operation.
- Researchers from various institutes under the Indian Council of Agricultural Research (ICAR) and Bidhan Chandra Krishi Viswavidyalaya found depleting trends in grain density of zinc and iron in rice and wheat cultivated in India.
- The Ministry of Agriculture and Farmers Welfare has provided an enhanced allocation of Rs. 2250 Crore for the year 2021-22 for 'Mission for Integrated Development of Horticulture (MIDH).
- Horticulture is the branch of plant agriculture dealing with garden crops, generally fruits, vegetables, and ornamental plants.
- CSIR Floriculture Mission has been approved for implementation in 21 States/UTs wherein available knowledgebase in CSIR Institutes will be utilized and leveraged to help Indian farmers and industry re-position itself to meet the import requirements.
- National e-Govemance Plan in Agriculture: The National e-govemance plan in agriculture (NeGPA) is a Centrally Sponsored Scheme.'! he scheme was initially launched in the year 2010-11 in 7 States. Recently, this scheme was extended up to March 31, 2021. The NeGPA guidelines were amended in 2020-2021 in order to infuse modern information technologies in the farm.
- Pradhan Mantri Fasal Bima Yojana (PMFBY): To ensure timely settlements of claims under Pradhan Mantri Fasal Bima Yojana (PMFBY), the Directorate General of Civil Aviation (DGCA) has approved the proposal of the department of agriculture for flying drones over 100 districts growing rice and wheat. The component schemes of Pradhan Mantri Kisan Sampada Yojana are as follows;
    - Mega Food Park
    - Integrated Cold Chain and Value Addition Infrastructure

- Infrastructure for Agro-Processing Clusters
- Creation /Expansion of Food Processing & Preservation Capacities
- Creation of Backward & Forward linkages
- Food Safety and Quality Assurance Infrastructure
- Human Resource and Institutions
- Operation Greens

The following community farming assets projects are eligible under the scheme: -
- Organic inputs production
- Bio stimulant production units
- Infrastructure for smart and precision agriculture.
- Projects identified for providing supply chain infrastructure for clusters of crops including export clusters.
- Projects promoted by Central/State/Local Governments or their agencies under PPP for building community farming assets or post-harvest management projects.

Eligible Beneficiaries
- Primary Agricultural Credit Society
- Agri-Entrepreneur
- Farmer
- Farmer Producers Organization
- Joint Liability Groups
- Self Help Group
- Start-up
- Marketing Cooperative Societies
- Multipurpose Cooperative Societies
- Aggregation Infrastructure Providers and
- Central/State agency or
- Local Body sponsored Public Private Partnership Project

- According to the second advance estimate of horticulture production released by the Ministry of Agriculture today, India is expected to have the highest ever horticulture production of 329.86 million tonnes in 2020-21, up by 2.93% over previous year.
- The increase in production has been registered in vegetables, spices, medicinal and aromatic crops. The Floriculture sector has been the hardest hit by the pandemic as production declined by 7.17%.
- Government of India had launched a campaign known as Garib Kalyan Rojgar Abhiyaan (GKRA) of 125 days on 20th June, 2020 to boost employment and livelihood opportunities for returnee migrant workers and similarly affected citizens in rural area, in the wake of COVID-19 pandemic.

- Deendayal Antyodaya Yojana - National Rural Livelihoods Mission (DAY-NRLM) aims to reduce poverty by organizing the rural poor women into Self Help Groups (SHGs), and continuously nurturing and supporting them to take economic activities till they attain appreciable increase in income over a period of time.
- Indian Farmers Fertilizer Cooperative Limited (IFFCO) has introduced the world's first Nano Urea Liquid for farmers across the world.
- As per 4th Advance Estimates, the estimated production of major crops during 2020- 21 is as under:
    - Foodgrains - 308.65 million tonnes, (record)
    - Rice - 122.27 million tonnes, (record)
    - Wheat - 109.52 million tonnes, (record)
    - Nutri / Coarse Cereals -51.15 million tonnes.
    - Maize -31.51 million tonnes, (record)
    - Pulses - 25.72 million tonnes, (record)
    - Tur - 4.28 million tonnes.
    - Gram - 11.99 million tonnes, (record)
    - Oilseeds - 36.10 million tonnes, (record)
    - Groundnut - 10.21 million tonnes (record)
    - Soyabean - 12.90 million tonnes
    - Rapeseed and Mustard - 10.11 million tonnes (record)
    - Sugarcane - 399.25 million tonnes
    - Cotton - 35.38 million bales (of 170 kg each)
    - Jute & Mesta - 9.56 million bales (of 180 kg each)
- Soyabean variety named MACS 1407 recently developed by Agharkar Research Institute of Pune along with ICAR. The variety contains 41% protein and 19.81% oil. This variety is also resistant to pests like leaf miner, leaf roller, girdle beetle, stem fly, white fly and aphids.
- International Labour Day or worker's day or May day- May. It was firstly hosted by the labour Kisan Party Madras in 1923.
- Decadal climate update is released by- World Metrological Organization. International Tea Day is celebrated on May 21 by the United Nations food & Agriculture Organization. It is celebrated since 2005. The theme was- "Tea and Fair Trade".

5
Memory Based Question (JRF, SRF, NET) 1997 to 2019

(1) The first Director of IARI was ——————.
    (a) M. S. Randhawa   (b) B. P. Pal
    (c) M. S. Swaminathan   (d) G. Kalloo

(2) In vegetable blanching is done for——————.
    (b) Removal of microorganism   (b) To activate the enzymes
    (c) To retention colour   (d) All of the above

(3) Viviparous refers to——————.
    (a) Germination of seed on mother plant
    (b) Germination of seed in the field
    (c) Germination of the seed on monoecious plant
    (d) Germination of seed on any place

(4) Sourness in radish is due to ——————.
    (a) Isosinate   (b) Isothiocinate
    (c) Anthocyanin   (d) All of the above

(5) Round the year grown variety of tomato is ——————.
    (a) Pusa Sadabahar   (b) Pusa Ruby
    (c) Pusa Uphar   (d) Pusa Divya

(6) CIPHET is situated at ——————.
    (a) Amritsar   (b) Ludhiana
    (c) Chandigarh   (d) Delhi

(7) Leading potato production state is ——————.
    (a) Uttar Pradesh   (b) M.P.
    (c) Gujarat   (d) Punjab

(8) Fire hazardous fertilizer is ——————.
    (a) Urea   (b) DAP
    (c) Ammonium Nitrate   (d) Ammonium sulphate

(9) Which of the following PGR is used for root formation ———————.
   (a) IBA   (b) ABA
   (c) NAA   (d) MH
(10) Stable Vegetable in our diet is ———————.
   (b) Tomato   (b) Potato
   (c) Cucumber   (d) Brinjal
(11) Main aim of pre- cooling in our diet is ———————.
   (a) To improve the quality   (b) To improve microbial load
   (c) To improve field heat (d) To fresh the product
(12) Plant hormone which is used for controlling of fruit drop ———————.
   (a) Auxin   (b) Cytokinin
   (c) Ethylene   (d) Gibberellins
(13) First time cytoplasmic genetic male sterility is used ———————.
   (c) Onion   (b) Carrot
   (c) Pea   (d) Brinjal
(14) TCA cycle is occurred in ———————.
   (a) C_3 plant   (b) C_4 plant
   (c) CAM plant   (d) All of the above
(15) DNA synthesis is occurring in which phase ———————.
   (a) Prophase   (b) Interphase
   (c) GI phase   (d) S phase
(16) 9: 7 ratio of gene interaction is ——— ———————.
   (a) Complementary   (b) Supplementry
   (c) Inhibitory   (d) Polymeric
(17) Sourness in radish is due to ———————.
   (d) Isosinate   (b) Isothiocinate
   (c) Anthocyanin   (d) All of the above
(18) Flower inducing hormone in cucurbits is ———————
   (a) NAA   (b) IBA
   (c) Auxin   (d) GA3
(19) Which family is mostly attacked by lady bird beetle ———————?
   (a) Solanaceae   (b) Cucurbitaceae
   (c) Liliaceae   (d) Liliaceae
(20) Cucurbits having the highest Vitamin A is ———————.
   (a) Bittergourd   (b) Cucumber
   (c) Pumpkin   (d) All of the above

(21) PK-Ft -16 is a variety of —————————.
    (a) Tomato　　　　　　　　(b) Brinjal
    (c) Cucurbits　　　　　　　　(d) Chilli

(22) Carrot is a rich source of —————————.
    (a) Vitamin A　　　　　　　　(b) Vitamin B
    (c) Vitamin C　　　　　　　　(d) Vitamin D

(23) Botanical name of bitter gourd is —————————.
    (a) *Citrullus lanatus*　　　　(b) *Momordica charantia*
    (c) *Cucumis melo*　　　　　(d) None of the above

(24) Which of the following crop is not transplanted? —————————?
    (a) Tomato　　　　　　　　(b) Chilli
    (c) Brinjal　　　　　　　　(d) None of the above

(25) Little leaf of brinjal is caused by —————————.
    (a) Mycoplasma　　　　　　(b) Bacteria
    (c) Virus　　　　　　　　　(d) Fungi

(26) Which one of the following is a non-climacteric fruit? —————————?
    (a) Muskmelon　　　　　　(b) Cucumber
    (c) Watermelon　　　　　　(d) All of the above

(27) French bean belongs to family —————————.
    (a) Malvaceae　　　　　　　(b) Convolvaceae
    (c) Umbelliferae　　　　　　(d) Papilionaceae

(28) Sweet potato belongs to family —————————.
    (b) Malvaceae　　　　　　　(b) Convolvaceae
    (c) Umbelliferae　　　　　　(d) Papilionaceae

(29) Malathion is a —————————.
    (c) Fungicides　　　　　　　(b) Weedicides
    (c) Pesticide　　　　　　　　(d) Nematicide

(30) Main function of RNA is —————————.
    (a) DNA transfer　　　　　　(b) DNA replication
    (c) Protein synthesis　　　　(d) All of the above

(31) Chilli is a good source of —————————.
    (a) Fat　　　　　　　　　　(b) Protein
    (c) Vitamin C　　　　　　　(d) Vitamin K

(32) Which of the following crop can also be grown in acidic soil—————————.
    (a) Frenchbean　　　　　　(b) Cauliflower
    (c) Spinach　　　　　　　　(d) Watermelon

(33) In tomato red colour is due to ———————.
   (a) Anthocyanin   (b) Lycopene
   (c) Xanthophyll   (d) None of the above

(34) The crop showing greatest inbreeding depression is ———————.
   (a) Onion   (b) Lettuce
   (c) Tomato   (d) None of the above

(35) The hybrids of crop, mostly preferred in India are ———————.
   (a) Brinjal   (b) Cucumber
   (c) Onion   (d) Tomato

(36) Homozygosity and homogeneity is maximum in ———————.
   (a) Pure line   (b) Inbred line
   (c) Land races   (d) Synthetic varieties

(37) Greening in potato is due to ———————.
   (a) High temperature   (b) High light intensity
   (c) Low temperature   (d) All of the above

(38) White rust of crucifer is caused by ———————.
   (a) *Crostopus candida* (b) *Albugo candida*
   (c) *Peranospora*   (d) *Cleviceps*

(39) Seed rate of cauliflower is ———————.
   (a) 100g/ha   (b) 200g/ha
   (c) 350-600g/ha   (d) 1000g/ha

(40) In Brinjal, only those flowers set fruits which have ———————.
   (a) Short   (b) Long style
   (c) Medium   (d) Both (b) and (c)

(41) Browning in cauliflower is due to ———————.
   (a) Cu   (b) Bo
   (c) Ca   (d) Mo

(42) Good storage requires ———————.
   (a) High temperature and high humidity
   (b) Low temperature and low humidity
   (c) High temperature and low humidity
   (d) None of the above

(43) Daria method of cultivation is followed in which vegetables———————.
   (a) Cucurbits   (b) Cole crops
   (c) Tuber crops   (d) Rot crops

(44) Gynoecious lines are common in ———————————.
   (a) Watermelon   (b) Muskmelon
   (c) Cucumber   (d) Long melon
(45) Leguminous leafy vegetable is ———————————.
   (a) Cowpea   (b) Fenugreek
   (c) Spinach   (d) Palak
(46) Main constraints in commercial cultivation of spices is ———————————.
   (a) Lack of planting materials   (b) Lack of Market
   (c) Lack of knowledge   (d) None of the above
(47) Pink colour of onion is due to ———————————.
   (a) Quercetin   (b) Anthocyanin
   (c) Carotene   (d) None of the above
(48) Which growth regulator is commonly used as weedicide ———————————?
   (a) ABA   (b) IBA
   (c) NAA   (d) 2, 4-D
(49) Riceyness in cauliflower is due to ———————————.
   (a) Growing in rice field   (b) Unfavourable temperature
   (c) Mo deficiency   (d) None of the above
(50) Which one of the following is uded to enhance more fruit set———————————?
   (a) NAA   (b) 2.4-D
   (c) 2,4,5T   (d) Ethylene
(51) Acridity in colocassia is due ———————————.
   (a) Calcium oxalate   (b) Potassium
   (c) Sulphur   (d) None of the above
(52) Which type of can is used for packing of spinach, cauliflower and sweet potato ———————————?
   (a) AR cans   (b) SR cans
   (c) Plain   (d) All of the above
(53) Akashin, a physiological disorder observed in radish is due to deficiency of ———————————.
   (a) N   (b) Ca
   (c) K   (d) B
(54) ——————————— is one of the ripening hormone?
   (a) IBA   (b) NAA
   (c) Ethylene   (d) None of the above

(55) The crop showing high inbreeding depression is ——————.
   (a) Onion   (b) Lettuce
   (c) Tomato   (d) Lima bean
(56) A treatment of fruits and vegetables by slight temperature for small time is known as ——————.
   (a) Sterilization   (b) Blanching
   (c) Pasteurization   (d) None of the above
(57) Thiram is a ——————.
   (a) Weedicide   (b) Pesticide
   (c) Fungicide   (d) None of the above
(58) Whiptail of cauliflower is caused due to deficiency of ——————.
   (a) Magnesium   (b) Boron
   (c) Molybdenum   (d) Potassium
(59) Pungency of onion is due to ——————.
   (a) Betanin   (b) Allylpropyl disulphide
   (c) Solanin   (d) Isothiocyanate
(60) Which of the following vegetables belongs to the family Fabaceae ——————?
   (a) Spinach   (b) Fennugreek
   (c) Celery   (d) Lettuce
(61) The range of optimum night temperature for tomato fruit set is ——————.
   (a) 5-10°C   (b) 15-20°C
   (c) 25-30°C   (d) 35-40°C
(62) —————— is first Director of IIVR
   (a) G. Kalloo   (b) Mangala Rai
   (c) HP Singh   (d) Mathura Rai
(63) Which of the following vegetables is resistant to drought.——————.
   (a) Okra   (b) Brinjal
   (c) Palak   (d) Sweet potato
(64) Little leaf disease of brinjal is caused by ——————.
   (a) Virus   (b) Physiological disorder
   (c) Bacteria   (d) Mycoplasma
(65) Potato is good source of .——————.
   (a) Pectin   (b) Starch
   (c) Tannin   (d) Carotene

(66) The quickest method to develop pure line variety of heterozygous population is through ——————.
   (a) Selfing   (b) Sibmating
   (c) Doubling chromosome of haploid   (d) Doubling chromosome of diploids
(67) Obtaining a F_2 ratio of 9:3:3:1 in a dihybrid cross is an indicator of ——————.
   (a) Independence between the two genes
   (b) Linkage between the genes with less than 50% in combination
   (c) Linkage between the genes with zero recombination
   (d) Pleiotropy
(68) Complete genetic homogeneity and homozygosity is observed ——————.
   (a) Land race   (b) Hybrid
   (c) Synthetic population   (d) Pure line
(69) When two heterozygous individuals are crossed, the segregation is observed earliest in ——————.
   (a) Test cross   (b) Back cross -I
   (c) F_2 generation   (d) F_1 generation
(70) Uniform heterosis in the F_1 population is observed when ——————.
   (a) The parents are hohozygous inbred lines
   (b) The parents are F1,s themselves
   (c) The parents are progeny of a single self-fertilized plant
   (d) The parents are of different ploidy level
(71) Recombination of genes take place ——————.
   (a) During meiosis I   (b) During meiosis II
   (c) On completions of fertilization   (d) During all of above stage
(72) Maintenance of pure line requires regular sibmating in the case of ——————.
   (a) Pea   (b) Bean
   (c) Papaya   (d) Tomato
(73) For population improvement in any crop species the best method to be followed is ——————.
   (a) Pure line selection   (b) Pedigree selection
   (c) Mass selection   (d) Recurrent selection
(74) Inbreeding is deleterious in ——————.
   (a) Tomato   (b) Cucurbits
   (c) Onion   (d) Maize

(75) Once the hybrid vigour is obtained in the F_1 population further decline in the expression of traits in the subsequent generations is due to increase in the ―――――――――――.
   (a) Overdominance   (b) Homozygosity
   (c) Heterozygosity   (d) Segregation

(76) A back cross is a test cross in a dihybrid cross when ――――――――.
   (a) The F_1 is crossed to one of the parents where the genes are in repulsion phase
   (b) The F_1 is crossed to the parents with both genes in dominant form
   (c) The F_1 is crossed to the parents with both genes in dominant coupling phase
   (d) The F_1 is crossed to the recessive parent with both genes in coupling phase

(77) Which vegetable is dioecious and propagated by stem cutting? ――――――.
   (a) Pointed gourd   (b) Spinach
   (c) Sweet potato   (d) Tapioca

(78) Sweet potato is ――――――――――――――――――.
   (a) Allohexaploid   (b) Autohexaploid
   (c) Autotetraploid   (d) Allotetraploid

(79) Tomato fruits showing blossom end rot is a typical deficiency symptom of ――――――――――.
   (a) Calcium   (b) Magnasium
   (c) Phosphorus   (d) Boron

(80) F_1 hybrids in vegetables are commercially exploited ―――――――――.
   (a) Cucumber, onion   (b) Bitter gourd, *coccinia*
   (c) Radish, garlic   (d) cowpea lima bean

(81) Which type of self- incompatability is present in cole crop ――――――.
   (a) Sporophytic   (b) Gametophytic
   (c) Gamosporophytic   (d) Pseudo self- incompatability

(82) Isolation distance for self – pollinated crop is generally kept.―――――――.
   (a) 10-200m   (b) 1-2km
   (c) 10 feet   (d) No isolation distance

(83) Bacterial soft rot of vegetable is caused by ――――――――――――.
   (a) *Botrytis cinma*   (b) *Erwinia cartovora*
   (c) *Diplodia metalensis* (d) *Sclerotinia functioncola*

(84) The commercial seed production of which variety of cauliflower is not possible in plain. ―――――――.
   (a) Pusa Deepti   (b) Pusa Snowball-1
   (c) Pusa Komal   (d) Pusa Synthetic

(85) Which of the following potato variety is resistant to late blight ———?
  (a) Kufri Badshah (b) Kufri Sinduri
  (c) Kufri Bahar (d) Kufri Chandramukhi
(86) Onion bulbs show problem of bolting if the temperature remains below ———.
  (a) 30°C (b) 25°C
  (c) 20°C (d) 15°C
(87) In tomato, following dose of alachor (active ingredient) per hectare is found most suitable for weed control ———.
  (a) 1-1.5kg (b) 2-2.5kg
  (c) 3-3.5kg (d) 4-4.5kg
(88) Following seed rate is recommended for sowing one hectare area in case of cucumber———.
  (a) 1-1.5kg (b) 2.5-3.5kg
  (c) 4.5-6.0kg (d) 6.5-7.0 kg
(89) Highest quantity of carotene is available from.———.
  (a) Carrot (b) Spinach leaf
  (c) Colocassia leaf (d) Fenugreek leaf
(90) Daily requirement of Vitamin A for an healthy adult is ———.
  (a) 50-60mg (b) 100IU
  (c) 4000-5000IU (d) 0.003-0.005g
(91) A separate ministry of food processing in the Government of India was created in the year———.
  (a) 1968 (b) 1978
  (c) 1988 (d) 1998
(92) Potato contains following amount of protein(mg/100g) of edible portion———.
  (a) 1.6 (b) 1.8
  (c) 2.0 (d) 2.2
(93) Edible portion of turnip is ———.
  (a) Epicotyl (b) Root
  (c) Tuber (d) Stem
(94) Blancing is related to ———.
  (a) Freezing (b) Drying
  (c) Canning (d) Fermentation
(95) Which is progenitor of watermelon ———?
  (a) *C. edulis* (b) *C. vulgaris*
  (c) *C. colosnthis* (d) *C. citriodes*

(96) The first example for the use of biogent for controlling post-harvest diseases _____.

(a) *Candida oleophila* (b) *Pseudomonas synngae*
(c) *Bacillus subtilis* (d) *Trichoderma harzianum*

(97) CIMAP is located at _____.
(a) Lucknow (b) Pantnagar
(c) New Delhi (d) Gujarat

(98) Capacity of soil that resists sudden changes in pH of soil is called _____.
(a) Buffering capacity (b) Cation exchange capacity
(c) Anion exchange capacity (d) Biological oxygen demand

(99) Totipotency in plant is present in _____.
(a) Pholem (b) Cork
(c) Cambium (d) Meristem

(100) Free living bacteria involved in nitrogen fixation _____.
(a) *Rhizobium* (b) *Azotobacterial*
(c) *Anabaena* (d) *Azospirillum*

(101) The cultivation of crop from the re-growth of stubbles of crop after harvest is called _____.
(a) Arrowing (b) Reaping
(c) Ratooning (d) Seasoning

(102) Brinjal variety Arka Anand is resistant to _____.
(a) Fusarium wilt (b) Bacterial wilt
(c) Damping off (d) Phomopsis blight

(103) Chilli Hybrid -1 was developed at _____.
(a) IIHR, Bangalore (b) PAU, Ludhiana
(c) IARI, New Delhi (d) HAU, Hisar

(104) Self – incompatibility leads to cross – pollination is _____.
(a) Cucumber (b) Carrot
(c) Cauliflower (d) Okra

(105) The condition where a flower cannot be fertilized by the pollen of same is called _____.
(a) Dichogammy (b) Homogammy
(c) Unisexual (d) Self-incompatability

(106) In honey bees, the pollen basket is present in _____.
(a) Fore legs (b) Hinds legs
(c) Both legs (d) In between fore and hind legs

(107) The method of lye peeling is through————————————————.
  (a) Sulphuric acid   (b) Nitric acid
  (c) Caustic soda   (d) Peelers
(108) Sauerkraut is product prepared by————————————————.
  (a) Freezing   (b) Dehydration
  (c) Fermentation   (d) Canning
(109) Amino-ethoxyvinyl Glycine (AVG) is a potent inhibitor of————————.
  (a) Ethylene   (b) Cytokinin
  (c) Auxin   (d) Gibberellin
(110) Cell division is generally caused due to. ————————————————.
  (a) Auxin   (b) Cytokinin
  (c) Gibberellin   (d) Ethylene
(111) Chilling treatment for breaking dormancy is called————————————.
  (a) Scarification   (b) Stratification
  (c) Vernalization   (d) Leaching
(112) Leek is botanically known as.————————————————————.
  (a) *Allium fistulosum*   (b) *Allum porrum*
  (c) *Allium schoonpprasum* (d) *Allium cepa*
(113) National Horticulture Board (NHB) was established in————————.
  (a) 1984   (b) 1990
  (c) 1965   (d) 1978
(114) Heterobeltiosis is estimated over the. ————————————————.
  (a) Popular variety   (b) Mid parent
  (c) Better parent   (d) Popular hybrids
(115) ———————————————— is a autogamous vegetables.
  (a) Palak   (b) Onion
  (c) Radish   (d) Tomato
(116) A vegetable highly tolerant to soil acidity————————————————.
  (a) Okra   (b) Cauliflower
  (c) Asparagus   (d) Potato
(117) Rajendra Swathi is a variety of————————————————————.
  (a) Coriander   (b) Fennugreek
  (c) Black paper)   (d) Cardamom
(118) Scooping operation is adopted in ————————————————.
  (a) Cauliflower   (b) Knol- khol
  (c) Cabbage   (d) Kale

(119) The first tropical variety of cabbage which can set seeds in North Indian Plains _____.
   (a) Pusa Sambandh   (b) Pusa Ageti
   (c) Golden Acre   (d) Pusa Drum Head

(120) Main objective of vegetable blanching is _____.
   (a) Inactivation of bacteria   (b) Inactivation of enzymes
   (c) Fixation of colour   (d) Removal of tissue gas

(121) Pheromone traps are commonly used to control _____.
   (a) Aphids   (b) Fruit fly
   (c) Mites   (d) Red pumpkin beetle

(122) In cauliflower, which gene is responsible for assimilating beta carotene _____.
   (a) Mi   (b) B
   (c) Or   (d) Acr

(123) White brinjal recommended for _____.
   (a) Diabetes   (b) Heart disease
   (c) Skin disease   (d) Cancer

(124) Contribution of vegetables in horticulture sector in India is about _____.
   (a) 40%   (b) 50%
   (c) 60%   (d) 70%

(125) Indian Institute of Vegetable Research is located at _____.
   (a) New Delhi   (b) Varanasi
   (c) Bangalore   (d) Mumbai

(126) _____ enzyme produces bad smelling aldehydes in vegetables _____.
   (a) Polyphenol oxidase   (b) Lipooxidase
   (c) Amylase   (d) Protease

(127) The progenitor of cultivated cucumber is _____.
   (a) Cucumis *humifractus*   (b) Cucumis *hystibus*
   (c) Cucumis *matuliferus*   (d) Cucumis *hardwickii*

(128) _____ solanaceous vegetable has maximum seed productivity.
   (a) Tomato   (b) Chilli
   (c) Bell pepper   (d) Brinjal

(129) Pasteurization of fruits and vegetables juice is done at what temperature for 30 minutes _____.
   (a) 85-90°C   (b) 105-110°C
   (c) 45-50°C   (d) 65-70°C

(130) Name the tomato hybrid carrying resistance to three major diseases ————.
  (a) Arka Ananya (b) Arka Rakshak
  (c) Arka Alok (d) Arka Abha
(131) Storage life of fruits and vegetables is extended by keeping them ———— and ————.
  (a) High in CO_2 and low in oxygen (b) High in oxygen and low in CO_2
  (c) At low temperature (d) At relatively high humidity
(132) A triploid watermelon variety Pusa Bedana was released in ————.
  (a) 1970 (b) 1972
  (c) 1976 (d) 1982
(133) Which wild species of brinjal carries resistance to shoot and fruit borer ————?
  (a) *Solanum mammosum* (b) *Solanum sisymbrifolium*
  (c) *Solanum incannuum* (d) *Solanum torvum*
(134) Selfing and massing is useful for the improvement of which of the following vegetable crops ————.
  (a) Onion (b) Cabbage
  (c) Tomato (d) Spinach
(135) Pusa Parvati is a variety of French bean developed through ————.
  (a) Clonal Selection (b) Mass selection
  (c) Mutation breeding (d) Pedigree selection
(136) Drying recovery of trumeric is ————.
  (a) 10-15% (b) 15-30%
  (c) 20-30% (d) 5-10%
(137) Which of the following Indian State is popularly known as Garden of Spices ————.
  (a) Karnataka (b) Kerala
  (c) Andhra Pradesh (d) Tamil Nadu
(138) Coriander originated ————.
  (a) Mediterranean region (b) Indonesia
  (c) Western Ghats of India (d) South East Asia
(139) Edible part of tomato is ————.
  (a) Pericarp and placenta (b) Thalmus
  (c) Mesocarp (d) Endocarp
(140) Palam Triloki ia variety of ————.
  (a) Tomato (b) Garden pea
  (c) French bean (d) Broccoli

(141) The first KVK was established at Pondicherry during the year————————.
    (a) 1960   (b) 1974
    (c) 1980   (d) 1985

(142) Which year is the deadline for completion of PMGSY————————?
    (a) 2018   (b) 2019
    (c) 2020   (d) 2021

(143) Durga Kesar is a variety of————————.
    (a) Watermelon   (b) Pumpkin
    (c) Muskmelon   (d) Summer squash

(144) Powdery mildew disease is serious problem of ————————.
    (a) Root crops   (b) Cucurbits
    (c) Cole crops   (d) Leafy vegetables

(145) The sex form of commonly cultivated black pepper is ————————.
    (a) Monoecious   (b) Andromonoecious
    (c) Gynomonoecius   (d) Gynoecious

(146) ———————— a is wild species of tomato tolerant to salt.
    (a) *Solanum cheesmanii*   (b) *Solanum pimpinellifolium*
    (c) *Solanum chilense*   (d) *Solanum penneli*

(147) GMS in chilli is maintained by ————————.
    (a) Heterozygous pollinator   (b) Homozygous pollinator
    (c) Mutation   (d) Polyploidy

(148) *Curcuma* sp ———————— having high starch content
    (a) *C. augustifolia*   (b) *C. zedoana*
    (c) *C. aromatic*   (d) *Curcuma longa*

(149) Male sterility is common in ————————.
    (a) Carrot   (b) Beans
    (c) Okra   (d) Palak

(150) Monohybrids are derived by hybridizing two individuals which differ for ————————.
    (a) Single character   (b) Double character
    (c) Quadruple character   (d) Multiple character

(151) Sweet potato requires photoperiod of ————————.
    (a) 7 hours   (b) 13 hours
    (c) 9.5 hours   (d) 11 hours

(152) Corinder variety tolerant to whitefly and mites is ————————.
    (a) Sindhu   (b) Sadhana
    (c) Co-1   (d) Rajendra Swathi

(153) Palak variety tolerant to cercospora leaf spot is ─────────.
    (a) Pusa Jyoti   (b) Arka Harit
    (c) Pant Composite-1   (d) Jobner Green

(154) The major share of fruit and vegetable export from India ─────────.
    (a) UAE   (b) Russia
    (c) Europe   (d) USA

(155) ───────── is protandrous.
    (a) Tomato   (b) Cauliflower
    (c) Onion   (d) French bean

(156) Triploid cassava variety is ─────────.
    (a) Co_2 (b) Sree Harsha
    (c) Sree Savya   (d) Sree Ratna

(157) Where does the water vapour exit the plant ─────────?
    (a) Xylem   (b) Stem
    (c) Roots   (d) Stomata

(158) How many genes have been reported for powdery mildew resistance in pea ─────────.
    (a) Single recessive er
    (b) Two recessive er_1 and er_2
    (c) One recessive (er) and one dominant(Er)
    (d) Two recessive er1 and er2) and one dominant (Er3)

(159) Red flash pumpkins are rich source of ─────────.
    (a) Vitamin C   (b) Proteins
    (c) Vitamins A   (d) Acids

(160) First bioinsecticide developed on commercial scale was ─────────.
    (a) Quinine   (b) Sporeine
    (c) Organophosphate   (d) DDT

(161) Who proposed the biogenetic law ─────────?
    (a) Charles Darwin   (b) Ernst Haeckel
    (c) Morgan   (d) Watson

(162) ───────── is not ethylene inhibitor?
    (a) AVG   (b) MVO
    (c) NAA   (d) AOA

(163) Apical dominance in plants is related to ─────────.
    (a) Morpheclins   (b) Gibberellin
    (c) Auxin   (d) Cytokinin

(164) Bordeaux mixture was discovered in ——————————.
  (a) France	(b) UK
  (c) Australia	(d) USA

(165) Horticulture has been derived from the world *hortus* ——————————.
  (a) German	(b) Latin
  (c) English	(d) French

(166) Ecodormancy is due to——————————
  (a) Water stress	(b) Chilling
  (c) Photoperiod	(d) Seed coat

(167) Recalcitrant seeds are ——————————
  (a) Tolerant to moisture loss	(b) Tolerant to disease
  (c) Tolerant to insect attack	(d) Less storage life

(168) Good quality oleoresin can be extracted from ——————————.
  (a) Coriander	(b) Black pepper
  (c) Fennugreek	(d) Onion

(169) The botanical name of Chayote or chow-chow is ——————————.
  (a) *Sechium edule*	(b) *Mermecylon edule*
  (c) *Coccinia grandis*	(d) *Coccinia indica*

(170) Arka Lohit is a variety of ——————————.
  (a) Tomato	(b) Chilli
  (c) Carrot	(d) Beet

(171) —————————— PGR induces pistillate flower in cucumber.
  (a) Ethephon	(b) IAA
  (c) NAA	(d) GA_3

(172) Storage loss of onion and potato by sprouting can be prevented by using which of the following growth regulator as sprout inhibitor ——————————.
  (a) MH	(b) GA_3
  (c) CCC	(d) SADH

(173) Minimum isolation distance for the production of foundation seed in onion should be ——————————.
  (a) 200m	(b) 400m
  (c) 600m	(d) 1000m

(174) Anthracnose in chilli is caused by ——————————.
  (a) *Curvularia*	(b) *Colletrotrichum*
  (c) *Alternaria*	(d) *Phytophthora*

(175) All the cole crops have originated from which of the following single ancester ─────────────.

(a) *Brassica oleracea* var.*capitata* (b) *Brassica oleracea* var.*gemmifera*
(c) *Brassica oleracea* var. *sulvestris* (d) *Brassica oleracea* var. *botrytis*

(176) ───────── is a non- parasitic disorder due to Ca deficiency,

(a) Blossom end rot (b) Radial cracking
(c) Blotchy ripening (d) cat face

(177) Brown anther type male sterility is observed in ─────────

(a) Radish (b) Turnip
(c) Carrot (d) Beet

(178) The processing of producing several identical copies of a gene sequence is known as ─────────

(a) Gene cloning (b) Gene sequencing
(c) Gene splicing (d) Micro cloning

(179) Which compound imparts colour and flavour to chilli ─────────?

(a) Oleoresins (b) Capsanthin
(c) Solanin (d) Lycopene

(180) The process of synthesis of mRNA from a DNA template is known as ─────────.

(a) Translation (b) Transcription
(c) Transduction (d) Transformation

(181) Most efficient species of earthworm for vermicomposting in tropical region is ─────────.

(a) *Eisenia fetida* (b) *Eudrilus eugeniae*
(c) *Perionyx excavates* (d) *Bimastos parvus*

(182) The colour of tag used for breeder seed is ─────────.

(a) Golden brown (b) White
(c) Blue (d) Green

(183) ───────── is a good source of L- Dopa (Dopamine to cure Parkinson disease in humans.

(a) Broccoli (b) Broad bean
(c) Celery (d) Lettuce

(184) Core intact is one of the methods of the seed peoduction methods used in which of the following crops? ─────────

(a) Broccoli (b) Cabbage
(c) Knol- khol (d) Brussels sprouts

(185) In which of the following plants cellulose content is ―――――――.
   (a) Cotton   (b) Sugarcane
   (c) Radish   (d) Potato

(186) Tetrad is seen in ―――――――.
   (a) Diplotene   (b) Pachytene
   (c) Diakinesis   (d) Krebs cycle

(187) Anticancer properties in crucifers is due to the presence of ―――――――.
   (a) Indole-3 carbinol (b) Diadzein
   (c) Anthocyanin   (d) Carotenoids

(188) Technique of combining two or more major genes is known as ―――――――.
   (a) Genetic Engineering (b) Gene pyramiding
   (c) Gene map   (d) Combining ability

(189) The term climacteric was first used by ―――――――.
   (a) Gane   (b) Kidd and West
   (c) Cruess   (d) Bleaker

(190) Which of these compound is unable to under go mutarotation ―――――――?
   (a) Glucose   (b) Maltose
   (c) Lactose   (d) Sucrose

(191) Xanthosoma is a ―――――――.
   (a) Tuber crop   (b) Leafy crop
   (c) Fruit crop   (d) Edible flower

(192) In a bisexual flower, the phenomenon when gynoecious matures earlier than androecium is called ―――――――.
   (a) Protandry   (b) Autogammy
   (c) Heterogammy   (d) Protogammy

(193) Which of the mechanism exist in carrot ―――――――.
   (a) Heterostyly   (b) Protogyny
   (c) Protandry   (d) Decliny

(194) The main site for dark reaction of photosynthesis is ―――――――.
   (a) Stroma   (b) Grana
   (c) Chloroplast   (d) Chromoplast

(195) Pusa Early Bunching is a varidety of ―――――――.
   (a) Spinach   (b) Radish
   (c) Fennugreek   (d) Coriander

(196) Marsh spot in pea is due to the deficiency of ―――――――.
   (a) Cu   (b) Fe
   (c) Cl   (d) Mn

(197) C4 vegetable crop is ——————.
- (a) Tomato
- (b) Cucumber
- (c) Lettuce
- (d) Amranth

(198) The world auxin is derived from ——————.
- (a) Latin
- (b) Greek
- (c) English
- (d) French

(199) Term Golden revolution is related to ——————.
- (a) Mustard
- (b) Horticulture
- (c) Agroforestry
- (d) Sugarcane

(200) Intercropping of tomato with cabbage can reduce the damage done by ——————.
- (a) Diamond – back moth
- (b) Aphids
- (c) Butterflies
- (d) Leafhopper

(201) R lacquered cans are used for canning of ——————.
- (a) High acid foods
- (b) Non – acid foods
- (c) Low acid foods
- (d) Neutral foods

(202) End product of glycolysis is ——————.
- (a) Mallic acid
- (b) Citric acid
- (c) Pyruvic acid
- (d) oxalic acid

(203) Tag colour of certified seed is ——————.
- (a) Golden yellow
- (b) White
- (c) Purple
- (d) Blue

(204) Who gave student,s t test ——————.
- (a) Mendel
- (b) Gosset
- (c) Fisher
- (d) Karl Pearson

(205) Lincolin is an exotic variety of ——————.
- (a) Cowpea
- (b) Pea
- (c) French bean
- (d) Broad bean

(206) —————— is a good source of iodine in our diet.
- (a) Okra
- (b) Watermelon
- (c) Bitter gourd
- (d) Garlic

(207) Coated or pelleted seeds are used for ——————.
- (a) Low germination rate seeds
- (b) Bold seeds
- (c) To prevent odour
- (d) Dormant seeds

(208) Triploid water melon is ——————.
- (a) Tetra-2
- (b) Tetra -4
- (c) Arka Madhura
- (d) Pusa Rusal

(209) Kashi Haritima is a variety of ——————.
   (a) Cowpea   (b) Indian bean
   (c) Pea   (d) Jack bean
(210) Crossing over take place in which stage of meiosis——————.
   (a) Zygotene   (b) Diplotene
   (c) Pachytene   (d) Leptotene
(211) Naveen, Vaishali, Rupali, Avanish and Krishna are hybrids of——————.
   (a) Brinjal   (b) Tomato
   (c) Chilli   (d) Okra
(212) Arka Pitamber is ——————.
   (a) Yellow coloured onion   (b) White coloured onion
   (c) Red coloured onion   (d) Kharif onion
(213) Fruit cracking in tomato is due to ——————.
   (a) B deficiency   (b) Ca deficiency
   (c) K deficiency   (d) Mg deficiency
(214) Which variety of carrot is released from IARI Katrain Station——————.
   (a) Pusa Keser   (b) Pusa Meghali
   (c) Pusa Yamdagni   (d) Zeno
(215) Rio- De Janerio is a variety of ——————.
   (a) Black pepper   (b) Fenugreek
   (c) Ginger   (d) Turmeric
(216) Non- pungent part of chilli is ——————.
   (a) Seeds   (b) Pericarp
   (c) Placenta   (d) Tendril
(217) Hottest chilli of India is ——————.
   (a) Pusa Jwala   (b) NP-46A
   (c) Bhoot Jolokia   (d) Bhaskar
(218) Bitterness in bitter gourd is due to ——————.
   (a) Momordicidin   (b) Cucurbitacin
   (c) Allyl propyl di sulphide   (d) Cheratin
(219) Pungency in dradish is due to ——————.
   (a) Isothiocyanate   (b) Allyl propyl di sulphide
   (c) Diallyl di sulphide (d) Allyl iso thiocynate
(220) What is full form of PCR.——————?
   (a) Polymerase chain reaction   (b) Polymer chain reaction
   (c) Polymeric chain reaction   (d) Polymerised chain rection

(221) Which one of the physiological disorder of tomato ———————?
    (a) Puffiness   (b) Hollow stem
    (c) Black heart   (d) Spliting

(222) Indicator plant for B and Mo deficiency is ——————————.
    (a) Cucurbits   (b) Cauliflower
    (c) Tomato   (d) Okra

(223) Precursor of auxin is ——————————————.
    (a) Mn   (b) Mg
    (c) Tryptophane   (d) Lysine

(224) Precursor of ethylene is ——————————————.
    (a) Cystein   (b) Methionine
    (c) Leucine   (d) Lysine

(225) Edible part of clove is ——————————————.
    (a) Aril   (b) Unopened flower bud
    (c) Fruit   (d) Stigma

(226) ———————————— is used for increasing male floer in greenhouse cucumber.
    (a) NAA   (b) CCC
    (c) Ethylene   (d) $AgNO_3$

(227) AGMARK standard ——————————————.
    (a) Is related with agricultural products
    (b) Is approved by the Directorate of Marketing and Inspection
    (c) Both (a) and (b)
    (d) None of the above

(228) The method used to detect a viral disease is ——————————————.
    (a) AFLP   (b) RFLP
    (c) ELISA   (d) None of these

(229) First Agricultural University established in India was ——————————————.
    (a) IIHR, Bangalore   (b) PAU, Ludhiana
    (c) IARI, New Delhi   (d) GBPU&T, Pantnagar

(230) Term appetizing is related to ——————————————.
    (a) Packaging   (b) Canning
    (c) Storage   (d) None of these

(231) Lux is unit of ——————————————————.
    (a) Light   (b) Photosynthesis
    (c) Transpiration   (d) Respiration

(232) Pulsing is related to —————————.
 (a) Treatment   (b) Preservation
 (c) Handling   (d) Storage

(233) Cumin belongs to family—————————.
 (a) Ranunculaceae   (b) Apiaceae
 (c) Malvaceae   (d) None of these

(234) ————————— is wind pollinated vegetable.
 (a) Radish   (b) Beet
 (c) Amranthus   (d) Both (a) & (b)

(235) Progeny of foundation seed is —————————.
 (a) Certified seed   (b) Registered seed
 (c) Nucleus seed   (d) None of the above

(236) *Flavr savr* is an important variety of—————————.
 (a) Potato   (b) Tomato
 (c) Sweet Potato   (d) Beet

(237) Which of the following is ethylene absorbent —————————?
 (a) $KMNO_4$ (b) K_2HPO_4
 (c) KI   (d) None of the above

(238) Stomatal activity is controlled by—————————.
 (a) Cl   (b) K
 (c) Na   (d) Mg

(239) Rutabaga is a cross between —————————.
 (a) Cauliflower X cabbage   (b) Cauliflower and turnip
 (c) Cabbage and turnip   (d) Radish and turnip

(240) First Indian who won World Food Prize is —————————?
 (a) Vergese Kurein   (b) MS Swaminathan
 (c) G Kalloo   (d) RS Paroda

(241) First auxin identified by F W Went in the year 1928 was —————————.
 (a) IAA   (b) IBA
 (c) PAA   (d) None of the above

(242) Preservation of cells, tissue and organ in liquid nitrogen is known as —————————.
 (a) Cryopreservation   (b) *In-situ* preservation
 (c) Cold preservation (d) None of these

(243) Application of chemical fertilizers using modern irrigation techniques is known as ─────────────.
   (a) Chemigation (b) Fertigation
   (c) Foliar spray (d) None of above

(244) Photoblastism has been used in ─────────────.
   (a) Tomato (b) Watermelon
   (c) Potato (d) Lettuce

(245) Pulsing is performed to increase the shelf life of ─────────────.
   (a) Fruits (b) Flowers
   (c) Vegetables (d) All of the above

(246) Tomato is a ─────────────.
   (a) Long day plant (b) Short day plant
   (c) Day natural plant (d) None of the above

(247) Loss of vigour in productivity is due to ─────────────.
   (a) Genetic drag (b) Linkage
   (c) Luxuriance (d) Inbreeding depression

(248) Leafy vegetables are stored through ─────────────.
   (a) Air blasting (b) Vaccum cooling
   (c) Freezing (d) Room cooling

(249) Value added and fermentation product from white cabbage is ─────────────.
   (a) Sauerkraut (b) Pickle
   (c) Silage (d) All of the above

(250) Large cardamom is propagated by ─────────────.
   (a) Seeds (b) Rhizome
   (c) Roots (d) None of the above

(251) Spinach is generally ─────────────.
   (a) Wind pollinated (b) Insect - pollinated
   (c) Bird pollinated (d) All of the above

(252) Maximum amount of variability is present in ─────────────.
   (a) Cole (b) Cucurbits
   (c) Solanaceae (d) None of the above

(253) Total recommended requirement of vegetables per capita per day is ─────────────.
   (a) 300g (b) 245g
   (c) 120g (d) 220g

(254) Autoclave is mainly used for —————.
   (a) Sterilization of glassware
   (b) Culture of media
   (c) Sterilization of media
   (d) None of the above

(255) Globe artichoke is cultivated for —————.
   (a) Succulent leaves
   (b) Flower buds
   (c) Tuber
   (d) None of the above

(256) Asparagus belongs to —————.
   (a) Leguminosae
   (b) Liliaceae
   (c) Composite
   (d) None of the above

(257) Chemical formula of KMS is —————.
   (a) $K_2Cr_2O_7$
   (b) $KMnO_4$
   (c) $Kr_2Cr_3O_4$
   (d) $K_2S_2O_5$

(258) CH-1 cultivar of chilli is bred through —————.
   (a) CMS
   (b) CGMS
   (c) GMS
   (d) None of the above

(259) Protein synthesis takes place in —————.
   (a) Nucleus
   (b) Nucleolus
   (c) Ribosome
   (d) Mitochondria

(260) DNA synthesis take place in —————.
   (a) S- Phase
   (b) G-1 phase
   (c) G-2 phase
   (d) Telophase

(261) Leaf area is measured by —————.
   (a) Porometer
   (b) Barometer
   (c) Auxanometer
   (d) Planometer

(262) Specific gravity indicates —————.
   (a) Purity
   (b) Maturity
   (c) Harvesting index
   (d) All of above

(263) Commonly used bactericide is —————.
   (a) Mancozeb
   (b) Sulfex
   (c) Carboxin
   (d) Streptomycine sulphate

(264) Frozen material needs to be —————.
   (a) Rehydrated
   (b) Thawed
   (c) Dehydrated
   (d) Cooled

(265) The degree of freedom in 2X3 factorial experiment with 2 replications will be —————.
   (a) 21
   (b) 26
   (c) 28
   (d) 30

(266) Red tinge with pink flesh watermelon is due to ─────────.
    (a) Anthocyanin   (b) Lycopene
    (c) Both a and b   (d) None of above
(267) The multivitamins green vegetable is ─────────.
    (a) Winged bean   (b) Chekermanis
    (c) Amranthus   (d) French bean
(268) Aroma, taste and feel lead to─────────.
    (a) Rancidity   (b) Viscosity
    (c) Flavour   (d) None of the above
(269) Cucurmin is extracted from─────────.
    (a) Turmeric   (b) Kokum
    (c) Ginger   (d) Curry leaf
(270) Potato ploidy level and chromosome number is─────────.
    (a) 2n=2x=48   (b) 2n=4X=48
    (c) 2n=2X=24   (d) 2n=2x=24
(271) Autogamy is found in ─────────.
    (a) Brinjal   (b) Tomato
    (c) Chilli   (d) Okra
(272) In which vegetable crop curing is done? ─────────.
    (a) Garlic   (b) Potato
    (c) Onion   (d) Yam
(273) The head quarter of WTO is located at─────────.
    (a) Geneva   (b) New York
    (c) Brazil   (d) Canada
(274) AVRDC is situated at ─────────.
    (a) Brazil   (b) Taiwan
    (c) Geneva   (d) New York
(275) Which plant part shows symptom of Ca deficiency ─────────.
    (a) Lower leaves   (b) Terminal leaves
    (c) New leaves   (d) All of the above
(276) Hollow stem is more prominent in─────────.
    (a) Small size tubers   (b) Medium sized tubers
    (c) Over sized tubers   (d) All of these
(277) Indian Horticulture is published from─────────.
    (a) IIHR   (b) PAU
    (c) ICAR   (d) IARI

(278) Arka Alok is a variety of tomato which shows resistance against—————.
   (a) Bacterial wilt   (b) Nematode
   (c) TLCV   (d) All of above

(279) A cut brinjal when exposed to air instantly turn brown. Such browning is due to action of —————.
   (a) PPO   (b) Lipoxygenase
   (c) Catalase   (d) Ascorbic acid

(280) Randomization is used to remove—————.
   (a) Bios   (b) Degree of freedom
   (c) Probability   (d) LSD

(281) Range of correlation coefficient is —————.
   (a) 0-1   (b) -1 to +1
   (c) -1 to 0   (d) None of the above

(282) Golden nematode is a serious problem of—————.
   (a) Garlic   (b) Brinjal
   (c) Potato   (d) Tomato

(283) ————— chemical is used as preservative in tomato ketchup
   (a) Potassium meta bi sulphite (b) Sodium benzoate
   (c) Sorbic acid   (d) Glacial acetic acid

(284) ————— preservative is used to preserve coloured product.
   (a) Sodium benzoate   (b) Sorbic acid
   (c) Benlate   (d) Potassium meta bi sulphite

(285) Edible portion of broccoli is known as —————.
   (a) Head   (b) Leaves
   (c) Curd   (d) Flower

(286) ————— portion of brussels sprout is edible.
   (a) Stem   (b) Modified roots
   (c) Flower   (d) Small buds and sprouts

(287) In which plant part the process of photosynthesis takes place
   (a) Mitochondria   (b) Ribosomes
   (c) Peroxisome   (d) Chloroplast

(288) Indigenous vegetables among following —————.
   (a) Brinjal   (b) Tomato
   (c) Artichoke   (d) Potato

(289) Function of potassium is —————.
   (a) Osmotic regulation   (b) Closing and opening of stomata
   (c) N2 fixation   (d) All of the above

(290) ———————— is multipurpose vegetable.
   (a) Drumstick   (b) Curry leaf
   (c) Brinjal   (d) Radish

(291) ———————— is early variety of garlic.
   (a) Agrifound Parvati   (b) G-282
   (c) G-41   (d) All of the above

(292) Arka Pragati variety of onion is suitable for ————————.
   (a) Rabi season   (b) Kharif season
   (c) Both season   (d) Summer season

(293) Shape of Scarlet Globe variety of radish is————————.
   (a) Cylindrical   (b) Round
   (c) Square   (d) None of the above

(294) Which variety of turmeric have high dry recovery of curcumin ————————.
   (a) Rango   (b) Suroma
   (c) Sughandham   (d) All of the above

(295) Temperature for cryopreservation of liquid nitrogen is ————————.
   (a) -196°C   (b) -44°C
   (c) 196K   (d) 100°C

(296) Organization related to Agricultural Marketing is ————————.
   (a) NABARD   (b) UNESCO
   (c) FCI   (d) NAFED

(297) Harvesting of melons for distant market————————.
   (a) Half-slip stage   (b) Green mature stage
   (c) Full slip stage   (d) Immature stage

(298) ———————— is king of spices.
   (a) Cardamom   (b) Pepper
   (c) Clove   (d) Chilli

(299) ———————— is oxygenator plant.
   (a) Elodea   (b) Azolla
   (c) Trapa   (d) Eichornia

(300) Arka Ananya is a variety of ————————.
   (a) Tomato   (b) Chilli
   (c) Brinjal   (d) Potato

(301) N-53 is variety of————————.
   (a) Garlic   (b) Onion
   (c) Pea   (d) Cabbage

(302) The infloresence of colocassia is —————————.
 (a) Spadix   (b) Penicle
 (c) Raceme   (d) Cymose
(303) Function of B and Zn is —————————
 (a) Reduce absicission   (b) Reduce fruit set
 (c) Increase abscission   (d) None of the above
(304) ————————— is manmade vegetable.
 (a) Artichoke   (b) Rutabaga
 (c) Brinjal   (d) None of the above
(305) ————————— is autotetraploid vegetable.
 (a) Onion   (b) Garlic
 (c) Leek   (d) Brinjal
(306) ————————— is viviparron vegetable crop —————————.
 (a) Rutabaga   (b) Pea
 (c) Chow-chow   (d) Cucumber
(307) Triploids are generally —————————.
 (a) Fertile   (b) Sterile
 (c) Male sterile   (d) Male fertile
(308) Palam Priya variety of pea shows resistant to —————————.
 (a) DM   (b) PM
 (c) Blight   (d) All of the above
(309) Ribosomes are produced in —————————.
 (a) Mitochondria   (b) ER
 (c) Lysosome   (d) Protoplast
(310) In small samples which test is used —————————.
 (a) t test   (b) F test
 (c) Z test   (d) All of the above
(311) Leafy vegetable are rich source of —————————.
 (a) Fe   (b) Mg
 (c) Folic acid   (d) None of the above
(312) ————————— is antioxidant.
 (a) Vitamin C   (b) Sugar
 (c) Lipid   (d) Mg and Ca
(313) Sex reversal is found in —————————.
 (a) Cucumber   (b) Onion
 (c) Tomato   (d) Papaya

(314) *Brassica oleracea* var, acephala is botanical name of ——————.
    (a) Cauliflower   (b) Cabbage
    (c) Kale   (d) None of the above

(315) Chemical used as germicide ——————
    (a) Urea   (b) Glucose
    (c) $Ca(NO_3)_2$ (d) HQC

(316) In a DNA molecule, thymine always pair with ——————.
    (a) Adenine   (b) Guanine
    (c) Cystocine   (d) Urecil

(317) The non- preference insect resistance mechanism is known as ——————.
    (a) Antochlorosis   (b) Antibiosis
    (c) Antixenosis   (d) Antopheromonosis

(318) Abscisic acid promotes ——————.
    (a) Leaf division   (b) Bud emergence
    (c) Leaf senescence   (d) Shoot elongation

(319) The highest potato producing state in India is ——————.
    (a) UP   (b) MP
    (c) WB   (d) Bihar

(320) Pellagra is due to deficiency of ——————.
    (a) Naicin   (b) Pyrodoxine
    (c) Folic acid   (d) Biotin

(321) Kaveri is an improved variety of ——————.
    (a) Cauliflower   (b) Kokum
    (c) Passion fruit   (d) Coriander

(322) T & V programme was introduced in ——————.
    (a) 1964   (b) 1978
    (c) 1974   (d) 1982

(323) —————— is known as God saint vegetable.
    (a) Pea   (b) Tomato
    (c) Winged bean   (d) Brinjal

(324) Stamenless mutant is reported in ——————.
    (a) Cabbage   (b) Brinjal
    (c) Tomato   (d) Carrot

(325) The type of germination in garden pea is ——————.
    (a) Epigeal   (b) Hypogeal
    (c) Hypo-epigeal   (d) Epi-hypogeal

(326) Curd initiation and development in cauliflower is mainly influenced by _____.
    (a) Temperature   (b) Photoperiod
    (c) Relative humidity   (d) None of the above

(327) Black rot is a most common disease of _____.
    (a) Tomato   (b) Potato
    (c) Celery   (d) Cabbage

(328) _____ is ripening harmone.
    (a) Kinetin   (b) GA
    (c) MH   (d) Ethylene

(329) _____ is cleistogamous vegetable.
    (a) Garden pea   (b) Lettuce
    (c) Capsicum   (d) Tomato

(330) *Curcuma longa* is botanical name of _____.
    (a) Ginger   (b) Turmeric
    (c) Cardamom   (d) Black pepper

(331) Nutritional quality breeding deals with genetic improvement in _____.
    (a) Vitamins   (b) Oil
    (c) Protein   (d) All of the above

(332) South – west monsoon contributes about of _____ total rainfall.
    (a) 30%   (b) 20%
    (c) 50%   (d) 85%

(333) Self – pollination is a form of _____.
    (a) Out breeding   (b) Inbreeding
    (c) Random breeding   (d) None of the above

(334) In genome, each type of chromosome is represented _____.
    (a) Once   (b) Twice
    (c) Thrice   (d) Many time

(335) About 95 to 99.5 % portion of plant tissues are made _____.
    (a) C, H and O   (b) N, P and K
    (c) Ca, Mg and O   (d) Cu, Zn and Fe

(336) Cytoplasmic genes are found in _____.
    (a) Mitochondria   (b) Chloroplast
    (c) Both a and b   (d) Nucleus

(337) Alkali soils are reclaimed by _____.
    (a) Calcium chloride   (b) Gypsum
    (c) Lime   (d) All of the above

(338) *Su- generis* is a latin phrase which means ―――――.
    (a) A new system   (b) A secure system
    (c) A system of its own   (d) A conventual system

(339) Nutmeg plants are ―――――.
    (a) Monoecious   (b) Andromonoecious
    (c) Androdioecious   (d) Dioecious

(340) Nucleus was first discovered by ―――――.
    (a) Benda(1877)   (b) CamitoGolgi(1832)
    (c) Robert Brown(1833)   (d) Flemming(1882)

(341) A nullisomic individual is represented by ―――――.
    (a) 2n-1   (b) 2n+1
    (c) 2n-2   (d) 2n-2

(342) Colour blindness genes are located on ―――――.
    (a) Y chromosome   (b) X chromosomes
    (c) Autosomes   (d) All of the above

(343) In a DNA molecule, thymine always with ―――――.
    (a) Adenine   (b) Guanine
    (c) Cytosine   (d) Uracil

(344) Regeneration of a plant from a single cell in culture medium is called as―――――.
    (a) Protoplast culture   (b) Meristim culture
    (c) Organ culture   (d) Cell culture

(345) The number of chromosomes is reduced to half in plant obtained by ―――――.
    (a) Anther   (b) Ovule culture
    (c) Shoot tip culture   (d) Both a and b

(346) The term genotype and phenotype was coined by ―――――.
    (a) Correns (1900)   (b) Johanson(1903)
    (c) Shull(1908)   (d) East(1908)

(347) First intergeneric hybrid between radish and cabbage was made by―――――.
    (a) Hull(1945)   (b) Rimpu(1890)
    (c) Andrew Knight(1800)   (d) Karpechenko(1927)

(348) Central Agricultural Research Institute is located at ―――――.
    (a) Kolkata   (b) Portblair
    (c) Vishakhapatnam   (d) Thiruvananthapuram

(349) Self- incompatability results due to ―――――.
    (a) Genetic causes   (b) Morphological causes
    (c) Physiological and biochemical causes   (d) All of the above

(350) The principal of experimental design includes —————————.
    (a) Local control   (b) Replication
    (c) Randomization   (d) All of the above
(351) Denitrification is more in ——————————.
    (a) Light soils   (b) Water logged soils
    (c) Well drained soils   (d) Heavy soils
(352) The active principal of clove oil is ——————————.
    (a) Macine   (b) Oleum
    (c) Aldehyde   (d) Eugenol
(353) Symptom of calcium deficiency in plant is first noticed in ——————————.
    (a) Middle leaves   (b) Lower leaves
    (c) Terminal leaves   (d) All of the above
(354) —————— element is essential for protein synthesis ——————————.
    (a) Calcium   (b) Magnesium
    (c) Manganese   (d) Sulphur
(355) Pusa Rasraj is a F1 hybrid of ——————————.
    (a) Tomato   (b) Watermelon
    (c) Muskmelon   (d) Cucumber
(356) —————— vegetable crops can tolerate acid soil.
    (a) Cucumber   (b) Bean
    (c) Okra   (d) Watermelon
(357) —————— is often cross-pollinated vegetable crop ——————————.
    (a) Pea   (b) Tomato
    (c) Amranth   (d) Lima bean
(358) Indeterminate habit of tomato is associated with ——————————.
    (a) Lower provitamin A   (b) Lower total soluble solids
    (c) Poor keeping quality   (d) Higher total soluble solids
(359) —————— is an anemophilous vegetable.
    (a) Muskmelon   (b) Carrot
    (c) Amaranths   (d) Cauliflower
(360) The relationship between arthmetic mean (A), harmonic mean (H) and geometric mean (G) is——————————.
    (a) $H^2 = GA$ (b) $G^2 = AH$
    (c) $A^2 = GH$ (d) None of the above
(361) —————— p^H is the dividing line between the acid and non-acid foods ——————————.
    (a) 3.0   (b) 5.0 -4.0
    (c) 4.5   (d) 5.0

(362) Crucifer vegetables are value for —————.
    (a) Magnesium containing compound isothiocynates
    (b) Sulphur containing compounds glucosinolates
    (c) Molybdenum containing compound isoflavonones
    (d) Boron containing compound vanillin

(363) The immediate precursor for ethylene synthesis is —————.
    (a) SAM   (b) ACC
    (c) Polyamine   (d) Tryptophan

(364) Gametes are usually————— cells.
    (a) Diploid   (b) Haploid
    (c) Germ   (d) Polyploid

(365) Dormancy of seed can be due to—————.
    (a) Impermeability of seed coat   (b) Requirement of water
    (c) Presence of ABA in the seed   (d) All of above

(366) Gram contains about————— percent fat.
    (a) 2   (b) 4.5
    (c) 3   (d) 6

(367) Fusion of two cell is known as —————.
    (a) Cybrid   (b) Protoplast fusion
    (c) Hybrid   (d) Cell fusion

(368) Latest method of storage —————.
    (a) CAM   (b) MAP
    (c) Hypobaric   (d) Low temperature

(369) Geometic mean of a given series is always————— than its arithmetic mean.
    (a) Higher   (b) Lower
    (c) Equal   (d) None of the above

(370) ————— test is used for testing the significance of mean difference—————.
    (a) t-test   (b) z- test
    (c) x2- test   (d) f-test

(371) CRD is used when material is —————.
    (a) Heterogenous   (b) Homogenous
    (c) Specialized   (d) Both a and b

(372) Regression value always lie between —————.
    (a) 1and +1   (b) -1 to +1
    (c) +1 and -1   (d) None of the above

(373) —————— vegetable crop is commercially used for processing in India.
   (a) Tomato   (b) Ash gourd
   (c) Both a and b   (d) Tapioca
(374) Genetically modified vegetable crop recommended for commercial cultivation in India ——————.
   (a) Potato   (b) Brinjal
   (c) Tomato   (d) None of the above
(375) Ideal processing potato cultivar should have ——————.
   (a) High TSS and low sugar   (b) High dry matter and low sugar
   (c) High sugars   (d) High acidity
(376) —————— cultivation is mostly concentrated in peri- urban areas.
   (a) Fruit   (b) Vegetables
   (c) Medicinal crops   (d) Plantation crops
(377) Kaoline used as an antitranspirant ——————.
   (a) Reduces growth of the plants
   (b) Does reflect light from plant and leaf surface
   (c) Affects the clouser and opening of stomata
   (d) Forms thin layer on the leaf surface
(378) Pollination and fertilization before anthesis is called ——————.
   (a) Cleistogammy   (b) Chasmogammy
   (c) Homogammy   (d) Dichogammy
(379) Head office of NBPGR is located at ——————.
   (a) Raipur   (b) Hyderabad
   (c) New Delhi   (d) Manipur
(380) —————— is known as father of genetics.
   (a) Gregor Mendel   (b) Darwin
   (c) JC Kolreuter   (d) Lewig
(381) Polyphenol oxidase enzyme inactivate at the temperature of ——————.
   (a) 80-90°C   (b) 100°C
   (c) 50°C   (d) Above 100°C
(382) Indian Farming is a publication of ——————.
   (a) IARI, New Delhi   (b) CAZRI, Jodhpur
   (c) ICRISAT, Hyderabad   (d) ICAR, New Delhi
(383) Ideal packaging is ——————.
   (a) Minimizing post- harvest losses   (b) Makes commodity attractive for consumers
   (c) More fit for transportation (d) All of the above

(384) ELISA is ——————— linked test
   (a) Bacteria (b) Virus
   (c) Enzyme (d) Protozoa
(385) Molecular markers are used for determining variation in plants ———————.
   (a) Genetic (b) Phenotypic
   (c) Endoplasmic (d) Cytoplasmic
(386) RFLP is ———————.
   (a) Marker technique (b) Selection method
   (c) Mutation technique (d) Gene slicing technique
(387) ——————— is the source of Bt gene.
   (a) Fungi (b) Bacteria
   (c) Virus (d) Mycoplasma
(388) Early December is the improved variety of ———————.
   (a) Pea (b) Cauliflower
   (c) Cabbage (d) Carrot
(389) NRC for orchid is situated at ———————.
   (a) Gangtok (b) Pune
   (c) Delhi (d) Bikaner
(390) Genetic term is given by ———————.
   (a) Mendel (b) Galton
   (c) Jones (d) Batson
(391) ——————— is anti- gibberellin in action.
   (a) Cultar (b) GA
   (c) MH (d) all of the above
(392) In Plant breeding nearly homozygous lines are produced by continued self-fertilization accompanied by selection for ——————— generation.
   (a) 3-4 (b) 5-6
   (c) 7-8 (d) 9-10
(393) In onion ——————— is used as maintainer for the male sterile line.
   (a) Smsms (b) Nmsms
   (c) NMsMs (d) NMsms
(394) In cucumber production of male flowers can be induced by the application of ———————.
   (a) Boron (b) Ethephon 600ppm
   (c) GA2000ppm (d) Auxin

(395) In tomato variety Hisar Anmol, the donor for TLCV is ―――――――.
   (a) *Solanum peruvianum* (b) L. *pimpinellifollium*
   (c) *L. esculentum* (d) *L. hirsutum* f. sp. *glaberatum*
(396) Self- incompatability is overcome by ―――――――.
   (a) Bud pollination (b) Surgical techniques
   (c) Irridation (d) All of the above
(397) Single gene male sterility is transferred in ――――――― years.
   (a) 3-4 (b) 5-6
   (c) 7-8 (d) 4-5
(398) Hybrid onion programme originated in 1925 by ―――――――.
   (a) TW Whitaker (b) BD Dowker
   (c) V Swarup (d) HA Jones
(399) Pollen sterility in tomato can be identified by using ―――――――.
   (a) Acetocarmine (b) $KMnO_4$
   (c) Methylene blue (d) Phenolphthalein
(400) ――――――― is used as donor parent for salt tolerance in tomato breeding ―――――――.
   (a) sp *hirsutum* (b) sp *esculentum*
   (c) sp *cheesmanii* (d) sp *pennelli*
(401) ――――――― had introduced climate controlled greenhouse cultivation.
   (a) ICAR (b) Mahyco Seed Co.
   (c) Indo- American Hybrid Seed Co. (d) Sutton & Sons
(402) Protrays are related to ―――――――.
   (a) Post harvest (b) Nursery
   (c) Transport (d) None of the above
(403) International Institute of Horticulture is situated in ―――――――.
   (a) Brazil (b) Italy
   (c) USA (d) None of the above
(404) The most commonly used solidifying agent in plant tissue culture media is ―――――――.
   (a) Sucrose (b) Glucose
   (c) Lactose (d) Galactose
(405) The growth responsible for shoot elongation is ―――――――.
   (a) BAP (b) GA_3
   (c) 2-4-D (d) IAA

(406) —————— is constituent of nitrate reductase.
   (a) Mn   (b) Zn
   (c) Mo   (d) All of the above

(407) Auxanometer is used to ——————
   (a) TSS   (b) Transpiration rate
   (c) Acidity   (d) Growth of the plant

(408) —————— enzyme is responsible for fruit softening.
   (a) Catalase   (b) Oxidase
   (c) Phynyle ammonialyase   (d) Pectinase

(409) In Nutmeg, male and female flowers are planted in the ratio of ——————.
   (a) 1:10   (b) 1:15
   (c) 1:13   (d) 1:5

(410) Botanical Name of Bay leaf is ——————.
   (a) *Armoracia rusticana*   (b) *Petroseinum crisum*
   (c) *Laurus nobilis*   (d) *Satureia hortensis*

(411) Artificial pollination is essential in ——————.
   (a) Cinnamon   (b) Clove
   (c) Vanilla   (d) Nutmeg

(412) Multistyle flowers are common in ——————.
   (a) Okra   (b) Brinjal
   (c) Chillies   (d) Lima bean

(413) Mitochondria contain ribosome of ——————.
   (a) 60S type   (b) 70S type
   (c) 60S and 70S type   (d) 80S type

(414) Muriate of potash is called ——————.
   (a) K_2SO_4 (b) KCL
   (c) K_2O (d) $KMnO_4$

(415) DNA is polymer of ——————.
   (a) Nucleotides   (b) Nucleosides
   (c) Aminoacid   (d) None of the above

(416) Photorespiration involves ——————.
   (a) C_4 pathway   (b) Glyxylate cycle
   (c) TCA cycle   (d) Carbon cycle

(417) Yeast is used for making ——————.
   (a) Squash   (b) Cidar
   (c) Nectar   (d) Curd

(418) Poinsette is the variety of ———————————.
   (a) Logmelon   (b) Muskmelon
   (c) Watermelon   (d) Cucumber
(419) Pumpkin grown throughout in India is ———————————.
   (a) *C. moschata*   (b) *C. maxima*
   (c) *C. facifolia*   (d) *C. paradisi*
(420) Tomato sauce must have ———————————.
   (a) 10%   (b) 16%
   (c) 22%   (d) 30%
(421) Treatment of fruit of vegetables in boiling water or vapour for a definite period before their canning is known as ———————————.
   (a) Lye peeling   (b) Vapour heat treatment
   (c) Blanching   (d) Pasteurization
(422) FPO permits the use of only two chemicals as preservatives———————————.
   (a) Boric acid and sodium chloride
   (b) Acetic acid and citric acid
   (c) Potassium meta bisulphide and sodium benzoate
   (d) Potassium disulphide and sodium carbonate
(423) Onion bulbs show problem of bolting if the temperature remain below———————————.
   (a) 30°C   (b) 25°C
   (c) 10°C   (d) 15°C
(424) ——————— enhances seed germination.
   (a) PP333   (b) GA
   (c) NAA   (d) BA
(425) ——————— is recommended for maintain freshness in cut leaf vegetables.
   (a) Cytokinin   (b) ABA
   (c) Auxin   (d) GA
(426) Nantes is an important variety of ———————————.
   (a) Carrot   (b) Turnip
   (c) Beet root   (d) Radish
(427) TSS is measured by———————————.
   (a) Refractometer   (b) Thermometer
   (c) Colorimeter   (d) Spectrophotometer
(428) The book Horticultural Science has been written by ———————————.
   (a) TK Bose   (b) KL Chadha
   (c) Jules Janick   (d) JO Simmond

(429) The University of Horticulture and Forestry is situated at ———.
  (a) New Delhi (b) Solan
  (c) Shimla (d) Bangalore

(430) National Medicinal Plants Board is located at———.
  (a) Bangalore (b) New Delhi
  (c) Chennai (d) Mumbai

(431) Coconut Development Board is located at ———.
  (a) Patna (b) Kota
  (c) Cochin (d) Hyderabad

(432) In double helical structure of DNA, distance of one turn is ———.
  (a) 36A (b) 38A
  (c) 34A (d) 3.4A

(433) A gene located on Y chromosome and therefore, transmitted from father to son is known as ———.
  (a) Autosomal gene (b) Supplementary gene
  (c) Sex linked gene (d) Holandric gene

(434) The proportion of additive genetic variance is phenotypic variance is called ———.
  (a) Heritability (b) Heterobeltosis
  (c) Linkage relationship (d) Dominance relationship

(435) One gallon of water is equal to ———.
  (a) 4.55 l (b) 5.5 l
  (c) 3.5 l (d) 2.55 l

(436) Farmers can have a staggered harvesting of following as it does not set harvesting time ———.
  (a) Cabbage (b) Onion
  (c) Cassava (d) Cauliflower

(437) Amongst vegetable crops leaving aside potatoes, the maximum production in the world is that of ———.
  (a) Cucumber (b) Brinjal
  (c) Onion (d) Tomato

(438) At the following temperature, there is a sharp decline in tuberization in potato ———.
  (a) Above 21°C (b) Above 30°C
  (c) Above 40°C (d) None of the above

(439) Most varieties of onion grown in India are ———.
  (a) Chilling types (b) Day neutral
  (c) Short day types (d) Long day types

(440) Pusa Purple Cluster variety of brinjal is resistant to ———————.
 (a) Drought (b) Mites
 (c) Purple blotch (d) Bacterial wilt

(441) Seed borne diseses of cauliflower are generally managed by ———————.
 (a) Methyl Bromide application (b) Soil sterilization
 (c) Formaldehyde (d) Hot water treatment of seeds

(442) Fruits and vegetables are generally rich in dietry fibre which protects from ———————.
 (a) Fat deficiency (b) Protein deficiency
 (c) Colon cancer (d) CHO deficiency

(443) Crossing over during meosis results in ———————.
 (a) Help mutation (b) Promoting
 (c) Breaking linkage (d) None of the above

(444) In heterosis breeding, one should capitalize on ———————.
 (a) In breeding depression (b) Over dominance
 (c) Higher see vigour (d) Homozygosity

(445) Mass selection is essentially a ———————.
 (a) Hybridization technique (b) Population improvement
 (c) Mutation breeding (d) Clonal selection

(446) ——————— is known as vegetable of twentieth century ———————.
 (a) Winged bean (b) Cluster bean
 (c) Chekkurmanis (d) Amranthus

(447) The part used as vegetable in gerusaleum artichoke ———————.
 (a) Flower bud (b) Root tuber
 (c) Stem (d) Leaf stalk

(448) In a diploid species generally following number of the chromosomes are involved in pollen mitosis ———————.
 (a) n (b) 2n
 (c) 3n (d) 4n

(449) DNA contains following number of nitrogenous bases repeated in various sequences———————.
 (a) One (b) Two
 (c) Four (d) Ten

(450) Linkage between genes effects ———————.
 (a) Vernalization (b) Fertilization
 (c) Anaphase (d) Independent assortment

(451) Swarn Rekha is a variety of ―――――――.
    (a) Bottle gourd   (b) Bitter gourd
    (c) Pointed gourd   (d) None of the above

(452) The ability of cell to generate in to a whole plant is known as ―――――――.
    (a) Autolysis   (b) Regeneration capacity
    (c) Totipotency   (d) None of the above

(453) In India, leading cardamom growing state is ―――――――.
    (a) Assam   (b) Karnataka
    (c) Maharashtra   (d) Kerala

(454) The large cardamom is propagated by ―――――――.
    (a) Seeds   (b) Suckers
    (c) Cutting   (d) All of the above

(455) Arka Abir is a variety of ―――――――.
    (a) Tomato   (b) Brinjal
    (c) Cucurbits   (d) Chilli

(456) Oxalates are present in ―――――――.
    (a) Root crops   (b) Bulb crops
    (c) Leafy vegetables   (d) None of the above

(457) Central Arid Zone Research is Located at ―――――――.
    (a) Jhansi   (b) Jabalpur
    (c) Ranchi   (d) Jodhpur

(458) ――――― is used as a surface sterilant.
    (a) $HgCl_2$ (b) KNO_3
    (c) KCl   (d) None of the above

(459) The FPO licence is given by ―――――――.
    (a) Central Government   (b) State Government
    (c) Ministry of Agriculture   (d) None of the above

(460) Term epistasis was given by ―――――――.
    (a) Bateson   (b) Halman
    (c) Mendal   (d) None of the above

(461) Non – bulb forming member of onion family is ―――――――.
    (a) Garlic   (b) Leek
    (c) Onion   (d) None of the above

(462) Gerkhins belongs to ――――― genus
    (a) *Cucurbita*   (b) *Cucumis*
    (c) *Citrulus*   (d) None of the above

(463) Who among the following is associated with vegetable Research ————.
　　(a) Swaminathan　　　　　　(b) G. Kalloo
　　(c) TK Bose　　　　　　　　(d) All of the above
(464) IIHR, Bangalore was established in ————.
　　(a) 1968　　　　　　　　　　(b) 1967
　　(c) 1961　　　　　　　　　　(d) 1970
(465) Pusa Kiran is a variety of ————.
　　(a) Palak　　　　　　　　　　(b) Chauli
　　(c) Kaddu　　　　　　　　　 (d) Guar
(466) Edible part of Sweet corn is ————.
　　(a) Kernel　　　　　　　　　(b) Spike
　　(c) Seeds　　　　　　　　　 (d) Stem
(467) The toxic substances in spinach is ————.
　　(a) Saponin　　　　　　　　 (b) Solanine
　　(c) Sinigrin　　　　　　　　　(d) Serotonin
(468) Phule Shubhangi variey of cucumber is cross of ————.
　　(a) Himangi X China Long　　(b) Poinsette X Kalyanpur Agethi
　　(c) Pusa Sanyog X Poinsette　 (d) Straight Eight
(469) Baby potato is variety of ————.
　　(a) Potato　　　　　　　　　(b) Lima bean
　　(c) Sweet potato　　　　　　(d) Chinese potato
(470) The plant which loves sun shine are known as ————.
　　(a) Helotrops　　　　　　　　(b) Pagiotrops
　　(c) Geotrops　　　　　　　　(d) Psezotrops
(471) Amariophallus Yam is used as ————.
　　(a) Anti sterility　　　　　　　(b) Blood purifier
　　(c) Apetizer　　　　　　　　 (d) Cardian stimulant
(472) Pillow is a serious disorder of ———— due to deficiency of Ca.
　　(a) Coriander　　　　　　　　(b) Chekurmanis
　　(c) Cucumber　　　　　　　 (d) Pumpkin
(473) Harvesting index is a ratio of ————.
　　(a) Biological yield : Economic Yield　(b) Economic Yield : Biological Yield
　　(c) Total weight : Dry weight　　　　 (d) Dry weight : Total weight
(474) The term hyrdroponic was derived by ————.
　　(a) Boyle　　　　　　　　　 (b) Setchell
　　(c) Jemson　　　　　　　　 (d) Licon

(475) Tree culture is a synonym of ————————————.
 (a) Arboriculture   (b) Silviculture
 (c) Horticulture   (d) Floriculture

(476) ———————— are phosphate solubilizes.
 (a) *Penecillum*   (b) *Aspergilus*
 (c) Azospirillum   (d) Both a and b

(477) Nityashree and Navashree are cultivar of————————.
 (a) Cumin   (b) Cinnamon
 (c) Cardamom   (d) Fennel

(478) Corn flower is national flower of ————————.
 (a) Germany   (b) Japan
 (c) Wales   (d) India

(479) The minimum degree of freedom required for good statistical analysis is ————————.
 (a) 8   (b) 12
 (c) 15   (d) 20

(480) ———————— fertilizers are not produced in India ————————.
 (a) Nitrogenous   (b) Phosphatic
 (c) Potassic   (d) All of the above

(481) Prevention of food adulteration act was passed in————————.
 (a) 1956   (b) 1954
 (c) 1915   (d) None of the above

(482) ———————— variety shows resistant against fruit fly of pumpkin.
 (a) Arka Chandan   (b) Arka Bahar
 (c) Arka Suryamukhi   (d) None of the above

(483) Which of the following is dioecious ————————?
 (a) Date palm   (b) Asparagus
 (c) Both a and b   (d) None of the above

(484) The recommended dose of Vitamin C per day is ————————.
 (a) 40mg   (b) 20mg
 (c) 60mg   (d) 80mg

(485) Osteoporasis occurs due to deficiency of ————————.
 (a) Iodine   (b) Calcium
 (c) Phosphorus   (d) Iron

(486) The functional unit of gene is ————————.
 (a) Cistron   (b) Muton
 (c) Recon   (d) None of the above

(487) Fruit and vegetable processing was started in ―――――――.
  (a) 1927 (b) 1932
  (c) 1925 (d) 1940

(488) Stigma is used as a spice of―――――――.
  (a) Vanilla (b) Saffron
  (c) Clove (d) None of the above

(489) Aspexginase is found in ―――――――.
  (a) Green chillies (b) Bell pepper
  (c) Red chillies (d) Bird chilli

(490) Perkin Long Log Green is a variety of ―――――――.
  (a) Brinjal (b) Cluster been
  (c) Okra (d) Cucumber

(491) Lysosomes was discovered by ―――――――.
  (a) Duve (b) Disy
  (c) Buckner (d) None of the above

(492) 16 July is celebrated as ―――――――.
  (a) Flag day (b) IARI day
  (c) ICAR day (d) CIMAP day

(493) Functional male sterility is controlled by ―――――――.
  (a) Dominant gene (b) Recessive gene
  (c) Cytoplasmic gene (d) None of the above

(494) Linkage between gene effect ―――――――.
  (a) Vernalization (b) Fertilization
  (c) Independent assortment (d) Sex – linked characters

(495) Punjab Surakh is a multiple disease resistant variety of ―――――――.
  (a) Capsicum (b) Brinjal
  (c) Chilli (d) Pea

(496) Conversion factor for organic carbon to organic matter is ―――――――.
  (a) 1:9 (b) 1:18
  (c) 1:2.7 (d) 1:24

(497) Artificial gene synthesis is done by ―――――――.
  (a) B. Chattopadhyaya (b) H. Khurana
  (c) C. Mukherjee (d) CT Patel

(498) Piwla Dhotra is used for reclamation of ―――――――.
  (a) Saline soil (b) Acidic soil
  (c) Alkali soil (d) Calcareous soil

(499) Succulometer is used to measure maturity of ―――――――.
   (a) Sweet corn (b) Radish
   (c) Sweet potato (d) Carrot

(500) Central Institute of Medicinal and Aromatic Plant is located at ―――――――.
   (a) Rajasthan (b) AP
   (c) Kerala (d) Lucknow

(501) Chairman of Block level Panchayat Samiti is ―――――――.
   (a) MLA (b) BDO
   (c) Mayor (d) Progressive farmers

(502) London Flag is a variety of ―――――――.
   (a) Celery (b) Cassava
   (c) Leek (d) Turnip

(503) Hemaglutine is found in ―――――――.
   (a) Broad bean (b) Cluster bean
   (c) French bean (d) Pumpkin

(504) Cobalt is present in which of following vitamins ―――――――.
   (a) Vitamin B1 (b) Vitamin B12
   (c) Vitamin B5 (d) Vitamin D

(505) The term appetizing is used for ―――――――.
   (a) Canning (b) Dehydration
   (c) Syruping (d) Sterilization

(506) Agriculture produce act is also known as ―――――――.
   (a) PFA act (b) FPO act
   (c) AGMARK act (d) ISI act

(507) Central Food Laboratory is located at ―――――――.
   (a) New Delhi (b) Mysore
   (c) Kolkata (d) Mumbai

(508) First Indian Horticulture Congress was held in ―――――――.
   (a) 1942 (b) 2004
   (c) 2005 (d) 1950

(509) Spice crop largely exported from India is ―――――――.
   (a) Black pepper (b) Cardamom
   (c) Turmeric (d) Ginger

(510) Plantation and spice are main crops at ―――――――.
   (a) Eastern Plateau and Hill zone (b) Western plateau and hill zone
   (c) Southern plateau and Hill zone (d) Central plateau and Hill zone

511. State accounting maximum area and production of cardamom is ——————.
    (a) Tamil Nadu   (b) Kerala
    (c) Karnataka   (d) Meghalaya
512. Indian Central Spices and Cashewnut Committee was established in——————.
    (a) 1968   (b) 1961
    (c) 1955   (d) 1980
513. The leading produce of spices in the world is ——————.
    (a) India   (b) Brazil
    (c) Mexico   (d) Indonesia
514. National Research Center of Spices was created in ——————.
    (a) 1965   (b) 1975
    (c) 1980   (d) 1986
515. Spices Cess Act came in to existance in——————.
    (a) 1966   (b) 1970
    (c) 1980   (d) 1992
516. Directorate of Arecanut and Spices Development (Calicut) was established in ——————.
    (a) 1960   (b) 1964
    (c) 1966   (d) 1968
517. All India Spices Development Council was established in ——————.
    (a) 1963   (b) 1966
    (c) 1971   (d) 1990
518. Black pepper is produced from ——————.
    (a) Runner   (b) Laterals
    (c) Plagiotrophic shoot (d) Hanging shoot
(519) Spices and condiments are group under——————.
    (a) Industrial crops   (b) Commercial crops
    (c) Food adjuncts   (d) Field crops
(520) Glycolysis occurs in ——————.
    (a) Cytoplasm   (b) Mitochondria
    (c) Nucleus   (d) Only a and b
(521) Soaking up of H_2O by dry substances due to hydrophyllic condition is ——————.
    (a) Imbibition   (b) Plasmolysis
    (c) Osmosis   (d) All of the above

(522) Photosynthetic rate is highest in —————————.
    (a) C4 plants　　　　　　　　(b) C3 plants
    (c) CAM plants　　　　　　　 (d) None of the above

(523) ————————— variety of radish was developed through mass selection.
    (a) Japanese White　　　　　 (b) Punjab Safed
    (c) Pusa Himani　　　　　　　(d) Arka Nishant

(524) ————————— variety of okra is suitable for cultivation in hilly areas of north India
    (a) Varsha Uphar　　　　　　 (b) Perkins Long Green
    (c) Hisar Barsati　　　　　　　(d) Pusa Sawani

(525) ————————— variety of brinjal is green fruited.
    (a) Arka Sheel　　　　　　　　(b) Arka Nidhi
    (c) Arka Shirish　　　　　　　 (d) Pusa Bindu

(526) National Research Center for Seed Spices is situated at —————————.
    (a) Ajmer　　　　　　　　　　(b) Bikaner
    (c) Junagarh　　　　　　　　　(d) Jaipur

(527) ————————— are trimonoecius vegetables.
    (a) Cucumber　　　　　　　　(b) Ridgegourd
    (c) Muskmelon　　　　　　　 (d) All of the above

(528) ————————— has capitulatum type of inflorescence.
    (a) Taro　　　　　　　　　　　(b) Globe Artichoke
    (c) Lettuce　　　　　　　　　　(d) Beet

(529) Pusa Shandar is a variety of —————————.
    (a) Sponge gourd　　　　　　　(b) Snap melon
    (c) Muskmelon　　　　　　　 (d) None of the above

(530) ————————— is a iodine rich vegetable —————————.
    (a) Okra　　　　　　　　　　　(b) Garlic
    (c) Beet　　　　　　　　　　　 (d) All of the above

(531) Bermuda yellow is a introduced variety of —————————.
    (a) Carrot　　　　　　　　　　(b) Onion
    (c) Tomato　　　　　　　　　 (d) Turnip

(532) ————————— is highly sensitive to photoperiod vegetable.
    (a) Watermelon　　　　　　　 (b) Muskmelon
    (c) Cucumber　　　　　　　　(d) Bottle gourd

(533)—————— vitamin I is found only in animal foods.
    (a) Vitamin A   (b) Vitamin B$_1$
    (c) Vitamin B$_{12}$ (d) Vitamin E
(534) Hooking of leaf tip is the symptom of —————.
    (a) Cu deficeincy   (b) Mg deficiency
    (c) K deficiency   (d) Ca deficiency
(535) PGR act as substitute to stratification —————.
    (a) Auxin (NAA)   (b) Gibberellin (GA$_3$)
    (c) ABA   (d) Cytokinin
(536) Soil moisture tension at field capacity is —————.
    (a) 31 atm   (b) 15 atm
    (c) 1/3 atm   (d) 0.001 atm
(537) In India, first gene sanctuary established in Garoo hill in Assam —————.
    (a) Wild mango   (b) Wild citrus
    (c) Wild grape   (d) Ber
(538) —————— is called universal cell organelle.
    (a) Mitochondria   (b) Lysosome
    (c) Ribosome   (d) Centriole
(539) Zero tillage was first adopted in —————.
    (a) Japan   (b) USA
    (c) China   (d) India
(540) Local control principle of experiment design is not used in —————.
    (a) CRD   (b) RBD
    (c) RCBD   (d) LSD
(541) The relation between mean deviation and standard deviation is —————.
    (a) Mean deviation is less than standard deviation
    (b) Mean deviation is =1/2 standard deviation
    (c) Mean deviation= standard deviation
    (d) None of the above
(542) —————— is antiethylene chemical
    (a) AgNO$_3$ (b) KMNO$_4$
    (c) CaCO$_3$ (d) Both a and b
(543) —————— is double trisomy.
    (a) 2n-1-1   (b) 2n+2
    (c) 2n+1+1   (d) 2n+1

(544) —————— is a white flower gourd.
    (a) Bottle gourd   (b) Bitter gourd
    (c) Ridge gourd   (d) Snake gourd

(545) —————— year is declared as international potato year.
    (a) 2004   (b) 2006
    (c) 2009   (d) 2008

(546) Many fold effect of single gene is known as ——————.
    (a) Multiple allele   (b) Pleotrophy
    (c) Polygene   (d) All of the above

(547) International fiber year is ——————.
    (a) 2004   (b) 2006
    (c) 2008   (d) 2009

(548) Sadashiva is a variety of ——————.
    (a) Cabbage   (b) Cauliflower
    (c) Knol-Khol   (d) Radish

(549) —————— is generally used for preparation wine ——————.
    (a) Broad bean   (b) Lima bean
    (c) Indian bean   (d) None of the above

(550) Thermodormancy is found in ——————.
    (a) Lettuce   (b) Amranthus
    (c) Parseley   (d) Celery

(551) —————— is the quantity of colour permitted in the preservation of food.
    (a) 200ppm   (b) 400ppm
    (c) 600ppm   (d) 800ppm

(552) IIVR is situated at ——————.
    (a) Varanasi   (b) Lucknow
    (c) Kanpur   (d) New Delhi

(553) CMRS is situated at ——————.
    (a) Lucknow   (b) Calicut
    (c) Nagpur   (d) Pune

(554) NRCO is situated at ——————.
    (a) Putlur   (b) Nasik
    (c) Ajmer   (d) Gangtok

(555) CRI for Chikoo is located at——————.
    (a) Lucknow   (b) Bihar
    (c) Bikaner   (d) Chennai

(556) NRC for Cashew is at —————————.
   (a) Nagpur   (b) Calicut
   (c) Putur   (d) Vital
(557) ————————— variety of carrot used for canning purposes.
   (a) Pusa Kesar   (b) Pusa Lal
   (c) Pusa Meghali   (d) Chantaney
(558) Konkan Haritparni is a variety of —————————.
   (a) Spinach   (b) Arvi
   (c) Ridge gourd   (d) Fenugreek
(559) Solan Gola is variety of —————————.
   (a) Tomato   (b) Brinjal
   (c) Sweet potato   (d) None of the above
(560) ————————— is monocotyledonous vegetable crop.
   (a) Amranthus   (b) Cabbage
   (c) Cauliflower   (d) Yam
(561) Adhunik is a private sector hybrid of —————————.
   (a) Okra   (b) Brinjal
   (c) Bottle gourd   (d) None of the above
(562) Bitterness in carrot is due to —————————.
   (a) Isopentanol   (b) Carotene
   (c) Isocoumanin   (d) Sulphoraphane
(563) King of North is a variety of —————————.
   (a) Cabbage   (b) Knol-Khol
   (c) Cauliflower   (d) Brussels sprout
(564) ————————— variety is resistant to pithiness in turnip.
   (a) Pus Swati   (b) Pusa Swarnima
   (c) Pusa Chandrima   (d) Pusa Kanchan
(565) ————————— is developed seed plot technique in Potato.
   (a) G. Kalloo   (b) Harbhajan Singh
   (c) Puskarnath   (d) Kirti Singh
(566) Yerusseri product is prepared from ————.————————— vegetable.
   (a) Cabbage   (b) Pumpkin
   (c) Radish   (d) Carrot
(567) ————————— is present in vascular streaking.
   (a) Cassava   (b) Sweet potato
   (c) Yams   (d) None of the above

(568) Arka Samridhi is a variety of —————————.
    (a) Cowpea   (b) Brinjal
    (c) Tomato   (d) Guar

(569) Horse bean is another name of —————————.
    (a) Yam bean   (b) Winged bean
    (c) Broad bean   (d) Potato bean

(570) Florida Butter is variety of —————————.
    (a) Sem   (b) Cowpea
    (c) Pea   (d) Lima bean

(571) Dark Green is a cultivar of —————————.
    (a) Leek   (b) Lettuce
    (c) Chinese cabbage   (d) None of the above

(572) International Center of potato is situated at —————————.
    (a) Peru   (b) Mexico
    (c) India   (d) Brazil

(573) Frog eye leaf spot is a disease of —————————.
    (a) Cowpea   (b) Cucumber
    (c) Chilli   (d) Tomato

(574) ————————— is anther mutant variety of tomato.
    (a) Pusa Divya   (b) Kashi Aman
    (c) Kashi Chyaman   (d) Arka Rakashk

(575) Diploid apogamy is found in —————————.
    (a) Onion   (b) Tomato
    (c) Cabbage   (d) Chilli

(576) Clustering of flower in tomato is known as —————————.
    (a) Spear   (b) Truss
    (c) Spur   (d) Plaursaur

(577) Flavour in cardamom is due to —————————.
    (a) Sinigrin   (b) a-torpenyl acetate
    (c) 1,8-cineole   (d) Both b and c

(578) ————————— instrument is used to measure leaf area.
    (a) Alanometer   (b) Hydrometer
    (c) Tenderometer   (d) All of the above

(579) Sree Nandani is a variety of —————————.
    (a) Sweet potato   (b) Tinda
    (c) Cassava   (d) Colocassia

(580) Avinash, Vijetha -1 are varieties of —————————.
   (a) Small cardamom   (b) Large cardamom
   (c) Both a and b   (d) None of the above
(581) Konkan Tej is a variety of —————————.
   (a) Nutmeg   (b) Turmeric
   (c) Cinnamon   (d) Ginger
(582) ————————— has the highest cross pollination rate of 95%.
   (a) Broccoli   (b) Carrot
   (c) Radish   (d) Tomato
(583) Tip burn is a physiological disorder of —————————.
   (a) Pea   (b) Lettuce
   (c) French bean   (d) None of the above
(584) DWD-1 variety of curry leaf is sensitive to —————————.
   (a) Winter season   (b) Summer season
   (c) Rainy season   (d) None of the above
(585) Blossom drop and ovule abortion are common problem —————————.
   (a) French bean   (b) Lablab bean
   (c) Feba bean   (d) All of the above
(586) Root stock is used in crops like —————————.
   (a) Tomato   (b) Cucumber
   (c) Cauliflower   (d) Both a and b
(587) Phule Raja is a variety of —————————.
   (a) Sponge gourd   (b) Cucumber
   (c) Onion   (d) None of the above
(588) Pentamorphic sex is observed in —————————.
   (a) Beet   (b) Palak
   (c) Radish   (d) None of the above
(589) Vilayati Baigan is common name of —————————.
   (a) Brinjal   (b) Tomato
   (c) Yam bean   (d) Broad bean
(590) Toxic substances present in celery is—————————.
   (a) Cucurbitacin   (b) Dioscorine
   (c) Apiin   (d) Solasodine
(591) Potato unfit for comsumption if solanine is greater than —————————.
   (a) 25mg/100g   (b) 20mg/100g
   (c) 35mg/100g   (d) 45mg/100g

(592) Osmosis was discovered by ―――――――.
    (a) Sachs   (b) Pfferrfer
    (c) Nollet   (d) Strasburger

(593) In truck gardening, the world Truck has been derived from ―――――.
    (a) Latin   (b) French
    (c) Greek   (d) English

(594) ――――― is used for freezing ―――――.
    (a) Sprouting Broccoli   (b) Spinach
    (c) Lima bean   (d) All of the above

(595) Tomato sauce and ketchup contain――――――TSS.
    (a) 25-30%   (b) 30%
    (c) 30-35%   (d) 35-40%

(596) ――――――― is trade name of nitrogen.
    (a) Lasso   (b) Baslin
    (c) Machete   (d) Toke

(597) Arka Saguna is an improved variety of―――――――.
    (a) Amaranth   (b) Palak
    (c) Spinach   (d) Beet

(598) Welsh and Grimbell in 1947 first reported male sterility in―――――.
    (a) Radish   (b) Onion
    (c) Cucumber   (d) Carrot

(599) Ladybird beetle attack on family ―――――――.
    (a) Cucurbitaceae   (b) Solanaceae
    (c) Liliaceae   (d) None of the above

(600) ――――――― was discovered enzyme.
    (a) Buchner   (b) Funk
    (c) Kunain   (d) Symner

(601) Green back symptom of tomato is due to deficiency of ―――――.
    (a) B   (b) Ca
    (c) K   (d) N

(602) The inflorescence of Taro is called ―――――――.
    (a) Raceme   (b) Spadix
    (c) Cyme   (d) Panicle

(603) Sundry vegetable is ―――――――.
    (a) Potato   (b) Onion
    (c) Garlic   (d) Okra

(604) National Institute of Nutrition is situated at ———————————.
   (a) Hyderabad   (b) Madras
   (c) Karnataka   (d) Kerala

(605) Edible part of Endive is ———————————.
   (a) Root tuber   (b) Leaves
   (c) Tender pods   (d)

(606) The phenomenon paraheliotrophy is seen in ———————————.
   (a) Tomato   (b) Pea
   (c) Cucumber   (d) Muskmelon

(607) Update is a variety of ———————————.
   (a) Onion   (b) Potato
   (c) Tomato   (d) Sweet pepper

(608) ——————————— is one of the international markets yellow onions are preferred.
   (a) Bangladesh   (b) South East Asia
   (c) Gulf countries   (d) Japan

(609) Without crook – neck variety of bottle gourd is ———————————.
   (a) PSPL   (b) PSPR
   (c) Pusa Naveen   (d) Pusa Manjari

(610) Dandali is a variety of ———————————.
   (a) Pointed gourd   (b) Ash gourd
   (c) Ivy gourd   (d) None of the above

(611) Lacchi method of planting is followed in ———————————.
   (a) Ash gourd   (b) Cucumber
   (c) Chow-chow   (d) Pointed gourd

(612) Netting is associated with maturity of ———————————.
   (a) Watermelon   (b) Pumpkin
   (c) Muskmelon   (d) Wax gourd

(613) Yellow dwarf is a viral disease of ———————————.
   (a) Brinjal   (b) Chilli
   (c) Onion   (d) Garlic

(614) ——————————— is spice bowl of India ———————————.
   (a) Karnataka   (b) Kerala
   (c) Andhra Pradesh   (d) Tamil Nadu

(615) Union Council of Medical Research is situated at ———————————.
   (a) New Delhi   (b) Lucknow
   (c) Varanasi   (d) Kanpur

(616)——————— is the oldest vegetable research station———————.
   (a) Kalayanpur   (b) Pantnagar
   (c) Varanasi   (d) Agra

(617)——————— is leader producer of spices in the world.
   (a) India   (b) Brazil
   (c) Mexico   (d) Indonesia

(618)——————— is queen of spices.
   (a) Cardamom   (b) Black pepper
   (c) Saffron   (d) Nutmeg

(619)——————— is king of spices.
   (a) Black pepper   (b) Cardamom
   (c) Turmeric   (d) Ginger

(620) Chanchal is popular variety of ———————.
   (a) Capsicum   (b) Cauliflower
   (c) Radish   (d) Tomato

(621) ESP > 15 is in which soil ———————.
   (a) Saline   (b) Alkaline
   (c) Saline   (d) Alkaline

(622) The main constituent of cinnamon bark is ———————.
   (a) Enigenal   (b) Limone Cinnaldehyde
   (c) Eugenyl accetate   (d) None of the above

(623) In 2019, among following vegetable has been declared National Vegetable ———————.
   (a) Potato   (b) Topioca
   (c) Bottle gourd   (d) Brinjal

(624) Kulfa (Purslane , Portulaca oleracea) is sussessfully grown in ———————.
   (a) America   (b) India
   (c) Malaysia   (d) All of the above

(625) Shiranti (Joy weed *Alternanthera sessilis*) locally known as Shiranti in Bihar and Sanchi in West Bengal belongs family———————
   (a) Brassiceae   (b) Amrantheceae
   (c) Leguminaceae   (d) Solanaceae

(626) Keu (Crepe ginger, *Cheilocostus specious*) is native of India belongs to family ———————.
   (a) Cotaceae   (b) Leguminoaceae
   (c) Malvaceae   (d) Liliaceae

(627) Bichho grass (Neetle leaf, *Urtica dioica*) is native to ⎯⎯⎯⎯⎯⎯⎯⎯⎯⎯⎯⎯.
   (a) Brazil (b) India
   (c) China (d) Russia

(628) Sanai (Sunnhemp, *Crotalaria juncea*) is mainly known for ⎯⎯⎯⎯⎯⎯⎯.
   (a) Fiber (b) Fodder
   (c) Flower (d) All of the above

(629) *Nelumbo nucifera* is also known as ⎯⎯⎯⎯⎯⎯⎯⎯⎯⎯⎯⎯.
   (a) Padma (b) Kamal
   (c) Lily of Nile (d) All of the above

(630) Vegetable Humming Bird is also known as ⎯⎯⎯⎯⎯⎯⎯⎯⎯⎯.
   (a) Hog plum (b) Agathi
   (c) Spine gourd (d) None of the above

(631) Hog Plum (*Spondias pinnata*) is native of ⎯⎯⎯⎯⎯⎯⎯⎯⎯⎯⎯.
   (a) Malasia (b) India
   (c) China (d) None of the above

(632) Yard long bean is botanically known as ⎯⎯⎯⎯⎯⎯⎯⎯⎯⎯⎯.
   (a) *Vigna unguiculata* subsp. *sesquipedalis*
   (b) *Hibiscus sadariffa*
   (c) *Momordica dioica*
   (d) None of the above

(633) Mountain spine gourd closely related to ⎯⎯⎯⎯⎯⎯⎯⎯⎯⎯.
   (a) Bitter gourd (b) Pumpkin
   (c) Spine gourd (d) All of the above

(634) *Momordica charantia* var. *muricata* is ⎯⎯⎯⎯⎯⎯⎯⎯⎯⎯.
   (a) Kattupaval (b) Methipaval
   (c) A & b both (d) Sweet gourd

(635) *Sauropus androgymus* is botanically ⎯⎯⎯⎯⎯⎯⎯⎯⎯⎯.
   (a) Chekkurmanis (b) Oriental pickling melon
   (c) Spinach (d) Curry leaf

(636) Gongura pickle is prepared from ⎯⎯⎯⎯⎯⎯⎯⎯⎯⎯.
   (a) Roselle (b) Luffa
   (c) Pumpkin (d) Tomato

(637) Athalakkai is botanically known as ⎯⎯⎯⎯⎯⎯⎯⎯⎯⎯.
   (a) *Cucumis melo* var. *conomon* (b) *Luffa tuberosa*
   (c) *Momordica dioica* (d) None of the above

(638) Rajgheera is ethnic land race of ———————.
- (a) Spinach
- (b) Broccoli
- (c) Lettuce
- (d) Amranth

(639) Karuvachakka(*Solena amplexicaulis* is a wild gathered vegetable of drier tract of ———————.
- (a) Western ghats
- (b) Eastern ghats
- (c) Northern ghats
- (d) None of the above

(640) *Senna tora* syn: *Cassia tora* is a ———————.
- (a) Legume vegetable
- (b) Fruit vegetable
- (c) Annual crop
- (d) Both a and c

(641) *Alternanthera sessilis* is propagated by ———————.
- (a) Seed
- (b) Tuber
- (c) Stem cutting
- (d) Rhizome

(642) *Portulaca oleracea* or Common purslane is rich source of ———————.
- (a) Sugar
- (b) Vitamin C
- (c) Vitamin A
- (d) Omega 3 fatty acid

(643) Clove bean (*Ipomoea muricata*) is domesticated in ——————— as a vegetable.
- (a) Kerala and Karnataka
- (b) UP & Bihar
- (c) Gujarat & Rajasthan
- (d) None of the above

(644) Indian Pennywort is a ———————.
- (a) Spices
- (b) Tuber crop
- (c) Aromatic plant
- (d) Bulb crop

(645) Orcah is a kitchen garden vegetable cultivated in ———————.
- (a) Karnataka
- (b) Orissa
- (c) Gujarat
- (d) None of the above

(646) Mateera is generally known as the poor man,s vegetable and the common man,s fruit of ———————.
- (a) Rajasthan
- (b) Karnataka
- (c) Orissa
- (d) Bihar

(647) ——————— of the immature pods called as *sangri* and are used as vegetable.
- (a) Mateera
- (b) Khejari
- (c) Bitter apple
- (d) None of the above

(648) Bitter apple is ancestor of ———————.
- (a) Cucumber
- (b) Watermelon
- (c) Pumpkin
- (d) Wax gourd

(649) Phog (*Calligonum polygonoides*) belongs to family ─────────────.
    (a) Solanaceae　　　　　　　　(b) Liliaceae
    (c) Polygonaceae　　　　　　　(d) Malavaceae

(650) Arya (*Cucumis melo* var, *chate*) is ───────── in nature.
    (a) Dioecious　　　　　　　　(b) Gynoecious
    (c) Monoecious　　　　　　　(d) Andromonoecious

(651) Momordin is found in ─────────────.
    (a) Jhaar Kerela　　　　　　　(b) Sweet gourd
    (c) Spine gourd　　　　　　　(d) Bottle gourd

(652) Bitter cress (*Cardamine oligosperma*) belongs to family ─────────.
    (a) Cucurbitaceae　　　　　　(b) Malvaceae
    (c) Leguminoceae　　　　　　(d) Brassicaceae

(653) Stinging Nettle (*Urtica dioica*) is also known as ─────────────.
    (a) Kalami sag　　　　　　　　(b) Kandali Sag
    (c) Bichhu Ghas　　　　　　　(d) Both b and C

(654) Fiddlehead ferns belongs to family ─────────────.
    (a) Apocynaceae　　　　　　　(b) Polygonoceae
    (c) Onocleaceae　　　　　　　(d) None of the above

(655) Watercress is botanically known as ─────────────.
    (a) *Cichorium endiva*　　　　　(b) *Bidens pilosa*
    (c) *Nasturtium officinale*　　　　(d) None of the above

(656) BlackJack or Spinach needle is a ───────── medicine.
    (a) Unani　　　　　　　　　　(b) Japanese
    (c) Chinese　　　　　　　　　(d) All of the above

Answer Sheet

1. b　2. d　3. a　4. b　5. a　6. b　7. a　8. c　9. a　10. b　11 c
12. a　13. c　14 b　15. d　16. a　17. b　8 b　19. c　20 c　21. d　22. a
23. b　24. d　25. a　26. b　27. d　28. b　29. c　30 c　31. c　32 d　33. b
34 a　35. d　36. a　37. b　38. b　39. c　40. d　41. b　42. b　43. a　44. c
45. b　46. a　47. e　48. d　49. b　50. a　51. a　52. b　53. b　54. c　55. a
56. b　57. c　58. c　59. b　60. b　61. b　62. a　63. d　64. d　65. b　66. -
67. a　68. d　69. -　70. -　71. a　72. c　73. d　74. c　75. b　76. d　77. a
78. b　79. a　80. a　81. a　82. a　83. b　84. b　85. a　86. d　87. a　88. b
89. a　90. c　91. c　92. a　93. b　94. c　95. c　96. d　97. a　98. a　99. b
100. a　101. c　102. b　103. b　104. c　105 d　106. b　107. c　108. c　109. a　110. b
111. b　112. b　113. a　114. c　115. d　116. d　117. a　118. a　119. b　120. b　121. b
122. c　123. a　124. c　125. b　126. b　127. d　128. d　129. d　130. b　131. a　132. b
133. c　134. a　135. d　136. b　137. b　138. a　139. a　140. b　141. b　142. b　143. a
144. b　145. c　146. b　147. a　148. -　149. a　150. a　151. d　152. b　153. c　154. a

155. b	156. b	157 d	158. d	159. c	160. b	161. b	162. c	163. c	164. a	165. b
166. a	167. d	168. b	169. a	170. b	171. a	172 a	173. d	174. b	175. a	176. a
177. c	178. a	179. b	180. b	181. a	182. a	183. b	184. b	185. a	186. b	187. a
188. b	189. b	190. d	191. a	192. d	193. c	194. a	195. c	196. d	197. c	198. b
199. a	200. a	201. a	202. c	203 b	204. b	205. b	206. a	207. b	208. c	209. b
210. c	211. b	212. a	213. a	214. c	215. c	216. b	217. c	218. a	219. a	220. a
221. a	222. b	223. c	224. b	225. b	226. d	227. c	228. c	229. d	230. b	231. a
232. d	233. b	234. d	235. a	236. b	237. a	238. b	239. b	240. a	241. a	242. a
243. b	244. d	245. b	246. c	247. d	248. b	249. a	250. -	251. a	252. b	253. a
254. a	255. b	256. b	257. d	258. c	259. c	260. a	261. d	262. c	263. d	264. b
265. a	266. c	267. b	268. c	269. a	270. b	271 -	2 72. c	273. a	274. b	275. b
276. -	277. c	278. a	279. a	280. a	281. b	282. c	283. b	284. a	285. c	286. d
287. d	288. -	289. d	290. a	291. b	292. c	293. b	294. a	295. a	296. b	297. a
298. b	299. a	300. a	301. a	302. a	303 a	304. b	305. c	306. -	307. b	308. b
309. b	310. a	311. c	312. a	313. d	314. c	315. d	316. a	317. c	318. c	319. a
320. c	321. c	322. c	323. c	324. c	325. b	326. a	327. d	328. d	329. a	330. b
331. d	332. d	333. b	334. a	335. a	336. c	337. b	338. c	339. d	340. c	341. c
342. a	343. d	344. d	345. d	346. b	347. d	348. b	349. d	350. d	351. b	352. d
353. c	354. c	355. c	356. d	357. d	358. b	359. c	360. b	361. c	362. b	363. b
364. b	365. d	366. b	367. d	368. c	369. b	370. a	371 b	372. d	373. c	374. c
375. b	376. b	377. a	378. c	379. a	380. d	381. a	382. d	383. d	384. b	385. a
386. -	387. b	388. a	389. a	390. d	391. a	392. b	393. b	394. c	395. a	396. a
397. c	398. d	399. a	400. c	401. c	402. b	403 a	404. a	405. b	406. c	407. d
408. d	409. a	410. c	411. c	412. b	413. b	414. b	415. a	416. b	417. b	418. d
419. a	420. d	421. c	422. c	423. c	424. b	425. a	426. a	427. a	428. c	429. b
430. b	431. c	432. c	433. c	434. a	435. a	436. c	437. d	438. a	439. c	440. d
441. d	442. c	443. c	444. b	445. b	446. a	447. b	448. a	449. c	450. d	451. c
452. c	453. d	454. b	455. d	456. c	457. d	458. a	459. a	460. a	461. b	462. b
463. b	464. a	465. b	466. a	467. a	468. b	469. b	470. a	471 b	472. c	473. b
474. b	475. a	476. b	477. b	478 a	479. b	480. c	481. b	482. c	483. c	484. c
485. b	486. a	487. a	488. -	489. a	490. c	491. a	492. c	493. -	494. c	495. c
496. a	497. b	498. c	499. a	500. d	501. b	502. c	503 c	504. b	505. a	506. c
507. c	508. b	509. a	510. c	511. b	512. c	513. a	514. d	515. c	516. c	517. c
518. c	519. c	520. a	521. a	522. a	523. d	524. b	525. c	526. a	527. d	528. b
529. b	530. a	531. b	532. d	533. c	534. d	535. b	536. c	537. b	538. c	539. b
540. a	541. a	542. -	543. a	544. a	545. d	546. b	547. d	548. c	549. -	550. a
551. a	552. a	553. a	554. d	555. b	556. c	557. d	558. d	559. a	560. d	561. a
562. -	563. b	564. d	565. c	566. b	567. -	568. -	569. c	570. d	571 b	572. a
573. c	574. a	575. a	576. b	577. d	578 a	579. a	580. a	581. c	582. a	583. b
584. a	585. a	586. d	587. d	588. d	589. b	590. c	591. b	592. b	593. b	594. d
595. a	596. d	597. a	598. d	599. a	600. a	601. c	602. b	603 b	604. a	605. b
606. b	607. b	608. d	609. c	610. a	611. a	612. c	613. c	614. b	615. b	616. a
617. a	618. a	619. c	620. a	621. d	622. b	623. d	624. d	625. b	626. a	627. b
628. d	629. d	630. b	631. b	632. a	633. c	634. c	635. a	636. a	637. -	638. d
639. a	640. d	641. c	642. d	643. a	644. c	645. a	646. a	647. b	648. b	649. c
650. c	651. a	652. d	653. d	654. c	655. c	656. c				

Match the Pairs

S.No.	Colum 1	Colum 2	Answer Key
1.	(i) Hybrid seed production in onion	(A) Self- incompatability	D
	(ii) Micropropogation of orchid	(B) Cleistogammy	E
	(iii) Cabbage	(C) Anemophilus	A
	(iv) Spinach	(D) Male sterility	C
	(v) Groundnut	(E) Mericlone	B
2.	(i) Ripening hormone	(A) Glyphosate	B
	(ii) Growth retardant	(B) Ethylene	D
	(iii) Bioside	(C) Sucrose	E
	(iv) Herbicide	(D) Cycocel	A
	(v) Pulsing	(E) 8-HQC	C
3	(i) Tetramorphic	(A) Cucumber	C
	(ii) Dioecious	(B) Onion	D
	(iii) Self - incompatability	(C) Spinach	E
	(iv) Male sterility	(D) Asparagus	B
	(v) Parthenocarpy	(E) Loquat	A
4.	(i) VAM	(A) Symbattic fungi	A
	(ii) Pseudomonas	(B) Bacteria	B
	(iii) Trichodermavirdi	(C) Fungas+ biofungicide	C
	(iv) Braconbepets	(D) Parasitoids	D
	(v) Azobactor	(E) Free living bacteria	E
5	(i) 2-4-D	(A) Herbicide	B
	(ii) Glyphosate	(B) Weedicide	A
	(iii) Vivapary	(C) Ripener	D
	(iv) Ethylene	(D) Chow- chow	C
	(v) ABA	(E) Stomatal regulation	E
6	(i) CCC	(A) *HQC	C
	(ii) Biocide	(B) Sucrose	A
	(iii) Pulsing	(C) Growth retardant	B
	(iv) Insecticide	(D) Thiram	E
	(v) Fungicide	(E) Malathion	D
7	(i) Porometer	(A) Tissue water potential	C
	(ii) Psycometer	(B) Abiotic stresses	A
	(iii) Drought	(C) Stomatal aperture measurement	B
	(iv) Guava wilt	(D) Plant growth measurement	E
	(v) Auxanometer	(E) Alkaline soil condition	D
8	(i) Bathua	(A) Malavaceae	C
	(ii) Carrot	(B) Convolvulaceae	D
	(iii) Pea	(C) Chenopodiaceae	E
	(iv) Okra	(D) Umbelliferae	A
	(v) Sweet potato	(E) Papilionaceae	B
9	(i) PusaSambandh	(A) Carrot	E
	(ii) PusaYamdagini	(B) Cauliflower	A
	(iii) Whiptail	(C) Mo Deficiency	C
	(iv) Browining	(D) Cutting	B
	(v) Sweet potato	(E) Cabbage	D
10	(i) Ridge gourd	(A) 22	B

S.No.	Colum 1	Colum 2	Answer Key
	(ii) French bean	(B) 26	A
	(iii) Okra	(C) 12	D
	(iv) Spinach	(D) 130	C
	(v) Pea	(E) 14	E
11	(i) Chloroplast	(A) Protein synthesis	B
	(ii) Mitochondria	(B) Photosynthesis	D
	(iii) Ribosome	(C) Packaging of food materials	A
	(iv) Golgi bodies	(D) Cellular respiration	C
	(v) Lysosomes	(E) Digestive vacuoles	E
12	(i) Autoploidy	(A) Simple polyploidy	A
	(ii) Gametes	(B) Sex cells	B
	(iii) Landsteiner	(C) Abo blood group	C
	(iv) Albinism	(D) Digestive	D
	(v) Protein	(E) A polymer of aminoacid	E
13	(i) Transgenic plants	(A) Homozygosity	D
	(ii) Cross pollination	(B) Microcloning	C
	(iii) Self-incompatability	(C) Heterozygosity	E
	(iv) Self-pollination	(D) Foreign DNA	A
	(v) Micropropogation	(E) Deformed Stamen	B
14	(i) CAZRI	(A) Jhansi	E
	(ii) CPRI	(B) Delhi	D
	(iii) IGFRI	(C) Putur	A
	(iv) NRC on Cashew	(D) Shimla	C
	(v) IASRI	(E) Jodhpur	B
15	(i) ABA	(A) Shoot elongation	C
	(ii) Cross pollination	(B) Homozygosity	D
	(iii) GA3	(C) Drought	A
	(iv) Self - pollination	(D) Heterozygosity	B
	(v) NAA	(E) Rooting	E
16	(i) Snow pea	(A) 2n=32	E
	(ii) Ash gourd	(B) 2n=22	D
	(iii) Watermelon	(C) 2n=18	B
	(iv) Sugar beet	(D) 2n=24	C
	(v) Leek	(E) 2n=14	A
17	(i) Tomato	(A) Arka Niketan	D
	(ii) Cauliflower	(B) Poinsett	E
	(iii) Onion	(C) Scarlet	A
	(iv) Cucumber	(D) Sioux	B
	(v) Radish	(E) PusaDeepali	C
18	(i) Phyllody	(A) Carrot	D
	(ii) Black heart	(B) Cauliflower	E
	(iii) Blinding	(C) Tomato	B
	(iv) Catface	(D) Turnip	C
	(v) Forking	(E) Potato	A
19	(i) Lalbagh	(A) New Delhi	C
	(ii) Varindavan Garden	(B) Chandigarh	D
	(iii) Budha Garden	(C) Bangalore	A

S.No.	Colum 1	Colum 2	Answer Key
	(iv) Pinjore Garden	(D) Mysore	B
	(v) Shalimar Bagh	(E) Srinagar	E
20	(i) IIHR	(A) New Delhi	C
	(ii) CISH	(B) Srinagar	E
	(iii) IARI	(C) Bangalore	A
	(iv) CITH	(D) Calicut	B
	(v) IISR	(E) Lucknow	D
21	(i) Ethephon	(A) Fruit drop	C
	(ii) IBA	(B) Elongation	E
	(iii) GA3	(C) Ripening	B
	(iv) 2-4-D	(D) Growth retardant	A
	(v) MH	(E) Rooting of cutting	D
22	(i) *Allium cepa*	(A) Mountain range	B
	(ii) New world	(B) Onion	D
	(iii) Andes	(C) Cole crops	A
	(iv) Cliff cabbage	(D) Capsicum	C
	(v) Hindustan	(E) Brinjal	E
23	(i) DNA	(A) Growth harmone	B
	(ii) Glucose	(B) Nucleic acid	C
	(iii) Cellulose	(C) Sugar	E
	(iv) Metasystox	(D) Insecticide	D
	(v) GA_3	(E) Cell wall	A
24	(i) Ambilli	(A) Cucumber	E
	(ii) Arka Harit	(B) Bitter gourd	B
	(iii) Pusa Vikas	(C) Pumpkin	C
	(iv) Pusa Chikini	(D) Sponge gourd	D
	(v) Poinsett	(E) Bottle gourd	A
25	(i) Blssom end rot	(A) Mango	D
	(ii) Red rot	(B) Ground nut	C
	(iii) Whiptail	(C) Sugarcane	E
	(iv) Spongy tissue	(D) Tomato	A
	(v) Tikka Diseases	(E) Cauliflower	B
26	(i) Pea	(A) Leaf	C
	(ii) Cauliflower	(B) Head	E
	(iii) Rhubarb	(C) Seed	A
	(iv) Cabbage	(D) Spear	B
	(v) Asparagus	(E) Curd	D
27	(i) Ginger	(A) Eugenol	D
	(ii) Pepper	(B) Cineol	C
	(iii) Cardamom	(C) Piperin	B
	(iv) Turmeric	(D) Ginzebarine	E
	(v) Clove	(E) Curcumin	A
29	(i) CAZARI	(A) Lucknow	D
	(ii) CTCRI	(B) Trivendrum	B
	(iii) CSIR	(C) New Delhi	C
	(iv) NRCAMP	(D) Jodhpur	E
	(v) CIMAP	(E) Gujarat	A
30	(i) Ascorbic acid	(A) Yeast	E

S.No.	Colum 1	Colum 2	Answer Key
	(ii) Riboflvin	(B) Sunlight	D
	(iii) Thiamine	(C) Carrot	A
	(iv) VitaminA	(D) Bael fruit	C
	(v) Vitamin D	(E) Aonla	B
31	(i) Coriander	(A) Kerala>Karnataka>TamilNadu	C
	(ii) Black pepper	(B) AP>Orissa> West Bengal	A
	(iii) Turmeric	(C) AP>Rajasthan>TamilNadu	B
	(iv) Clove	(D) Kerala>Meghalaya> Orissa	E
	(v) Ginger	(E) TamilNadua>Kerala>Karnataka	D
32	(i) Black pepper	(A) 22	C
	(ii) Turmeric	(B) 24	D
	(iii) Ginger	(C) 52	A
	(iv)Cardamom	(D) 62	B
33	(i) Rutaceae	(A) Fenugreek	B
	(ii) Orchidaceae	(B)Currry leaves	D
	(iii) Papilionaceae	(C) Cinnamon	A
	(iv) Iridaceae	(D) Vanilla	E
	(v) Lauraceae	(E) Saffron	C
34	(i) Lettuce	(A) Day neutral	D
	(ii) Frenchbean	(B) Protandry	A
	(iii) Celery	(C) Balsam pear	B
	(iv) Bitter gourd	(D) Long day plant	C
	(v) Ash gourd	(E) White gourd	E
35	(i) Brinjal	(A) PusaSheetal, HisarArun	D
	(ii) Tomato	(B) Bhagyalaxmi, Andhra Jyoti	E
	(iii) Chilli	(C) PusaMuketa, PusaSambandh	B
	(iv) Onion	(D) PusaAnupam, Pusa Bihar	C
	(v) Cabbage	(E) PusaMadhvi, Early Grano	A
36	(i) Leaf curl resistant (Chilli)	(A) Bhaskar	D
	(ii) Fruit rot resistant	(B) K2	B
	(iii) F1 hybrid	(C) Agni	C
	(iv) Yellow anther chilli	(D) PusaJwala	A
39	(i) Tomato	(A) Seed rate:500-700g/ha	E
	(ii) Brinjal	(B) Seed rate: 1000-1200g/ha	A
	(iii) Chilli	(C) Seed rate: 350-500g/ha	B
	(iv) Cabbage early	(D) Seed rate: 10-15g/ha	C
	(v) Onion	(E) Seed rate: 500g/ha	D
40	(i) Bacterial wilt resistant (tomato)	(A) PubHybrid, HS-102	E
	(ii) Long brinjal	(B) ArkaNavneet	D
	(iii) Round brinjal	(C) Pant Samarat	B
	(iv) Nematode resistant (tomato)	(D) Pusa-120, Hisar Lalit	C
	(v) Hot set tomato (30°C)	(E) ArkaShirishArkaSheel	A
41	(i) Bottle gourd	(A) 24	C
	(ii) Ash gourd	(B) 16	A
	(iii) Carrot	(C) 22	D
	(iv) Garlic	(D)18	B
42	(i) Brussels sprout	(A) Moringaceae	B
	(ii) Celery	(B) Cruciferae	C

S.No.	Colum 1	Colum 2	Answer Key
	(iii) Palak	(C) Umbeliferae	D
	(iv) Drum stick	(D) Chenopodiacea	A
43	(i) Lesser yam	(A) *Cyamopsistetragonaloba*	C
	(ii) Cluster bean	(B) *Vignaungiculata*	A
	(iii) Cowpea	(C) *Dioscoreaesculenta*	B
	(iv) White Yam	(D) *Dioscorearotundata*	D
44	(i) CIPHET	(A) Kerala	D
	(ii) CITH	(B) Thiruvananthapuram	C
	(iii) CTCRI	(C) Srinagar	B
	(iv) CPCRI	(D) Ludhiana	A
45	(i) Term Heterosis	(A) Golden	B
	(ii) Progeny test	(B) Shull	C
	(iii) SSD Method	(C) Vilmorin	A
	(iv) CGMS in onion	(D) NilsonEhle	E
	(V) Bulk breeding method	(E) Jones	D
46	(i) Synthesis of tryptophan	(A) Cl	D
	(ii) Photolysis of water	(B) Mg	C
	(iii) Constituent of chlorophyll	(C) Mn	B
	(iv) Hooking leaf tip	(D) Zn	E
	(v) O_2 evaluation in primary reaction of photosynthesis	(E) Ca	A
47	(i) Nutmeg and Mace	(A) *Elettariacardamomum*	B
	(ii) Clove	(B) *Myristicafragans*	D
	(iii) Cardmom	(C) *Piper nigrum*	A
	(iv) Cinnamon	(D) *Syeygiumaromaticum*	E
	(V) Black pepper	(E) *Cinnamomumverum*	C
48	(i) Turmeric	(A) Bark spice	E
	(ii) Black pepper	(B) Bark spice	C
	(iii) Saffron	(C) Panniyuar-1	A
	(iv) Coriander	(D) Seed spice	D
	(v) Cinnamon	(E) Sudarshan	B
49	(i) Beri- beri	(A) Vitamin K	B
	(ii) Blindness	(B) Vitamin B1	D
	(iii) Scurvey	(C) Vitamin C	C
	(iv) Sterility	(D) Vitamin A	E
	(v) Blood coagulation	(E) Vitamin E	A

Appendices

1. Abbreviations used in Vegetable Science

APEDA	Agricultural and Processed Food Products Export Development Authority.
2,4,5-T	(2,4,5-trichlorophenoxy) acetic acid.
2,4-D	(2,4-dichlorophenoxy)acetic acid.
3-CPA	2-(3-chlorophenoxy) propionic acid.
4-CPA	4-chlorophenoxy-acetic acid.
ABA	Abscisic Acid.
AFLP	Amplified Fragment Length Polymorphism.
AGR	Absolute Growth Rate.
AOA	Agreement on Agriculture.
AOSA	The Association of Official Seed Analysis.
ARYA	Attracting and Retention of Youth in Agriculture
ATARI	Agricultural Technology Application and Research Institute
AVRDC	Asian Vegetable Research and Development Centre, Taiwan.
CBD	Convention on Biodiversity.
CFB	Corrguated Fibre Board.
CFTRI	Central Food Technological Research Institute, Mysore, Karnataka.
CGRCGIR	Crop growth rate.Consultative Group for International Agriculture Research
CHES	Central Horticultural Experiment Station.
Chlormequat (CCC)	2-chloroethyl-trimethyl ammonium chloride.
CIAH	Central Institute for Arid Horticulture, Bikaner, Rajasthan.
CIHNP	Central Institute of Horticulture for Horthern Plains, Lucknow, Uttar Pradesh.
CIPHET	Central Institute For Post Harvest Engineering and Technology, Luddhaina.
CITH	Central Institute for Temperate Horticulture, Srinagar, J & K.
CPP	Cow Pat Pit.
CPRI	Central Potato Research Institute , Shimla, Himachal Pradesh.
CPRS	Central Potato Research Station , Patna, Bihar.
CRL	Central Research Laboratory , IARI, New Delhi.
CSC	Central Seed Committee.
CTCRI	Central Tuber Crops Research Institute , Trivendrum, Kerla.
CVRC	Central Variety Release Committee.
Daminozide (B9, Alar)	Succinic acid-2, 2-demethyl hydrazine.
DIBA	Dot immunobinding assay.
DSC	Settling Trade Dispute.
DUS	Distinctness, uniformity and stability
ELISA	Enzyme Linked Immunosorbant Assay
EMC	Equilibrium Moisture Content.
Ethephon	2-chloroethyl Phosphonic acid
EVA	Ethylene Vinyl Acetate
FDA	Food and Drug Administration, USA.

FIRST	Farm Innovations Resource Science and Technology
FPO	Fruit Products Order, 1955.
FPTC	Food Porcessing and Training Centre.
GA_3	Gibberellic acid.
GATT	General Agreement on Tariffs and Trade.
GEAC	Genetic Engineering Approval Committee.
GI	Geographical Indication.
GOT	Grow Out Test.
GRAS	Generally Recognized As Safe.
GURT	Genetic use Restriction Technologies
HDP	High Density Polyethelene
HI	Harvest Index.
HOPCOMS	Horticultural Producers Co-operative, Marketing Society, Karnataka.
HPMC	Horticulture Produce Marketing and Processing Cooperative Limited.
HPMC	Horticultural Produce Marketing Co-operation, Himanchal Pradesh and J & K
IAA	Indole-3-acetic Acid.
IAAP	Intensive Agricultural Area Programme.
IADP	Intensive Agricultural Area Programme.
IAEA	International Atomic Energy Agency.
IAHS	Indo- American Hybrid Seeds Company.
IBA	Indole-3-butyric Acid.
IBSCs	Institutional Biosafety Committees.
IEF	Electrofocusing.
IFOAM	International Federation of Organic Agriculture Association.
IIHR	Indian Institute of Horticultural Research , Hessarghatta, Banglore , Karnataka.
IIVR	Indian Institute of Vegetable Research, Varanasi.
IME	Institute of Market Ecology.
INDOCERT	Indian Organic Certification Agency.
IOFGA	Irish Organic Farmers and Growers Association.
IPC	International Potato Centre, Peru.
IPCL	Indian Petro Chemical Corporation Ltd.
IPGRI	International Plant Genetic Resources Institute, Rome.
IQF	Individual Quick Freezing.
IEM	Immunosorbent Electron Microscopy.
ISHS	International Society for Horticultural Science, Belgium.
KIRAN	Knowledge Innovative Repository in Agriculture for North East
LAD	Leaf Area Duration.
LAI	Leaf Area Index.
LAR	Leaf Area Ratio.
LDPE	Low Density Polyethylene.
LLDPE	Linear Low-density Polyethylene.
LLP	Lab to Land Programme.
LWR	Leaf Weight Ratio.
MAHCO	Maharashtra Hybrid Seeds Company Ltd. Jalna.
Maleic hydrazide	1, 2-dihydro-3, 6-pyridazinedione.
MAP	Modified Atmospheric Packaging.
Mepiquat chloride	N, N-dimethyl Piperidinium chloride.
MOA	Mokichi Okada Association.

MSCS	Minimum Seed Certification Standards.
NAA	1-naphthalene Acetic Acid.
NAEP	National Agricultural Extension Project.
NAFED	National Agricultural Cooperative Marketing Federation.
NAR	Net Assimilation Rate.
NARP	National Agricultural Project.
NBPGR	National Bureau of Plant Genetic Resources, New Delhi.
NDDB	National Dairy Development Board.
NHB	National Horticulture Board, Gurgaon, Haryana.
NOSB	National Organic Standard Board.
NOVDB	National Oilseed and Vegetable Development Board.
NOXA BNOA	2-naphthalenyloxyacetic acid.
NPOP	National Programm for Organic Production .
NSKE	Neem Seed Kernel.
NVT	National Varietal Trial.
OFP	Organic Food Production.
OFPA	The Organic Foods Production Act.
ORP	Operational Research project.
Paclobutrazol	(2RS, 3RS)-1-(4-chlorophenyl)-4, 4-dimethyl-2-(1,2,4-triazol-1-yl)-pentan-3-ol.
PAGE	Polyacrylamide Gel Electrophoresis.
PCR	Polymerase Chain Rection.
PDV&FR	Protection of Plant Varieties and Farmer's Rights.
PEA	Plant Efficiency Analyzer
PET	Polyethylene trephthalate
PFAA	Prevention of Food Adulteration Act.
PSMEC	Priority Setting Monitoring and Evaluation Cell
RAFI	Rural Advancement Foundation International.
RAPD	Random Amplified Polymorphic DNA.
RCGM	Review Committee on Genetic Manipulation.
RDAC	Recombinant DNA Advisory Committee.
READY	Rural and Enterpreneurship Awareness Development Yougna
RASFF	Rapid Alert System for Food and Feed Specific Leaf Area
RFLP	Restriction Fragment Length Polymorphism.
RGR	Relative Growth Rate.
SWOT	Strength, Weakness, Oppertunity Threats
SLA	Specific Leaf Area.
SLW	Specific Leaf Weight.
SPS	Sanitary and Phytosanitary Measures.
SSDC	State Seed Development Corporation.
SSR	Simple Sequence Repeats.
SVRC	State Variety Release Committee.
TIBA	2,3,5-triiodo-benzoic acid.
TRIPS	Trade Related Intellectual Property Rights.
Uniconazol	(E)-(4-chlorophenyl)-4, 4-dimethyl-2-(1,2,4-triazol-1-yl)-1-penten-3-ol.
UPOV	Union International Pour La Protection Des Obtentions Vegetables.
UVT	Uniform Varietal Trial
WIPO	World Intellectual Property Organization

2. First in Vegetable Science

- First Agricultural University in India - **G.B.Pant, Pantnagar, Uttarakhand (On the pattern of Land Grant system of USA).**
- First Central Vegetable Breeding Station - **Katrain, Kullu Valley, Himachal Pradesh**
- First Project Coordinator of AICRP (VC) - **Dr.Vishnu Swarup.**
  First Project Director of IIVR - **Dr. Gautam Kalloo.**
- First mention of hybrid seeds in seed catalogue - **1945, USA.**
- First F_1 hybrid in public sector: **Pusa Meghdoot, Pusa Manjari 1971 by IARI.**
- First F_1 hybrid released in 1973 : **Tomato (Karnataka), Capsicum (Bharat) by IAHS, Bangalore.**
- First reported self- incompatibility- **J.G. Koelreuter in *Verbascum phoenicum.***
- First discovered CGMS **by Jonnes and Davis (1944).**
- First reported GCMS by **Jonnes and Emsweller in onion cv. Italian Red.**
- First F_1 hybrid in celery using CMS: **Green Giant (1982), Taki Seed Company, Japan.**
- First employed pure line selection **by Johannsen:** To improve seed weight in French Bean.
- First to produce single cross hybrid in maize: **Shull** and double cross by **D.F. Jones.**
- First time the word "Heterosis" was used by **Shull in 1914.**
- First single seed decent method was proposed by **Goulden.**
- First reported heterosis in tomato by **Hedrick and Booth** in **1907.**
- First report of hybrid vigor in brinjal -**Nagai and Kida (1926).**
- First interspecific hybrid **Sweet William and Carnation** by **Thomas Fairchild** in **1717.**
- First intergeneric hybridization **Bread Wheat and Rye** by **Rimpu in 1890.**
- First reported brown anther CMS in **Cv. "Tender Sweet"** by **Welch and Grimbell** in **1947.**
- First carrot fly resistant variety: **Flyaway in 1993.**
- First report of hybrid vigor of chilli: **1933 from IARI by Deshpande**
- First use of polyethylene as a greenhouse cover: **1948 by Emery Mayers Emmert.**
- First rank in grafted vegetable country: **Japan (81%) followed by Korea (54%).**
- First crop raised in hydroponics is **Tomato cv. Pipo (1973).**
- First transgenic variety: *"Flavr Savr"* **(Tomato) Calgene Company**(Enhanced Vase of life)
- First intergeneric hybrid radish & Cabbage by **Karpechenko** in **1927** in **Russia.**
- First proposed – *"Ideotype"* by **Donald** in **1968.**
- First coined term *"Homeostasis"* by **Lerner** in **1954.**
- First discovered CMS in onion – **1925** for seed production.
- First rank in productivity of garlic – **Egypt.**
- First indigenous variety of Leek – **Palam Paushtik.**

- First time reported heterosis in carrot – **Poole in 1937.**
- First discovered petaloid cytoplasm known as **Cornell Petaloid** by **Munger** in **1953.**
- First reported petaloid anther CMS in **Cv. "Cornell Wild"**
- First successful embryo culture **Cherry embryo (1993).**
- First molecular marker **RFLP (1980).**
- First concept of genetic map was presented by **Alfred H. Sturtevant 1913.**
- First genetic map was published in **1911** by **T.H.Morgan.**
- First bacterial genome sequenced - *Haemophilus influenza.*
- First multicellular organism sequenced - *Caenorhabditis elegans.*
- First plant sequenced - *Arabidopsis thaliana.*
- First black coloured variety of carrot in India : **Pusa Asita**
- First temperate hybrid developed using CMS: **Pusa Nayanjyoti.**
- First public sector hybrid using CMS: **Pusa Vasudha.**
- First genetically improved variety of Elephant foot yam in the world is **Sree Athira.**
- First genetically improved variety of Taro – **Sree Kiran.**
- First reported SI in cole crops by **Bateman (1954).**
- First cabbage hybrid - **Nagoka in Japan(1951).**
- First CMS based F_1 hybrid in cauliflower- **KTH-27.**
- First reported male sterility in radish-**Ogura 1968 in Japanese radish type.**
- First Purple Fleshed radish variety is **Pusa Sagarika.**
- First hybrid of knol-khol : **Roggli.**
- First sequenced vegetable crop – **Cucumber.**
- First Gynoecious line of Cucumber is- **M.SU. 713-5 (Shogoin X Wisconsin SMR 18)** developed by **C.E.Peterson(1960).**
- First gynoecious hybrid in cucumber: **Pusa Sanyog** in **(1971).**
- First tropical cucumber hybrid – **Phule Prachi** by **MPKV, Rahuri.**
- First extra early parthenocarpic variety of cucumber – **Pusa Seedless.**
- First true mini watermelon cultivar: **New Hampshire Midget Developed by A.F.Yeager.**
- First fusarium wilt resistant watermelon - **Conquerer Developed** by **W.A.Orton.**
- First horticultural classification of melons - **Naudin(1859); Modified** by **Munger and Robinson.**
- First F1 hybrid of muskmelon- **Punjab Hybrid (MS-1 X Hara Madhu)** in **1984- Resistant to powdery mildew.**
- First muskmelon variety grown under net-house in north Indian plains – **Pusa Sarda.**
- First Early Maturing variety of Round melon - **Pusa Raunak.**
- First early maturing variety of Long melon – **Pusa Utkarsh.**
- First improved variety of Ash Gourd – **Pusa Ujjwal** (Petha preparation).
- First inter-varietal hybrid of greater yam – **Shree Shilpa.**
- First Coleus Variety Grown in Kerala – **Shree Dhara.**

- First report of enation leaf curl virus in okra (Vector: White Fly) – **IIHR.**
- First machine-harvestable tomato cultivar was developed by **G.C.Hanna.**
- First male sterility in tomato was reported by **C.M.Rick.**
- First tomato resistant cv. **Pan America.**
- First triple disease resistant variety of tomato – **Arka Rakshak.**
- First hybrid of tomato available cultivation in **1973.**
- First root knot nematode resistant variety of tomato: **Sel-120.**
- First report of heterosis in Brinjal : **Kakisaki (1931).**
- First male sterility in brinjal : **Jasmin (1954).**
- First functional male sterile line in brinjal: **UGA1-MS (Cv. Florida High Bush), Pathak and Jaworski (1989).**
- First reported little leaf of brinjal : **Thomas and Krishnaswamy.**
- First discovered CMS in chilli by **Peterson** in **PI164835.**
- First president of Central Potato Research Institute – **Dr. S.Ramanujan.**
- First International Flower Auction Centre was launched in **Bangalore.**
- First rank foliage plant in global market- **Dieffenbachia.**
- First rank cut flower in global Market – **Rose.**
- First rank pot plant in global market – **Hedera.**
- First rank dry flower in global market – **Helichrysum.**
- First Spice Park in India – **Chhindwara, M.P.**
- First fruit product research laboratory in India : **Kodur.**
- First Fruit & vegetable processing industry in india : **Mumbai (1935).**
- First Fruit preservation & canning institute: **Lucknow (1949).**
- First carotenoid pigment was found in **Carrot.**
- First Bioagent for controlling post-harvest diseases : *Bacillus subtillis.*
- First climate controlled greenhouses established in India by **Indo-American Hybrid Seed Company in 1965.**
- First Fruit &Vegetable processing started in **1857.**
- First International Okra Workshop at **NBPGR, NewDelhi 1990.**
- First to develop heat tolerant cabbages by **Japanese breeders varieties (i.e) KK Cross & KY Cross – popular in South East Asia.**
- First YVMV tolerant variety in okra: **Pusa Sawani:** Presently it has become susceptible.
- First *in vitro* culture of pepper: **Smith & Heiser (1957).**
- First used the term Regression by **Francis Galton.**
- First Chi-square test was used by **Karl Pearson (1900).**
- First report on double stranded DNA in cauliflower mosaic virus – **Sheperd** *et al.***(1968).**
- First Auxin was discovered by **F.W.Went in 1926.**
- First commercial level production of plantlets in spice crops was achieved in **Cardamom.**
- First largest seed spice is **Coriander.**

- First wilt resistant variety of cumin is **Gujarat Cumin-1**.
- First cultivated food plant in France is **Celery** in **1623**.
- First quality vanilla is **Bourbon like vanilla: Mexican vanilla**.
- First used the term Regression by **Francis Galton**.

3. Latest Released Varieties/ Hybrids

Cabbage

KTCBH-81: This hybrid has been developed from ICAR-IARI, RS, Katrain and released in 2013. The plants are dark green. Head is round shaped, compact and covered with outer leaf, ready for harvest in 60-65 DAT. The yield potential is 40.0-45.0 t/ha. It is recommended for Jammu&Kashmir, HimachalPradesh and Uttarakhand.

KTCBH-822: This hybrid has been developed from ICAR-IARI, RS, Katrain and released in 2019. The plants are dark green and waxy leaves. Head is flat shaped, compact and covered with outer leaf, ready for harvest in 70-75 DAT with good field staying capacity. The yield potential is 40.0 t/ha. It is recommended for cultivation in states like Jammu &Kashmir, H.P. and Uttarakhand, Rajasthan, Gujarat, Haryana, Delhi.

Cauliflower

Sabour Agrim (SBECF-102): This variety has been developed from BAU, Sabour and released in 2014. It is early, form curd at 22-27°C. Plants are erect to semi- spreading. Curds are ready for harvesting in 65 - 68 days for 50% curd (450g) maturity. The yield potential is 15.0-20.0t/ha. It is recommended for cultivation in Madhya Pradesh, Maharashtra and Goa.

DC-31 : This variety has been developed from IARI, Pusa, New Delhi and released 2014. This variety is suitable for July transplanting and reaches marketable maturity during October. Curds are compact with retentive white colour. Curd weight is 500-600g with yield potential of 16.0-18.0t/ha. It has been recommended for cultivation in states like Punjab, Uttar Pradesh, Bihar and Jharkhand.

KTH-301: This hybrid has been developed from IARI, Katrain and released in 2019. It is suitable for cultivation in the mid-season with the harvesting of curd in the month of November-December. An average yield is 39.0 t/ha. It has been recommended for cultivation in states like Jammu & Kashmir, Himachal Pradesh,. Uttarakhand, Rajasthan. Gujarat, Haryana and Delhi.

Kashi Gobhi-20: This variety has been developed from IIVR, Varanasi. This variety belongs in tropical cauliflower group matures .Curds are white compact, hemispherical and free from riceyness, leafiness and fuzziness. Marketable curd weight is 600-700 g and yield ranged from 25 to28 t/ha.

Bitter Gourd

Pusa Aushadhi (Sel-1): This variety has been developed from IARI, New Delhi and released in 2013. The fruits are light green in color, medium long (16.5). The yield potential is 20.0-22.0 t/ha. It is recommended for cultivation in state like Rajasthan, Gujarat, Haryana and Delhi.

NBIH-2009: This hybrid has been developed from Nuziveedu Seeds Pvt. Ltd. and released in 2018. The fruits are dark green in colour, sharp and dense tubercles. The yield potential is 15.0-20.0 t/ha. It is recommended for cultivation in states like Punjab, Uttar Pradesh, Bihar and Jharkhand.

Kashi Mayuri: This variety belongs in medium maturity group. Fruit are 15-18 cm long, with uniform shape-size and higher shelf life. It is tolerant to anthracnose, downy mildew and mosaic diseases in field condition. An average yield is 19 t/ha.

Bottle Gourd

NDBG-619: This variety has been developed by NDUA&T, Faizabad and released in 2009. The plants have long cylindrical attractive fruits. The yield potential is 63.5 t/ha. It has been recommended for cultivation in states like Madhya Pradesh, Maharashtra and Goa.

Santosh-20 (KBGH-20): This hybrid has been developed by Krishidhan Vegetable Seeds Pvt. Ltd. and released in 2010. The plants are tolerant to both powdery and downy mildews in field. The yield potential is 35.0-40.0 t/ha. It has been recommended for cultivation in Punjab, Uttar Pradesh, Bihar and Jharkhand.

PBOG-89: This variety has been developed by GBPUAT, Pantnagar. The fruits of this variety are cylindrical and attractive.The yield potential is 35.0-40.0 t/ha. It has been recommended for cultivation in states like Punjab, UttarPradesh, Bihar, Jammu&Kashmir, Himachal Pradesh, MadhyaPradesh, Maharashtra and Jharkhand.

Anurag: This hybrid has been developed by Nuziveedu Seeds Pvt. Ltd. This hybrid is suitable for long distance transport.The yield potential is 30.0-35.0 t/ha. It has been recommended for cultivation Punjab, Uttar Pradesh, Bihar and Jharkhand.

NDBG-10: This variety has been developed by NDUA&T, Faizabad. The plants are tolerant to downy mildew. The yield potential is 30.0-50.0 t/ha. It is recommended for Punjab, UttarPradesh, Bihar and Jharkhand.

Kashi Kirti: This variety has been developed at IIVR, Varanasi. Fruits are green, small, cylindrical shape (Gutka type), resistant to downey mildew, early maturing and high yielding. This variety is suitable for distant marketing due to longer pos tharvest life. An average yield is 45 t/ha.

Kashi Kundal: Attractive pear-shaped green fruits, medium size, 12-14 fruits/plant, weight of 1.3 to 1.5 kg, resistant to downey mildew, yield (47.5 t/ha).

Kashi Kiran: Light green attractive round fruits, tolerant to downey mildew, Fruit weight 600-700 g, number of fruits 13-14/plant, average yield 45-48 t/ha.

Cucumber

Rajani-Cu-05: This hybrid has been developed from Syngenta Seeds Pvt. Ltd. In 2013. Fruits are cylindrical, dark green and medium long. The yield potential is 20-25 t/ha with seed rate of 1.0-2.0 kg/ha. It is recommended for J&K, HP and Uttarakhand.

DC-83: This variety has been developed from ICAR-IARI, New Delhi in 2018. Fruits are straight, rudimentary vines, soft skinned with tender and crispy flesh. The yield potential is 20-25 t/ha with seed rate of 1.0-2.0 kg/ha. It is recommended for Punjab, Bihar and Jharkhand.

DGCH-18: This hybrid has been developed from ICAR-IARI, New Delhi in 2019. It is gynoecy based hybrid with predominantly gynoecious behaviour, earliness. The yield potential is 24.0 t/ha. It is recommended for Jammu & Kashmir, HimachalPradesh and Uttarakhand.

Pointed Gourd

Kashi Suphal: This variety was developed from IIVR, Varanasi. Plant spread of this variety is 3.5-4.0 meter. Harvesting starts 90-95 days after planting. Fruits are attractive light green in colour, less seeded and good flesh than other variety with better keeping quality. This variety yielded 18.0-21.5 t/ha. This variety is suitable for cultivation in Uttar Pradesh, Bihar, Jharkhand and West Bengal.

Kashi Amulya: Less seeded (5-8 seed/fruit as compared to 25-30 seeds in normal variety), suitable for confectionary purpose, more fleshy, attractive light green fruit with sparsely distributed white stripes, yield 20-22 t/ha. Planting of 8000-10000 cuttings /ha in 9:1 ratio of female: male.

Pumpkin and Squashes

Swarna Amrit (HAPK-10): This hybrid has been developed from ICAR-ICER, RC, Ranchi in 2009. Fruits are flat and are able to harvest 75-80 days after sowing. The yield potential is 45-50 q/ha with seed rate of 3.5-4.0 kg/ha. It is recommended for J&K, HP and Uttarakhand.

Pusa Vikas: It is developed from IARI, New Delhi. Flesh is yellow. It is suitable for growing in spring-summer season (February-March to June) in northern plains of India.

Kashi Shishir: Early maturing hybrid with small round and mottled green fruits (2 - 2.25 kg), 3-4 fruits/plant, suitable for cultivation in summer as well as Kharif; yield (38-42 t/ha); seed rate : 4-4 kg/ha.

Sponge Gourd

VR-1 (Kashi Divya): This variety has been developed from ICAR-IIVR, Varanasi in 2010. The plants have cylindrical fruits of 20-25 cm length. The yield potential is 13-16 t/ha with seed rate of 3.5-4.0 kg/ha. It is recommended for Punjab, UP, Bihar and Jharkhand.

VRSG-1 (Kashi Shreya): This variety has been developed from ICAR-IIVR, Varanasi in 2018. The plants have dark fruits of 20-25 cm length. The yield potential is 12-15 t/ha with seed rate of 3.5-4.0 kg/ha. It is recommended for Punjab, UP, Bihar and Jharkhand.

Kashi Rakshita: This hybrid has been developed from ICAR-IIVR, Varanasi in 2018. The plants have dark fruits of 20-25 cm length. The yield potential is 12-15 t/ha with seed rate of 3.5-4.0 kg/ha. It is recommended for Punjab, UP, Bihar and Jharkhand.

Kashi Saumya: Medium maturing hybrid, dark green fruits, resistant to SGMV and tolerant to downy and powdery mildew under field condition, yield (18-19 t/ha).

Kashi Jyoti: Light green fruits of 100-150 g, resistant to sponge gourd mosaic virus tolerant to downy mildew and root knot nematode, yield 15-18t/ha.; Seed rate : 4-5 kg/ha.

PUSA Supriya: It is released from IARI, New Delhi. Flowers are monoecious, campanulate. Fruits become ready for picking 50-55 days after sowing in spring-summer and 44-48 days after in kharif season.

Ridge Gourd

Kashi Kushi: This variety has been developed from ICAR-IIVR, Varanasi. Plant bearing light green fruits with 10 dark green superficial & continuous longitudinal ridges. Single plant bears 140 fruits/ plant in cluster of 5-6 fruits. An average yield is 5.24 -6.27 kg/ plant.

Pusa Nutan: Its variety was released from IARI, New Delhi. Fruits are 25-30 cm long, straight and attractive green, An average fruit weight is 105 g, flesh tender, suitable for spring summer and *kharif* season. Maturity is 45-50 days after seed sowing. An average yield is 16.0 t/ha.

Pallavi: This hybrid has been developed from Sungro Seed Pvt.Ltd. in 2009. It has green long fruits with high ridge. The yield potential is 13.8-18.5 t/ha with seed rate of 3-4 kg/ha. It is recommended for Chhattisgarh, Orissa, AP and Telangana.

Nhyrgh-5HB (Haritham): This hybrid has been developed from APHU, Hyderabad in 2009. Fruits are light in color and can be harvested after 65-68 days. The yield potential is 20.1-26.0 t/ha with seed rate of 3-4 kg/ha. It is recommended for MP, Maharashtra and Goa.

Kashi Shivani (VRRG-27): This variety has been developed from ICAR-IIVR, Varanasi in 2016. Fruits are green long and straight of 20-30 cm with diameter of 3-4 cm. The yield potential is 18-20 t/ha with seed rate of 3-4 kg/ha. It is recommended for Punjab, UP, Bihar and Jharkhand.

Watermelon

Kashi Surabhi: This is open-pollinated variety developed from ICAR- IIVR, Varanasi and recommended for cultivation in Uttar Pradesh, Bihar, Jharkhand and Punjab. This variety is suitable for petha preparation. Fruits can be stored up to 4 months after harvesting at ambient temperature. Plant is tolerant to leaf minor& resistance to anthracnose disease. This variety is yielded 60-70 t/ ha.

Mudliar: This is a popular variety of Tamil Nadu, whose fruits are big with light green colour.

Pusa Ujwal: This variety has been released from IARI Pusa, New Delhi and recommended for commercial cultivation in Delhi, Punjab and Haryana. Fruits are green in immature stage and white in full mature stage. Fruits are cylindrical, rind colour green, immature fruits are pubescent, cylindrical (Av. length-23.50 cm and girth55.0 cm).Mature fruits have a dense pelt of minute hairs and covered with a white waxy bloom. The average fruits weight is 7 kg and average number of fruits per plant is 3. The fruit flesh is white and seeds are creamy white in colour. Fruits ellipsoid, ideal for packing and long-distance transportation. Maturity 120 days after seed sowing.This variety is resistant to cucumber mosaic virus and water melon mosaic virus. An average yield is 45.0 t/ha.

KAG-1: This variety has been developed from CSAU&T, Kanpur in 2011, Vines are medium long. The yield potential is 53.9-78.7 t/ha and seed rate are of 2.0-2.5 kg/ha. It is recommended for Karnataka, Tamil Nadu, Kerala and Pondicherry.

DAGH-16 (PUSA URMI): This hybrid has been developed from IARI, New Delhi in 2013. Fruits are oblong ellipsoid with greenish white rind and white flesh. The yield potential is 47.5 t/ha and seed rate are of 2.0 Kg/ha. It is recommended for Rajasthan, Gujrat, Haryana, Delhi, Karnataka, Tamil Nadu and Kerala.

DAGH-14 (PUSA SHREYALI): This hybrid has been developed from IARI, New Delhi in 2013. Fruits are cylindrical with white rind and white flesh. The yield potential is 52.0 t/ha and seed rate are of 2.0 Kg/ha. It is recommended for Punjab, UP, Bihar and Jharkhand.

DAG-12: This variety has been developed from IARI, New Delhi in 2013. Fruits are cylindrical with greenish white rind and white flesh, easy long-distance transportation. The yield potential is 52.0 q/ha and seed rate are of 2.0 Kg/ha. It is recommended for Karnataka, Tamil Nadu and Kerala.

Brinjal

Utkal Tarini: This is a bacterial wilt resistant variety, developed at OUA & T, Bhubaneshwar, derived through pedigree method from the cross Pusa Kranti x Gopa Local.

Fruits are oblong, medium sized, deep purple with cremish white flesh, 6-7 days shelf life. It is recommended for cultivation West Bengal, Assam, Chhattisgarh, Orissa and Andhra Pradesh.

PB-67: This variety is resistant to bacterial wilt and *Phomopsis*, early maturity with yield potential of 41.0 t/ha. It is developed by GBAU&T, Pantnagar in 2009 and recommended for cultivation in states like Punjab, UP, Bihar and Jharkhand.

Rsika: This hybrid has been developed by Bejo Sheetal Pvt. Ltd, Jalana in 2009. The fruit length is 16.3 cm with yield of 40.0-58.0 t/ha. It is recommended for cultivation in states like Punjab, UP, Bihar and Jharkhand.

Shamli: This hybrid has been developed by Seminis Seeds Pvt. Ltd. in 2009. It is long fruited hybrid with yield of 35.0-65.0 t/ha. It is recommended for cultivation in states like Punjab, UP, Bihar and Jharkhand.

VNR-51C: This hybrid has been developed by VNR Seeds Pvt. Ltd. Raipur in 2009. It is a small round fruited hybrid with yield of 45.0-50.0 t/ha. It is recommended for cultivation in states like Punjab, UP, Bihar and Jharkhand.

HBAH-8: This hybrid has been developed by ICAR-ICER, RC, Ranchi in 2009. It is a small round fruited hybrid with yield of 37.5-54.4 t/ha. It is recommended for cultivation in states Karnataka, Tamil Nadu, Kerala and Pondicherry.

PB-70: This variety is resistant to bacterial wilt, fruit and shoot borer and *Phomopsis*, early maturity with yield potential of 40.0 t/ha. It is developed by GBAU&T, Pantnagar in 2010 and is recommended for cultivation in states like Punjab, UP, Bihar, Jharkhand, Chhattisgarh, Orissa, AP, MP, Maharashtra, Goa, Karnataka, Tamil Nadu, Kerala and Pondicherry.

DBL-02: This variety has been developed by IARI, New Delhi in 2010. It is long fruited hybrid with yield of 37.0-39.0 t/ha. It is recommended for cultivation in states like Punjab, UP, Bihar and Jharkhand, J&K, HP, Rajasthan, Gujarat, Haryana and Delhi.

DBHL-20: This hybrid has been developed IARI, New Delhi in 2009. It is long fruited hybrid with yield of 45.0-67.5 t/ha. It is recommended for cultivation in states like Punjab, UP, Bihar and Jharkhand.

PHBL-51: This hybrid has been developed by PAU, Ludhiana in 2012. It is long fruited hybrid with yield of 55.0-65.0 t/ha. It is recommended for cultivation in states like Punjab, UP, Bihar and Jharkhand.

PBHSR-31: This hybrid has been developed by PAU, Ludhiana in 2012. It is long fruited hybrid with yield of 55.0-65.0 t/ha. It is recommended for cultivation in states like Punjab, UP, Bihar and Jharkhand.

VNR-218: This hybrid has been developed by VNR Seeds Pvt. Ltd. Raipur in 2012. It is a small long fruited hybrid resistant to bacterial wilt.. It is recommended for cultivation in states like West Bengal and Assam.

HABR-21: This variety has been developed by ICAR-ICER, RC, Ranchi in 2013. Its fruit are oblong with yield of 55.0-60.0 t/ha. It is recommended for cultivation in states like Punjab, UP, Bihar and Jharkhand.

PBHL-52: This hybrid has been developed by PAU, Ludhiana in 2014. Fruits are medium long with yield of 675 q/ha and seed rate of 150-200 g/ha. It is recommended for Punjab, UP, Bihar and Jharkhand.

Nishant: This hybrid has been developed by Advanta Seeds Pvt. Ltd. in 2015. It is long fruited hybrid with green calyx cluster and yield of 30.0-35.0 t/ha. It is recommended for cultivation in states like Punjab, UP, Bihar and Jharkhand.

DBL-175: This variety has been developed by IARI, New Delhi in 2018. It is long fruited hybrid with yield of 35.0-40.0 t/ha. It is recommended for cultivation in states like MP, Rajasthan, Gujarat, Haryana, Maharashtra, Goa and Delhi.

PBL-232: This variety has been developed by PAU, Ludhiana in 2019. Fruits are medium long, early maturing, green calyx with yield of 36.0 t/ha. It is recommended for cultivation in states like MP, Rajasthan, Gujarat, Haryana, Maharashtra, Goa and Delhi.

IVBL-21: This variety has been developed by ICAR-IIVR, Varanasi in 2019. This variety is tolerant to *Phomosis* blight and Fusarium wilt, Fruits are medium long, early maturing with yield of 40.0 t/ha and seed rate of 400-500 g/ha. It is recommended for Punjab, UP, Bihar and Jharkhand.

PusaVaibhav (DBPR-23): This variety has been developed by IARI, New Delhi in 2019. The plants are tall with round fruits (15 cm) and non-spiny green calyx with yield of 41.0 t/ha. It is recommended for cultivation in states like Punjab, UP, Bihar and Jharkhand.

Kashi Himani: This variety has been developed by ICAR-IIVR, Varanasi in 2019. Fruits are medium long, shiny white fruits with less seeds, suitable for kharif season, An average yield 40.0-43.0 t/ha.

Chilli

HH-41786: This hybrid has been developed from Syngenta Seeds Pvt. Ltd. in 2011. The plants are long fruited. The yield potential is 10.0-13.0 t/ha. It is recommended for cultivation in states like MP, Maharashtra, Goa.

Garima (BSS-378): This hybrid has been developed from Bejo Sheetal Seeds Pvt. Ltd. in 2012. The plants are long fruited (14-16 cm). The yield potential is 15.0 t/ha. It is recommended for cultivation in MP, Maharashtra and , Goa.

Vidya: This hybrid has been developed from VNR Seeds Pvt. Ltd. Raipur in 2013. The plants are tolerant to Fusarium wilt. The yield potential is 20.0-22.0 t/ha. It is recommended for cultivation in states like Punjab, UP, Bihar and Jharkhand.

LCA-620: This variety has been developed from Dr YSRHU, RS, Lam in 2014. The fruits of this plants are dry color. The yield potential is 13.8 t/ha. It is recommended for cultivation in states like Chhattisgarh, Orissa and AP.

CH-27: This hybrid has been developed from PAU, Ludhiana in 2019. The plants are spreading tall, pungent (0.8% capsaicin), resistant to leaf curl virus, fruit rot and root knot nematodes. The yield potential is 14.5 t/ha. It is recommended for cultivation in states like Punjab, UP, Bihar, Jharkhand.

Kashi Abha: Fruits short stout with blunt apex (bullet type), highly pungent, tolerant to biotic (anthracnose, CLCV, thrips and mites) and abiotic stress (low and high temperature), yield 15 t/ha,

Kashi Ratna: CMS based F_1 hybrid suitable for green chillies, tolerant to anthracnose and thrips, fruit contains 0.82% (93450 SHU) capsaicin and 175.6 mg/100g vitamin C. Yield potential is 20-22 t/ha.

Kashi Nutan: Early maturing hybrid, cylindrical long, light green colour with mottling at peduncle end, resistant to downy mildew and suitable for Kharif and summer seasons. Yield is 17.5 t/ha.

Capsicum

PRCH-101: This hybrid has been developed by UUHF, Ranichauri in 2013. Plants bear fruits of 80g (average fruit). Yield potential is 30.0-32.0 t/ha. It is recommended for cultivation in states like Jammu&Kashmir, HP and Uttarakhand.

DARL-70: This variety has been developed by DIBER, Pithoragarh in 2013. Plants are tolerant to fusarium wilt and powdery mildew. The yield potential is 20.0-22.0 t/ha. It is recommended for cultivation in states like Jammu & Kashmir, Himachal Pradesh and Uttarakhand.

KTC-1: This variety has been developed by IARI (RS), Katrain in 2019. The plants bear 6-7 fruits/plants with average fruit weight of 70 g. The yield potential is 200 q/ha and seed rate are 1.5-2.0 kg/ha. It is recommended for cultivation in Jammu&Kashmir, Himachal Pradesh and Uttarakhand.

Okra

JOL-2K-19 (GJO-3): This hybrid has been developed from JAU, Junagadh in 2011. The plants are moderately resistant to Jassids, whitefly and pod borer damage and field resistant to YVMV, attractive green pods and good shelf life. The yield potential is 15.0-24.0 t/ha. It is recommended for cultivation in states like Chhattisgarh, Orissa and Andhra Pradesh.

OH-597: This hybrid has been developed from Syngenta India Limited, Pune in 2011.The plants have green tender pods and resistant to YVMV disease. The yield potential is 12.0-15.0 t/ha. It is recommended for cultivation in Madhya Pradesh, Maharashtra and Goa.

AROH-631: This hybrid has been developed from Ankur Seeds Pvt. Ltd in 2012. The plants have cylindrical fruits, smooth dark green capable of harvesting 40-50 DAS. The yield potential is 12.5-16.0 t/ha. It is recommended for cultivation in states like Chhattisgarh, Orissa and A.P., and M.P., Maharashtra and Goa.

JOH-0819: This hybrid has been developed from JAU, Junagadh in 2014. The plants have medium long fruits, smooth light of 10-15 cm long with fruit weight of 10-15 g. The yield potential is 16.0 t/ha. It is recommended for Rajasthan, Gujarat, Haryana and Delhi & M.P.,Maharashtra and Goa.

Kashi Vardan (VRO-25): This variety has been developed from ICAR-IIVR, Varanasi in 2014. This is an early variety with short internodes along with 2 to 3 branches. It takes 42-44 days for first flowering. The yield potential is around 14.0-15.0 t/ha. It is recommended for cultivation in states like Punjab, U.P., Bihar and Jharkhand.

Kashi Chaman: Plants medium tall (120-125 cm), flowering 39-41 DAS, fruiting period 45-100 days, fruits dark green, length 11-14 cm, resistant to YVMV and OLECV under field conditions, suitable for summer and rainy seasons. Yield potential is 15-16 t/ha.

Kashi Lalima: Reddish purple fruits, tolerant to YVMV and OLCV, medium tall and short internodes, rich in anthocyanin and phenolics, suitable for both summer and Kharif seasons. Yield potential is 14-15 t/ha.

Kashi Shristi: F_1 hybrid, medium tall, short internodes, narrow angled 2-3 branch, tolerant to YVMV, dark green fruits, medium fruit length, suitable for both summer and Kharif seasons. Yield potential is 18- 19 t/ha.

Tomato

DARL-68: This variety has been developed by DIBER, Pithoragarh in 2014. Plants are indeterminate growth habit with yield of 32.0 t/ha.

Punjab Ratta: This variety has been developed by PAU University, Ludhiana in 2014. Plants are of determinate growth habit with yield of 56 t/ha. The seed rate is 350-400 g/ha. The fruits are suitable for processing purposes.

IIHR-H-240 (Arka Samrat): This hybrid has been developed by ICAR-IIHR, Bangalore in 2015. Hybrids are of semi-determinate growth habit with yield of 80.0-85.0 t/ha. This hybrid is resistant to ToLCV, bacterial blight, early blight. This variety is suitable for processing purposes.

ATL-08-21: This variety has been developed by AAU University, Anand in 2016. Plants are of determinate growth with yield of 45.0-50.0 t/ha. The seed rate is 35.0-40.0 g/ha. This variety is resistant to fruit borer and leaf minor.

Kashi Amul: This variety has been developed by ICAR-IIVR, Varanasi in 2016. Plants are semi-determinate in growth with yield of 50.0-60.0 t/ha.

Kashi Adarsh: This variety has been developed by ICAR-IIVR, Varanasi in 2016. Plants are of semi-determinate in growth with yield of 60.0 t/ha.

Kaveri-304: This hybrid has been developed by private sector in 2018. Plants are of semi-determinate in growth with yield of 90.0 t/ha.

TODHYB-05: This hybrid has been developed by private sector in 2017. Plants are of semi-determinate in growth with yield of 32.5 t/ha.

TOLCVRES-5: This variety has been developed in 2018. Plants are of semi-determinate in growth habit with yield of 35.0 t/ha. This variety is resistant to ToLCV.

BT19-1-1-1: This variety has been developed by OUAT, Bhubaneswar in 2019. Plants are of semi-determinate in growth with yield of 30.0 t/ha. This variety is tolerant to bacterial wilt.

Kashi Chayan: This variety has been developed by ICAR-IIVR, Varanasi in 2019. Plants are of indeterminate growth habit with yield of 60.0-70.0 t/ha.. This variety is resistant to ToLCV carrying Ty3 gene.

CTH-1: This hybrid has been developed by TNAU, Coimbatore in 2019. Plants are of determinate in growth with yield of 80.0-90.0 q/ha. This hybrid has flat round fruits and thick pericarp with shelf life of 10 days at room temperature.

Amranths

Kashi Suhavani: Soft succulent green leaf, delayed flowering and high yield potential (30-33 t/ha). Suitable for growing during summer and rainy seasons. Sowing time is February-March and June-July in northern India.

Bathua

Kashi Bathua 2: Upright growth, green, plant 180-195 cm long at 160 DAS, excellent source of folic acid, minerals, vitamin A, vitamin C, phenolics and antioxidants, yield 30-32 t/ha.

Kashi Bathua 4: Upright growth habit with purplish green leaf and petiole, excellent source of vitamin C, vitamin A, folic acid, minerals, phenolics and antioxidants, yield - 35 t/ha.

POI/Indian Spinach

Kashi POI-1: Soft, lush green shoots, early picking with delayed flowering, green vines/stem with green succulent leaves, tolerant to water lodging, excellent source of antioxidant with low oxalate content, picking starts 40 DAT and continues up to 220-230 days at 20-30 days interval, high yield 50.7 t/ha.

Kashi POI-2: Fast growing bush type, green stems and leaves, suitable for making saag, pakode, soup etc. excellent source of antioxidant with slightly lower oxalate content, first picking starts 38-40 DAT and continues up to 140-150 days at 20- 30 days interval, yield 63.5 t/ha.

Kashi POI-3: Fast growing plant with twinning growth habit, red stem and mid ribs, high betalain content, suitable for year-round cultivation, rich source of carotenoids with lower oxalate content, picking starts 40 DAT and continues up to 240-250 days at 20-25 days interval, yield 61.3t/ha.

Cowpea

Ankur Gomti: This variety has been developed from Ankur Seeds Pvt. Ltd. in 2011. This is a bush type variety. The yield potential is 8.0-9.0 t/ha. It is recommended for cultivation in states like Chhattisgarh, Orissa and AP.

CP-55: This variety has been developed from IARI, New Delhi in 2018. The pods of this variety are smooth, slender and straight. The yield potential is 12.0-15.0 t/ha. It is recommended for cultivation in states like Rajasthan, Gujarat, Haryana and Delhi.

French Bean

Kashi Sampann: This variety was developed at IIVR, Varanasi through hybridization (Arka Komal X Contender). Plants are bushy and multi-branched. Flowering occurs in axial of the leaf. Pods are light green in round shape. Sometimes 75 pods are harvested in one plant. Pods are 11.2-14.0 cm in length, 0.7-1.0 cm width and 0.6-0.9 cm in thickness consisting 5-7 seeds per pod. Per plant yield harvested 400 g green pods/ plant in 4-5 pickings. Flowering fruiting continue in day temperature upto $38^{\circ}C$. Seeds are bright reddish black.

Kashi Rajhans: This variety was developed at IIVR, Varanasi from a Canadian line EC-595960 exclusively for vegetable type. This has ability to give flowering – fruiting at temperature of day temperature $32-34^{0}C$ in Varanasi condition (heat tolerance). Peak fruiting period is second week of February and continue up to first week of April. An average green pod yield is 23.8t/ha green pods. This variety yielded 33.09% more pod yield with national check Arka Komal.

Kashi Baigani (VPFBP-14): This variety has been developed from ICAR-IIVR, Varanasi in 2019. It is a pole type variety, pods are purple in color, slightly curved and free from parchment. The yield potential is 16.0 t/ha. It is recommended for cultivation in states like J&K, HP, Uttarakhand, Chhattisgarh, Orissa, AP, MP, Karnataka, Tamil Nadu, Kerala and Pondicherry.

Garden Pea

Vivek Matar 11: This variety is an indigenous collection, the seed sample was collected from Srirampur village, Agra developed at Vivekananda Parvatiya Krishi Anusandhan Sansthan (ICAR),Almora, Uttarakhand. Plants are dwarf, vigorous in growth with green foliage. Pods are long, dark green and curved.Seed is wrinkled and greenish in colour.This variety is resistant to powdery mildew and wilt (5-8% incidence), white rot (2-3% incidence) and leaf blight. Less incidence of pod borer has also been recorded. An average yield is 10.0-11. 0 t/ha.

VP- 434: This variety has been developed by ICAR-VPKAS, Almora in 2012. The plants are of medium maturity group and resistant to powdery mildew. The yield potential is 11.0-12t/ha and seed rate are 140-150 kg/ha. It is recommended for J&K, HP and Uttarakhand.

Pusa Agrani (GP-17): This variety has been developed by ICAR-IARI, New Delhi in 2013. The plants have dark green color pods with 6-7 seeds. The yield potential is 9.0-10.0 t/ha. It is recommended for cultivation in states like Jammu&Kashmir, Himanchal Pradesh and Uttarakhand.

Arka Priya (IIHR-1): This variety has been developed by ICAR-IIHR, Bangalore in 2016. The plants are resistant to powdery mildew and rust. The yield potential is 12.0 q/ha and seed rate are 140-150 kg/ha. It is recommended for cultivation in Karnataka, Tamil Nadu, Kerala and Pondicherry.

Indian Bean

Kashi Khushhaal: This variety was released and notified through SVRC in 2019 for commercial of cultivation. This is cluster bearing semi-pole type promising line collected from Ramana village of Varanasi from farmer field. This line was collected from Uttrakhand. It is the earliest variety which flowering starts in 66 days and pods are ready for harvesting in 95-100 days after seed sowing. Fruiting starts from last week of September and continued second week February in agro-climatic condition in Utar Pradesh, Varanasi. Fruits are dark green in colour measured 12.5-14.6cm in length, 1.3-2.4cm in width with 07-1.0cm thickness containing 4-6 seeds per pod. Per plant yield is 6.6 kg/plant green pods. Pod seeds are brown in colour measured 1.16-123cm and 0.78-0.84cm in length and width, respectively.

Kashi Sheetal: This variety was released and notified through SVRC in 2019 for commercial cultivation in Uttar Pradesh. An average pod yield is about 359.0 q/ha green pods which was 30.6% more green yield in compare with national check variety Swarn Utkrisht. This variety has been found as tolerant to DYMV in field condition.

DB-10: This variety has been developed from ICAR-IARI, New Delhi. The plants are pole type resistant to common bean mosaic virus and cercospera disease. The yield potential is 30.0-35.0 t/ha and seed rate of 30 kg/ha. It is recommended for cultivation in states like MP, Maharashtra and Goa.

Kale

KTK-64: This variety has been developed from ICAR-IARI, RS, Katrain in 2013. Highly serrated, purplish green leaves, 40-50 cm in length and 15-20 cm in width, plant height is 50-60 cm. Leaves available throughout the winter season in multiple harvestings, higher contents of phenols, anthocyanin, ascorbic acid, chlorophyll-a, chlorophyll-b, total chlorophyll, lycopene and total carotenoids, high tolerance to cold and frost conditions. Average leaf yield over locations is 35.0 t/ha. It is recommended for cultivation in J&K, H.P. and Uttarakhand.

Mustard Green

UHF VR-12-1: This variety has been developed from Ranichauri in 2018. Leaves green to purplish green. About 3-4 times leaves can be picked in direct sown crop of Rabi (Oct-April). Leaves are for harvest at 45-50 DAS and has an average leaf yield 35-40 t/ha. Seed rate: 2 – 2.5 kg/ha. Sowing time: Rabi (Oct.-Nov.). It is recommended for cultivation states like Sikkim, Meghalaya, Manipur, Nagaland, Mizoram, Tripura, Arunachal Pradesh and Andaman & Nicobar Islands.

Carrot

Pusa Vasudha: This hybrid has been developed from IARI Pusa, New Delhi. This is first public sector tropical carrot hybrid developed using CMS system. Self-red colored carrot hybrid. High in total carotenoids, lycopene, TSS and minerals.

Pusa Vrishti: Its variety is released from IARI Pusa, New Delhi. It is a new heat tolerant tropical carrot variety. It is suitable for early sowing beginning in July under north Indian plains. Maturity is 85-90 days. An average yield is 25.0 t/ha.

Pusa Yamdagni: It is developed by hybridization between EC-9981 x Nantes and released by IARI Regional Station, Katrain. It combines the earliness of EC-9981 and self-colored core character of Nantes. Roots are 15-16 cm long, orange, self-colored core and slightly tapering stumpy to semi-stumpy ending. It's top is medium size and quick growing in comparison with other temperate types. It is high yielder and richer in carotene content.

Kashi Krishna: This variety has been developed by ICAR-IIVR, Varanasi in 2019. Attractive black self-core colour roots, tolerant to bolting, fewer secondary roots/scars, suitable for salad, juice, halwa, kanji (fermented juice) and nutraceutical purpose, rich source of anthocyanin (285 mg/100g FW), phenolics and antioxidants, root yield (20-22 t/ha).

Radish

Pusa Mridula: This variety has been developed from IARI Pusa New Delhi. Roots are globular with bright red skin, mildly pungent. Maturity is within 25 days. An average yield is 13.0 t/ha.

Pusa Reshmi: It is a main season variety in Asiatic group and suitable for mid-September to early October sowing. Roots are 30-35 cm long, tapering white with green shoulder, mildly pungent, tolerant to slightly higher temperatures medium light green top with cut leaves.

Kashi Aadra: This variety has been developed from ICAR-IIVR, Varanasi in 2019.

(VRRAD-150): Suitable for sowing during mid-September to mid-December; Roots are attractive white in colour, long and icicle in shape. Roots are ready to harvest in 40-45 days after sowing. Root yield (24-40 t/ha). Seed rate: 8-10 kg/ha. Sowing time: Rabi. It is recommended for West Bengal and Assam.

UHFR-12-1: This variety has been developed from Ranichauri, in 2019. The roots are round to slightly tapering, 10-12 cm in length and width, white in colour. Weighing 180-200 g at edible maturity. Roots are ready to harvest in 55-60 days after sowing. Root yield (35 t/ha). Seed rate: 7-8 kg/ha. It is recommended for J&K, H.P. and Uttarakhand.

Kashi Mooli-40: This variety has ability to tolerate day temperature of 35-42 °C and suitable for summer cultivation in eastern part of Uttar Pradesh. It has delayed bolting, less pithiness, attractive white colour, having root yield of 30-35 t/ha in winter season and 20-23 t/ha during summer season.

Kashi Lohit: Attractive red colour roots, suitable for salad dressing, good source of anthocyanin and phenolic content, tolerant to pithiness, root yield 40-45.0 t/ha in winter.

Glossary

A

Abiotic stress: Adverse conditions for crop growth and production caused by environment factor such as deficiency or excess of nutrition, moisture, temperature and light, the presence of harmful gases or toxicants, and abnormal soil condition such as salinity, alkanety and acidity.

Abnormal seedling: Seedling which are unable to develop in to normal plants.

Acclimatization: The process of introduced plants to adapt or adjust to the new or changed environment.

Acentric chromosomes: A chromosomes without centromere.

A-chromosomes: Normal member of chromosomes complements of a species which are essential for normal growth and development.

Acicular: Long, narrow and cylindrical; i.e., needle-shaped as the leaves of onion, etc.

Acid foods: Foods having pH 4.5 to 3.7 which are usually spoiled by non-spore forming aciduric, butyric anaerobes, etc. e.g., products of tomato, etc.

Acrocentric chromosomes: A chromosomes in which centromere is located very near to one end or has sub terminal position.

Active collection: Germplasm which is meant for medium term storage (10 to 15years). Such collection is subjected to regeneration, multiplication, evaluation, distribution and documentation after every 10 to15 years.

Adaptation: The process by which individuals (or part of individuals), population or species change in form of function in such a way to better survive under given environmental conditions.

Additive gene effect: Gene action in which the effect on a genetic trait are enhanced by each additional gene, either an allele at the same locus or gene at different loci.

Additive variance: That portion of genetics variance which is produced by the average effects of genes at all segregating loci.

Additive X additive epistasis: Interaction between two loci each exhibition lack of dominance individually.

Additive X dominance epistasis: Interaction between two loci, one exhibiting lack of dominance individually.

Adjacent 1-segrigation: Segregation of one normal chromosomes with one translocated ($t_1 + n_2$ and $t_2 + n_1$).

Adjacent 2- segregation: Segregation of $t_1 n_1$ to one pole and $t_2 n_2$ to another pole.

Adult resistance: Resistance exhibited by young seedling.

Adventive embryony: Development of embryo from the diploid cells of ovule lying outside the embryo sac belonging to either nucellus or integument.

Alien addition: Addition of one chromosome of yield species to the normal compliments of a cultivated species.

Aline substitution: Replacement of one pair of chromosomes of cultivated species with those of wild donor species.

A-line: The male sterile line.
Alkylating agents: Chemical mutagens which cause mutation by adding alkyl group at various positions in DNA.
Allele: An allele is an alternative form of a gene.
Allelochemical: Chemical substances which are liberated in allelopathy and inhibit the growth of another species growing together.
Allelopathy: Suppression of plant growth on the same piece of land in a year multiplied by hundred.
Alliin: A colorless, odorless, water soluble amino acid present in the uninjured bulb of garlic which, on crushing, breaks down in presence of the enzyme alliinase to allicin, the principal ingredient of the odoriferous diallyl disulfide.
Allogamy: When pollen grains from flowers of one plant pollinate the flower of other plants.
Allohaploids: Polyhaploid which develop from a autopolyploid species.
Allosomal linkage: Linkage of genes which are located in allosomes or sex chromosomes.
Allosomes: Those chromosomes which differ in number and morphology in male and female sex, also known as sex chromosomes.
Alternate segregation: At anaphase movement of two normal chromosomes (n_1 and n_2) towards one pole and that of two translocated chromosomes (t_1 and t_2) to another pole.
Amino acid: Organic: compound which contains carboxyl (COOH) or acidic group and an amino (NH_2) or basic group. Amino acids are of two types, essential and non-essential.
Amphidiploid: An allopolyploid combining genomes of two diploid species.
Analysis of covariance: The statistical procedure which splits simultaneously the variation of two variables in various components.
Aneuhaploids: Haploid which develops from a aneuploid species. Aneuploids are four types, viz. disomic haploid, nullisomic haploids, substitution haploid and miss division haploid.
Aneuploidy: The change in chromosomes number which involves one or few chromosomes of the genome. Aneuploids are of three type, viz., .monosomic, nullisomic, and polysomic.
Anther culture: The culturing of anthers *in vitro* for the purpose of generating haploid plantlets.
Anthesis: Full flower expansion including anther extrusion.
Antibiosis: Adeverse effects of the host on feeding, development and reproduction of insect - pests.
Antibody: Substances in a tissue or fluid of the body that acts in antagonism to a foreign substance (antigen).
Antigen: A substance, usually a protein, introduced into a living organism that elicits antibody formation.
Antimutator gene: Gene which decreases the frequency of spontaneous mutation of other genes in the same genome. Such gene has been reported in bacteria and bacterio-phases.
Apical dominance: In plants, the inhibition of lateral buds by high level of auxins, produced in the lead shoot or apical meristem.
Apogamy: Development of embryo either from synergids or antipodal cells of embryo sac.
Apomixis: Development of seed without sexual fusion (fertilization). Apomixis are 4 types, parthenogenesis, apogamy, apospory, and adventive embryony.
Apospory: Development of another embryo sac without reduction from the cell of ovule outside the embryo sac may develop either from archesporium (generative apospory) or from nucellus or integument (somatic apospory).
Arithmetic mean: Sum of all observation in a sample divided by their number.
Aroids: A group of vegetable crops under the family Araceae where edible plant parts are corm and cormels e.g., *Colocasia* spp., *Amorphophallus* sp., etc.
Artificial selection: The practice of choosing individuals from a population for reproduction, usually because these individuals possess one or more desirable traits.

Aseptic conditions: Pathogen free environment.

Asexual reproduction: Any processes of reproduction that does not involve that formation and union of gamets from the different sexes or mating types.

Asiatic carrot: Carrot cultivars which do not require any low temperature treatment for flowering and produce seed freely in the plains of India. e.g., Pusa Kesar.

Asnapsis: The failure or partial failure in the pairing of homologous chromosomes during the meiotic prophase.

Assortive mating: Mating in which the partners are chosen because they are phenotypically similar.

Autogamy; Transfer of pollen grains from the anther to the stigma of the same flower, also called self-pollination various machanism which promote autogamy include, bisexuality, homogamy, cleistogamy etc.

Autohaploids: Polyploids which develop from an autohaploid species.

Autopolyploid: A polyploid that has multiple and identical or nearly identical sets of chromosomes (genomes). A polyploidy species with genomes derived from the same original species.

Autoradiograph: A record or photograph prepared by labelling a substance such as DNA with a radioactive material such as tritiated thymidine and allowing the image produced by decay radiations to develop on a film over a period of time.

Autosomal linkage: Linkage of autosomal genes.

Autosomes: Those chromosomes which do not differ in number and morphology in male and female sex.

Autozygote: A diploid individual in which the two genes of a locus are identical by descent from an ancestral gene.

Auxotroph: A mutant microorganism (e.g., bacterium or yeast) that will not grow on a minimal medium but that requires the addition of some compound such as an amino acid or a vitamin.

Avoidence: Escape of a variety from insect attack either due to earliness or its cultivation in the season when insect population is very low.

B

B- line: The fertile counterpart of a line. This line does not have fertility restorer genes and used as the male parents to maintain the A- line.

Back cross method of breeding: It is method of breeding in which the desirable character (s) of a non-recurrent (donor) parent is added to the genetic back ground of recurrent (recipient) parent through subsequent generations of back crossing and selection.

Back cross: Crossing the F_1 hybrids with either of the parents. This may be done to test the genotypic ratio of F_1 or to transfer specific gene complex from one species to another.

Balanced heterosis: It is a type of true heterosis, which occurs from the balanced combinations of genes in a hybrid.

Balanced polymorphism: Regular occurrence of several phenotypes in a genetic population due to superiority of heterozygote over homozygotes.

Base collection: Plant materials which are meant for long term storage (up to hundred years). Seed of such material is stored at -18^0 to -20^0c.

Base station: The RTK-GPS receiver and radio that are placed in a stationary position, functioning as the corrections source for roving tractor units in an area. These stations can be either portable or permanently installed systems and their coverage can range from 5 to 10 miles depending on topographic conditions, antenna height, and radio-transmit power.

Base temperature: The threshold temperature level below which plant do not develop. Each plant has its own base temperature. e.g., pea (4.4°C), French bean (10°C), asparagus (5.5°C), spinach (2°C), pumpkin (13°C), tomato (15°C), etc.

Basic number: The number of chromosomes in ancestral diploid ancestors of polyploids, represented by x.

B-chromosomes: Chromosomes which are found in addition to normal chromosomes compliments of a species and are not essential for normal growth and development. Also known as accessory, supernumerary or extra chromosomes.

Beaded root: Root possessing swellings at frequent intervals, seen in *Basella, Momordica*, etc.

Bhasinda: The underground stem of *Nelumbium*, a popular vegetable in North India.

Bi-directional replication: Replication of DNA in both directions from the point of origin.

Biennial: Plant having a two-year life cycle, vegetative in the first season and reproductive in the second season, and this transition from vegetative to reproductive stage often requires environmental trigger such as vernalization or photoperiod. e.g., cabbage, onion, carrot, etc.

Biological yield: Total dry matter production per plant.

Biochemical mutation: A mutation that alters biochemical function of an individual.

Biometrical genetics: A branch of genetics which utilizes various statistical concepts and procedures in the study of genetics. It includes quantitative genetics and population genetics.

Biometrical pathway: A definite sequential path of biochemical reaction.

Biometrics: The science dealing with the application of statistical methods to biological problems.

Biotechnology: The application of recombinant DNA, cell and tissue culture, and other methods used to develop new and improved plants and plant products.

Biotic stress: Adverse condition for crop growth and production caused by biological factors such as diseases, insects, and parasitic weeds.

Biotype: Distinct physiological race or strain within morphological species. A population of individuals with identical genetic constitution. A biotype may be made up of homozygotes, of which only the former would be expected to breed true.

Biparental cross: Crossing of randomly selected plants in F_2 or sub-sequent generation of a cross between two pure lines in a definite fashion. Concept of biparental mating was originally developed by Comstock and Robinson (1948, 1952). There are three mating designs of biparental cross, viz North Carolina Design 1 and 2 North Carolina Design 3, also called NCD1, NCD2, and NCD3.

Bisexuality: A system of producing both stamen and carpel in the same flower on a plant.

Bit: A binary digit, 0 and 1.

Black leaf speck of cabbage: Small, sharply sunken brown or black specks on leaves which occur under refrigeration in transit and storage and under sharp temperature drops in the fields, also found in Chinese cabbage and cauliflower.

Blanching (cultural): Exclusion of light from the edible parts of salad crops like asparagus, leek, cauliflower, etc., which makes the crop crisp, reduces acrid flavor, improves flavor and tenderness.

B-line: The fertile counterpart of a line. This line does not have fertility restorer genes and used as the male parents to maintain the A-line.

Bolter: Sporadic occurrence of abnormally big sized tuber in potato.

Bolting: Significant stem elongation that proceeds flowering; also includes the case of premature emergence of flower stalk.

Bract: A more or less modified subtending a flower or belonging to an inflorescence.

Breeder seed: Seed increased by the originating, or sponsoring, plant breeder or institution, used as the source for production of foundation seed.

Breeder's rights: Legislation giving the equivalent of patent right for plant cultivars. The effect is to give the developer of a cultivar sole legal possession of that cultivars and a legal basis for compensation for its use by others. Breeder's rights may be referred to as "Plant cultivar protection".

Breeding methods: Various procedures (selection, hybridization, mutation etc.) which are used for genetic improvement of crop plants, also called breeding procedures. Breeding methods which are commonly used in autogamous species include, introduction, pure line selection, mass selection, pedigree method, back cross method, bulk method, single seed descent method, multiline breeding, mutation breeding etc.

Broad sense heritability: Ratio of total genetic variance phenotypic variance.

Bud pollination: Pollination of flowers before they attain maturity (open and shed pollen grain) with the pollen grains collected from the same plant/ genotype.

Bulb crops: Vegetable crops under the genus *Allium* which include onion, garlic, leek, shallot and chive, whose bulbs are eaten raw or cooked or they and their leaves are used to flavor other vegetables, meat, fish and sauces.

Bulb scale: Fleshy 'leaves' that together form the bulb.

Bulb tunic: The dead, papery, leathery or fibrous covering that surrounds most bulbs.

Bulb: A specialized underground organ consisting of a short, fleshy, usually vertical stem axis bearing at its apex a growing point or a flower primordium enclosed by thick, fleshy scales, e.g., onion, etc.

Bulbil: Vegetative part that is actually the modification of flower(s). It develops into plant directly without formation of seeds; seen in onion, garlic, etc.

Bulblets: Miniature bulbs produced around the base of the mother bulb due to development of meristem in the axil of scale leaves.

Bulb-to-seed method: A method of seed production of bulb crops where the bulbs harvested during warm weather are selected, stored and again replanted in winter for seed production.

Bulk breeding: A selection procedure is segregating population of self-pollinated species in which material is grown in bulk plots from F_2 to F_5 with or without selection; next generation is grown from bulk seed and individual plant selection is practiced in F_6 or later generations.

Bus: A group of lines used to transfer bits between micro procedure and the components of computer system.

Byte: A group of 8 bits.

C

Callus: A mass of unorganized callus in culture medium (plural calli).

Capsanthin: A carotenoid pigment responsible for the characteristic orange-red coloration of ripe chilli.

Caruncle: An outgrowth near the hilum of the seeds as seen in dolichos bean

Celery lettuce: Stem type cultivar of lettuce, grown for its thick stem which is eaten after peeling.

Cell culture: Regeneration of a whole plant from a single cell in nutrient medium. It may include somatic cell or germinal cell (pollen).

Cell cycle: The period in which one cycle of cell division is completed. It consists of interphase and mitotic phase.

Cell division: The process of reproduction of new cells from the pre- existing cell.

Cell: A basic unit of structure and function in all living organisms.

Centers of diversity: A placed, region or area where maximum variability of crop plants is observed, also called centers of origin. Center of origin include, China, India, Indo-malaya, Central Asia, Near East, Mediterranean, Abyssinia, South Mexico and Central America, Chile and Brazil.

Certified seed: The progeny of a foundation or registered seed, which maintains the satisfactory genetic identity and purity and has been certified approved by the certifying agency.

Chasmogamy: Opening of flower after the completion of pollination.

Chemical dormancy: Type of seed coat dormancy in which germination inhibiting chemicals *viz.,* various phenols, coumarin and abscisic acid accumulated in the fruit as well as in the seed coverings strongly inhibit germination; found in cucurbits, tomato, etc.
Chiasma terminalization: The movement chiasma away from the centromere towards the end of tetrad.
Chiasma: The point of exchange of segment between non- sister chromatids of homologous chromosomes during pachytene.
Chlorophyll: Containing plastids, sites or photosynthesis in green plants.
Chloroplasts: Plastid of green color that is associated with photosynthesis.
Chromatine: A partially clumped and tangled mass of nuclear chromosomes.
Chromoplast: Plastid with other than green color.
Chromosomal DNA: This is found in the chromosome.
Chromosome models: The pattern of organization of chromatin fibers in a chromosome.
Chromosomes maps: Line diagrams which depict position of various genes on chromosomes and recombination frequency between them also known as genetic maps or linkage maps.
Chromosomes: Darkly staining nucleoprotein bodies that are observed in cells during division. Each chromosome carries a linear array of genes.
Circular Chromosomes: A chromosomes with circular shape and structures, found in bacteria and viruses.
Circular DNA: DNA which has a ring or circular shape such DNA is usually found in prokaryotes, chloroplast and, mitochondria.
Cis position: Presence of two yield allele in one homologous chromosomes (++/ab) also called coupling phase of alleles.
Cistron: The largest elements within a gene which is the unit of function.
Cleistogamy: A built-in breeding mechanism where flowers remain closed at the time of pollination which favour self-pollination, as seen in lettuce.
Clonal selection: A procedure of selecting superior clones from the mixed population of asexually propagated crop such as sugarcane, potato etc.
Clone: A group of individuals derived by a single original plant propagated by vegetative means.
Closed anther mutant: A mutated anther type which through produces viable pollen grains but does not shed them due to non- rupturing and causes bareness in plants particularly where self-pollination is the rule. Closed anther mutants can be maintained by hand pollination with the pollens collected from such flowers by force opening at full maturity.
Co-repressor: A combination of repressor and metabolite which parent's protein synthesis. Such process is termed as co repressor.
Codominance: Expression of both the alleles in the heterozygote or F_1.
Codominant genes: Alleles, each of which produce an independent effect in F_1.
Codon: Triplet sequence of RNA bases which codes for a particular amino acid.
Co-heritability: Ratio of genotypic covariance to the phenotypic covariance. It measures simultaneous inheritance of two characters.
Colchicine: Alkaloids extracted from seeds or corms of *Colchicum autumnale,* which induces pollyploid by arresting, spindle formation during mitosis.
Colchiploidy: Polyploid which is induced by colchicine treatment.
Cole crops: A group of vegetable crops which originated from wild cabbage of Mediterranean region and belonging to genus Brassica (Brassicaceae) which include cabbage, cauliflower, Broccoli, Brussels sprouts, knol-khol, Chinese cabbage, etc., whose leaves, unopened flower buds, inflorescences or swollen stems are used as cooked or raw vegetables.
Combinational heterosis: The heterosis in quantitative characters resulting from the overall combination of the favorable cumulative effects of a number componential character.

Compatible: Capable of fertilization.

Complementary genes: Those genes which have more or less similar phenotypic expression individually but when they come together they interact to produce a new character expression. If two such genes are complementary for a dominant effect, a 9:7 ratio results in F_2; if two are complementary for a recessive effect, a 15:1 ratio results in F_2.

Complete diallel: All possible single crosses among n parents that are n (n-1) /2.

Complete linkage: Linkage in which crossing over does not occur.

Complete penetrance: Expression of a gene in all the individuals which carry it.

Completely randomized design: Experimental design which is used when the experimental plot is small and homogenous such as pot culture.

Composite variety: In cross-pollinated species, a variety developed by mixing the seed of various genotypes which are similar in maturity, height, seed size, seed colour, etc.

Composites: The advanced generations seed mixture of an interval or interracial cross.

Computer: An electronic device that can transmit, store and process information or data. Computer are of four types; micro, mini, main frame and super computer.

Conical roots: When the root is broad at the base and gradually tappers towards the apex like a cone as in carrot.

Conservation: The protection of genetic diversity from genetic erosion either under natural condition or by storing in gene banks.

Conservative replication: DNA replication in which one new DNA molecule has parental strands and other contains both newly synthesized strands there is no experimental proof for this method also.

Constitutive enzyme: An enzyme whose production is constant irrespective of metabolic state of cell.

Consumptive use of water (CUW): Water used to meet the evapotranspiration (ET) need and metabolic activities of plant is collectively known as consumptive use of water.

Contrasting character: Feature of an individual with marked (observable) phenotypic differences, such as red and white, tall and dwarf.

Contributing alleles: Those alleles which contribute to continuous variations also referred to as effective alleles.

Convergent improvement: A system of double back crossing for the purpose of improving each of two inbred lines without greatly modifying the yield of their F_1 cross.

Coordinate projection: Refers to a coordinate system using a specific model of the Earth. UTM coordinates would be considered to be a coordinate projection as it uses a model of the Earth that is cylindrical. UTM's are projected onto a map based on latitude and longitude.

Coreless carrot: Good quality cultivars of carrot in which the core or xylem is small and deeply pigmented so that the cortex or phloem and the core is evenly colored.

Corm: Bulky, short and vertical underground modified stem in which foods are stored as in Elephant's foot (*Amorphophallus* sp.), etc.

Correction factor: The square of grand total divided by number of observation in the analysis of variance.

Correlation: A statistical measure which is used to find out the degree and direction of relationship two or more variable. It is of three types viz. simple, partial, and multiple.

Coupling: Linkage between dominant (AB) or recessive (ab) genes.

Cover crops: Crops that are grown both for the protection of the soil from erosion and for soil improvement. e.g., cowpea.

Cris –cross inheritance: Inheritance of sex linked gene from grandfather to grandson through daughter.

Critical differences: Least significance differences greater than which all the difference is significant.

Critical moisture periods: Critical periods of irrigation needs can best be defined as that time when soil moisture stress can reduce yield most in an otherwise healthy crop. This is not to say that it is the only time in the life of the crop that moisture stress reduces yield. It is, however, the time when stress has the greatest effect.

Crop ideotype: A plant model which is expected to yield greater quantity of grains, fibers, oil or other useful product when developed as a cultivar.

Cropping index: Number of crops grown on the same piece of land in a year multiplied by hundred.

Cross incompatibility: In ability of a functional pollen of one species or genus to effect fertilization of the female gametes of another species or genus of the same family.

Cross pollinated crops: An assembly of genetically heterozygous individuals under commercial cultivation which share a common gene pool and in which each individual takes new genotype generation after generation.

Cross- pollination: Transfer of pollen from the anther of one plant to the stigma of another plant. It affects the union of genetically dissimilar gametes.

Cross: The products of the mating between two or more parents are dissimilar genetic constitution. The various type of crosses utilized in heterosis breeding programs are single, three-way, double crosses, top crosses, multiple crosses etc.

Crossing over: Inter change of parts between no sister chromatids of homologous chromosomes during pachytene.

Crossing: Artificial mating of two or more parents of unlike genetical constitutions.

Cucurbits: A large and diverse group of vegetable crops under cucurbitaceae, used as vegetables (pumpkin, different gourds, etc.) pickles (cucumber) and as desert fruits (muskmelon, watermelon).

Cultigroup: An intraspecific category below subspecies which includes cultivated types such as *Vigna unguiculata* cultigroup sesquipedalis (vegetable cowpea).

Cultivar: An assemblage of cultivated plants which is clearly distinguished by any character and which, when reproduced, sexually or asexually, retain the distinguishing characters.

Culture medium: A nutrient medium which contains all essential micro and macronutrients, carbohydrate vitamins and hormones.

Curd size index: A curd character of cauliflower which is the equatorial × polar diameter of the curd.

Curd: Edible part of cauliflower which is actually the repeatedly branched prefloral fleshy apical meristem.

Cyclic selection: Selection in one direction for one generation or season and in opposite direction in next generation or season.

Cytogenetic/cytonuclear male sterility: Male sterility is determined by interaction of gene and cytoplasm but none of them singly can control sterility.

Cytokinesis: The process of division of cytoplasm.

Cytoplasmic and genic male sterility: Sterility of pollen grains which is governed by the interaction between sterile cytoplasm and recessive gene, seen in onion, beet, carrot, etc.

Cytoplasmic DNA: The DNA which is found in cytoplasm either in chloroplast or in mitochondria.

Cytoplasmic genic male sterility (CGMS): Pollen sterility which is controlled by both cytoplasmic and nuclear genes.

Cytoplasmic inheritance: Inheritance which is governed by cytoplasmic gene or plasma gene, by chloroplast or mitochondrial DNA. also known as extra chromosomal inheritance or extra nuclear inheritance or organeller inheritance or non- mendelian inheritance.

Cytoplasmic male sterility (CMS): Pollen sterility which is caused by cytoplasmic genes.
Cytoplasmic mutation: A mutation in cytoplasmic gene.
Cytoplasmic sterility: Transmission of male sterility by the cytoplasm.
Cytoplasmic: Concerned with the cytoplasm in the cell.

D

D2-statistics: Statistical procedures which measures forces of differention at intra and inter cluster levels and determines the relative contribution of each component trait to the total divergence. This technique was developed by P.C.Mahalanobils (1928) and first used for assessment of variability in plant breeding by C.R.Rao (1952).
Data layer (in GIS): A layer of information on a GIS map. A map can have many layers to present different types of information. For example, the first layer of a map may be a satellite image of an area. The next layer may have only lines that represent roads or highways. The next layer may contain topographic information and so forth.
Day-neutral plant: Plant in which flowering is not influenced by day length. e.g., tomato, cucumber, okra, asparagus, capsicum, snap bean, etc.
Decompound leaf: When the leaf is more than thrice pinnate as in carrot, etc.
Defective seed: Seed which are broken, disease, infested, insect damaged, undeveloped and unfit for germination.
Degreening: The process of decomposing the green pigment in fruits by applying ethylene (1000-2000 ppm) or similar metabolic inducers to give a fruit its characteristic color as preferred by consumers, generally followed in citrus fruits but also practiced in banana, mango, tomato, etc.
Dehaulming: Removal of the top portion (haulm) of potato in the seed crop to avoid the infestation of virus carrying insect vectors.
Dehiscence: Splitting open of a fruiting structure or anther.
Dehiscent fruit: Fruit whose peri carp bursts to liberate the seeds at maturity as seen in okra, etc.
Deletion: Loss of a segment from a chromosome, also called deficiency. It is of two types, viz. terminal and interstitial.
Denaturation: The process of separation of DNA strands on heating of DNA molecule at high temperature.
Diakinesis: A sub stage of meotic prophase 1st in which bivalents are distributed throughout the cell.
Diallele cross: A diallele cross can be defined as all possible combinations i.e.; n (n-1)/2 of single crosses among one parents.
Dichogamy: The maturing of male and female gametes at different time when the male gametes matures earlier than the female, it is called protandry, but when female matures first, it is called protogyny, this situation is called dichogamy.
Dicliny: A situation, where plants produce unisexual flowers.
Differential Global Positioning System (DGPS): This system operates using the same GPS satellites, with the addition of a differential corrections source (WAAS satellite or Coast Guard Beacon) to increase the accuracy of the system. In both cases, multiple ground stations provide the information about satellite error. Accuracy is typically better than 10 feet (3 meters) and can be better than 40 inches (1 meter).
Dihaploid: A haploid which develops from a tetraploid species.
Dioecious: Plant species in which unisexual flower, staminate or pistillate, is borne on separate plants, as in pointed gourd, etc.
Dioecy: Where male and female flower or borne singly and different plants.
Dipeptide: A product of union of two different amino acids.

Diploid number: The somatic chromosome number of a true diploid species.
Diploids: Individual with 2x somatic chromosome number.
Directional selection: Selection in favor of extreme types, viz; earliness and lateness or tallness and dwarfness.
Disomic haploid: A haploid which develops from a tetrasomic species (n+1).
Dispersion: The degree to which numerical data tend to spread about the mean value. It is a measure of variation in a sample.
Dispersive replication: DNA replication in which the new DNA molecules have old and new DNA in patch. This method is not accepted as it could not be proved experimentally.
Displaced: Presence of duplication away from the original segment but on the same arm of chromosomes.
DNA probes: The small segments of DNA with known base sequences, origin and function.
DNA replication: The process by which a DNA molecule makes it identical copies.
DNA: Deoxyribonuclic acid; the information –carrying genetic material that comprises the genes.
Dominance hypothesis: Heterosis due to superiority of dominant alleles over recessive alleles. Heterosis is directly proportional to the number of dominant genes contributed by each parent.
Dominance variance: That portion of genetic variance which arises due to deviation from the additive scheme of gene action resulting from intra –allelic interaction. It is due to the deviation of heterozygote (Aa) from the average of two homozygotes (AA and aa).
Dominance: The phenomenon in which the dominant gene has an overriding effect on its allele in such a way that heterozygote (Aa) is phenotypically indistinguishable from the dominant homozygote (AA).
Dominant (inhibitory) epistasis: Gene interaction in which a dominant allele at one locus can mask the expression of both (dominant and recessive) alleles at second locus resulting in 13:3 ratios, also known as inhibitory.
Dominant gene: When two parents contrasting characters are crossed, the character of one parent appears in F_1 generation to the inclusion of the character of the other parent even through both the genes are present in the hybrid.The gene, which expresses itself ,is said to be dominant over the other recessive.
Dominant: The character which expresses in F_1.
Donar parent: The parent which donates desirable genes also called non- recurrent parent, because it is used once in the crossing programme.
Double cropping: Cultivation of crops one after another on the same field in a year.
Double cross hybrid: A hybrid obtained by crossing two single crosses i.e; (Ax B) x (C x D).
Double crossing over: The formation of two chiasmata between non- sister chromatids of homologous chromosomes.
Double fertilization: A phenomenon in angiosperm where by one male nucleus unites with the egg nucleus to form zygote (2n) which develops into the embryo and the second male nucleus unites with two polar nuclei in the embryo sac to form the endosperm(3n).
Double stranded DNA: DNA which has spirally arranged double strands. It is found in all plants, animals, and bacteria.
Double tetrasomic: Addition of two chromosomes to two different pairs (2n+2+2).
Double top cross: A cross-obtained from crossing single cross with an open pollinated variety.
Drought avoidance: Ability of plants to the maintain a favorable internal water balance under moisture stress.
Drought hardening: Improvement in drought tolerance ability of a genotype through various seeds and seedling treatment.

Drought tolerance: Ability of crop plants to grow, develop and reproduce normally under moisture deficit conditions. In other words, survival of plants under water deficit condition without injury.

Drought: Condition of soil moisture deficiency or water scarcity. There are four mechanism of drought resistance, viz. drought escape, drought avoidance, drought tolerance and drought resistance.

Duplicate dominant epistasis: Gene interaction in which recessive alleles at either of two loci can mask the expression of recessive alleles at the two loci, resulting in 15:1 ratio, also referred to as duplicate gene interaction.

Duplicate recessive epistasis: Gene interaction in which recessive alleles at either of two loci, resulting in 9:7 ratio, also called complimentary epistasis.

Durable resistance: Long lasting resistance. It may be vertical or horizontal.

E

Earthing up: The process of putting the soil just near the base of stems of certain crops like potato, cassava, banana, etc. to provide support and to prevent root exposure.

Effective root zone: It is the depth where the most of the active roots of mature plant are concentrated and are capable of extracting soil moisture.

Effector: The molecule which acts as an inducer or co-repressor in the Operon model of *E.coli.*

Electrophoresis: The migration of suspended particles in an electric field.

Electroporation: A process whereby cell membranes are made permeable to DNA by applying an intense electric current.

Emasculation: In bisexual flowers, the removal of stamens before they burst and shed pollen grains. Its purpose to check self–pollination and is done before effecting cross-pollination.

Embryo culture: The cutting of an immature embryo on a sterile nutrient medium.

Embryo: The portion of seed, which contains the dominant, miniature, rudimentary plant. It arises from the zygote.

Emigration: Outgoing of alleles from a population.

Endoplasmic reticulum: A vast network of membrane enclosed tubules, vesicles and sacs found in cytoplasm.

Endosperm: The nutritive tissue formed inside the embryo sac in the seed. It arises from the triple fusion of sperms nuclei with polar nuclei of embryo sac.

Environmental correlation: The association between two variables which is entirely due to environmental effects. It is estimated from error variances and co-variances.

Enzyme: A protein that accelerates a specific chemical reaction in a living system.

Epidemic: Wide spread uncontrolled incidence of a disease.

Epistasis: The phenomenon in which a non- allelic or gene combination exert a dominant effect over another gene or combination of genes (non- allelic interaction).

Epistatic variance: That portion of genetic variance which arises due to deviation as a consequence of inter-allelic (inter genic) interaction.

Essential amino acids: Amino acids which cannot be synthesized in human body and their requirement has to be met through dietary intake. These are methionine, isoleucine, leucine, lysine, threonine, tryptophan, valine phenylalanine, histidine, and agrinine.

Euchromatin: Lightly staining region of chromosomes during inters phase. Usually found in the middle of chromosomes, genetically active and takes part in transcription.

Eugenics: A branch of genetics which deals with frequencies of genes and genotypes in a population, and also with various forces which tend to alter gene frequencies in a population leading to evolutionary changes.

Euhaploid: Haploid which develop from a euploid species. Euhaploid are of two types, viz, monohaploids, and polyhaploids.
Eukaryotes: Organism whose cells contain well defined nucleus.
Euploidy: The change in the chromosomes number which involves entire set. Euploidy includes monoploids, diploids and polliploids.
European carrot: Carrot cultivars which are biennial in nature and require low temperature (4.8-10°C) treatment for certain periods for flowering, hence do not produce seeds in plains of India, e.g., Nantes, Chanteny, Imperator etc.
Evaporation: It is the loss of water from soil surface of a particular area during a certain period.
Evapotranspiration: It is the total loss of water due to transpiration from a crop plants and evaporation from the soil.
Evolution: The process of the origin of the organisms (varieties, species, genera, families etc.).
Exons: Coding sequence of DNA in split genes.
Exotic collection: The germplasm which is collected or received from other country.
Explant: Plant part which is used for regeneration. It may be a cell, a protoplast, a tissue or an organ.
Ex-situ conservation: The preservation of germplasm in gene banks.
Extinction: Permanent loss of a crop species due to various reasons.

F

F_1: An abbreviation to designate the first hybrid generation. It is composed of progenies raised by showing the seeds obtained from cross between two genetically unlike parents.
F_2: Progeny of F_1 plants obtained by selfing.
Fasciculated roots: Swollen tuberous roots which are developed in a cluster or fascicle at the base of the stem as in asparagas, etc.
Fertility restorer gene: Usually a gene when put into a cytoplasmic male sterile background is able to bring back the production of normal functional pollen grains.
Fertility: The ability of an organism to form viable off spring.
Field application efficiency: The field application efficiency is the fraction of the applied water that is used by the crop. Provided there are no runoff losses, the field application efficiency (%) is the required irrigation depth (mm), divided by the average applied irrigation depth (mm), and multiplied by 100%.
Field capacity (FC): Field capacity is the amount of water that a well-drained soil holds against gravitational forces, or the amount of water remaining when downward drainage has markedly decreased. This situation usually exists one to three days after the field has thoroughly been wetted by rain or irrigation. The field capacity is the upper limit of available soil moisture to the plants. The soil moisture tension at field capacity generally varies from 0.1-0.3 atmospheres.
Field resistance: Resistance which gives an effective control of a parasite under field condition.
Five parameter models: A model of generation mean analysis which provides information about five parameters, viz. m, d, h, i and l involves P_1, P_2, F_1, F_2, and F_3 generations of a cross in analysis.
Floating garden: A type of vegetable garden found in the lakes of Kashmir valley where vegetables are grown on a floating base prepared with some grass (*Typha*), compost and other organic matters.
Floppy disk: A thin plastic-coated disk with magnetic oxide which is used for information storage in computers.
Foliaceous stipules: A large paired leafy outgrowth as seen in pea, etc.

Foundation seed: The seed stocks that are so handed as to most nearly maintain the specific genetic identity and purity of the original stock and provide the source for the production of called certified and registered seed.

Founder effect: Establishment of a new population in the main population by single or few individuals.

Frame shift mutations: Mutation which arises due to addition or detection of nucleotides in mRNA.

Fresh under germinated seeds: Viable seeds which can abort water but do not germinate and remain fresh in germination test.

F-test: A test of statistical significance which is used to compare the differences among several means.

Full diallel: All possible both away (direct and reciprocal) crosses among n genotype.

Full slip: A harvesting index of muskmelon or cantaloups for local market when the fruit can easily be removed (slip) with a slight pressure from the stem leaving a clean stem cavity.

Functional male sterility: Where plants produce normal pollen grains but anthers remain closed and caused male sterility.

Fusiform root: When the root is swollen in the middle and gradually tappers towards the apex and the base, being more or less spindle-shaped in appearance as in radish.

F-value: The ratio of treatment variance to error variance.

G

G1: A pre-DNA replication phase it lise between telo phase s- phase.

G2: A post DNA replication phase during which protein and RNA synthesis take place.

Gamete selection: It is a type of selection for detecting and combining of desirable gamete a genetically variable heterozygous population into the back ground of an inbred line of known performance and combining ability.

Gamete: A matured sex cell, capable of fusing with another to form a zygote.

Gametocide: A chemical substance destructive to gametes.

Gametophyte: A phase of life cycle of plants, which has haploid nuclei, during it the sex cells are produced. It arises from a spore produced by meiosis from a sporophyte (diploid).

Garden for vegetable processing: A type of vegetable farming where vegetables are produced with a soe objective of supplying them to the processing factories.

Gene action: The manner in which genes control phenotypic expression of various characters in an organism.

Gene deployment: Planned geographical distribution of major genes for specific resistance to a pest for use in varietal devlopment and production.

Gene expression: The hereditary properties of an organism as represented by the gene and are expressed in generation under a set of environmental factors.

Gene flow: The spread of genes from one breeding population to another by migration, possible leading to allele frequency changes.

Gene for gene hypothesis: This hypothesis states that for each gene controlling pathognecity in the pathogene, also called Flor hypothesis after the name of scientist who developed this concept.

Gene frequency: Proportion of a gene or allele or its series present in a population or a sample thereof. It is usually expressed as number between 0 to 1.

Gene pool: Some total of all genes in a breeding population.

Gene pyramiding: Incorporation of two or more major genes in a variety for specific resistance to a pest.

Gene symbol: Various symbol which are used to represent genes or alleles.

Gene: A heredity determinant of a specific biological function; a unit of inheritance (DNA) located in a fixed place on the chromosome.

General combining ability: The comparative ability of a line or a genetic stock to combine with a tester or a group of testers.

Genetic advance: The expected gain in the mean of the population for a particular quantitative character by one generation of selection of a specified percent of the highest-ranking plant conditions. Genetic resistance is of two types viz. vertical and horizontal.

Genetic architecture: A term used to denote the general genetic structure of a species.

Genetic break down: A term used to indicate the loss of vigour and often the early death of F_2 plants which lack the necessary adaptive complexes of either or both the original parents.

Genetic code: The relationship between the sequences of amino acid in a polypeptide chain.

Genetic engineering: Genetic manipulation that use recombinant DNA methods (gene splicing) to change the genetic makeup of an organism.

Genetic equilibrium: In a random mating population, the stage in which genotype frequencies do not change from one generation to another.

Genetic erosion: Gradual disappearance of various forms of a cultivated species and its wild relatives.

Genetic homeostasis: The ability of a random mating population to equilibrate its genetic composition so as to resist sudden environmental changes.

Genetic male sterility: Sterility of male gametes in a flower governed by genes and is heritable.

Genetic resistance: Ability of some genotypes to give higher yield of good quality than other varieties at the same initial level of disease or insect infestation under similar environmental conditions. Genetic resistance is of two types viz. vertical and horizontal.

Genetic RNA: The RNA which act as DNA or genetic material.

Genetics: The science of heredity and variation.

Genic male sterility: Sterility of pollen grains which is governed by a single recessive gene as found in squash, pumpkin, muskmelon, Brussels sprouts, cabbage, cauliflower, lettuce, sprouting broccoli, etc.

Genome: A basic or monoploid set of chromosomes. In a genome, each type of chromosomes is represented only once.

Genotype X environmental interaction: The interplay in effect of the genetic and non-genetic factors on the development of an organism.

Genotype: The hereditary properties of an organism as represented by the gene constituents. They may be expressed or latent.

Genus (pl.genera): A taxonomic category that includes group of closely related species.

Geographic (spatial) data – Data that contains information about the spatial location (position) and the attribute being monitored such as yield, soil properties, plant variables, seed population, etc.

Geographical diversity: The diversity of the biological population produced by the presence of a geographical barrier such as mountains, rivers, canyons.

Germination: Emergence of normal seed ling from the seeds under ideal condition of light temperature, moisture, oxygen, and nutrients.

Germplasm complexes: The advanced generation of mixed seeds obtained by a purpose ful intermixing of a number of genotypes or hybrids among them.

Germplasm: In plant breeding sense, germplasm is the sum total of genetic stocks of a particular crop species.

GIS – Geographical computer system that records, measures, manages, or analyzes geographically referenced information or data.

Global navigation satellite system (GNSS)- It is the standard generic term for satellite navigation systems that provide geo-spatial positioning with global coverage using time signals transmitted from satellites. The United States GPS and the Russian GLONASS are the only two fully operational GNSS. Top of the line GNSS receivers can communicate with both GPS and GLONASS satellites effectively doubling the available reference satellites at any given time

Global positioning system (GPS)- A system using satellite signals (radio-waves) to locate and track the position of a receiver/antenna on the Earth. GPS is a technology that originated in the U.S. It is currently maintained by the U.S. government and available to users worldwide free of charge. There are 30 satellites in the GPS constellation.

Gourd: Generally, it refers to the fruit of cucurbits. Actually, this epithet refers to the fruit character: hard and tough rind upon complete maturation as in bottle gourd, pumpkin or summer squash, even though the term gourd is applied to other fleshy fruits like bitter gourd or snake gourd whose skin do not become tough when ripe.

Grana: Small cylindrical structures found inside the inner memberane of a chloroplast.

Green pepper: Tender, semi-mature green pepper (*Piper nigrum*) spike which is used commercially in pickles.

Gross irrigation requirement (GIR): GIR is the total quantity of water applied to field including losses due to leakage, seepage and evaporation from open channels.

Growth crack of sweet potato: Longitudinal or transverse splits and fissures due to irregular on interrupted growth.

Guar gum: The mucilaginous seed flour of guar (*Cyamopsis tetragonolobus*) is valued as guar gum (Galactomannan) used in textile, paper, cosmetic and oil industries throughout the world. It is also a useful absorbent for explosives.

Gynandro morphs: Individuals with sex mosaic also called gynanders.

H

Hakuran: An artificial amphidiploid of cabbage, as Chinese cabbage produced through embryo culture technique, which is a good leafy vegetable.

Half diallel: All possible one-way crosses among n genotypes, i.e. (n-1)/2.

Half sib: Progeny having one parent in common.

Half-hardy vegetable: Vegetable crops which can thrive well in cool weather condition but can not tolerate frost. e.g., beet, carrot, cauliflower, lettuce, spinach, etc.

Haploid number: The gametic chromosome number of a species.

Haploids: Individual with gametic (half) chromosomes number. Haploids are of different types.

Hard disk: A disk which is permanently fixed in a computer, also called Winchester disk. Hard disks have more storage capacity and are faster in reading and writing.

Hardware: The physical components of the computer such as key board, processing unit, monitor and printer.

Head shape index: A head character of cabbage, Chinese cabbage, calculated by dividing mean head length (cm) with mean head width (cm).

Head: Edible portion of cabbage, Chinese cabbage and head lettuce which is a structurally distinct, compact leafy portion made up of numerous overlapping leaves covering the terminal bud.

Head-to-seed method: A method of seed production practiced in cabbage where the selected plants with fully matured heads are lifted prior to snowfall, stored and again replanted at the onset of spring for seed production.

Heredity: Resemblance among individuals related by descent; transmission of traits from parent to off spring.

Heritability: Degree to which a given trait is controlled by inheritance.

Herkogamy: Hinderance of self-pollination due to some physical barriers such as presence of a hyline membrane around anther.

Hermaphrodite: An individual with both male and female reproductive organs.
Heterobeltiosis: Heterosis expressed over the better parent of the cross.
Heterochromatin: Chromatin staining darkly even during inter phase, often containing repetitive DNA with few genes.
Heterogametic sex: Sex with dissimilar type of sex chromosomes such as xy, xo, and zw.
Heterogeneous population: A population which is composed of genetically, dissimilar plants such as land races, mass selected populations, composites, synthetics and multilines.
Heterokaryons: Hybrid cell combining protoplast of two different species.
Heteroploidy: Any change in the chromosomes number from the diploid state. It is of two types, viz. Euploidy and aneuploidy.
Heterosis: Hybrid vigour such that an F_1 hybrid fall out side the range of parents with respect to same characters. Usually applied to size, rate of growth or general anthocyanin absent fitness.
Heterostyly: Different lengths of styles and filaments in a flower.
Heterozygosity: The phenomenon in which the homologous chromosomes of an organism possess different genes of the same allelic series.
Heterozygote: An organism with one or more heterozygous pairs of genes or unlike alleles at one or more corresponding loci .As a result of heterozygosity the organism will not breed true.
Heterozygotic potential variability: The variability which is stored in heterozygotes, e.g. AaBb.
Heterozygous: Individual having dissimilar alleles on the corresponding locus of homologous chromosomes.
Holokinetic chromosomes: A chromosomes with diffused centromere.
Homeostasis: The buffering capacity of a genotype to environmental fluctuation. Adaptability is a result of homeostasis.
Homogametic sex: Sex with similar type of sex chromosomes such as xx or zz.
Homogamy: Maturation of anthers and stgma of a flower at the same time.
Homogeneous population: A population of genetically similar plants such as a pure line, F_1 between two pure lines and progeny of a clone.
Homokaryons: Hybrid cell combining protoplasts of the same species.
Homologous chromosomes: Chromosomes that cover in pairs and are generally similar in size and shape, one having come from the male parent and the other from the female parent. Such chromosomes contain the same array of genes.
Homozygosity: The proportion of homozygous individuals in a segregating population. Homozygosity is equal to $[(2m-1)/2m]^n$, where m is the number of selfing generations and n is the number of gene pairs segregating .
Homozygote: An organism with identical genes at corresponding loci on homologous chromosomes.
Homozygotic potential variability: Variability which is stored in homozygotes, viz., AAbb or aaBB.
Homozygous: Individual having similar alleles on the corresponding locus of homologous chromosomes.
Horticultural traits: Character of economic importance in horticultural crops.
Hub crop: Crop which has the greatest comparative advantage over other crops in a sequential cropping system e.g. vegetable crops.
Hybrid inviability: In ability of zygote to grow into a normal embryo under the normal conditions of development.
Hybrid sterility: In ability of a hybrid to produce viable off spring.
Hybrid vigour: Increase in vigor, growth yield or function of a hybrid over the parents that result from the crossing of genetically unlike organisms.

Hybrid. The progeny of a cross between two or more individual plants of unlike genetic constitution.

Hybridization: A method of crop improvement in which two or more plants of unlike genetical constitution differing in one or more characters are crossed together.

Hydroponics: Growing plants in nutrient solution.

Hypersensitivity: A host pathogen reaction which leads to death of infested tissues.

Hypogeous germination: A pattern of germination where the lengthening of the hypocotyl does not raise the cotyledons above the ground and only the epicotyl emerges, as seen in pea, etc.

I

Ideotype breeding: A method of crop improvement which is used to enhance genetic yield potential through genetic manipulation of individual plant character.

Ideotype: A biological model which is expected to perform or behave in a predictable manner within a defined environment.

Immigration: In coming of new alleles in a population.

In situ: From the Latin, meaning in the natural place. Refers to experimental treatments performed on cell or tissues rather than on extracts from them.

In vitro: From the Latin, meaning within glass; biological processes made to occur experimentally outside the organism in a test tube or other container.

Inbred line: A line produced by continued in crop breeding a nearly homozygote line usually originating from continued self-fertilization accompanied by selection.

Inbred: In cross pollinated species, a true breeding line obtained by continuous in breeding.

Inbreeding depression: The loss of vigor as a consequence of inbreeding. It is primarily due to the breakdown of specific gene system governing the expression of vigour, governing a particular trait or traits and is often accompanied by reduction in yield, size, fecundity etc.

Inbreeding: Mating of closely related individuals .example –self-pollinated crops.

Incompability: Failure to obtain fertilization and seed formation after self- pollination, usually due to failure of pollen tube to penetrate stigma, or to reduced growth of the pollen tube in the style tissues.

Incomplete dominance: Partial resemblance of F_1 with one of its parents.

Incomplete linkage: Linkage in which some frequency of crossing over occur.

Incomplete penetrance: Expression of gene in less than 100% of its carries.

Independent assortment: Random or free segregation of chromosomes and gene during gamete formation i. e. during meiosis.

Indigenous collection: The germplasm which is collected within the country.

Induced mutation: Mutation in which are produced by the use of mutagenic agent.

Inducer: The substances which allows initiation of transcription (lactose in lac operon). Such process is known induction.

Inducible enzyme: An enzyme whose production is enhanced by adding the substrate in culture medium. Such system is called inducible system.

Inert matter: Non- living materials such as sand, pebbles, soil particles, straw, etc.

Inflorescence: (1) A flower clusters (2) the arrangement and mode of development of the flowers on a flower axis.

Interference: The tendency of one cross over to reduce the change of another crossover in adjacent region.

Inter-generic hybridization: Crossing between two different genera of the same family. Triticale and raphanobrassica are the outcome of inter generic crosses.

Internal browning of tomato: Gray-brownish discoloration of internal tissues in green fruit which extend to the surface and form lesion that remain greenish or yellow in ripe fruit; caused by water imbalance and high temperature and/or nutrient imbalance.

Inter-specific hybridization: Crossing between two different species of the same genus, also called intra-generic hybridization. Such crosses are called inter specific crosses.

Introgression: Transfer of some genes from one species into the genome of another species.

Introgressively hybridization: A type of hybridization in which a number of genes or gene block of one species are added to the genetic background of another species by crossing and often by back crossing.

Inversion: Structural changes in which a segment is oriented in a reverse order. Inversion are of two types, viz. paracentric and pericentric.

Irradiation: Exposure of plants or plant parts to X-rays or other radiations to increase mutation rate.

Irrigation frequency: It refers to the number of days between irrigation during periods without rainfall.

Irrigation period: It is the number of days that can be allowed for applying one irrigation to a given design area during the peak consumptive use period of the crop being irrigated.

Irrigation requirement: Irrigation requirement for crop production is the amount of water, in addition to rainfall, that must be applied to meet a crop's evapotranspiration needs without significant reduction in yield.

Irrigation: An artificial application of water to the soil or plant for the purpose of crop production is known as irrigation.

Isoallele: An allele which is similar in its phenotypic expression to that of other independently occurring allele.

Isogenic lines: Lines differing from each other genetically at one locus only i.e. lines identical in all traits but one.

Isolation: The condition in which individuals of common ancestry is separated into two or more mating groups that mating between or among groups is prevented.

Isosine: A newly discovered nucleotide which is found in third position in a codon and can pair with A, U, and C. It is of three types, dispersive conservative and semiconservative.

J

J. shape chromosome: A chromosome which assumes J shape at anaphase.

Jumping genes: The genes which keep on changing their position in chromosomes and also between the chromosomes in a genome. Also called transpose or transposable elements. The first case of jumping gene was reported by Mc.Clintock in 1950 in maize.

K

Karyokinesis: The process of the division of nucleus.
Karyotype: The characteristic feature of chromosomes of a species.

L

Lamp brush chromosome: A chromosome having lamp brush appearance.

Landraces: Traditional cultivars with sufficient genetic integrity to be morphologically identifiable and differing in adaptation and to cultural practices, but genetically variable.

Latitude – A global standard coordinate used to identify a position on earth given in degrees, minutes and seconds, indicates the north/south position above/below the equator, positive is in the northern hemisphere and negative is in the southern hemisphere.

Lattice design: Incomplete block design in which the number of varieties or treatments form a square.
Lethal gene: Gene which causes death of its carrier when in homozygous condition.
Leucoplasts: Colorless plastids of green color that is associated with storage of starch, protein and fat.
Line breeding: A system of which a number of genotypes which have been progeny tested in respect to some character or group of characters are composite to form a variety.
Line X tester analysis: A system of lines for combining ability in the genetic background of a number of proven testers.
Line: This term refers to a group of individuals obtained from a common ancestry. In maize programme; it is synonym to inbred line.
Linear DNA: DNA which has a thread like structure with both the ends free. Such DNA is found in eukaryotes.
Linkage: Association of two or more non-allelomorphic genes so that they tend to passed from generation to generation as an inseparable unit and fails to show independent assortment. The potential types appear in greater frequency than expected in F_2. This is due to locating of linked factors on the same chromosome.
Lipase: An enzyme that joins the ends of two strands of nucleic acid.
Locus: A position on chromosomes which is occupied by an allele.
Long-day plant: A plant which requires a day longer than its critical day length for flowering. e.g., lettuce, radish, onion, cabbage, carrot, spinach, beet, etc.
Longitude – A global standard coordinate used to identify a position on earth given in degrees, minutes and seconds, indicates the east/west position around the globe from a reference point which overlays Greenwich, England. Negative values are east of Greenwich and positive values are west.
Luxuriance: It refers to the phenomenon in which the crossing of two parental forms bring an excessive, accidental, un adaptable and often unbalanced expression of a number of an attribute. Or the superiority of F_1 over its parents in vegetative growth, but not in yield and adaptation, also called pseudo- heterosis.
Lycopene: Red pigment found in ripe tomato which is a straight chain derivative of carotene with no vitamin activity. Its chemical composition ($C_{40} H_{56}$) is same as that of carotene.
Lysimeter: Cemented micro plots of various sizes used for the study of roots and salt tolerance.
Lysosomes: Cellular particles which contain several digestive enzymes.

M

Mainframe computers: Computers which is large storage capacity and very high speed of processing as compared to micro and mini computers.
Maintainer line: A genotype use to maintain the male sterility of cytoplasmic- genic system.
Male gametocides: Chemical which are used for induction of male sterility.
Male sterility: A condition in which either pollen is absent or non- functional in flowering plants.
Marker gene: Common gene differences which assort independently from all other readily testable gene loci are called marker genes e.g. potato leaf in tomato, brown seed in onion, anthocyanin pigment in cotyledons of chilli/ brinjal etc.
Market gardening: Vegetable farming for supply of vegetables to the consumers in the local market; one of the most intensive types of vegetable farming.
Mating system: The system in which individuals are arranged in pairs leading to sexual reproduction.

Matric potential of soil water: It is defined as the amount of work that a unit quantity of water is capable of doing (in equilibrium soil water system) when it moves to another equilibrium system identical in all respects except that there is no matrix present.

Meiosis: Two successive spindle using divisions which reduce the chromosome number from diploid to haploid.

Memory: A medium that stores binary information such as instructions and data and provides that information to the microprocessor whenever necessary. Memory is of two types, viz; read only memory and read and write memory.

Meristem culture: Culture of apical meristems, particularly shoot apical meristem, for production of shoots and plantlets.

Messenger RNA: The RNA which carries information from nuclear DNA to cytoplasm for protein synthesis

Metacentric chromosomes: A chromosome in which centromere is located is in the middle portion. Such chromosomes assume V shape at anaphase.

Microcomputer: A computer that is designed using a microprocessor as its central processing unit (CPU). IT includes four components: microprocessor, memory, input, and output.

Microprocessor: A semiconductor device manufactured by using large scale integration technique. It includes arithmetic logic unit (ALU), register arrays and control circuit on a single chip.

Minicomputer: A medium size computer which is more costly and powerful than microcomputer.

Mitochondria: A rod like cytoplasmic organelle which is the main site of cellular respiration.

Mitosis: The spindle using nuclear division which produces two identical daughter cells from a mother cell.

Modern cultivars: The currently cultivated high yielding varieties.

Moisture content (% by vol.) = Moisture content (% by wt.) x Bulk density of soil

Momeostatic: A genotype giving better performance even in adverse environments because of wide genetic base.

Monoallelic SI: Self- incapability which is controlled by single gene. It is found in some species of the family leguminosae, solanaceae, and cruciferae.

Monoculture: Repetative growing of the same sole crop on the same field.

Monoecious: The condition in which both male and female sex organs are produced separately but on the same plant.

Monoecy: Where male and female flower are unisexual but produced on the same plant.

Monohaploids: Haploids which develop from a normal diploid species.

Monohybrid crosses: A cross involving one gene pair affecting one character.

Monoploids: Individuals with basic chromosomes number.

Multiple cropping: Cultivation of two or more crops on the same field in a year.

Multivitamin green: Chekurmanis (*Sauropus androgymus*), a perennial leafy vegetable crop is called so because of the availability of various vitamins like A, B, C, D, F and K from this vegetable crop.

Mutable gene: A gene which exhibits higher mutation rate than other gene.

Mutagen: Physically or chemical agents which greatly enhance the frequency of mutation.

Mutant: Any plant, which has originated or acquired a heritable variation as a result of mutation.

Mutation: A sudden heritable changes in the phenotype of an individual.

Mutational heterosis: It is a type of true heterosis, which results from the occurrence of the balanced types of mutations plants.

Muton: The smallest element within a gene, which can give rise to a mutant phenotype or mutation.

N

Napiform root: Root which when swollen become spherical at the upper part and sharply tapering at the lower part, as in beet, turnip, etc.

Narrow sense heritability: Ratio of total genetic variance to phenotypic variance.

Negative mass selection: Removal of off type plants from a mixed population allowing rest of the plants to grow further.

Net irrigation requirement (NIR): NIR is the irrigation water which is delivered to the field and available for the crop to use. This is primarily water that is stored in crop's root zone.

Non-contributing alleles: Those alleles which do not contribute to continuous variation, also called non-effective alleles.

Non-essential amino acids: Amino acid which can be synthesized by human body and they need not be supplied through diet.

Non-genetic RNA: RNA which does not acts as genetic materials. It is found in higher organisms where DNA is the genetic material.

Non-recurrent parent: The donor parent in the back crossing programme. The desire character of this parent is added in the genetic background of recurrent parent.

Non-sense mutations: Mutation with codons which do not code any amino acid.

Normal isoallele: An isoallele which acts within the phenotypic range of a wild character.

Nucleic acid: A macro molecule composed of phosphoric acid, pentose, sugar, and organic bases; DNA and RNA.

Nucleic acid: A macromolecule composed of phosphoric acid, pentose, sugar, and organic bases; DNA and RNA.

Nucleoside: A combination of deoxyribose sugar and nitrogen base.

Nucleotide: A unit of DNA and RNA molecule containing a phosphate, a sugar, and an organic base.

Nucleus: The part of a eukaryotic cell that contains the chromosomes, separated from the cytoplasm by a membrane.

Nullisomic haploid: A haploid which develops from a nullisomic (n-1).

Nullisomic: An individual lacking one pair of chromosomes from a diploid set (2n-2).

Nursery bed: A prepared area where seed is sown or into which transplants or cuttings are planted.

O

Obtuse: Blunt pointed.

Okazaki Fragments: Short segment of nucleotide synthesized in lagging strand of DNA as result of discontinuous replication.

Oleoresin: A natural combination of resinous substances and essential oils present in the fruits of certain crop plants like chilli.

Oligogenic traits: Characters which are governed by one or few genes, also called qualitative characters.

Open pollinated: The progeny of individuals where pollination took place automatically in flowers in nature.

Orbicular: Circular, Round.

Organogenesis: The process of differentiation of shoot and root from somatic embryos.

Orthodox seed: Seed which can dried to low moisture content and stored at low temperature without losing their viability.

Osmotic potential of soil water: Osmotic potential can be defined as the amount of work that quantity of water in an equilibrium soil water system is capable of doing when it moves to another equilibrium system identical in all respects except that there is no solution.

Out breeding: Mating of unrelated individuals.

Out cross: A type of natural cross obtained from the crosses of a number of unknown genotypes.
Ovate: Egg shaped; broadest below center (contrast ovate and elliptic).
Over dominance hypothesis: Heterosis due to superiority of heterozygote over both the homozygous.
Over dominance: An effect the heterozygote (Aa) that is greater than the effect of homozygous dominant (AA).
Overlapping Genes: Genes which code for more than one protein. In such gene, the complete nucleotide sequence codes for one protein and part of such nucleotide sequence for another protein. Such genes have been reported in tumor producing viruses such as phi X 174, Sv40 and G4.
Ovule: The term generally applied to the whole seed forming apparatus inside an ovary i.e; nucellus plus the integument.

P

Parietal placentation: When placentae bearing the ovules develop on the inner wall of the one-chambered ovary corresponding to the confluent margins of carpels as seen in radish, etc.
Parthenocarpy: The development of fruit without fertilization and the formation of normal seeds.
Parthenogenesis: The development of an individual from the female gamete without fertilization.
Pedicels: The flower stalks, which grow from the upper branches of the peduncle.
Pedigree: Record of ancestry of an individual selected plant for its various segregations.
Peduncle: The main flower stalk, often branched once or twice (or more) in its upper part and bearing pedicles near the end of the branches.
Pentagonal: In which the lobes are rather broad than long giving, the corolla a pentagonal or 5-pointed appearance.
Pepo: Fleshy, many seeded fruit which develops from an inferior, one-celled or spuriously three-celled, syncarpous pistil with parietal placentation, e.g., cucumber, melons, squash, gourds, etc.
Perianth: When the calyx and corolla do not differ much in shape and color, they together are said to form the perianth as seen in onion, garlic, etc.
Pericarp : The outer wall of the fruit .
Peripheral embryo: Embryo which encloses endosperm or peri sperm tissue as seen in *Amaranthus*, etc.
Permanent wilting point (PWP): The amount of moisture left in the soil after a plant has permanently wilted is called 'wilting coefficient' or permanent wilting point. At this stage film of water around soil particle are held so tightly that roots in contact cannot take water. The moisture tension at this point varies from 7 to 32 atmospheres, but 15 atmospheres is commonly used tension for this point.
Petiole: The stalk of a leaf .
Petiolate: With petioles.
P^F of Soil: It is a negative pressure of soil moisture expressed in cm (based on the height of water column above free water level in cm). pF= log 10^h , where h = soil moisture tension in cm of water.
Phenotype: The visible manifestation of the genotype produced as a consequence of growth and development.
Phenotypic disassortative mating: Mating of individuals with contrasting phenotypic character. Example –crossing between tall × dwarf plants.
Phenotypic assortative mating: Mating of individuals with similar phenotypes. Example-crossing between tall individual with Aa x AA or Aa or Aa x AA and aa x Aa genotypes.

Phenotypic stability: The stability of a genotype (or genotypes) over a spectrum of environmental condition. It is also known as developmental homeostasis.

Photo dormancy: A type of physiological dormancy of seed where germination of seed is sensitive to light i.e., seeds of some plants require light to germinate whereas others require darkness. e.g., lettuce seed require light and *Allium*, *Amaranthus*, etc. seed require darkness for germination.

Phytoalexin: A phenolic substance having antifungal principle, synthesized by plants in response to parasite invasion or infection by certain fungi, e.g., pisatin, phaseolin, trifolirhizin, orchinol and isocumarin from pea and bean pods, and carrot root, respectively.

Pie plant: Rhubarb (*Rheum raponticum*), one of the oldest cultivated vegetable crops, is commonly known as "pie plant".

Pinnate: A leaf divided into terminal and lateral leaflets.

Pinnatipartite: When the incision of leaf margin is more than half way down towards the midrib, as in radish.

Planting ratio: The male and female plants when planted in a certain proportion to ensure proper pollination and fertilization e.g., 10:1 (female: male) ratio in pointed gourd.

Pleiotropy: Phenomenon of a single gene affecting two or more different characters.

Poi: The pressure-cooked taro (*Colocasia esculenta*) corms after being passed through strainer are allowed to ferment which gives an acidic product called 'poi'.

Polyhaploid: Haploid which develop from polyploid species. Polyploids again are of two types, viz. allohaploids and autohaploids.

Polycross: An isolated group of plants or clones arranged in same fashion to facilitate random inter pollination.

Polymerase chain reaction (PCR): A procedure involving multiple cycles of denaturation, and polynucleotide synthesis that amplifies a particular DNA sequence.

Polymerization: Chemical union of two or more molecules of the same kind to form a new compound having the same elements in the same proportions but a higher molecular weight and different physical properties.

Polymorphism: Two or more kinds of individuals maintained in a breeding population.

Polynucleotide: A linear sequence of joined nucleotides in DNA or RNA.

Polypeptide: A linear molecule with two or more amino acids and one or more peptide groups. They are called dipeptides, tripeptides, and so on, according to the number of amino acids present.

Polypetalous: When the petals remain free from each other as seen in cabbage, radish, etc.

Polyploid: An organism with more than two sets of chromosomes (2n diploid) or genomes (e.g., triploid (3n), tetraploid (4n) and so on.

Positional sterility: A type of male sterility, also called functional sterility, where pollens are functional but anthers fail to dehisce, found in some mutants of tomato.

Post-harvest: After harvest.

Pot herbs: A group of vegetable crops whose foliage and sometimes immature stem are used as cooked vegetables; also called leafy vegetables or green e.g., palak, amaranthus, spinach, basella, etc.

Potential evapo-transpiration (PET): PET is defined as the evapotranspiration that occurs when the ground is completely covered by short actively growing vegetation in large area and where there is no limitation in soil moisture.

Pricking: A method of raising secondary nursery for the crops having very small seeds; in the case of high density sowing, it 1elps to develop a thinner stand in the nursery. In Cole crops and lettuce, pricking is done when first pair of true leaves develop.

Progeny testing: The practice of ascertaining the genotype of an individual by mating it to an individual of known genotype and examining the progeny.

Progeny: The off springs of a particular mating.
Program: A set of instructions written in a specific sequence for the computer to accomplish a given task.
Prokaryote: A number of a large group of organisms (including bacteria and blue green algae) that lack true nuclei in their cells and that do not undergo meiosis.
Pseudo – heterosis: See luxuriance.
Pseudo- dominance: The phenomenon of the apparent dominance of a recessive gene in the area opposite a chromosome deficiency.
Pseudo- incompatibility: Incompatibility due to physiological and physical reasons only and is not heritable e.g. physiological factors like temperature, light and physical factors like heterostyly, protandry, closed anther, etc.
Pure line: The descendants obtained from self-fertilization of a single homozygous or an inbred homogenic strain.

Q

Qualitative characters: The characters, which show discrete variation and easily identified by visual observations. Examples, flower color, fruit shape etc.
Quantitative characters: The characters, which show continuous variation such that visual identification of individual genes segregation is not possible. These are usually governed by the cumulative effect of polygenes and are highly influenced by environmental conditions. Examples are yield per plant, fruit per plant etc.
Quarantine: The prophylactic measure which is used to prevent the entry of new diseases, insects and weeds along with plant introduction from other countries.
Quercetin: A pigment which imparts coloration to the outer skin on onion bulb.
Quiescence: Describes the condition in which the seed can germinated immediately upon the absorption of water in the absence of any internal germination barriers. The embryo (or seed) is called to be quiescent.

R

Radiation genetics: A branch of genetics which deals with effects of various types of radiations on chromosomes and genes.
Random mating: Arrangements of pairs is by chance i.e. each individual has equal chances to mate with another. Examples- cross-pollinated crops.
Range: Difference between the lowest and the highest values present in the observation in a sample.
Read only memory (ROM): A memory that stores binary information permanently. The information can be read from this memory but cannot be altered.
Read/Write memory (R/WR): A memory that stores binary information during the operation of computer. This is used as writing pad to write user's program and data. The information stored in this memory can be read and altered easily.
Readily available water (RAW): Soil moisture lying between field capacity and permanent wilting point (15 atmospheres) is referred as readily available moisture. As the water content above field capacity can not be held against the forces of gravity and will drain out, and plant roots can not extract water content below permanent wilting point. The fraction of RAW that a crop can extract from the root zone without suffering water stress is the readily available soil water.
Recalcitrant seeds: The seeds which show drastic loss in viability with decrease in moisture content below 12 or 13 %. Such species include coconut, mango, tea, coffee, rubber, jackfruit, oil palm etc. Such seeds can not be conserved in seed banks.

Recessive epistasis: Gene interaction in which recessive alleles at one locus mask the expression of both the alleles at another locus resulting in 9:3:4 ratio. Also called supplementary epistasis.

Recessive gene: When parents with contrasting characters are crossed, the character of one parent does not appear in F_1. even though genes for both the factors are present in the hybrid. The gene, which does not appear, is said to be recessive.

Reciprocal hybrids: Two hybrids produced by crossing the same parents but the male of first is used as female in the another and similarly the female of first is used as male in another such as (A× B) and (B × A).

Reciprocal translocation: Mutual exchange of segments between non –homologous chromosomes.

Recombinant DNA: The DNA which contains genes from different sources and can combine with DNA of any organism.

Recons: The regions (units) within a gene between which recombination's can occur, but the recombination can not occur within a recon.

Recurrent selection: Reselection generation after generation with inter mating of selected plants to provide for genetic recombination.

Registered seed: The progeny of foundation seed that is so handled as to most nearly maintain the satisfactory genetic purity and has been approved by certifying agency.

Relative heterosis: Usually refers to the heterosis expressed over the mid parental value of a cross.

Renaturation: Union of separated (denatured) DNA strands on cooling.

Replication: Repetition of treatments under investigation.

Repulsion: Linkage between dominant and recessive alleles.

Restitution: Union of broken chromosomes segments which restores original gene sequence.

Restriction enzyme: An endonuclease that recognizes a specific short sequence in DNA and cleaves the DNA molecule.

Restriction map: A linear or circular diagram of a DNA molecule showing the sites that are cleaved by different restriction enzyme.

Reverse transcriptase: An enzyme that catalyzes the synthesis of DNA using an RNA template.

Reversion: Restitution of a mutant gene to the wild type condition, or at least to a form that gives the wild phenotype. More generally, the appearance of trait expressed by a remote ancestor; a throwback; atavism.

RFLP (Restriction fragment length polymorphism): A genetic difference among individuals that is decided by comparing DNA fragments released by digestion with one or more restriction enzyme.

Ribosome: Cytoplasmic organelle on which proteins are synthesized.

Ring chromosomes: A physically circular chromosome, usually found in prokaryotes such as bacteria and viruses.

RNA (Ribonucleic acid): The information – carrying material in some viruses. More generally, a molecule derived from DNA by transcription that may carry information (messenger or mRNA), provide subcellular structure (ribosomal or rRNA), transport amino acids (transfer or T RNA), or facilitate the biochemical modification of itself or other RNA molecules.

Rod shaped chromosomes: A chromosomes which assumes rod shaped at anaphase.

Rouging: Process of removal of off types (phenotypically different) plants from the field of an improved variety to avoid contamination.

Root crops: A group of vegetable crops whose swollen tap roots and in some cases hypocotyl along with tap root such as carrot, beet, radish, turnip, rutabaga etc. are cooked or eaten raw.

Root forking: Branching of tap roots in the root crops, particularly in radish and carrot, due to the presence of impediment, undecomposed organic matter or plant refuse in the soil.

Root tuber: The fleshy root of a herbaceous perennial plant with buds or eyes in the upper regions. e.g., sweet potato.

Root-to-seed method: A method of seed production in root crops where the fully matured roots are harvested, selected and after giving proper root and shoot cuts, they are replanted for seed production.

S

Sagittate: Arrow shaped with the basal lobes directed downwards. e.g., leaves of some aroids.

Salad crops: Green leafy vegetables which are usually consumed raw with oil, vinegar and various other condiments, e.g., lettuce, endive, celery, chicory, parsley, etc.

S-allele: A set of alleles controlling sporophyte self-compatibility.

Satellite: A communications vehicle orbiting the Earth. Satellites typically provide a variety of information from weather data to television programming. Satellites send time-stamped signals to GPS receivers to determine the position on the Earth.

Saturation point: When all the pores of soil are filled with water, the soil is said to be saturated or having maximum water holding capacity. At this point soil moisture tension is almost zero.

Scheduling of Irrigation: It is the decision-making process of determining when to irrigate and how much water to apply in each irrigation.

Scooping: Removal of central portion of the curd for easier initiation of flower stalk in cauliflower.

Secondary gene pool: The genetic material that leads to partial fertility on crossing with primary gene pool. It includes genotypes of related species and is designated as GP_2.

Seed parent: The female (pistillate parent of a hybrid).

Seed potato: Potato tubers used for planting.

Seed: The matured ovule having all the essential structures to produce a new plant.

Seedless watermelon: It refers to auto triploid (3x) watermelon which is both male and female sterile due to unequal chromosomal distribution in meiosis resulting in seedless condition in the fruit.

Seed-to-seed method: A method of seed production where the plants are allowed to produce seed in its original place of growing.

Segregating generations: The F_2 and onward generation of a hybrid where separation of parental from material chromosomes at meiosis and consequent separation of genes takes place leading to the possibility of recombination in the off spring.

Selection: The process in which a number of individuals with certain desirable characteristics are favored for further reproduction.

Self– pollinated crops: An assembly of homozygous plants. These crops often have one single genotype and reproduce it as such from generation to generation.

Self- sterile: The organism which fails to fertilize and set seeds after self – pollination.

Self-incompatibility: Failure of fertilization even though both male and female parts of the bisexual flower are fully functional as seen in cabbage, cauliflower, radish, etc.

Selfing: The process of putting the pollen of the same flower on its stigma i.e. enforcing self-pollination. It is extensively practiced in maize and other cross-pollinated crops to obtain inbred lines for utilization in the hybrid programme.

Semi – sterility: A state of only partial fertility in plant zygotes usually associated with chromosomal translocations.

Semigamy: Abnormal fertilization in which the male gamete fertilizes the egg, but does not fuse with the egg nucleus.

Sex determination: The process of sex differentiation which utilizes various genetical concepts to decide whether a particular genotype will develop into male or female sex.

Sex influenced genes: Gene whose expression depends on the sex of an individual such as baldness in humans.

Shoot apex culture: A tissue culture procedure for eliminating virus or other pathogens from plant parts where excision and aseptic culture of the small segment of the terminal growing points done because the terminal growing point of a plant is often free from virus and other pathogens even if the rest of the plant infected; practiced in potato, sweet potato, cassava, etc.

Shuttle vector: A plasmid capable of replicating in two different organisms such as yeast and *E. coli*.

Sib- mating: Sibling or crossing at random the two or more-individual obtained from the same parentage. It is form inbreeding and refers to brother – sister mating.

Sib-pollination: Pollination between closely related biotypes.

Silencer: A DNA sequence that helps to reduce or shut off the expression of a nearby gene.

Simla mirch: Bell-shaped, non-pungent, mild and thick fleshed.

Single cross hybrid: A hybrid involving two cross-compatible lines.

Single cross: A cross between two inbred, A x B.

Single crossing over: Formation of single chiasma between non-sister chromatids of homologous chromosomes.

Software: A group of programmes.

Soil moisture content: The moisture content of a sample of soil is usually defined as the amount of water lost when dried at constant weight. It is usually expressed in per cent.

Soil moisture content (% by wt.) = Wt. of moist sample — Wt. of oven dry sample X 100 Wt. of oven dry sample

Moisture content (% by vol.) = Moisture content (% by wt.) x Bulk density of soil

Soil moisture stress: It is the sum of the soil moisture tension and osmotic pressure of soil solution. The osmotic pressure developed by the soil solution retards the uptake of water by plants. Plants growing in a soil with a soil moisture tension is say 1 atmosphere, apparently can extract enough moisture for growth. But if the osmotic pressure of the soil solution is, say 10 atmospheres, the total stress are 11 atmospheres. Under such condition, plant cannot extract sufficient water for good growth.

Soil moisture tension: It is a measure of the tenacity with which water is retained in the soil. The tenacity is measured in terms of the potential energy of water in the soil measured usually with respect to free water. Soil moisture tension is expressed in atmospheres or bar or kPa (1 atmospheres=1036 cm of water, and 1 bar = 1023 cm of water column or 100 kPa).

Sole cropping: Raising one crop alone in pure stands, also called solid planting.

Somaclonal variation: The variation which is generated by the use of tissue culture.

Somatic hybridization: Crossing of plants through fusion somatic cells (protoplast).

Somatopalstic sterility: A type of sterility, which results as a consequence of the collapse of fertilized ovules or zygote during the embryonic or early development stages due to disturbance in embryo endosperm relations.

Southern blot: The transfer of DNA fragments from an electrophoretic gel to a cellulose or nylon membrane by capillary action.

Spadix: A spike with a fleshy axis which is enclosed by one or more large, often brightly colored, bracts called spathes, as in aroids, etc.

Spear: The shoot which is the edible part of asparagus.

Special chromosomes: Chromosomes which significantly differ in structure and function from normal chromosomes such as lamp brush chromosomes, polytene chromosomes and *B*-chromosomes.

Species: A unit of taxonomic classification containing group of individuals enough alike so that it may be reasonably assumed that they have arisen from a common ancestor.

Specific combining ability: The deviation in the performance expected on the basis of general combining ability.

Spike: The inflorescence in which the main axis is elongated and the lower flower opens earlier than the upper ones as in raceme, but the flowers are sessile, as seen in amaranth us, etc.

Splicing: The process that covalently joins exon sequence of RNA and eliminates the intervening intron sequences.

Sporophytic incompatibility: Failure of fertilization due to genetic abnormalities.

Stamens. Organ of flower, which produces pollens.

Standard deviation: The square root of the arithmetic mean of the squares of the deviation measured from the mean. It is the square root of variance.

Standard heterosis: Heterosis expressed over the standard or check variety. It is also known as useful heterosis.

Statistics: A branch of applied mathematics which deals with collection, presentation, analysis, and interpretation of numerical data tend to spread about the mean value. It is a measure of variation in a sample.

Steckling: Matured root of carrot, radish, turnip, etc. to be replanted after over-wintering for seed production in root-to-seed method.

Sterility: The phenomenon in which an organism is infertile and is incapable of reproducing.

Stolon: Slender, underground lateral stems arising from buds on the underground portion of the stem which enlarge at its tip to produce the tuber, as in potato.

Strain: The mating group within a variety or species with distinct morphological or physiological features.

Stump method: A method of seed production in cabbage where the head after full maturity is cut off just below the base, keeping the stem with outer whorl of leaves intact.

Substitution haploid: A haploid which develops from a substitution line (n-1+1).

Super computer: A computer with extremely large storage capacity and at least 10 times faster computing speed than other computers. Such computers are used in scientific and engineering disciplines.

Supermarket on a stalk: the winged bean (*Psophocarpus tetragonolobus*) plant is described so, as six different foods are supplied by this plant: leaves like spinach, succulent shoots resembling large thin asparagus, fried flowers for making a sweet garnish, tender pod as vegetable, the seed and the underground tubers are exceptionally rich in proteins.

Sweet pepper: It is the chilli of commercial value (*Capsicum annuum*).

Syneptonemal complex: A protein frame work which is found between paired chromosomes.

Synthetic varieties: These are the open pollinated advanced generation population of a number of hybrids obtained by crossing a number of tested lines grown in isolation.

T

Tabasco pepper: Small fruited, very pungent chilli (*Capsicum frutescens*).

Tandem: Duplication with normal sequence (similar to original segment) of genes.

Tautomerization: The process of sift of hydrogen atoms from one position to another either in a purine or in a pyrimidine base.

Taxa (Sing. Taxon): A general term for taxonomic classification, irrespective of rank.

Template: A macromolecule which provides information for the synthesis of another complimentary macro- molecule.

Tender vegetable: Vegetable crops which cannot withstand frost and some of them even do not thrive in cool weather, e.g., brinjal, okra, chilli, cucurbits, sweet potato, cowpea, tomato, cassava, beans, etc.

Tenderometer: An instrument by which toughness of the seed coat and firmness of pulp is determined and is mostly used to determine the seed quality of pea where high value of tenderometer indicates low quality.
Terminal delation: Loss of either terminal segment of a chromosome.
Tertiary gene pool: The genetic materials which leads to production of sterile hybrids on crossing with primary gene pool. It is designated as GP_3.
Tertiary trisomic: A trisomic in which the additional chromosome is translocated one.
Test cross: The cross of the F_1 with recessive homozygous parent i. e; Aa x aa.
Tetrasomic: Addition of two chromosomes to one pair or two different pairs.
Three-way cross hybrid: Hybrid progeny between a single cross and an inbred, i.e. (A x B) xC.
Three-way cross: The cross between a single cross and an inbred line. With three lines (A, B, C), it could be represented as (Ax B) x C. It has been genetically used in the maize-breeding programme for the production of commercial sweet maize hybrids.
Thrum: In primula, flowers with short style and high anthers.
Tissue culture: The development of an entire organism from plant cells and tissues *in vitro* or artificial media.
Top cross: An out cross of selections clones or inbred, to a common pollen parent. In corn, a commonly an inbred – variety cross.
Total available water (TAW): The total available water in the root zone is the difference between the water content at field capacity and wilting point (TAW= FC–WP). TAW is the amount of water that a crop can extract from its root zone, and its magnitude depends on the type of soil and the rooting depth.
Totipotency: Single cell culture develops in a plant.
Transformation (bacteria): Genetic alteration of bacteria brought about by the incorporation of foreign DNA in the bacterial cells.
Transgenic: A term applied to organisms that have been altered by introducing DNA molecules into them.
Transgressive segregation: The appearance of individuals showing an extreme development of a character than either of the parent in F_2 or later generations. This usually occurs because of cumulative and complementary effect of genes contributed by the parents of the cross
Transition: A mutation caused by the substitution of one purine by another purine or one pyrimidine by another pyrimidine in DNA or RNA.
Translocation: One way or reciprocal exchange of segments between non-homologous chromosomes.
Transpiration ratio: It refers to the quantity of water (in gram) transpired by the plant to accumulate 1 g of dry matter. This ratio varies from 200 to 1000 depending upon crop species, cultivation conditions and crop growth stage.
Transpiration: It is water loss through living plants in the form of water vapour is called transpiration.
Transposition: Existance of one wild and one mutant allele in each homologous chromosome (+a / +b) also called repulsion phase of alleles.
Triple cross: The cross between two three-way crosses. With 6 lines (A, B, C, D, E. F), it could be represented as (Ax B) x C) x (D x E) x F).
Trisomic: Addition of one chromosome to one pair in a diploid set.
Tristyly: Style having two positions, viz., low and high.
Tuber: A special kind of swollen modified stem structure that functions as an underground storage organ as in potato, Jerusalem artichoke, etc.
Tubercle: Small aerial tuber produced in the leaf axils as seen in Yam (*Dioscorea alata*).
Tunic: The papery or fibrous coats covering bulbs and corms.

Tunicated bulbs: Type of bulb where the outer leaves are usually thin, membranous and dry and completely ensheath the inner portion and the central axis of the bulb like a tunic as the bulb of *Allium*.

U

Undeveloped embryo: Partially developed torpedo-shaped embryos that may attain a size up to one half that of the seed cavity at maturity as seen in carrot, etc.

Unidirectional replication: Replication of DNA in one direction only from the point of origin.

Uniform expressivity: Similar on or uniform expression of gene in all the individuals that carry such genes.

Unique DNA: The DNA segments (nucleotides) having only single copy per genome.

Utilization index: Ability of root and tuber crops for better accumulation of photosynthates in the storage organ which can be judged by root: shoot ratio.

V

V- Shaped chromosomes: A chromosomes which assumes V –shaped at anaphase.

Vacuum cooling: A technique of cooling vegetable (leafy vegetables, asparagus, Brussels sprout, etc.) having a high surface to volume ratio, rapidly and uniformly by boiling off some of their water at 1ºC and at low pressure (5 mm mercury) into a sealed container. The produce is cooled by evaporation of water from the tissue surface and is more rapid than hydrocooling.

Variable expressivity: Differential or variable expression of a gene in the individuals that carry it.

Variance: The average of the squared deviations from the mean or the square of the standard deviation.

Variation: Differences amongst the individuals arising as a consequence of differences in the genetic makeup; the effect of the environment or interplay of both. It is the chief characteristics of living organisms and is the basis of evolution.

Varietal blend: Mechanical mixtures of seed of two or more varieties.

Varietal deterioration: permanent reduction either in the genetic or agronomic value of a released variety.

Variety: An agricultural variety is a population of similar individuals having common identifiable plant fruit or seed characteristics along with a good agronomic base. This is generally utilized for commercial cultivation.

Vector. An animal or insect, which transmits parasites e.g. white fly, is the vector of viral diseases of tomato okra, chili, etc.

Vegetable forcing: A specialized type of vegetable farming where vegetables are grown out of their normal season. Vegetable forcing requires some special structures like glasshouse, hot bed, cold frame, etc.

Vegetable: An edible plant or plant part eaten cooked or raw as a main part of a meal, side dish, or appetizer.

Vegetative apomixis: A type of apomixis where flowers are replaced in the plants by bulbils or buds which fall to the ground like seeds, found in *Allium cepa* var. *Viviparum*, *Poa bulbosa*, etc.

Vertical resistance: Resistance conferred by few genes, specific to few known races.

Vexiliary: Of the five petals, when the posterior one is the largest and almost covers the two lateral petals as in pea, dolichos bean, etc.

Viability: The capability to live and develop normally.

Virulent: A race of a pathogen capable of attacking a host with specific resistance.
Vital mutation: Mutation in which all mutants survive.

W

Water Management: It is an efficient and planned used of water for crop production.
Water Requirement: It is the quantity of water needed for raising a crop in a given period. It includes; consumptive use, economically unavoidable losses and water needed for some special operation such as land preparation, transplanting, leaching of salt etc.
Water Stress: When the potential energy of the soil water drops below a threshold value, the crop is said to be water stressed.
Water use efficiency (WUE): It is the ratio of crop yield (Y) to the amount of water depleted by the crop in the process of evapotranspiration (ET). WUE (crop) = Y/ ET
Waxing: A short term storage technique of fresh fruits and vegetables under ambient conditions by applying wax emulsion containing paraffin wax, triethanol and aleic acid which provide a thin, discontinuous layer on the fruit surface and thus curtails the respiration and transpiration resulting increase in shelf-life. It also helps to keep away the microbes when fungicides like, Benlate 50 are used in wax emulsion.
Waxy blister of tomato: A disorder where white to cream colored irregular blisters, 3 to 6 mm in diameter and often more than 3 mm high occur which become light to dark brown, depressed and crack as the fruits ripen.
White heart: Whiteness at the central portion of watermelon instead of uniform development of pink color from center to rind, indicating poor quality.
Wide crossing: Mating between different genera of the same family; also called distant hybridization.
Wobble base pairing: The pairing of mRNA codon with tRNA anticodon in which first two bases of a codon have normal pairing and third base has abnormal base pairing.
Word: A group of bits the computer recognizes and processes at a time.
Working collection: The germplasm which is meant for short term storage (3 to 5 year). Seed of such material are stored at 5 to 10^0C.
WUE (crop) = Y/ ET

X

X- rays: separately ionizing and highly penetrating radiations generated in x-rays tubes and used for induction of mutations.
X^2-Test: A test of statistical significance which is used to test the significance of differences between observed and expected frequencies. It is used for the analysis of oligogenic characters.
Xenia: The immediate effect of pollen (i.e.; male gamete) on the endosperm. It is frequently observed in open pollinated maize with respect to kernel color.

Y

Yearling: One-year-old bulblets which have been formed on the scale.

Z

Z- DNA: The DNA in which sugar and phosphate linkages follow a zig –zag pattern. Such DNA has left handed double helical model.
Zoning of beet: Under unfavorable conditions, particularly in hot weather, beetroot show alternate white and colored circles when sliced, called zoning.

Z-test: A test of significance which is used to compare two means when the sample size is large (more than 30).
Zygote: The cell formed by the fusion.

Printed and bound by CPI Group (UK) Ltd, Croydon, CR0 4YY
22/04/2026

14866396-0003